BASIC DRAFTING TECHNOLOGY

Rotmans
Horton
Good

BASIC DRAFTING TECHNOLOGY

Rotmans
Horton
Good

10 9 8 7 6 5

LIBRARY OF CONGRESS CATALOG CARD NUMBER: 78-050424
ISBN: 0-8273-1293-8

Printed in the United States of America
Published simultaneously in Canada
by Nelson Canada,
A Division of International Thomson Limited

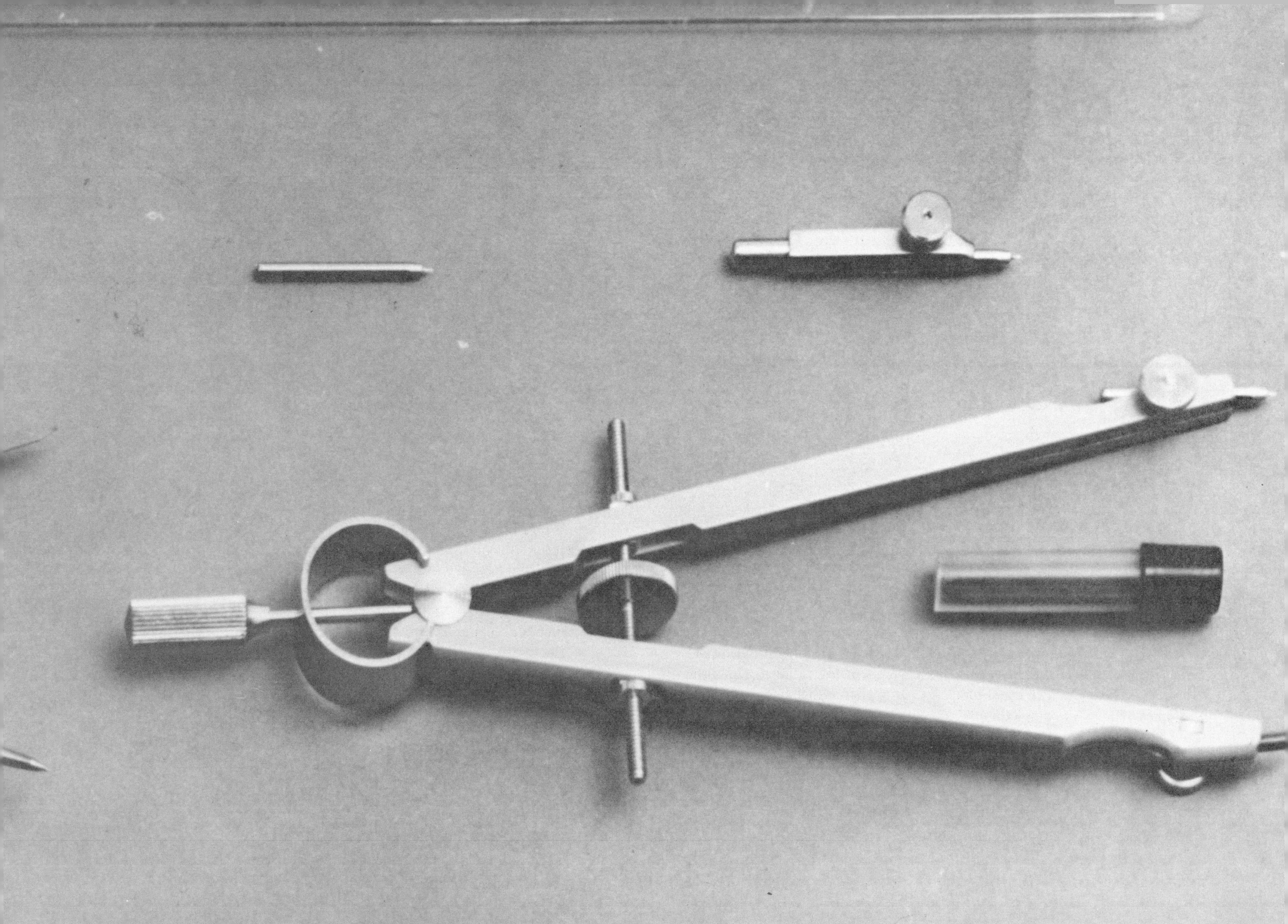

DELMAR PUBLISHERS INC. • ALBANY, NEW YORK

PREFACE

BASIC DRAFTING TECHNOLOGY provides a well-rounded course in drafting which is ready for use without further preparation on the part of the instructor. Each unit includes a progressive series of instruction sheets and activities to develop fundamental skills and an understanding of up to date drafting practices.

This book is designed to include the best tested teaching methods of motivation, presentation, application, and testing. Each unit includes a detailed discussion of each new drafting principle or terms, illustrations and explanations to show how these practices are applied, assignments, and step-by-step directions for completing the assignments.

This technique of individualized instruction enables each student to know what is expected, permitting them to work on their own assignments with a minimum of assistance according to their own ability. At the same time, the instructor has more freedom to supervise and give effective individual attention to those students who need it.

The use of a projection box and models is recommended for the first few problems dealing with the theory of projection. Using this technique, the student is better able to translate the text drawing to the graphic representation asked for in the assignment. The test material and drawings conform to the American Standard Drawings and Drafting Room Practice and to other related standards approved by the American National Standards Institute.

This text contains the following major features:

- All the material is up to date, presented in logical sequence in the order it should be learned.
- The material is based upon the author's actual use of the information over many years of teaching drafting.
- Assignments are spread throughout the units to allow the students to practice technique as it is taught rather than at the end of a unit.
- The metric system is introduced as a part of the text and used throughout the book in dual-dimensioning assignments.
- A second color is used to clarify many of the illustrations and add emphasis when necessary.
- Review questions throughout the text evaluate the student's comprehension before going on to new material.
- Technical terms are explained in the text when they are introduced.
- An Instructor's Guide is available for the teacher to speed the marking of each drawing as well as supply helpful information for the teaching of specific units.

The revising author, Mary Good, has been a teacher since 1959. She is presently teaching drafting at Hilton Central High School, Hilton, New York. A graduate of the University of Utah, Mrs. Good has written or illustrated articles for local and state technical journals. She is presently a member of the NYSIATA, the AIAA, and the Monroe IA Teachers Association.

CONTENTS

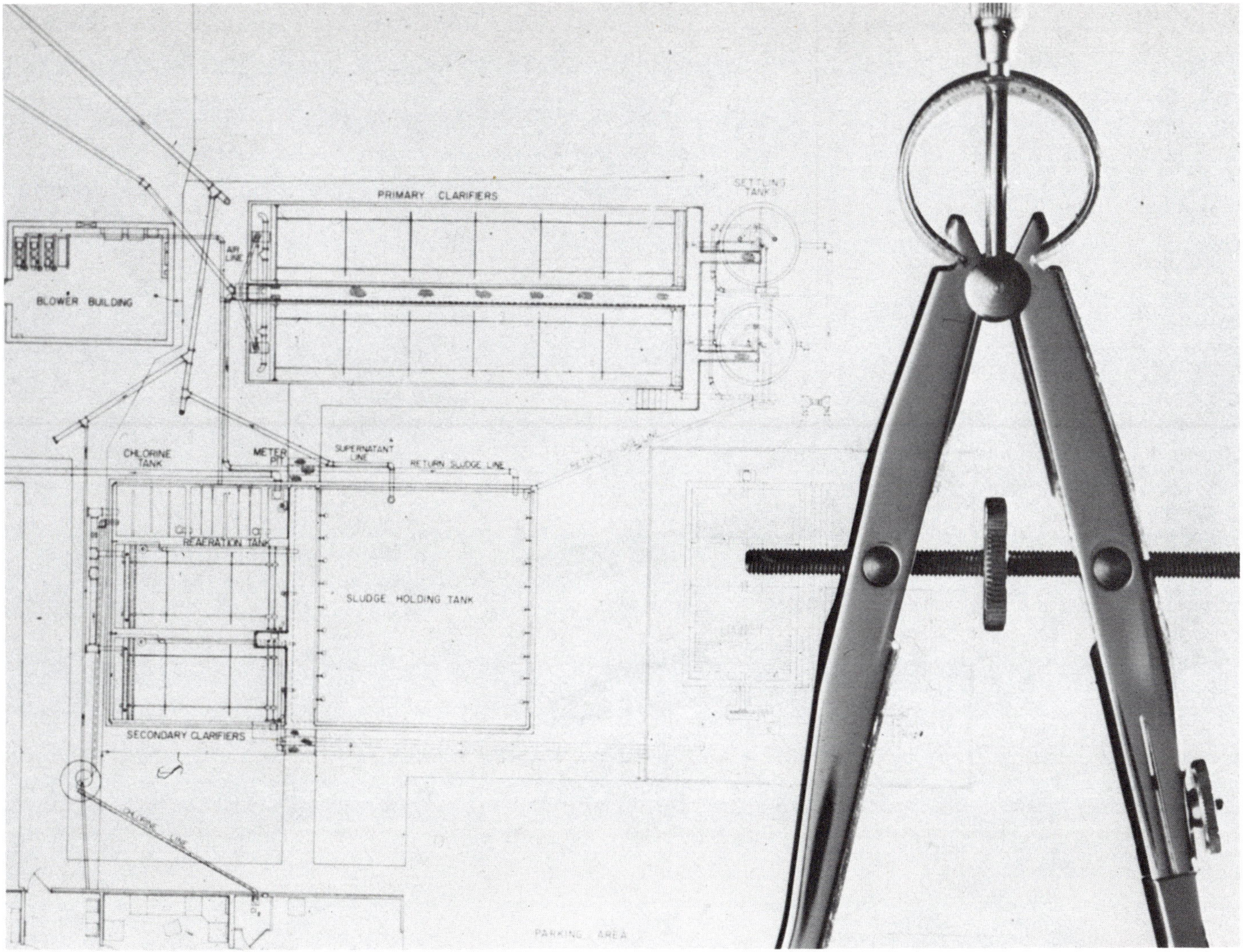

UNIT 1 TOOLS AND EQUIPMENT OF DRAFTING

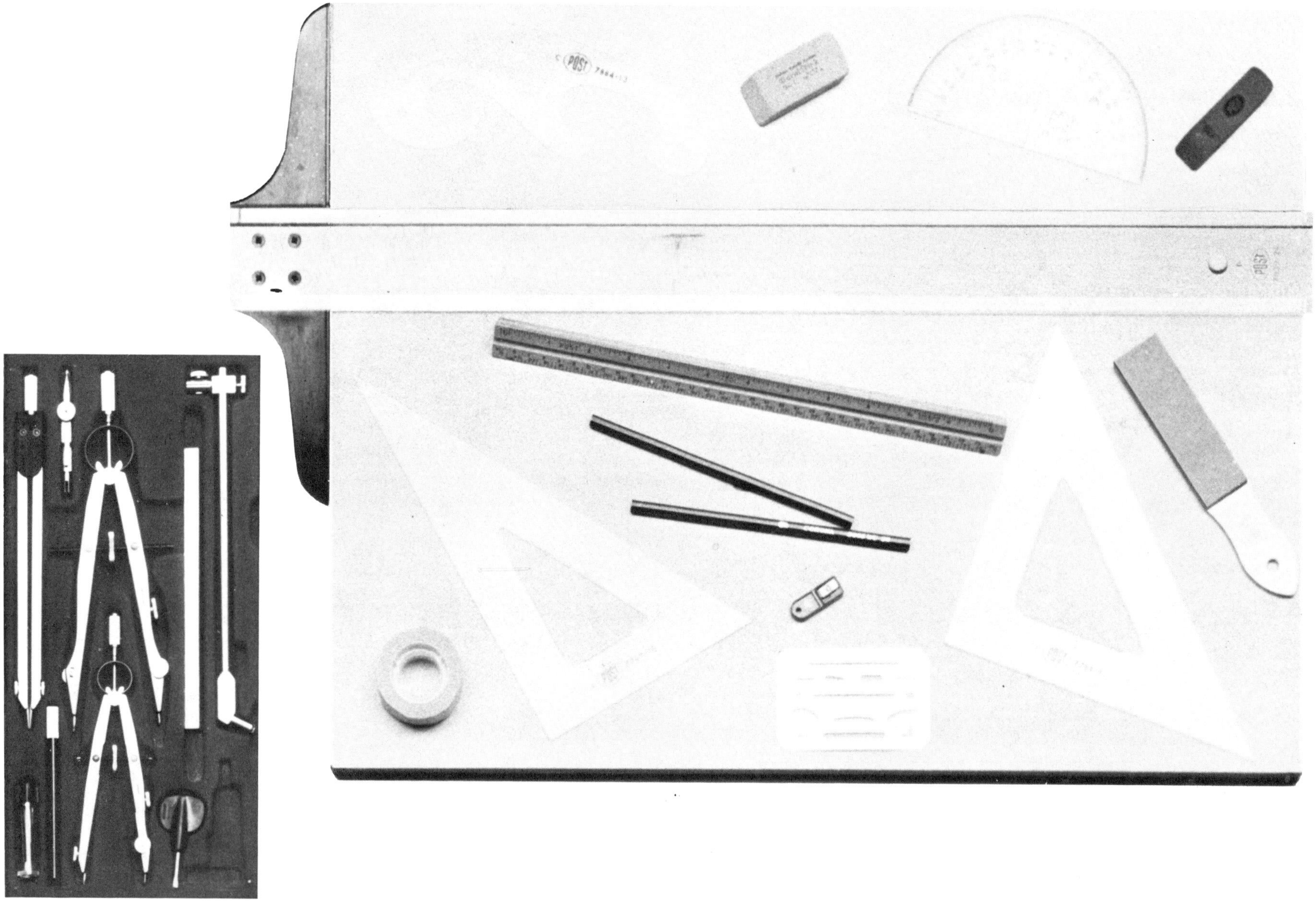

Fig. 1-1 Most commonly used student drafting equipment

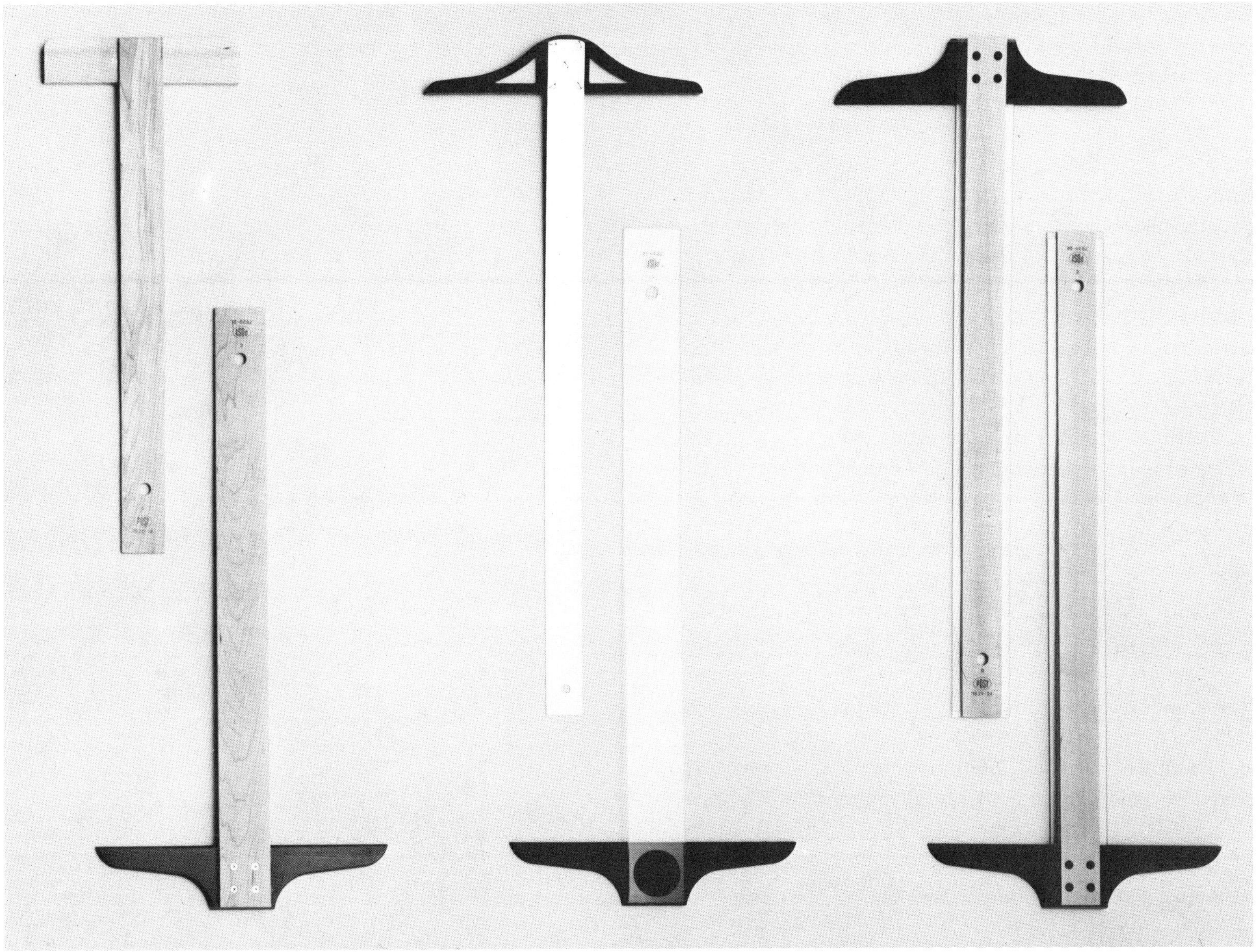

Fig. 1-2 T squares of various styles and materials

Speed and accuracy in drafting are constantly improved by new drafting techniques and technological advances in equipment and mechanical aids. Proper use of the correct tools make it easier to draw with precision. Some of the more commonly used tools and practices are explained in this unit, figure 1-1.

THE DRAWING SURFACE

The *drawing board* has a smooth flat surface on which the drawing paper is placed. It has a perfectly straight edge on each side to guide the T square accurately. The head of the T square moves along the left edge for right handed drafters, and along the right edge for left handers.

Portable drawing boards are usually made of soft wood. They have metal sides to keep the edges accurate longer. They can be used flat on a table or tilted for better vision. Drawings are usually attached directly to the surface of drafting tables. Industrial drafting tables come in various heights and designs. The tops can be raised and lowered or tilted to any angle. A drafter may sit or stand while working at a drawing board.

T SQUARES

A *T square* is used to draw horizontal lines. It has two parts, a head and a blade, fastened at right angles, figure 1-2. T square blades are made of wood, metal, plastic, and combinations of these materials, such as aluminum or wood with plastic edges.

Right handed drafters hold the head tightly against the left edge of the drawing surface with the left hand. Left handers use the right hand to hold the head of the T square tightly against the right edge of the drawing surface. Horizontal lines are drawn along the upper edge of the blade. The lines are drawn from the head towards the blade of the T square.

PARALLEL STRAIGHTEDGE

The *parallel straightedge* eliminates the need to control the head of the straightedge along the edge of the drawing surface as with the T square, figure 1-3. The wires that attach the straightedge to the surface keep it in a parallel position as it moves up and down the board. This allows the drafter more freedom in drawing parallel horizontal lines.

Fig. 1-3 The parallel straightedge does not need to be held against the working edge as does the T-square.

TRIANGLES

Two right angle triangles are regularly used in drafting. One has angles of 90, 60, and 30 degrees, figure 1-4. The other has one angle of 90 degrees and two angles of 45 degrees, figure 1-5.

The triangles are used for drawing vertical and inclined lines. They are supported along the upper edge of the T square or parallel straightedge. Vertical lines are drawn from the bottom towards the top. The angles in figures 1-6, 1-7, and 1-8 are drawn using these two triangles.

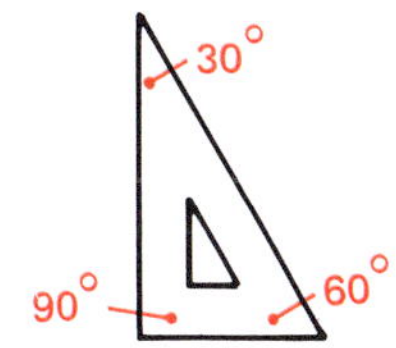

Fig. 1-4 The 30° - 60° triangle

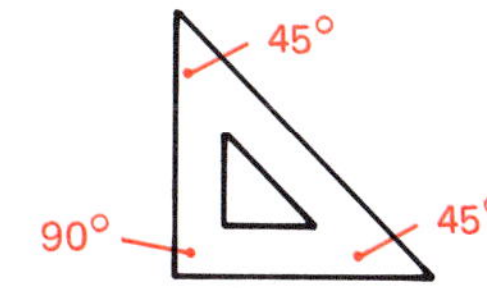

Fig. 1-5 The 45° triangle

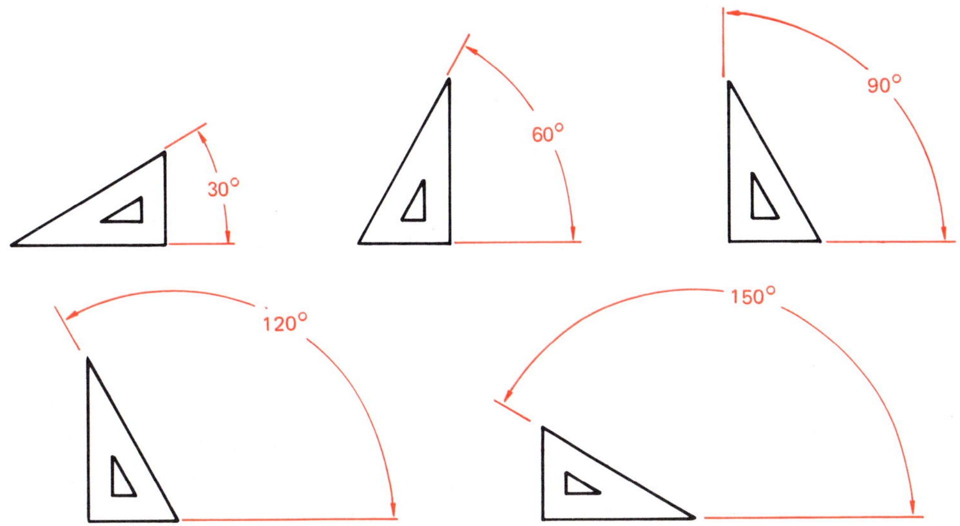

Fig. 1-6 Angles drawn with the 30° - 60° triangle

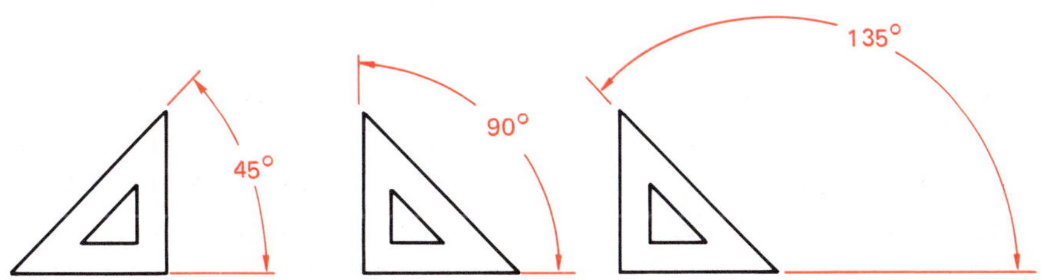

Fig. 1-7 Angles drawn with the 45° triangle

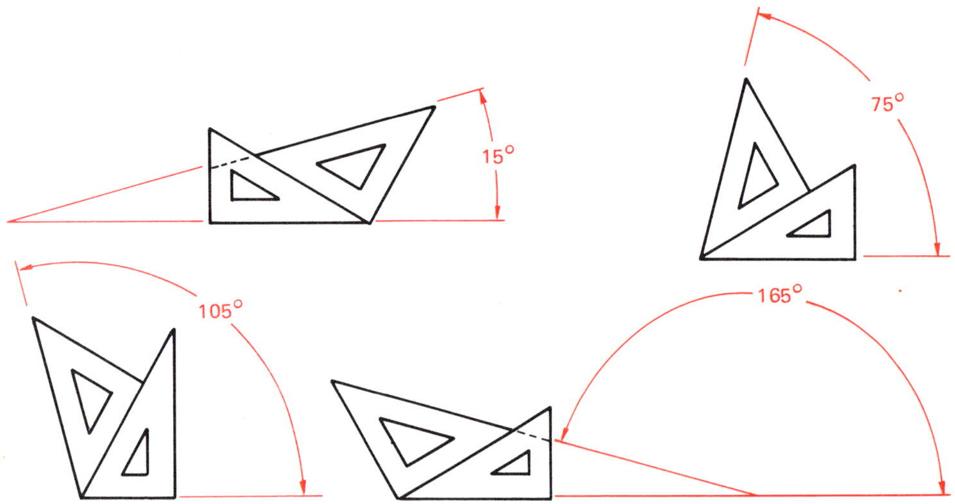

Fig. 1-8 Angles drawn with the combined triangles

PROTRACTOR

The protractor, figure 1-9, is used for laying out angles which cannot be drawn easily by using the triangles.

SCALES

Scales are used for measuring actual size, or enlarging or reducing the drawing of an object in proportion. They are made in different shapes, figure 1-10, and of different materials, such as boxwood, plastic, and metal. The shape determines the number of scales on each tool. The two-bevel and opposite bevel shapes each have two different scales. The four-bevel shape has four scales. The triangular scales, figure 1-11, provide six faces with room for eleven different scales.

There are four kinds of scales used in engineering drawing. These are the architect's scale, mechanical engineer's scale, civil engineer's scale, and metric scale.

The architect's scale, figure 1-12, is used in working drawings of buildings, some electrical, mechanical and engineering drawings, and machine parts. This scale is open divided along the face with each division representing one foot. The fully divided end unit is similar to a 12-inch ruler. It permits direct measurements on architectural drawings ranging from full size to 1/128th size.

The mechanical engineer's scale, figure 1-13, is used for measuring machine drawings. The conventional scale has inches and fractions of an inch divisions representing inches. The decimal-inch system, used in the automotive industry and other engineering fields, has inches divided into decimals. For full size, one inch is divided into 50 units.

The civil engineer's scale is used to lay out maps, roads, and public projects that need to be plotted, graphed, and charted. It is divided into decimals.

The metric scale is increasing in use as the International System of measuring becomes accepted by American industries as the official dimensioning system. Metric scales come in a variety of shapes. They are divided into the metric increments of centimetres and millimetres. The scales indicate the relative size of the

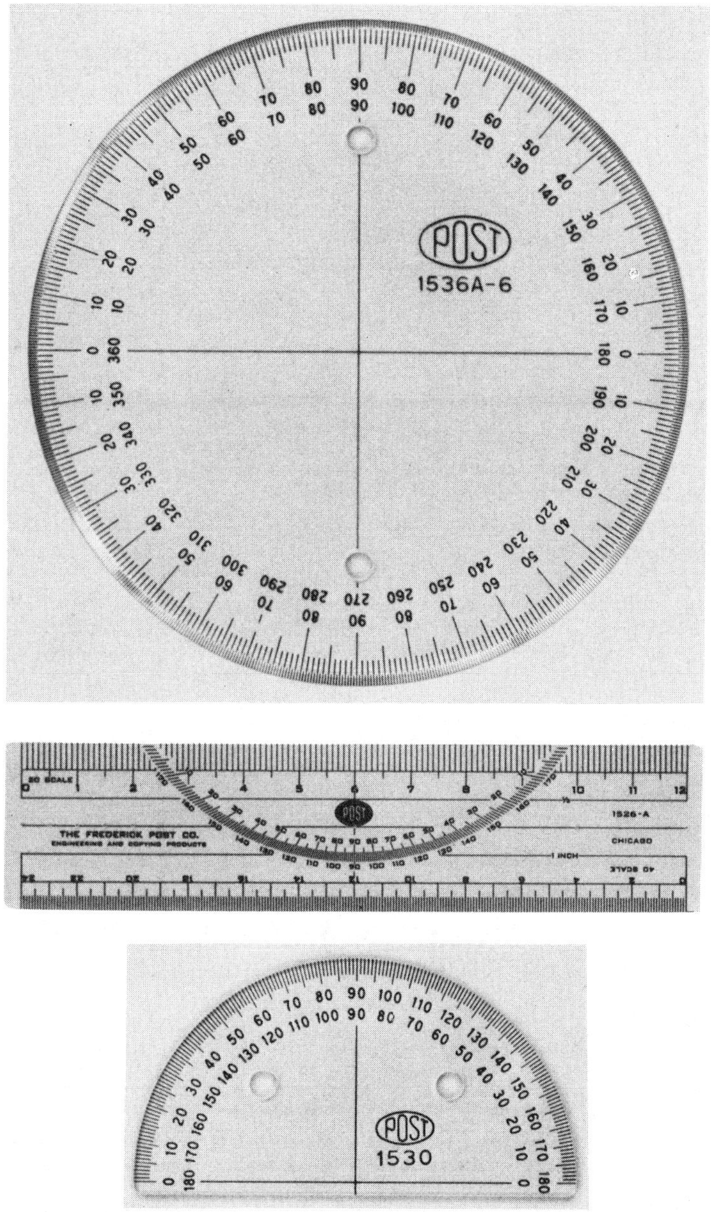

Fig. 1-9 Three types of protractors

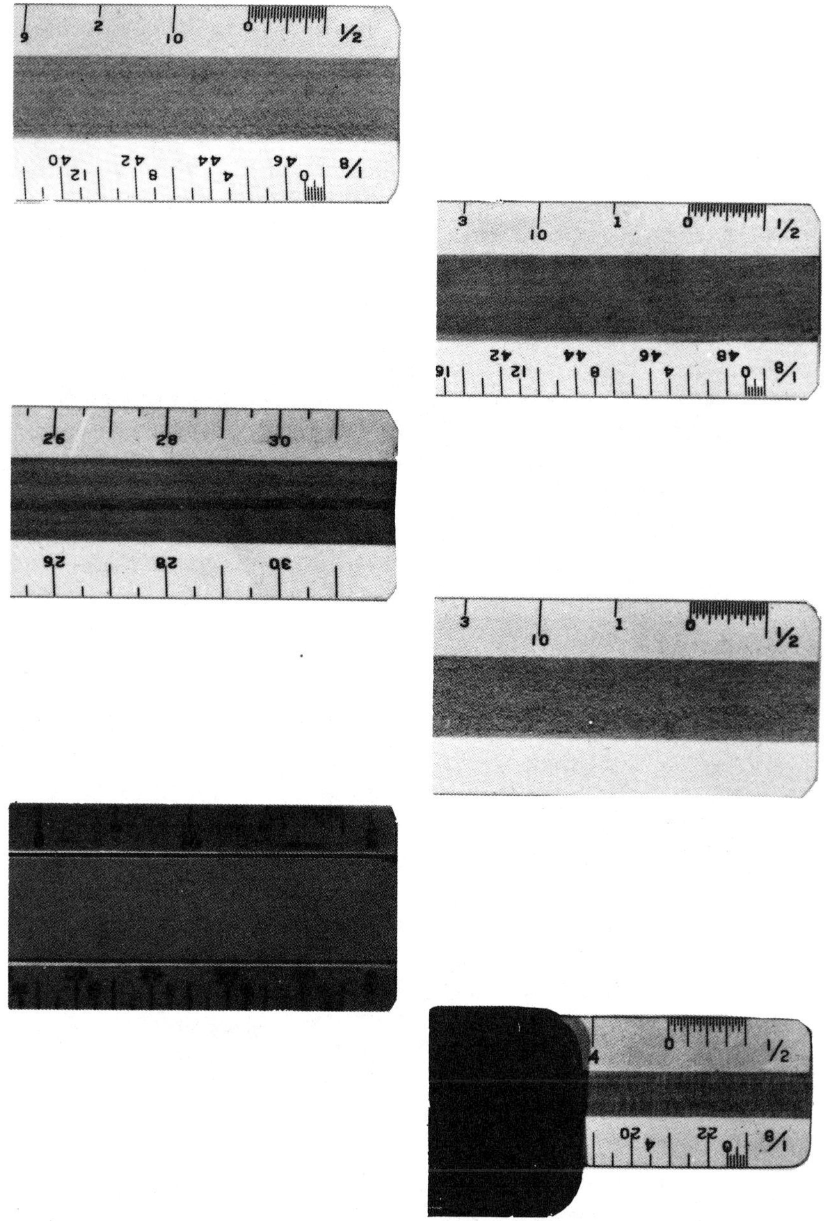

Fig. 1-10 Scales differ in shape and type of measurement.

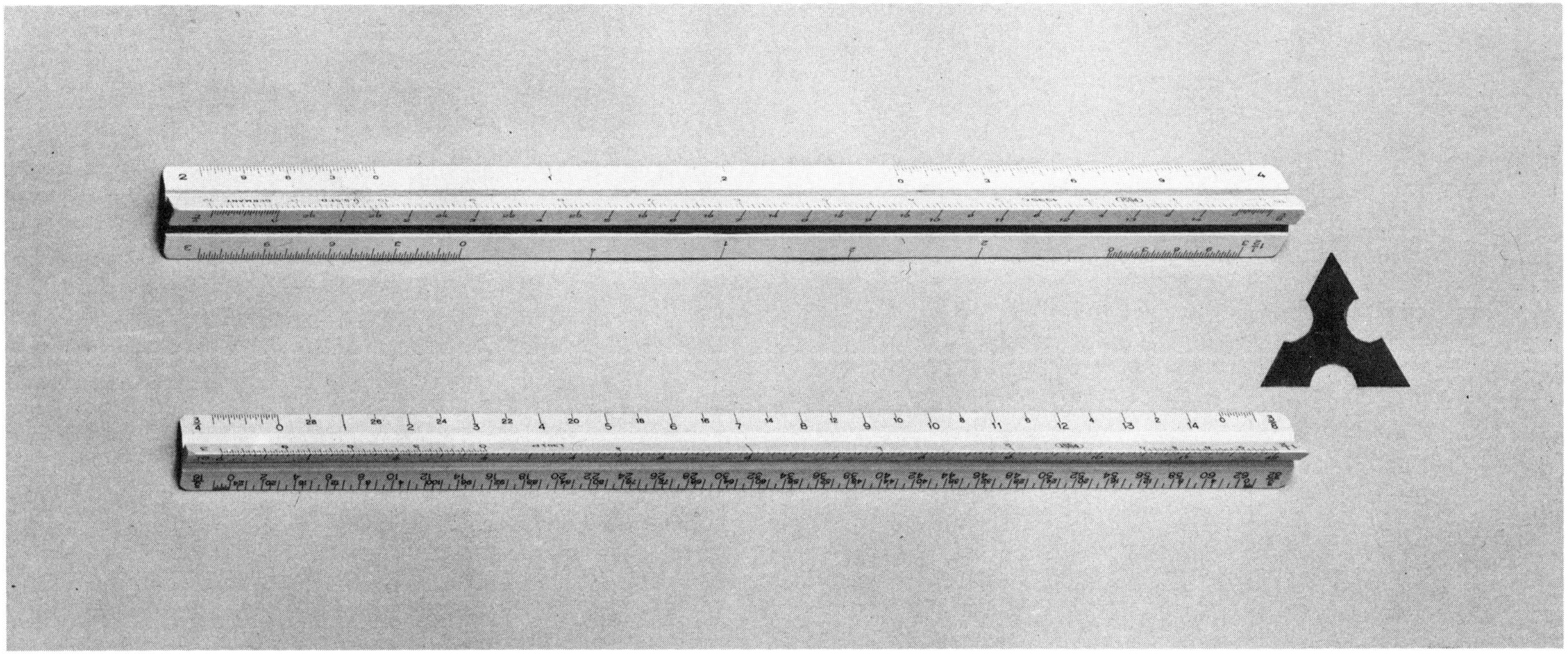

Fig. 1-11 Triangular scales have six faces

Architect's Scales:

1.	Scale:	12″ = 1′-0″	(full size)
2.	Scale:	6″ = 1′-0″	(1/2 size)
3.	Scale:	3″ = 1′-0″	(1/4 size)
4.	Scale:	1 1/2″ = 1′-0″	(1/8 size)
5.	Scale:	3/4″ = 1′-0″	(1/16 size)
6.	Scale:	3/8″ = 1′-0″	(1/32 size)
7.	Scale:	3/16″ = 1′-0″	(1/64 size)
8.	Scale:	3/32″ = 1′-0″	(1/128 size)
9.	Scale:	1″ = 1′-0″	(1/12 size)
10.	Scale:	1/2″ = 1′-0″	(1/24 size)
11.	Scale:	1/4″ = 1′-0″	(1/48 size)
12.	Scale:	1/8″ = 1′-0″	(1/96 size)

Fig. 1-12 Recommended forms to indicate several scales

drawing as compared to the actual size of the object being drawn. A scale of 1:20 indicates the drawing is done 1/20th the size of the object. It is referenced as 5 centimetres = 1 metre on the drawing. The scale 1:100 is referenced 1 centimetre = 1 metre on the drawing. The scale 1:100,000 is 1 centimetre = 1 kilometre on the drawing.

Measuring with the Architect's Scale

The divisions for feet are along the face of the scale. The divisions at the end of the scale divide the feet into inches or fractions of an inch. To lay out a distance of 2′-2″, start at 0, figure 1-14. Count two full feet divisions along the face plus number of divisions needed to measure two inches from the zero toward the end of the scale. These divisions change depending on the scale being used. Each mark on the 1/8-inch scale equals 2 inches, while each mark on the 3-inch scale represents 1/8-inch. Become familiar with these differences to avoid errors in measuring.

Mechanical Engineer's Scales:

1.	Scale:	1″ = 1″	or	1 in. = 1 in.	(full size)
2.	Scale:	1/2″ = 1″	or	1/2 in. = 1 in.	(1/2 size)
3.	Scale:	1/4″ = 1″	or	1/4 in. = 1 in.	(1/4 size)
4.	Scale:	1/8″ = 1″	or	1/8 in. = 1 in.	(1/8 size)
5.	Scale:	2″ = 1″	or	2 in. = 1 in.	(double size)

Fig. 1-13 The first number refers to the drawing and the second refers to the object.

Laying Off Measurements with a Scale.

The scale is placed on the drawing paper with the face to be used up and away from the drafter. The measuring is done from directly above the scale. Distances are marked off with a short dash at right angles to the scale made by a sharply pointed lead.

Successive distances on the same line are marked without moving the scale. They are double checked by measuring individually after marking.

DRAFTING MACHINES

Drafting machines are used throughout the industry to reduce drawing time considerably. There is no longer any need to reposition the five basic drafting tools; the T square, triangle, straightedge, protractor, and scale. All their operations are combined into one precision instrument. There are two types of manual drafting machines currently in use. One is the track type, figure 1-15. The other is an arm type, figure 1-16. The basic operating principles are the same for either machine.

Machines are made for right- or left-handed drafters. The head is held in the nondrawing hand. A release button allows rapid setting and resetting of angles for drawing and measuring as the arms of the machine move up and down and across the drawing board. A protractor, precisely graduated in degrees and seconds, is built into the head. Scales for the machines are made in various graduations, styles, and materials. The track-type machine can be used with a computer attachment capable of "reading" a drawing onto numerical control tapes. It is important to follow the manufacturer's directions for the type of drafting machine used, but the following general procedures are used for aligning:

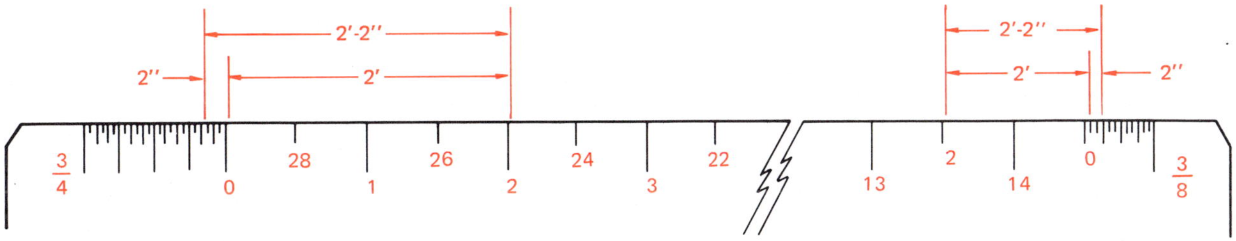

Fig. 1-14 Measuring with the architect's scale

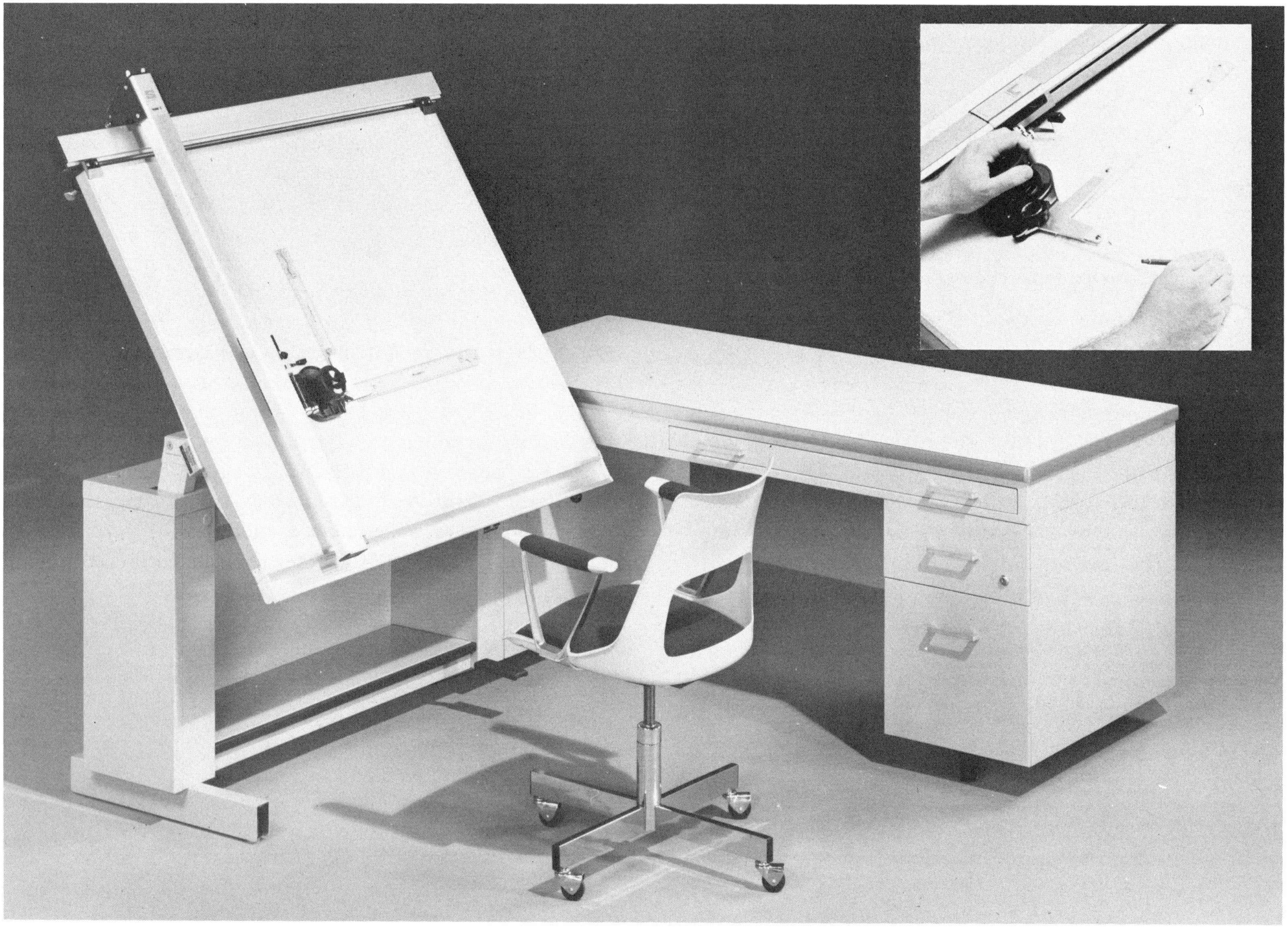

Fig. 1-15 Track-type drafting machine on adjustable drafting table

Fig. 1-16 Arm style drafting machine

Base Line Alignment

1. Set the protractor indexing head at zero. Release the wing nut or lever that secures the protractor.
2. Adjust the horizontal scale to the desired base line. Make sure the screws holding the scale chucks are tight, and the scales are firmly inserted in the base plates.
3. Tighten the protractor wing nut or screw to hold the base setting.

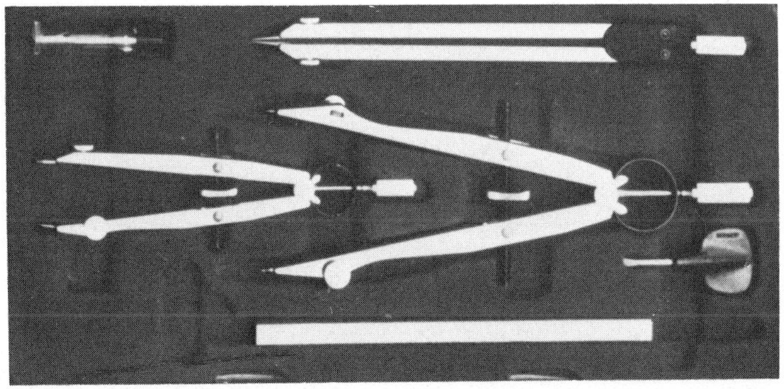

Fig. 1-17 Example of a student drafting set

Vertical Scale Alignment

1. Set the protractor head at 0 and draw a horizontal line along the top of the scale. Always hold the scale firmly with one hand when drawing so the pencil pressure does not move the free end of the scale.
2. Rotate the vertical scale through 90 degrees and check it against that horizontal line.
3. If it does not line up perfectly, loosen the chuck clamping screws and make the minor adjustment so that the scales are exactly 90 degrees apart. Tighten the screw.

DRAWING INSTRUMENTS

Drawing sets are available in a wide variety of instrument combinations. They are as simple as the combination in figure 1-17, or as elaborate as a combination of different sizes and types of specialized tools for professional drafters, engineers, and illustrators. Each tool is designed to increase accuracy and efficiency in drawing, and must be handled and respected as a precision instrument.

Compasses

The *compass* is used for drawing circles and arcs. The *bow compass* is the most commonly used style, figure 1-18. It has a

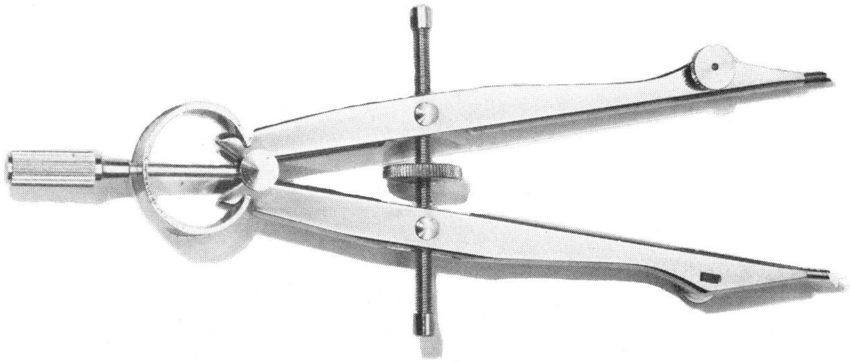

Fig. 1-18 Bow compass used to make circles and arcs

Fig. 1-19 A beam compass makes circles larger than those a bow compass can handle.

center wheel adjustment. One leg has a removable point to accept a pen, pencil, extender bar, or divider ends. The legs are solid or hinged to make them knee adjustable. The knees are bent to bring the legs perpendicular to the paper when making a large circle or when inking. The bow compass holds a radius more securely than a friction compass.

A *friction compass* is opened by pulling directly on the legs. It is closed by pushing the legs together. The knees are solid or bent like the bow compass.

To draw a circle or arc, the compass is inclined slightly in the direction in which the line is drawn. It is turned by rolling the handle or thumbscrew between the thumb and forefinger. The pencil point is kept sharpened to a chisel point and turned to the outside.

Very large circles are drawn with a *beam compass*, figure 1-19, or with an extender bar attached to the bow compass. One leg has a pin point. The other has attachments for either lead or ink. The pin is held at the center by one hand and the beam rotated by the other.

Dividers

Dividers look like compasses with two pin points, figure 1-20. While the bow style is the most commonly used type of compass, the friction divider is the most preferred style of divider. It is easily adjusted with the fingers of one hand. A divider is used to transfer and mark off distances. Expert use of the divider increases speed and improves accuracy in drawing.

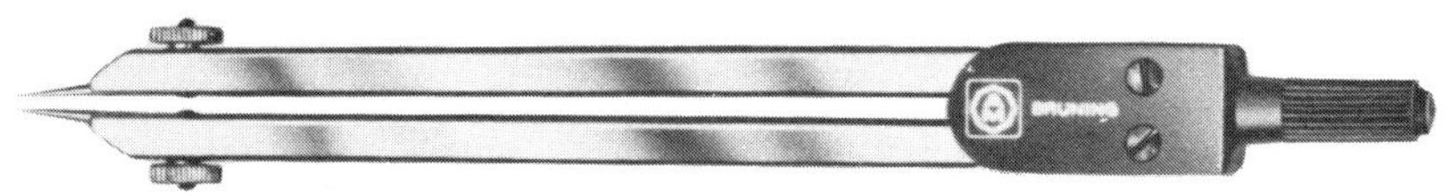

Fig. 1-20 A divider aids in making accurate measurements.

IRREGULAR CURVES

Irregular curves, sometimes referred to as French curves, are used to draw any noncircular curve. There are many sizes and shapes available, figure 1-21. There are also some specialized

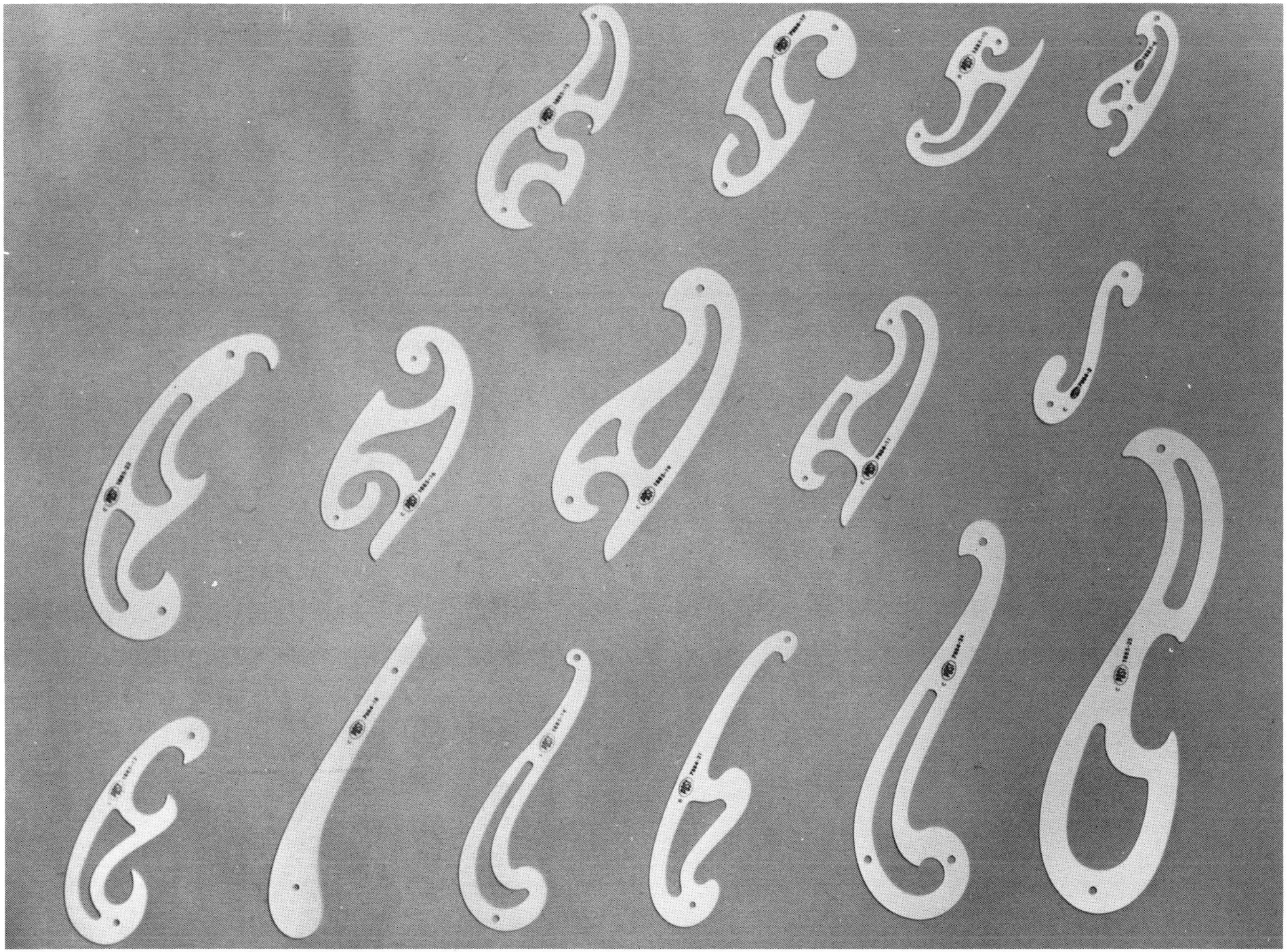

Fig. 1-21 A variety of irregular curves

classifications, such as ship's curves, rule curves, and railroad curves. Irregular curves connect a series of points in a noncircular pattern. The irregular curve is adjusted until three or more dots line up with a portion of the edge of the curve. Draw the line short at the first and last points to avoid making humps in the line. A segment of the curve is drawn this way until the line is the desired length, figure 1-22.

FLEXIBLE CURVES AND SPLINES

Flexible curves and splines are made of metal, plastic, rubber, or any combination of these materials. They are easily flexed into any position so a smooth line can be drawn along the edge, figure 1-23. Some curves use lead weights to hold them in position as the curve is drawn.

TEMPLATES

Thin sheets of plastic cut with openings in various shapes and symbols are called *templates*. General templates include shapes

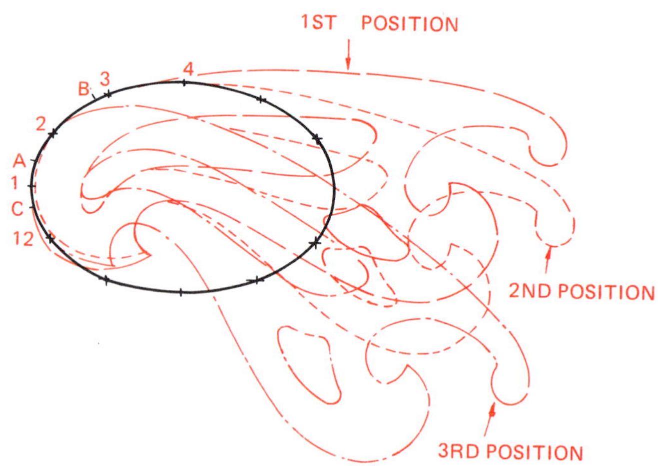

Fig. 1-22 Using the irregular curve

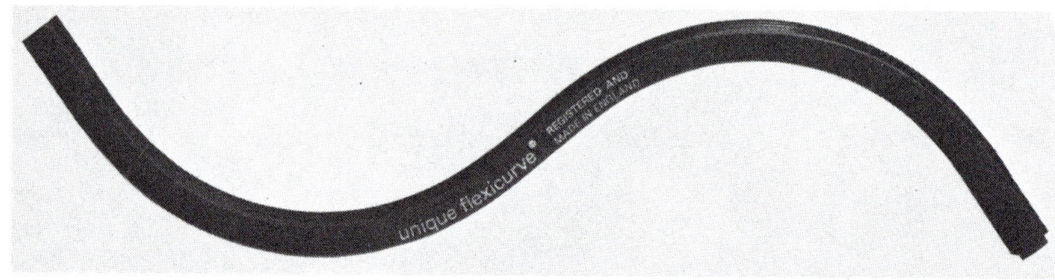

Fig. 1-23 A flexible curve aids in drawing irregular curves.

such as circles, figure 1-24; ellipses, figure 1-25; and lettering guides. Specialized templates include symbols for architectural, electrical, chemical, mathematical, structural, mechanical engineering, isometric, and metric drafting. Custom templates are also made for specialized jobs. They are useful, time saving devices.

DRAFTING MEDIA

The choice of material for drawing sheets was once limited to paper or cloth. Now there are a wide variety of papers, cloths, and films each having its own advantages and limitations. The charac-

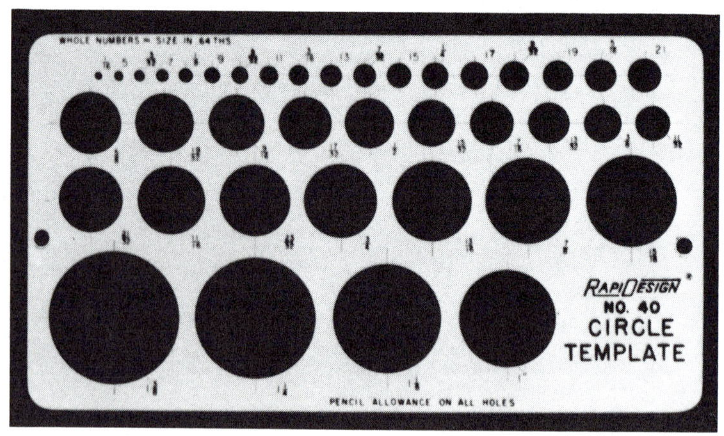

Fig. 1-24 Circle template

teristics of strength, transparency, erasability, permanence, and stability vary widely. Since new improvements are being made all the time, experience is the only reliable basis for selecting the best medium for a particular job.

Wood Pulp Papers

Opaque and translucent tracing papers made totally of wood pulp are now used only where a cost factor is of major concern, such as in beginning drafting classes. The papers are generally tinted so they do not soil as easily as some other media. The tracing paper is not as transparent as those with special agents added.

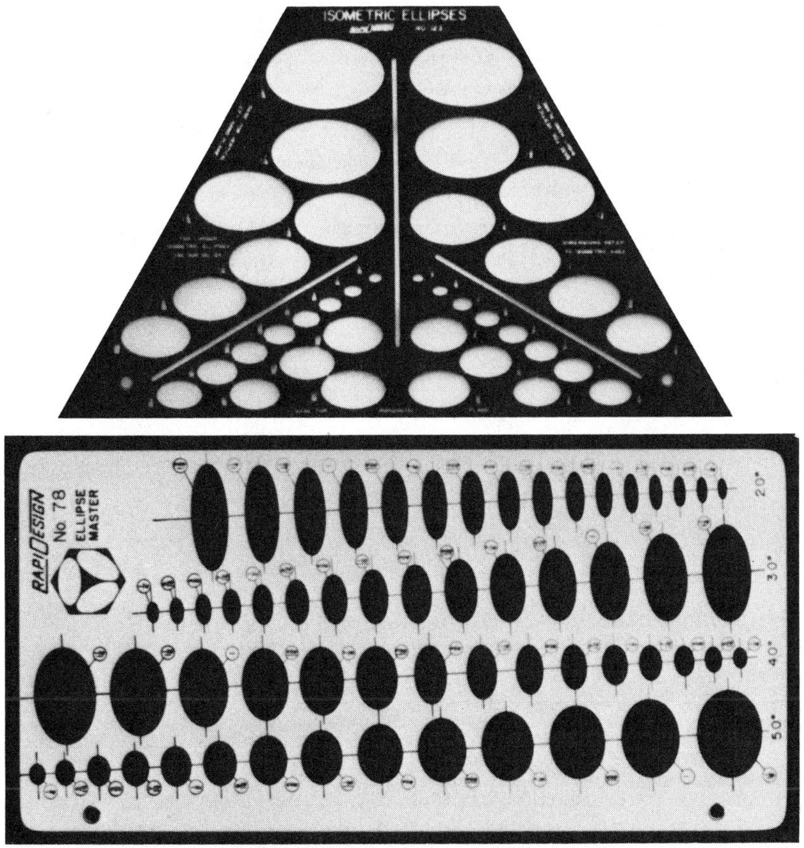

Fig. 1-25 Elliptical templates

Rag Content Papers

High quality, rag content papers and illustration boards are used for illustrations, pictorials, maps, and charts that must be reproduced by camera processing. The surface textures are varied to accept many different types of inks, paints, and shading media as well as pencil. The heavier weight boards are used when presentation quality is important.

Vellum

Vellum is transparent rag content paper coated with a synthetic resin. The resin increases its strength, transparency, durability, and erasability. Paper treated in this manner does not discolor, become brittle, or "ghost." *Ghosting* is the erased lines that show through when a drawing is reproduced. Vellum is the most widely used drawing material in industry. Although the processes and chemicals used to make vellums vary with the manufacturer, they are constantly being improved with testing and experimentation.

Tracing Cloth

Tracing cloth is a finely woven cotton fiber material that is sized with starch to make a surface that accepts both pencil and ink. The working side is dull or "frosted." The erasability is excellent. The white cloth is used for pencil; blue for ink. Its transparency and permanence are greater than paper, but it is unstable and expands or shrinks depending on the humidity.

Film

Polyester drafting film is the best medium for drawings requiring high transparency, durability, stability, and easy erasing.

Its surface is waterproof, does not discolor, or become brittle with age. The thickness and quality of this material is controlled in the manufacturing process. Its only drawback is its high cost.

Standard Sizes of Drawing Sheets

Standard sheet sizes are as follows:

ENGLISH SYSTEM			METRIC SYSTEM	
A	8 1/2 x 11	9 x 12 (inches)	A4	210 x 297 (millimetres)
B	11 x 17	12 x 18	A3	297 x 420
C	17 x 22	18 x 24	A2	420 x 594
D	22 x 34	24 x 36	A1	594 x 841
E	34 x 44	36 x 48	A0	841 x 1189

PENCILS AND LEAD HOLDERS

Drawing pencils are available in different degrees of hardness ranging from 9H (hard) to 6B (black or soft). The most commonly used drafting leads are the hard (2H, 3H, 4H) or medium (HB, F, H). The softer leads are used for sketching or rendering. They wear down faster and smudge easier than the hard leads used for instrument drawing. Special leads marked K, L, and E are made in the same range of hardness, but are used on plasticized papers and films.

The drafter must know the degree of hardness best suited for a particular job. The selection of a wood-cased pencil or a mechanical lead holder is another personal decision. There are a variety of styles available, figure 1-26. Some hold .035 inch thick leads and some hold .5 mm or .8 mm extra fine leads, figure 1-27. The lead holders are refillable. Refill leads come in all degrees of hardness, as well as plastic leads for use on drafting films.

SHARPENERS AND SHARPENING TECHNIQUES

Wood-cased pencils are sharpened with a knife or a drafting pencil sharpener. Only the wood is removed leaving about 3/8 inch of lead exposed. The lead is then pointed on a sandpaper pad or a lead pointer in the same manner as the graphite used in the lead holder is shaped.

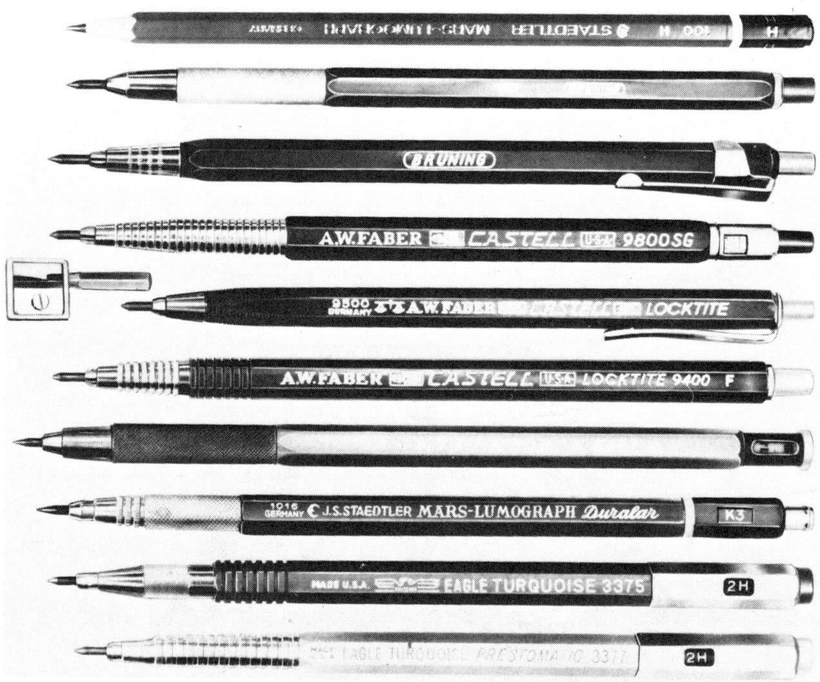

Fig. 1-26 Drafting pencils and lead holders

Various styles and shapes of lead pointers, figure, 1-28 are used to shape the lead on a pencil or a lead holder. They are electric or manual, pocket-size or attached to the table. The faster and more accurately the point is produced, the more convenient the pointer is to use.

A conical point is preferred for lettering and measuring. It is made by rotating the pencil between the fingers while rubbing it along the sandpaper pad in one direction, figure 1-29. A wedge or chisel point is sometimes preferred for drawing lines. The wedge is formed by rubbing opposite sides of the lead along the

Fig. 1-27 Lead holders designed for either thick or thin leads

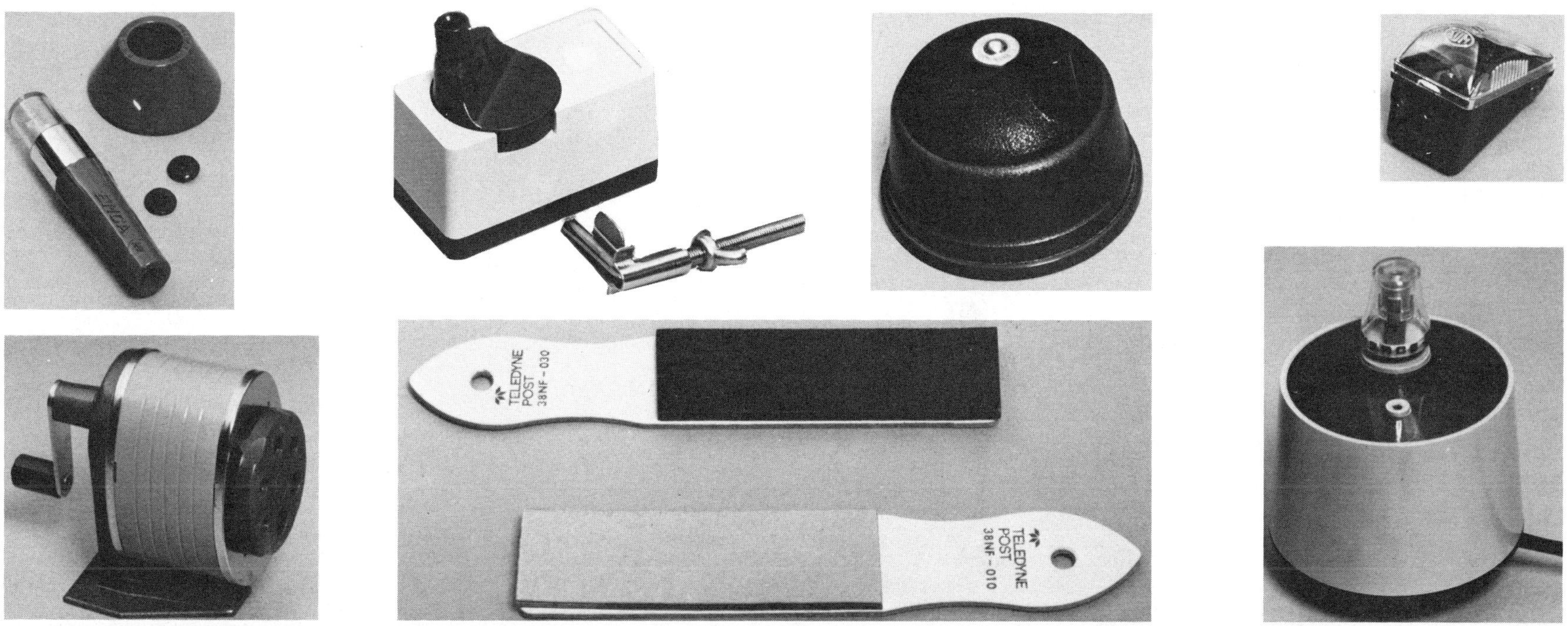

Fig. 1-28 Pencil pointing devices

sandpaper pad, shaping the point similar to the end of a screwdriver. A chisel point is made by rubbing only one side of the lead on the sandpaper to form a point like the edge of a chisel. This is the point used in a compass.

Excess graphite is removed from the point after sharpening by using a cloth or a prepared point cleaner disc.

ERASERS

Erasers vary almost as much as the drawing media they clean, figure 1-30. They are available in block or pencil form. A soft rubber eraser removes light lines and cleans up smudges. A vinyl eraser works best to remove plastic lead from polyester drawing film. It is also excellant for general cleanup of vellums. A moldable, kneaded rubber eraser removes excess graphite without damaging the surface on any drafting medium. It does not remove heavy lines.

Heavy lines that must be removed to change a drawing or correct an error are best erased with an electric eraser, figure 1-31. Soft rubber and vinyl erasers are made for use in the electric eraser.

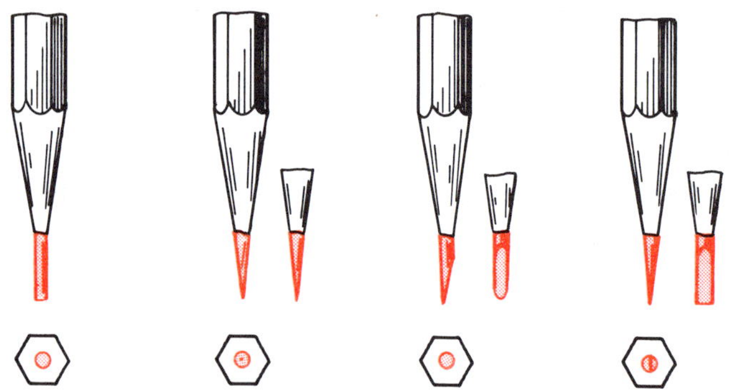

Fig. 1-29 Drafting pencils are sharpened to various shapes.

Fig. 1-30 Erasers vary in shape, composition and hardness.

Avoid too much pressure when erasing. This causes overheating or smudges. Hard rubber erasers remove ink lines only. Ink can also be removed from drafting film with special chemical erasers and solvents.

ERASING SHIELD

An erasing shield is a thin, metal plate with various shaped openings in it, figure 1-32. The shield is placed on the drawing so that the line to be erased appears in one of the openings. As the line is erased, the shield protects other lines near it.

DUSTING BRUSH

The drafter's dusting brush is used frequently to clean away eraser crumbs and dust, figure 1-33.

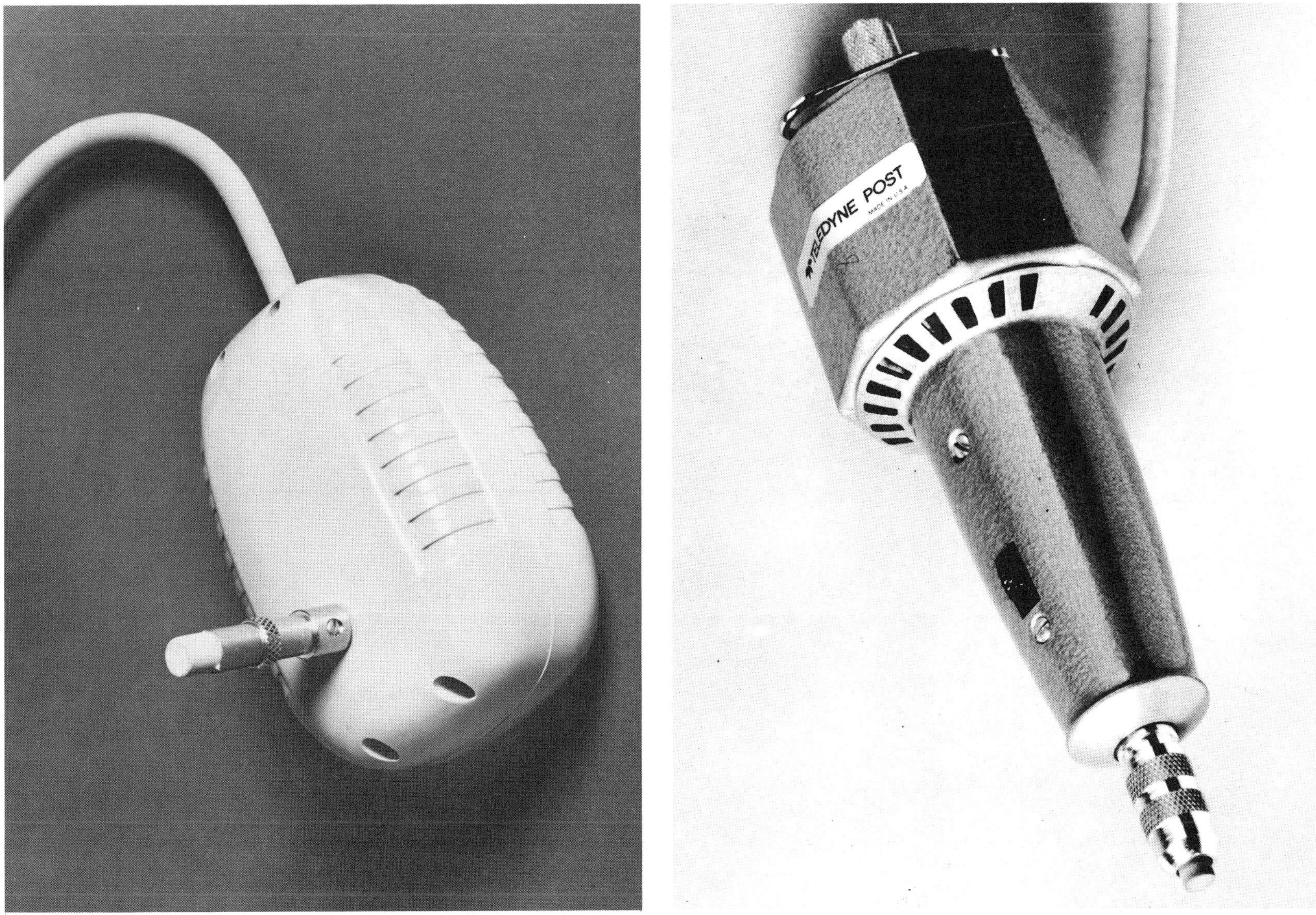

Fig. 1-31 Two styles of electric erasers

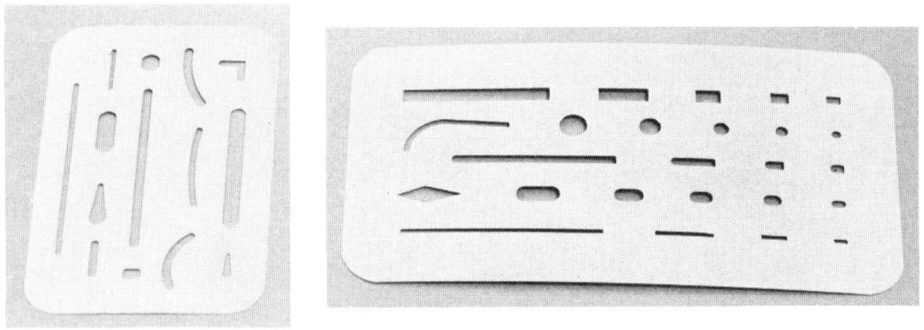

Fig. 1-32 Erasing shields

SPECIAL CLEANING AIDS

A dry cleaning pad filled with erasure materials can be used to keep the entire drawing surface clean, figure 1-34. A powdered substance called *pounce* is sometimes sprinkled over vellum, drafting cloth, or film to keep the surface clean. It also improves the quality of the lines drawn on the medium. Excess powder is removed with a dusting brush after the drawing is finished.

DRAFTING TAPE

Used to fasten the drawing sheet to the drafting surface, *drafting tape* is specially treated to stick to the surface without tearing the sheet when it is removed. Tapes come in rolls or in a convenient dot form for easy dispensing.

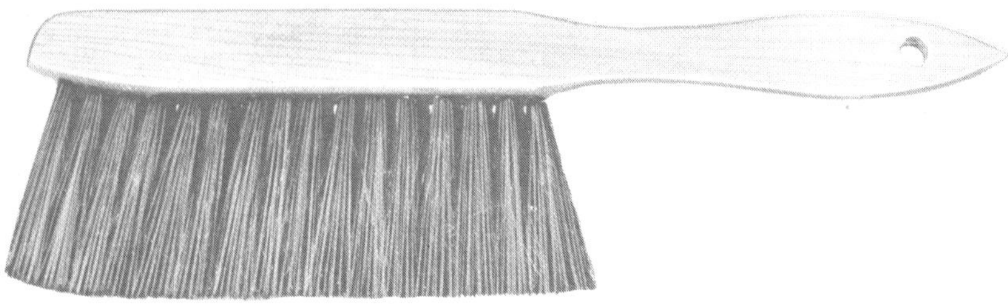

Fig. 1-33 A dusting brush is used to keep the drawing surface clean.

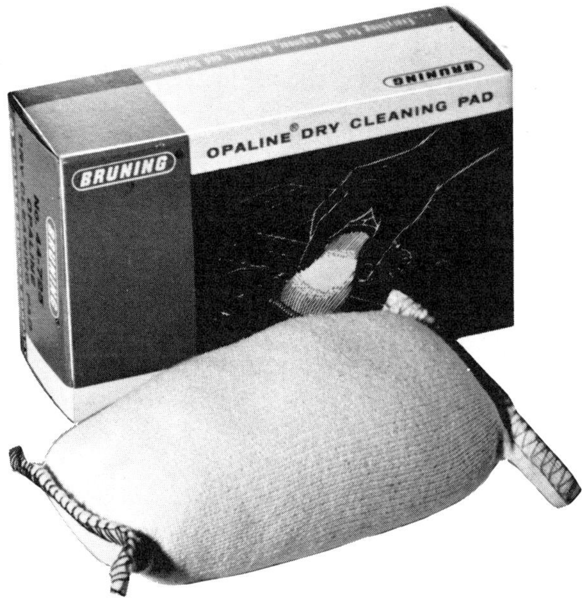

Fig. 1-34 A dry cleaning pad will assist in keeping the entire drawing clean.

ASSIGNMENT A

The drawings made from the instructions in this book are to be drawn on 11 by 17 inch paper. This is a standard "B" size sheet. This assignment demonstrates how to layout the border and set up the title strip.

Procedure

1. Place the T square against the working edge near the bottom of the drawing surface. Rest the drawing sheet against the top of the blade near the working edge.
2. Fasten the upper corners of the sheet with drafting tape or thumbtacks. Move the T square. Smooth the sheet down and fasten the lower corners.

3. When using a drafting machine, the drawing sheet is aligned to the machine horizontally, or on any angle that is comfortable to the drafter.

Steps 4 through 6 are necessary only if the paper is larger that 11 by 17 inches. If the sheet is 11 by 17 inches, proceed with step 7.

4. Put a dot 11 inches up from the bottom of the paper and draw a light horizontal line through that mark.
5. Holding a triangle against the T square and drawing from bottom to top, figure 1-35, draw a light vertical line close to the working edge of the drawing sheet to connect the horizontal line to the bottom of the paper.
6. Measure 17 inches to the right of the vertical line and place a dot. Draw a vertical line through this mark connecting the horizontal line to the bottom of the paper.

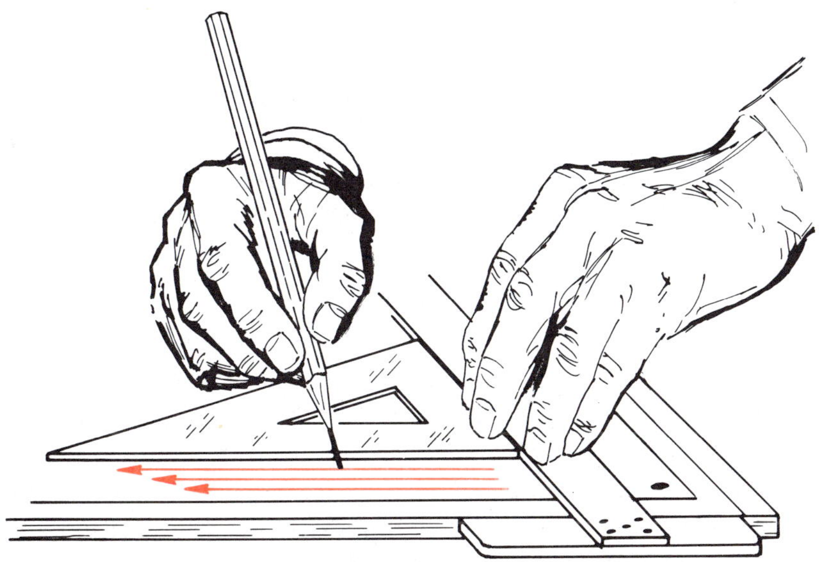

Fig. 1-35 Position of triangle for drawing vertical lines

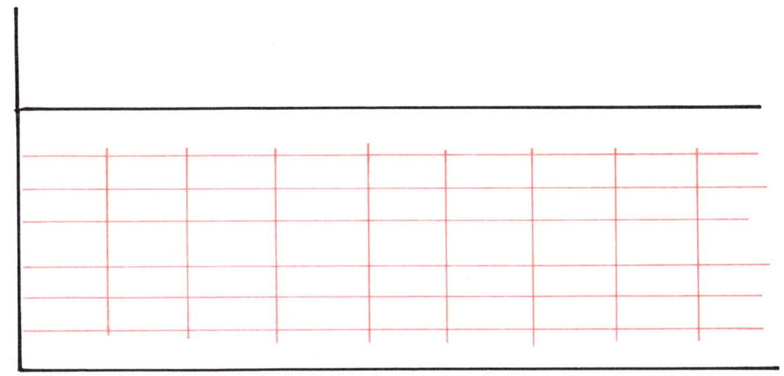

Fig. 1-36 Guide lines for vertical lettering

The lines just drawn are trimming lines along which the sheet is cut when the drawing is complete. The drawing sheet will then be 11 by 17 inches.

7. Measure and mark 1/4 inch toward the center from each edge of the drawing sheet, or trim lines, and draw lines through these marks parallel to the edges. These lines are margin lines and are darkened when the drawing is completed.
8. Measure up 3/4 inch from the bottom margin line and make a horizontal line. This will be the title strip. It is called a title block when the shape is a rectangular block in the lower right corner of the sheet.
9. Measure from the left 4 1/2 inches, 4 inches, and 4 inches. Draw vertical lines through these points within the title strip.
10. From the top of the title strip, measure down 1/8 inch, 3/16 inch, 1/8 inch, and 3/16 inch and draw light horizontal guidelines for lettering. Draw a series of light vertical guidelines about 1/4 inch apart to aid in freehand lettering, figure 1-36.

The sheet is now ready for the drawing. This standard layout is used on each sheet in this course.

Blocking in the Views

When several details are done on one sheet, the views are spaced to give a balanced appearance to the drawing. This is accomplished by first selecting the details to be drawn. Determine the length and width of the space required for the views of each detail including the necessary space for dimensioning between the views. After the overall length and width of each view is found, they are blocked in with light lines. The blocks are shifted on the drawing until the desired spacing is attained before any views are drawn. Figure 1-37 shows how the views are blocked in. Layout solutions for each assignment are given in the procedure directions for drawings.

Parts List

Assembly drawings, unit assemblies, group assemblies, or subassemblies usually contain a list showing the number and name of each part included in the assembly. This list is called a *parts list.* The parts are identified in the assembly drawing by a detail number. The detail number is enclosed in a circle attached to the part with a leader.

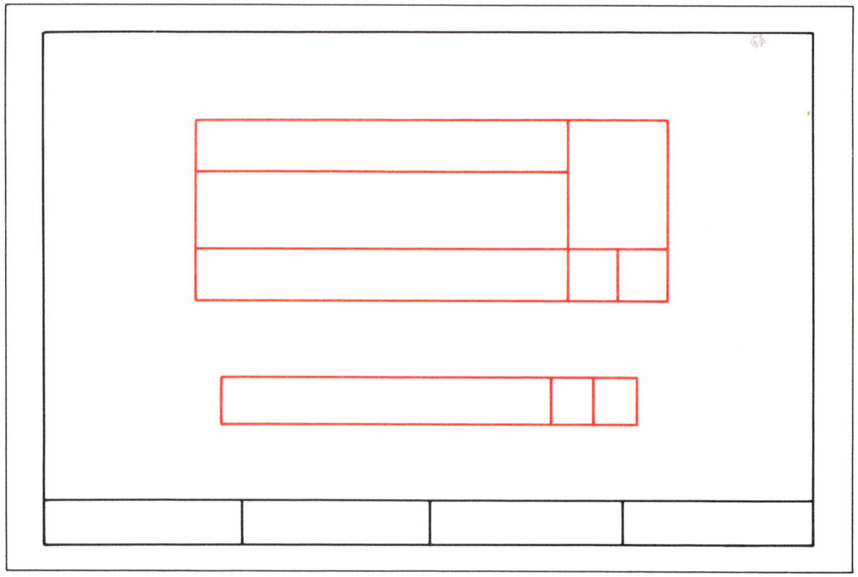

Fig. 1-37 The views blocked in

Material List

A bill of materials, or *materials list,* is placed on detail drawings that show more than one part. The material list usually specifies:

- the part number.
- the size of the unfinished or raw material.
- the allowance for machining.
- a description of standard parts such as bolts, screws, and nuts.
- an identifying name or number of special parts that are purchased.
- the number of each part required.
- the kind of material each part is made of.

The material list is used as a guide when the necessary raw material and standard parts are ordered from the stock department. A typical material list is illustrated in figure 1-38. It is usually placed in the lower right corner of a drawing sheet above the title strip.

LETTERING

Drafting is a language that uses lines, symbols, dimensions, and notations to describe an object. The lines and symbols are drawn with drawing instruments and tools, but the dimensions, notes, titles, bills of materials, and specifications are usually lettered freehand on the drawing. Lettering is used instead of writing because it is more legible. Capital, single-stroke, gothic letters are generally used because they can be drawn fast and easily. These letters are formed by straight and curved strokes of uniform width.

5	DIE CASTING D-18	ZINC ALLOY	1
4	NO. 2 TAPER PIN	STD.	2
3	1/2 - 13 HEX. HD. CAP SCR.	STD.	4
2	PATT. NO. 4718-L	C.I.	1
1	5/8 x 1 x 6 1/8 LONG	C.D.S.	1
PART NO.	SIZE OF MATERIAL	MAT.	NO. REQ.

Fig. 1-38 A material list

The lettering sheet in figure 1-39 shows the shape of the different letters and numerals. It also shows the sequence and direction of the strokes used to form them.

Horizontal guidelines give a uniform height for the letters. Vertical guidelines at the beginning keep letters straight. With practice, the vertical lines become unnecessary.

An H, 2H, or 3H pencil with a conical point is best for lettering on vellum.

ASSIGNMENT B

Good lettering adds much to the attractive appearance of a drawing. Careless or poor lettering spoils an otherwise good drawing. Skill and speed in lettering is acquired through practice. This assignment shows how to form the letters properly.

Procedure

1. Use the sheet already prepared with margin lines and title strip.
2. From the top margin line, measure down 3/4 inch and place a dot. From this dot, continuing downward, point off two more dots 3/16 inch apart. Draw a light horizontal line through each dot.
3. Measure down 1/2 inch and place a dot. Point off two more dots 3/16 inch apart. Draw a light horizontal line through each dot.
4. Repeat Step 3 four more times. When completed there are six sets of horizontal guidelines.
5. Measure down 3/8 inch and place a dot. From this dot point off two more dots 3/32 inch apart. Draw a light horizontal line through each dot.
6. Repeat Step 5 four more times. When completed, there are five more sets of horizontal guidelines.
7. Measure down 3/8 inch and place a dot. From this dot point off 1/8 inch, 1/4 inch, and 1/8 inch. Draw a light horizontal line through each dot.
8. On the top line of the title strip beginning 1 inch from the left end, point off 1/2 inch spaces all the way across the line. From each of these points draw a light vertical line upward to the top margin line.
9. Check for accuracy and proceed with lettering.
10. The first group of letters, figure 1-40, are formed with vertical and horizontal strokes. Study the form of each letter and the order in which the strokes are made. Note that stroke 4 of the letter E and strokes 3 of the letters F and H are slightly above the center of the guideline. Except for the letter I, the normal widths of the letters shown are a little more than 2/3's of their height. For good spacing, the area of the spaces between the letters is approximately equal as shown by the shaded portion in figure 1-41. On the first set of horizontal guidelines, beginning 1 inch from the left margin line, make five of each letter in figure 1-40. The space between

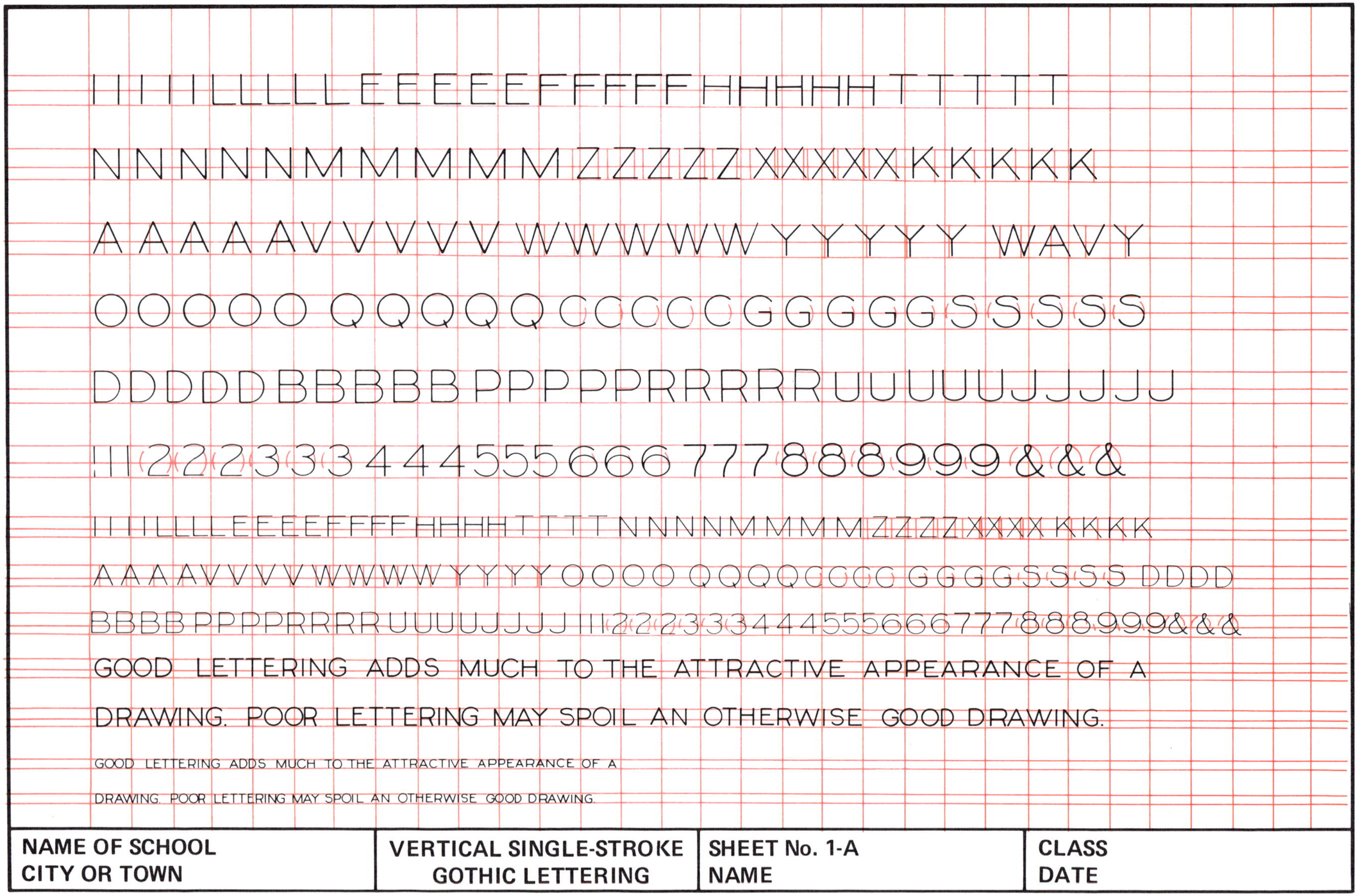
GOOD LETTERING ADDS MUCH TO THE ATTRACTIVE APPEARANCE OF A
DRAWING. POOR LETTERING MAY SPOIL AN OTHERWISE GOOD DRAWING.
GOOD LETTERING ADDS MUCH TO THE ATTRACTIVE APPEARANCE OF A
DRAWING. POOR LETTERING MAY SPOIL AN OTHERWISE GOOD DRAWING.
NAME OF SCHOOL
CITY OR TOWN
VERTICAL SINGLE-STROKE
GOTHIC LETTERING
SHEET No. 1-A
NAME
CLASS
DATE

Fig. 1-39

these letters is approximately the same as those in figure 1-41. The downward strokes are parallel to the vertical guidelines.

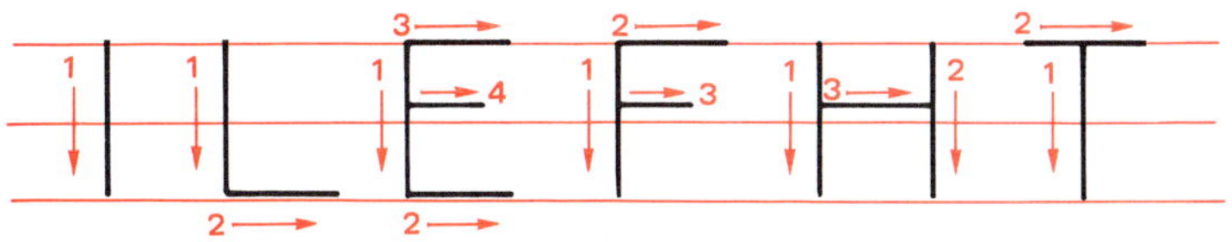

Fig. 1-40 Letters formed with vertical and horizontal strokes

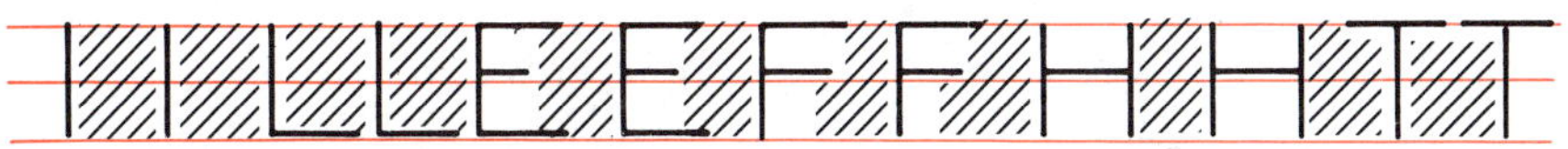

Fig. 1-41 Spacing lettering

11. The second group of letters, figure 1-42, are combinations of vertical, horizontal, and oblique strokes. For the letters Z, X, and K, a light freehand vertical guideline is first sketched at each side as an aid in forming them. The widths of the top of these three letters are slightly less that the widths at the bottom. The width of the letter M is greater that its height. On the second set of guidelines, make five of each of these letters.

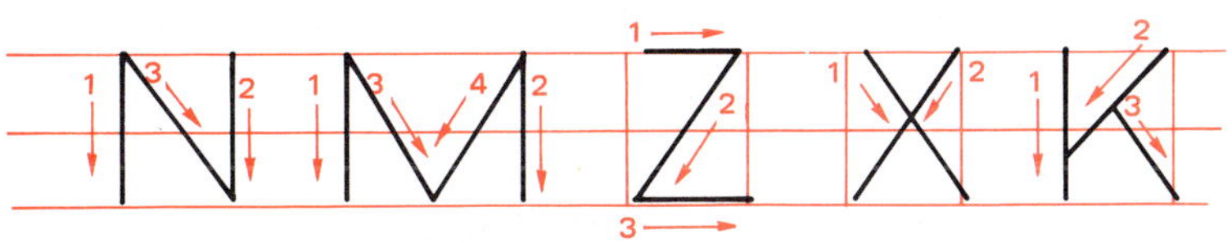

Fig. 1-42 Letters formed with vertical, horizontal and oblique strokes

12. The next group of letters is shown in figure 1-43. A light freehand vertical guideline is first sketched to mark the center of these letters. The widths are pointed off and the inclined strokes are made. The width of the letter W is greater than its height. Notice that stroke 3 of the letter A is well below the center guideline. On the third set of guidelines make five of each letter.

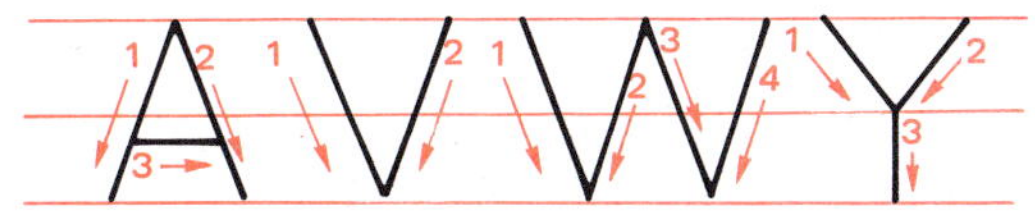

Fig. 1-43 Letters formed with oblique strokes

13. Figure 1-44 shows a group of letters formed with curved strokes. First sketch the approximate circles freehand, then form the letters with the strokes as indicated. Stroke 5 of the letter G is placed on the center guideline. On the fourth set of guidelines make five of each letter.

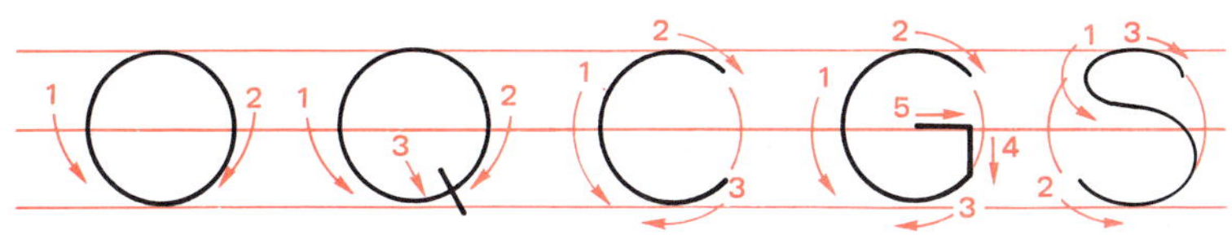

Fig. 1-44 Letters formed with curved strokes

14. The final group of letters are combinations of straight and curved strokes, figure 1-45. Notice that stroke 3 of the letter B is slightly above the center line, while stroke 2 of the letters P and R is slightly below the center line. The letter U is normal width. The letter J is slightly less than normal width. On the fifth set of guidelines make five of each of these letters.

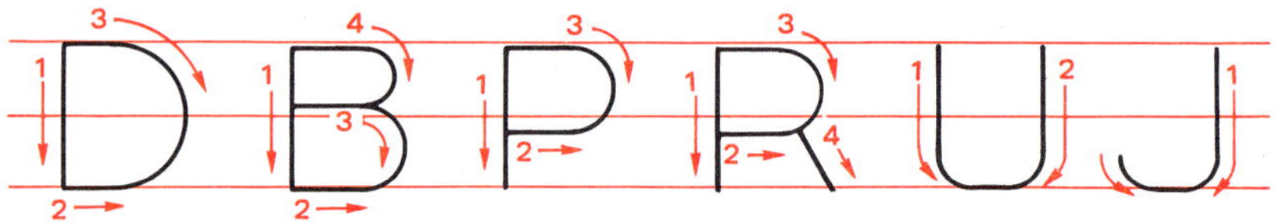

Fig. 1-45 Letters formed with combined straight and curved strokes

15. The construction of numerals is given in figure 1-46. The symbol at the right is called the *ampersand.* It is sometimes used in place of the word "and." Make three of each numeral on the sixth set of guidelines.

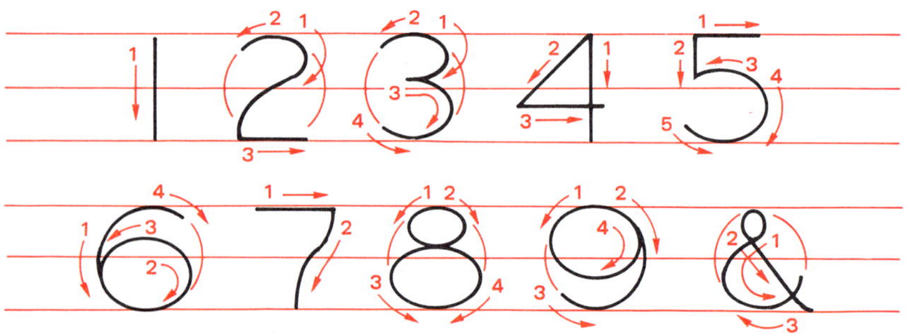

Fig. 1-46 Vertical numerals

16. The remainder of the letters and numerals on the sheet are made 3/16 inch or 1/8 inch high. Letters of this height are used in the title strip and for notes on the drawings. Complete the lettering as shown in figure 1-39. Carefully form each letter with the proper strokes.
17. Letter the title strip.
18. Clean up the sheet.

REVIEW QUESTIONS

1. List two ways to keep drawings clean and neat.
2. Explain how to use an erasing shield.
3. List some industries that commonly use a decimal-inch, mechanical engineer's scale.
4. Why is it poor practice to use the edge of a scale to draw a line?
5. What type of lines are the heaviest (darkest) on a drawing?
6. What are the requirements for a good drawing board?
7. Explain what equipment is used to make a series of parallel vertical lines.
8. Explain why construction lines are light weight.
9. What type of eraser cleans a pencil line off a plasticized sheet best?
10. What is the standard size of the drawing sheets to be used in this course?
11. What is the purpose of trimming lines?
12. What is the purpose of the margin lines?
13. How wide is the recommended margin?
14. What are the dimensions of the title strip?

UNIT 2
GEOMETRY IN DRAFTING

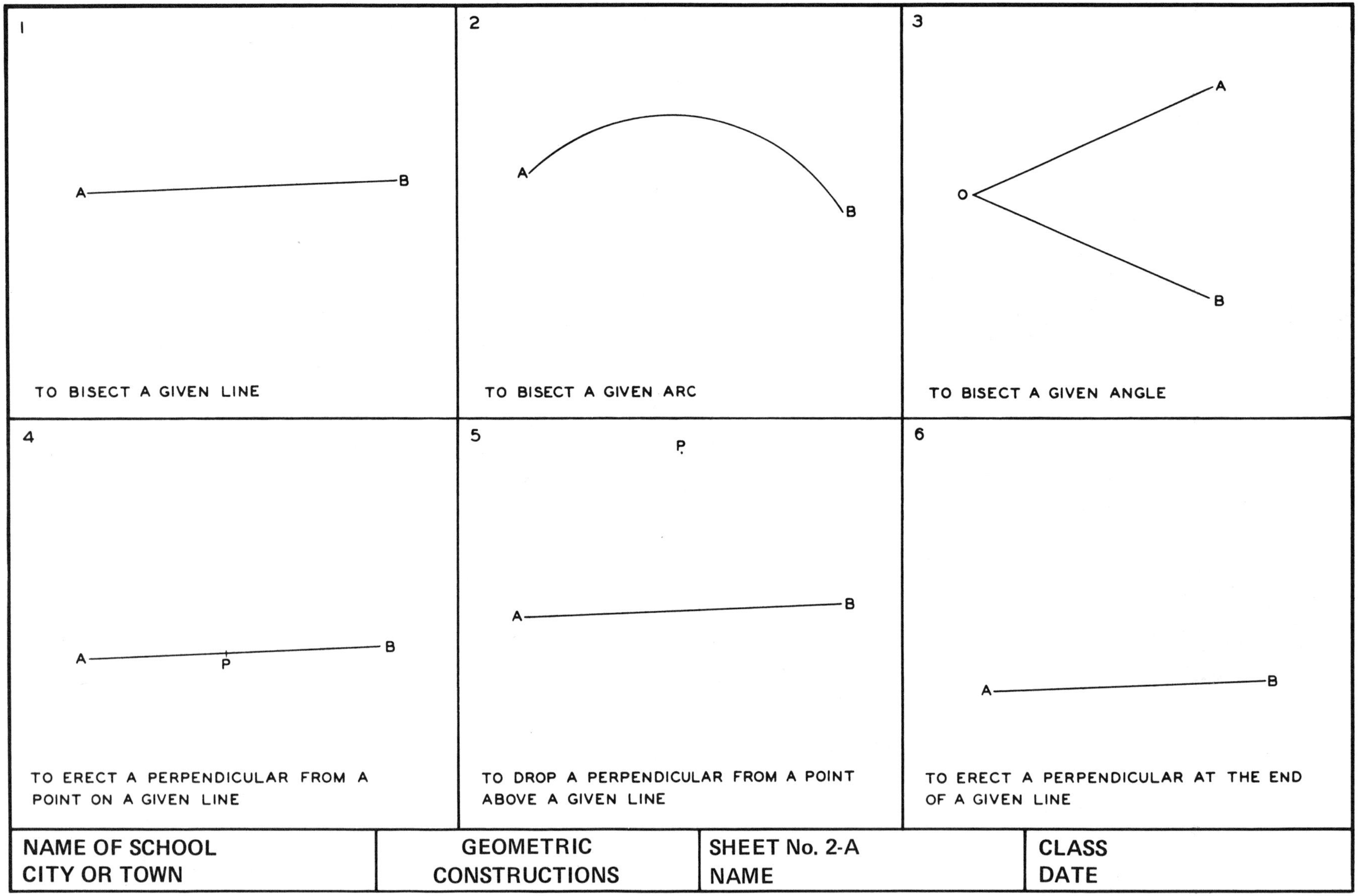
1
A
B
TO BISECT A GIVEN LINE
2
A
B
TO BISECT A GIVEN ARC
3
A
O
B
TO BISECT A GIVEN ANGLE
4
A
P
B
TO ERECT A PERPENDICULAR FROM A POINT ON A GIVEN LINE
5
P
A
B
TO DROP A PERPENDICULAR FROM A POINT ABOVE A GIVEN LINE
6
A
B
TO ERECT A PERPENDICULAR AT THE END OF A GIVEN LINE
NAME OF SCHOOL
CITY OR TOWN
GEOMETRIC CONSTRUCTIONS
SHEET No. 2-A
NAME
CLASS
DATE

Fig. 2-1

The geometric constructions in this section show some short, accurate methods of drawing parallel lines, perpendiculars, and tangent arcs; and the construction of triangles and polygons under various conditions.

Some of these constructions are used frequently by drafters and other craftsmen in their daily work. Many are useful in laying out large sheet metal patterns, airplane work, and in other applications where the T square and triangles cannot be used in the conventional manner. For this reason, the geometric constructions which follow should be made without a T square. Only a straightedge, compass, and dividers should be used.

Each assignment has six geometric constructions or problems. An illustration shows step-by-step how each problem is drawn. Each problem has some part given, such as a line, an arc, or an angle. The given portion of each problem is drawn with a solid or unbroken line of medium thickness. The lines used in the construction of the problem are drawn with long dashes and are thin lines. The lines which are required to complete the problem are solid or unbroken lines of medium thickness. The lines used in the construction of the problem are drawn with long dashes and are thin lines. The lines which are required to complete the problem are solid or unbroken and are drawn heavier than the given lines.

For example, in the first problem, the given line to be bisected is of medium weight. The construction lines are long thin dashes. The line which bisects the given line is made heavier than the given line.

TERMS

- A *point* has no dimensions. It has position only.
- A *straight line* has position, direction, and length but not width or thickness.
- A *radius* is a straight line from the center of a circle to the circumference or to an arc.
- An *angle* is the space or opening between two straight lines which intersect.
- A *perpendicular* is a straight line at a right angle to another straight line.
- An *arc* is a part of a circle.
- A *right angle* contains 90 degrees.
- To *bisect* is to divide into two equal parts.

ASSIGNMENT A

Draw the six geometric constructions in figure 2-1. Use only a straightedge (the edge of a triangle) and compass in their construction.

Procedure

1. Lay out the margin lines on the drawing sheet and letter the title strip. Use the T square and triangles for this purpose.
2. Divide the sheet into six equal spaces.
3. Use letters 1/8 inch high to name each problem in the proper space.
4. Put the T square away.
5. To bisect a line, figure 2-2, draw the given line A-B along the edge of the triangle in the first space. Draw this line any convenient length and in the approximate position. Then proceed with the problem step-by-step as shown. Place letters on each part of the construction. These letters are in alphabetical sequence to show the order of the construction.
6. To bisect an arc, figure 2-3, draw the given arc A-B with any convenient radius and in the approximate position. Proceed step-by-step to complete the problem.
7. Proceed as shown in figures 2-4 to 2-7 to complete the remaining problems.
8. Check each construction carefully.

PROBLEM 1

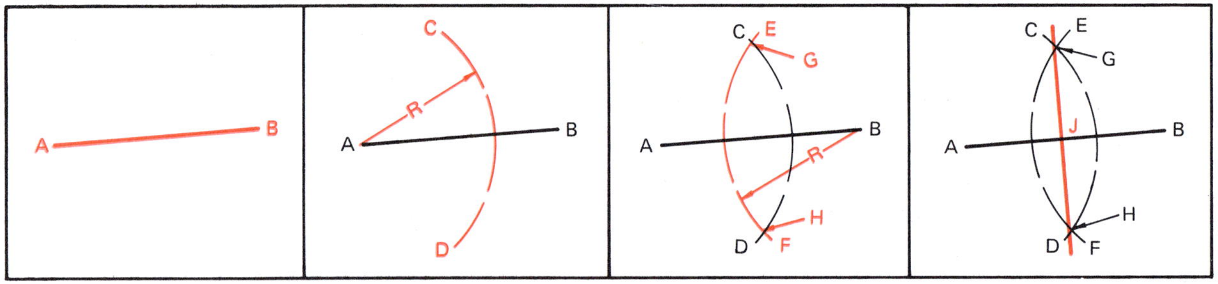

DRAW LINE A-B

WITH R GREATER THAN HALF DRAW ARC C-D

WITH SAME R DRAW ARC E-F TO LOCATE G AND H

CONNECT INTERSECTIONS G AND H TO BISECT LINE AT J

Fig. 2-2 To bisect a given line

PROBLEM 2

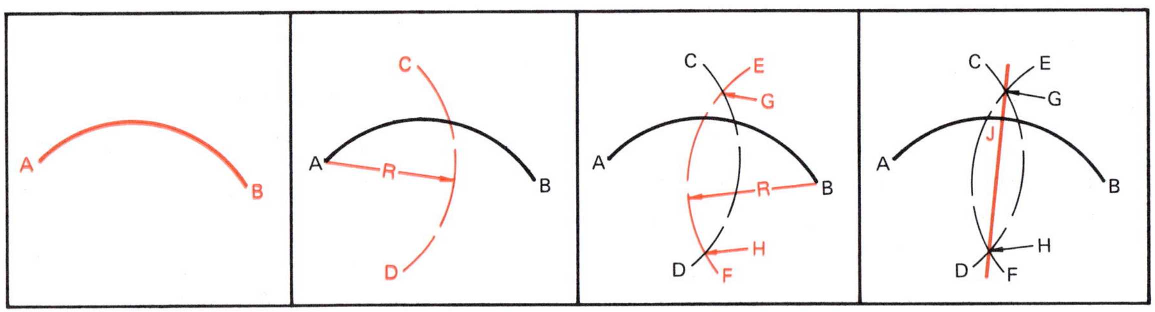

DRAW GIVEN ARC A-B

WITH R GREATER THAN HALF DRAW ARC C-D

WITH SAME R DRAW ARC E-F TO LOCATE G AND H

CONNECT INTERSECTIONS G AND H TO BISECT GIVEN ARC AT J

Fig. 2-3 To bisect a given arc

PROBLEM 3

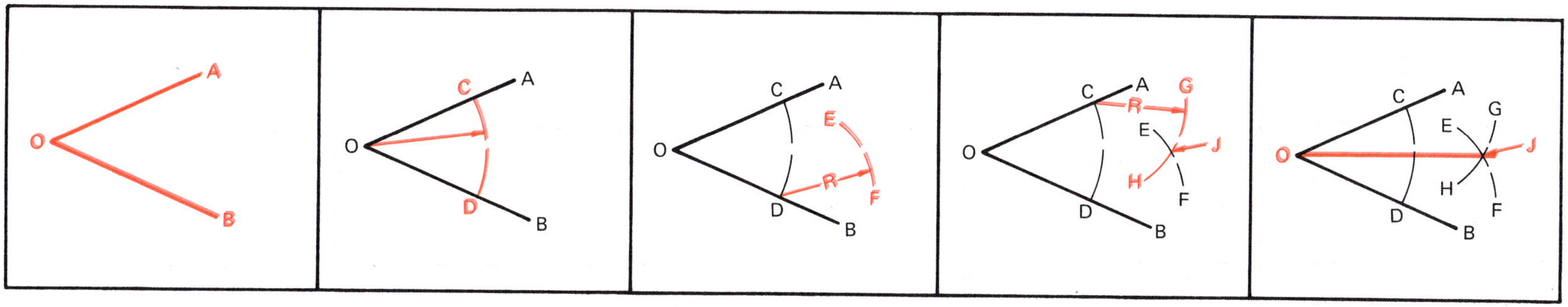

DRAW GIVEN ANGLE A-O-B | DRAW ARC C-D | DRAW ARC E-F | DRAW ARC G-H TO LOCATE J | CONNECT J AND O TO BISECT GIVEN ANGLE

Fig. 2-4 To bisect a given angle

PROBLEM 4

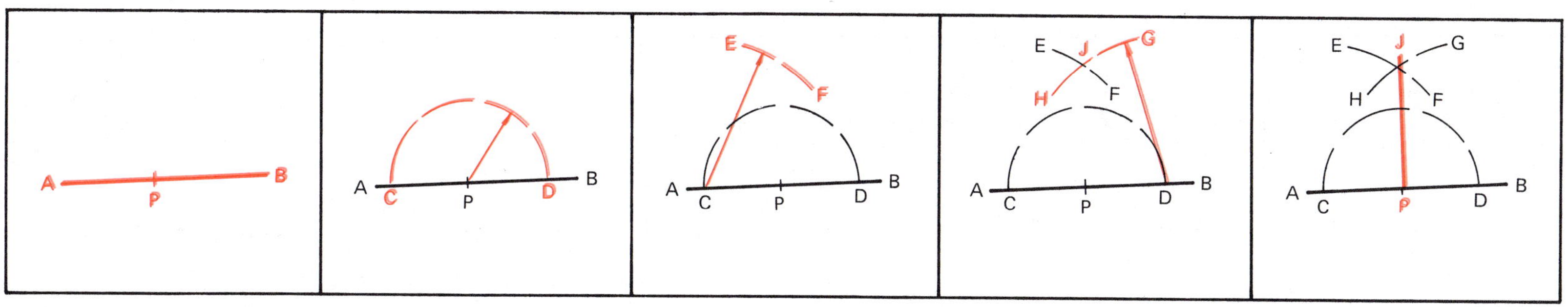

DRAW GIVEN LINE A-B AND SELECT POINT P | DRAW SEMI-CIRCLE C-D | DRAW ARC E-F | DRAW ARC G-H TO LOCATE J | CONNECT J AND P TO ERECT PERPENDICULAR

Fig. 2-5 To erect a perpendicular from a point on a given line

PROBLEM 5

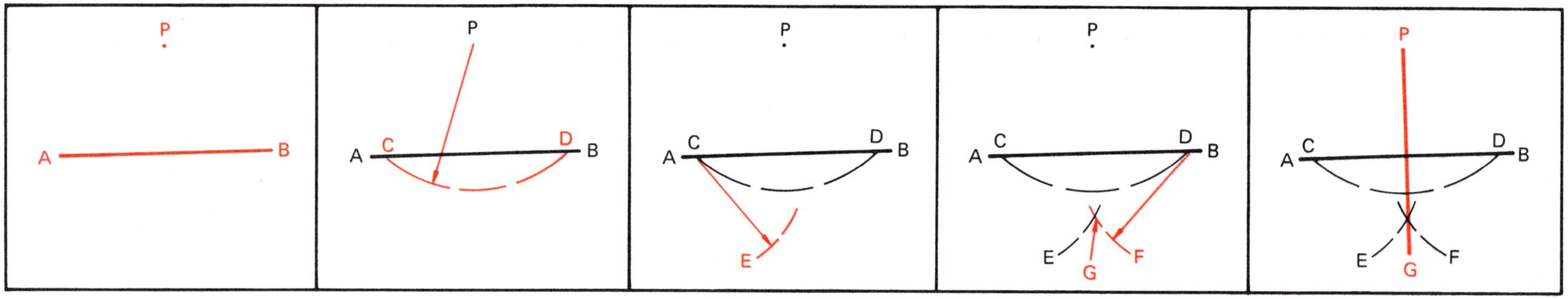

DRAW GIVEN LINE A-B AND SELECT POINT P

DRAW ARC C-D

DRAW ARC E

WITH SAME R DRAW ARC F TO LOCATE G

CONNECT P AND G TO DROP PERPENDICULAR

Fig. 2-6 To drop a perpendicular from a point above a given line

PROBLEM 6

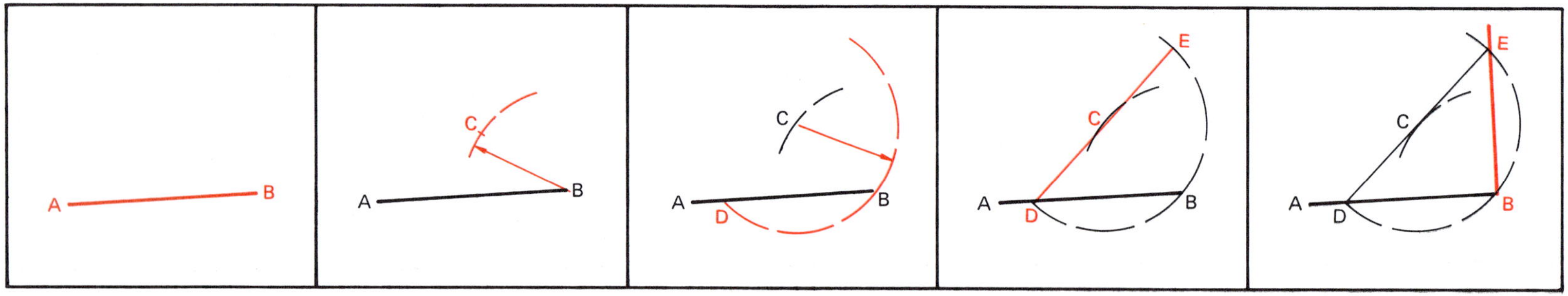

DRAW GIVEN LINE A-B

DRAW AN ARC AND SELECT POINT C

WITH SAME R DRAW A SEMI-CIRCLE AND LOCATE D

DRAW A LINE FROM D THRU C TO LOCATE E

DRAW PERPEN-DICULAR FROM B TO E

Fig. 2-7 To erect a perpendicular at the end of given line

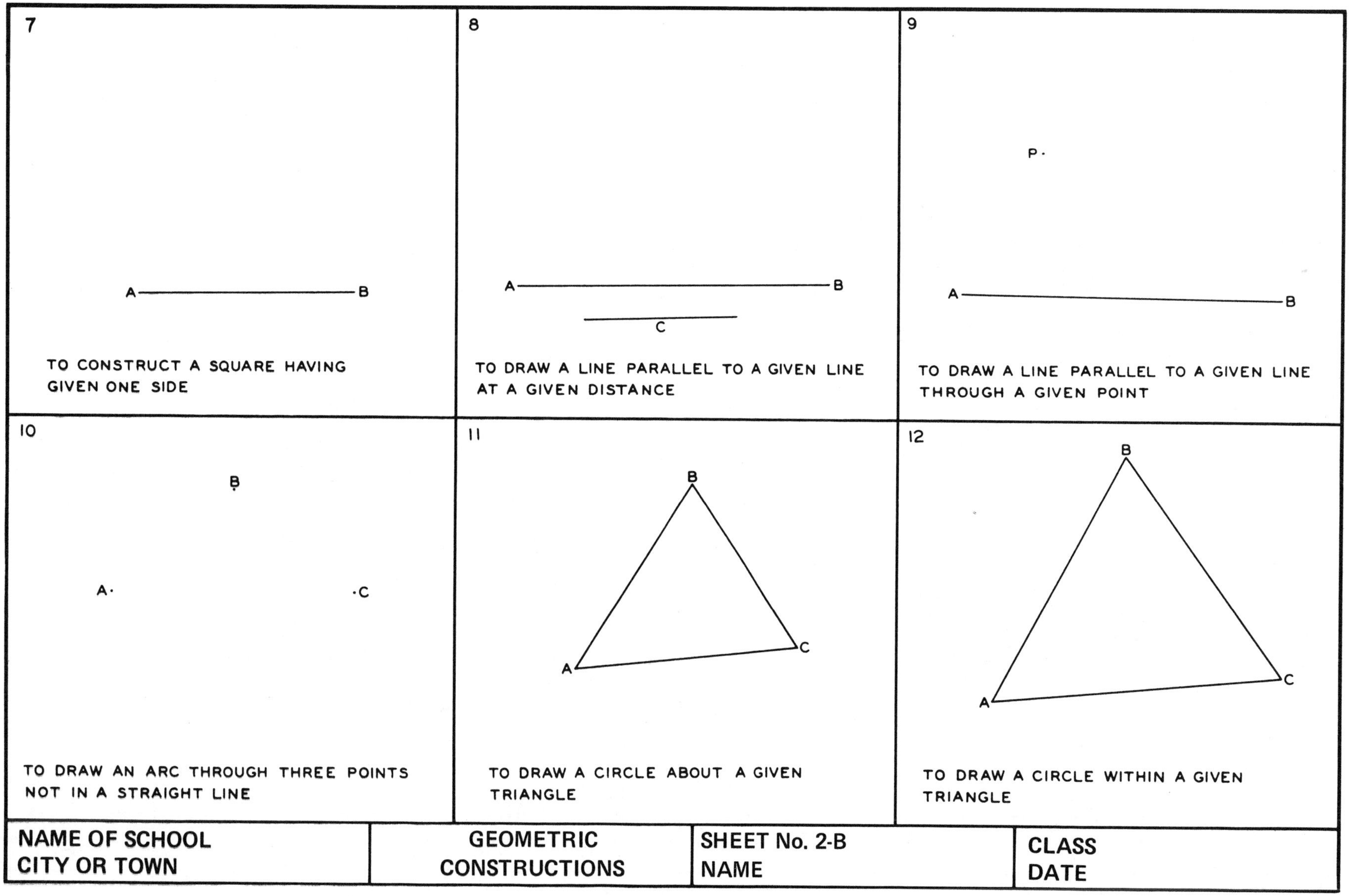
7
A
B
TO CONSTRUCT A SQUARE HAVING GIVEN ONE SIDE
8
A
B
C
TO DRAW A LINE PARALLEL TO A GIVEN LINE AT A GIVEN DISTANCE
9
P
A
B
TO DRAW A LINE PARALLEL TO A GIVEN LINE THROUGH A GIVEN POINT
10
B
A
C
TO DRAW AN ARC THROUGH THREE POINTS NOT IN A STRAIGHT LINE
11
B
A
C
TO DRAW A CIRCLE ABOUT A GIVEN TRIANGLE
12
B
A
C
TO DRAW A CIRCLE WITHIN A GIVEN TRIANGLE
NAME OF SCHOOL
CITY OR TOWN
GEOMETRIC CONSTRUCTIONS
SHEET No. 2-B
NAME
CLASS
DATE

Fig. 2-8

ASSIGNMENT B

Six geometric constructions are made in figure 2-8. The first problem is the construction of a square when one side is given. Problems two and three require the construction of lines parallel to given lines. The fourth problem shows the method of drawing an arc through three given points. The next shows the method of drawing a circle through three given points, or to draw a circle about a given triangle. The sixth problem describes the method of drawing a circle within a given triangle so that the circle is tangent to the sides of the triangle.

Draw the six geometric constructions specified in figure 2-8. Use only a straightedge and compass in their construction.

Procedure

1. Lay out the margin lines on the drawing sheet and letter the title strip.
2. Divide the sheet into six equal spaces and letter the name of each problem in the proper space with lettering 1/8 inch high.
3. The construction of each problem is shown in figures 2-9 to 2-14. Proceed step-by-step to complete the problems.

TERMS

- A *square* is a figure having four equal sides and four right angles.
- *Parallel lines* are an equal distance apart throughout their entire length.
- A *circle* is a closed curved line on a plane such that every part is an equal distance from a point within called the center.
- A *diameter* is a straight line or distance from one side of a circle through the center to the other side.
- A *triangle* is a figure having three angles and three sides.

PROBLEM 7

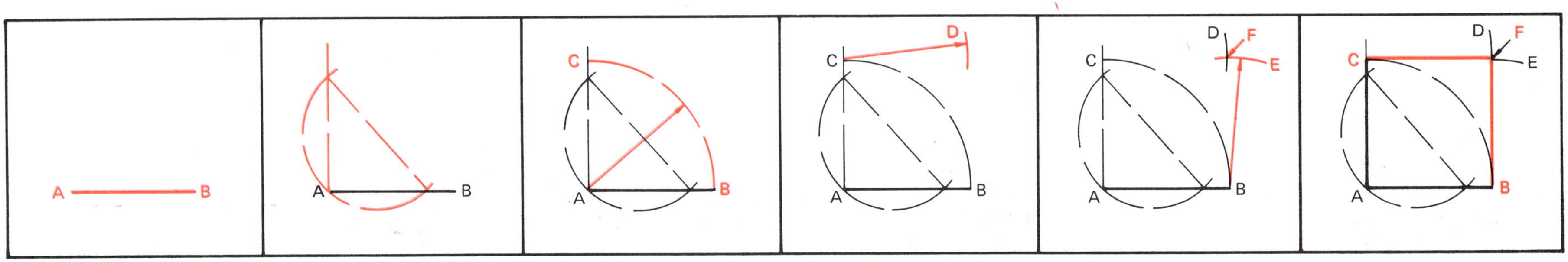

DRAW GIVEN SIDE A-B

ERECT A PERPENDICULAR AT A. PROB. 6

DRAW ARC B-C

WITH SAME R DRAW ARC D

WITH SAME R DRAW ARC E TO LOCATE F

DRAW LINES C-F AND B-F TO COMPLETE SQUARE

Fig. 2-9 To construct a square having given one side

PROBLEM 8

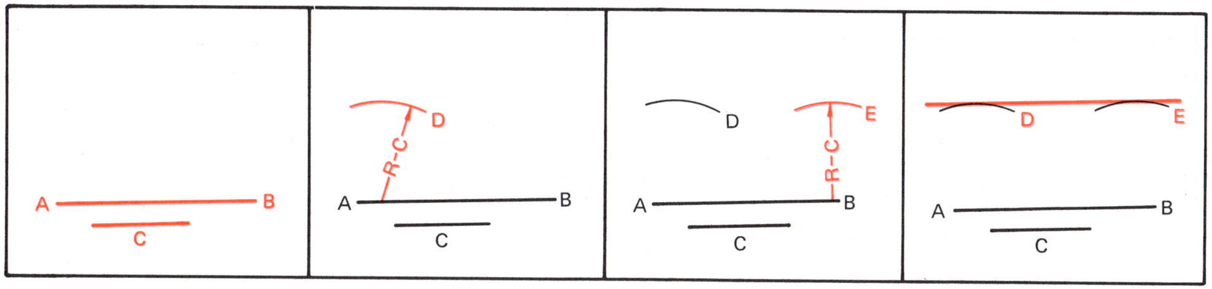

DRAW GIVEN LINE A-B AND DISTANCE C

DRAW ARC D

DRAW ARC E

DRAW PARAL-LEL LINE TAN-GENT TO ARCS D AND E

Fig. 2-10 To draw a line parallel to a given at a given distance

PROBLEM 9

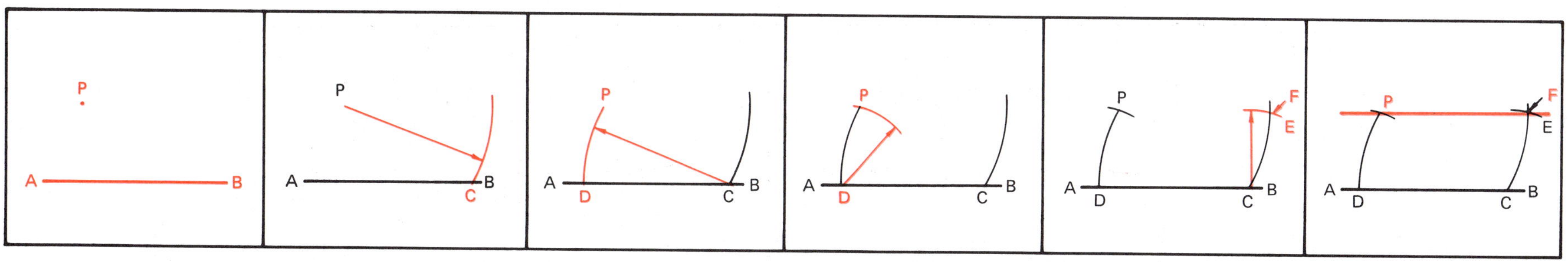

DRAW GIVEN LINE A-B AND SELECT POINT P

DRAW AN ARC TO LOCATE C

DRAW ARC P-D

MEASURE CHORD D-P WITH COMPASS-ES

WITH RADIUS D-P DRAW ARC E TO LOCATE F

DRAW PARALLEL LINE THROUGH P AND F

Fig. 2-11 To draw a line parallel to a given line through a given point

PROBLEM 10

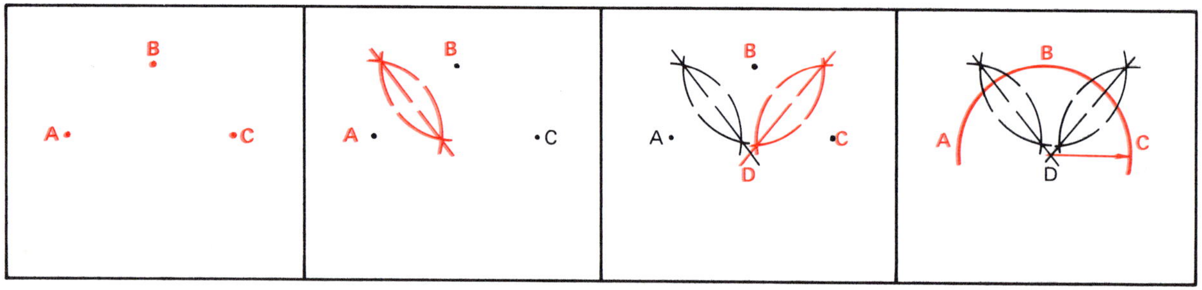

SELECT POINTS A, B AND C	BISECT A-B PROB. 1	BISECT B-C TO LOCATE CENTER D	DRAW ARC THRU POINTS A, B AND C

Fig. 2-12 To draw an arc through three points

PROBLEM 11

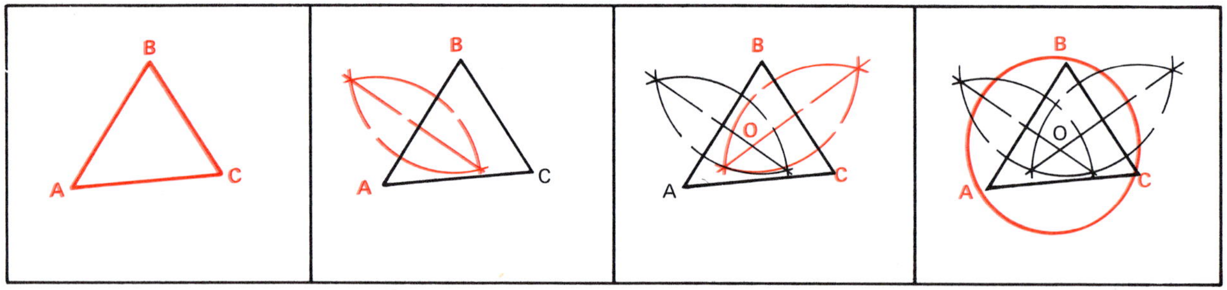

DRAW GIVEN TRIANGLE A-B-C	BISECT SIDE A-B PROB. 1	BISECT SIDE B-C TO LOCATE CENTER O	DRAW CIRCLE THRU A-B-C

Fig. 2-13 To draw a circle about a given triangle

PROBLEM 12

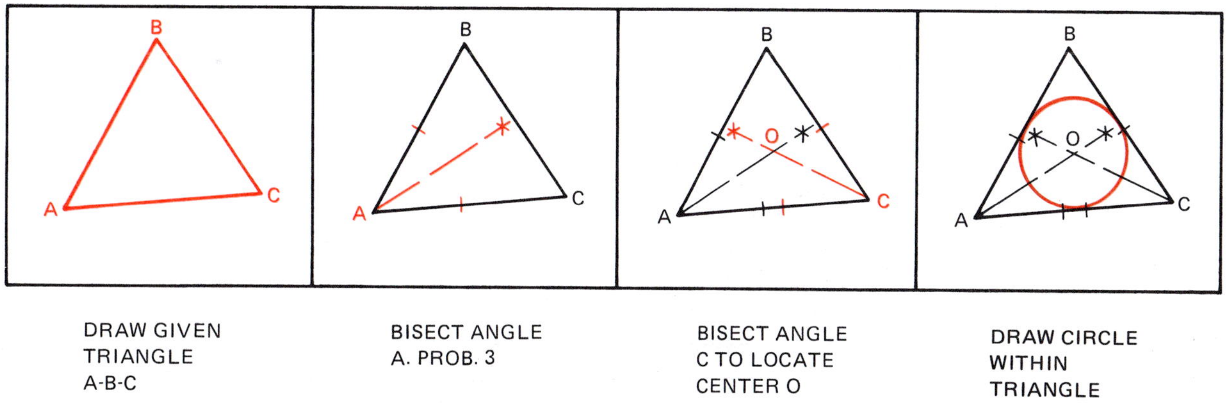

Fig. 2-14 To draw a circle within a given triangle

ASSIGNMENT C

The first problem in figure 2-15 shows the construction of an angle equal to a given angle. The other constructions show five methods of drawing triangles.

Draw the six geometric constructions specified in figure 2-15.

Procedure

1. Lay out the margin sheet and letter the title strip.
2. Divide the sheet into six equal spaces and letter the name of each problem in the proper space.
3. The construction of each problem is shown in figures 2-16 to 2-21. Proceed step-by-step to complete the problems.
4. Check each construction carefully.

TERMS

- A *scalene triangle* has unequal sides and unequal angles.
- An *isosceles triangle* has two equal angles and two equal sides.
- An *equilateral triangle* has three equal sides and three equal angles.
- A *right triangle* has one right angle.

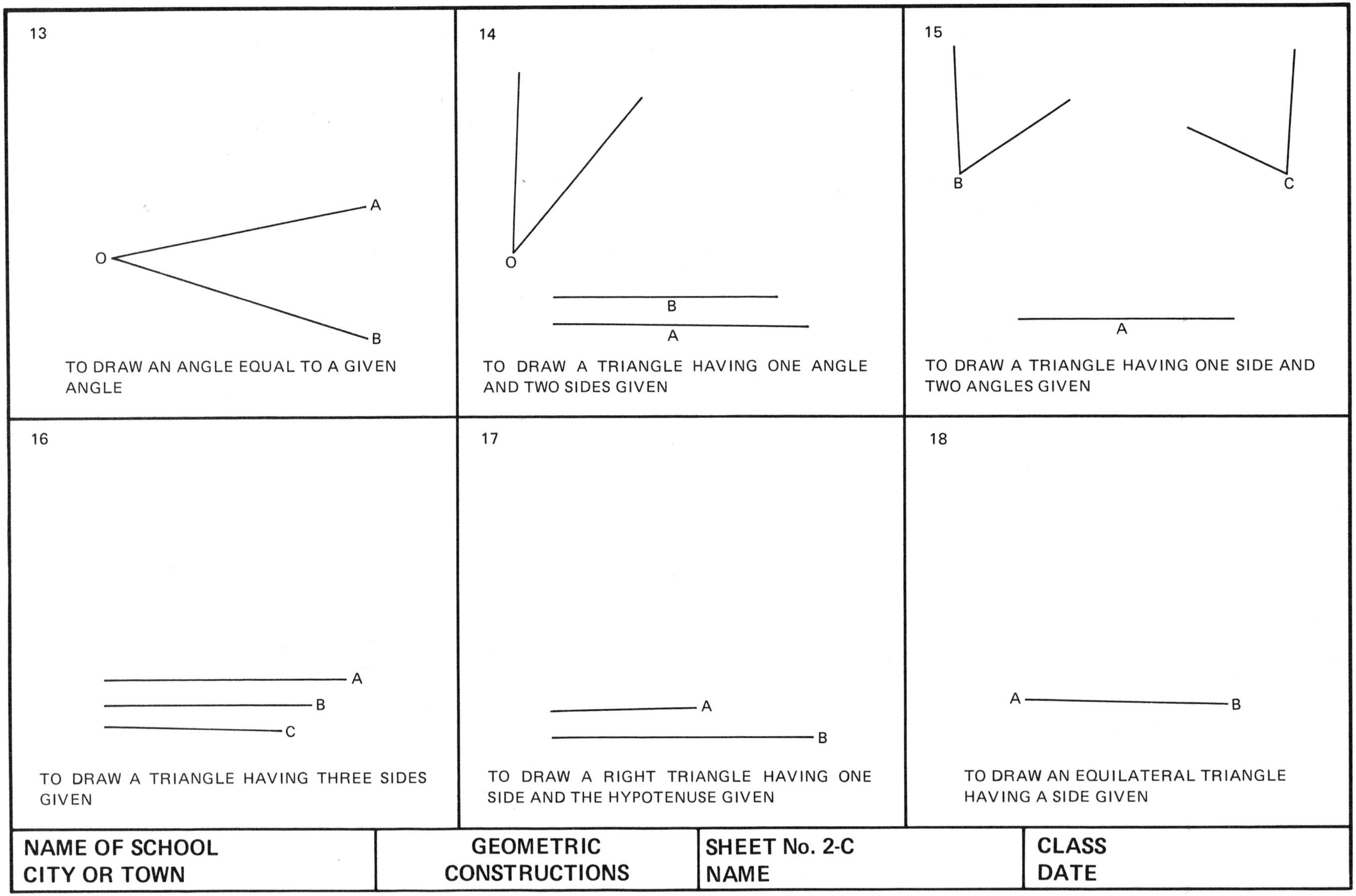
13
A
O
B
TO DRAW AN ANGLE EQUAL TO A GIVEN ANGLE
14
O
B
A
TO DRAW A TRIANGLE HAVING ONE ANGLE AND TWO SIDES GIVEN
15
B
C
A
TO DRAW A TRIANGLE HAVING ONE SIDE AND TWO ANGLES GIVEN
16
A
B
C
TO DRAW A TRIANGLE HAVING THREE SIDES GIVEN
17
A
B
TO DRAW A RIGHT TRIANGLE HAVING ONE SIDE AND THE HYPOTENUSE GIVEN
18
A
B
TO DRAW AN EQUILATERAL TRIANGLE HAVING A SIDE GIVEN
NAME OF SCHOOL
CITY OR TOWN
GEOMETRIC CONSTRUCTIONS
SHEET No. 2-C
NAME
CLASS
DATE

Fig. 2-15

PROBLEM 13

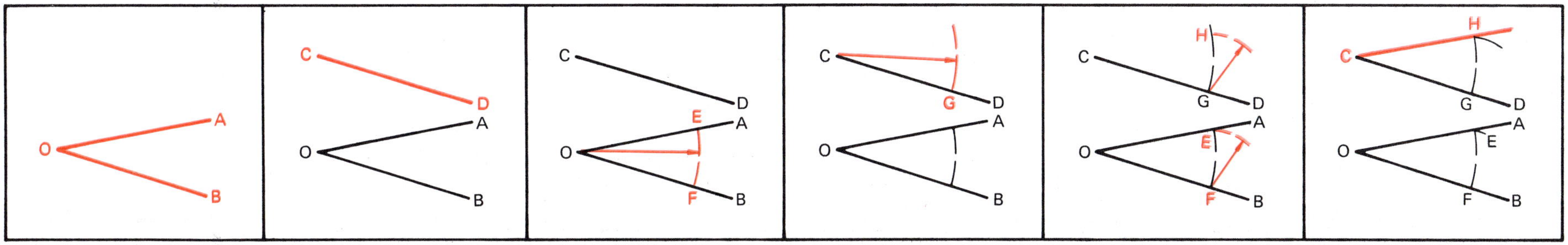

DRAW GIVEN ANGLE A-O-B	DRAW SIDE C-D OF ANGLE TO BE CONSTRUCT-ED	DRAW ARC E-F	WITH SAME R DRAW ARC FROM G	WITH RADIUS E-F LOCATE H	DRAW LINE FROM C THRU H TO COMPLETE ANGLE

Fig. 2-16 To draw an angle equal to a given angle

PROBLEM 14

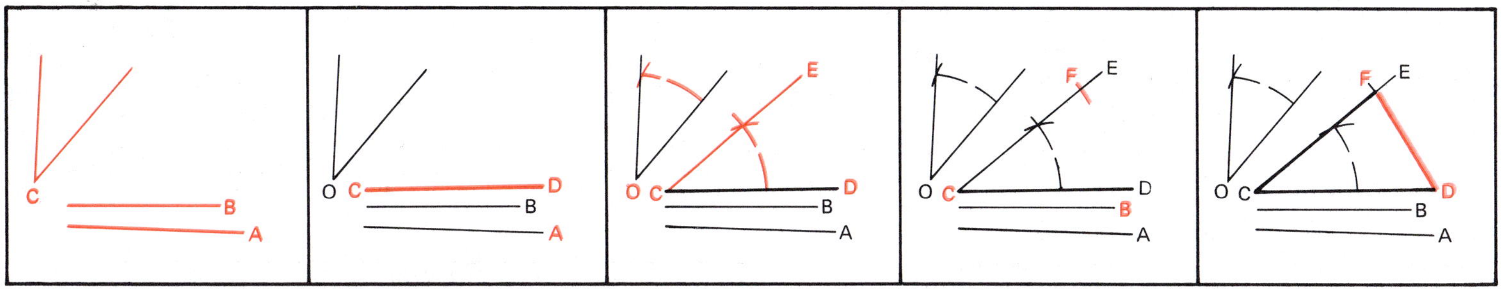

DRAW GIVEN SIDES A AND B. AND ANGLE O	DRAW SIDE C-D EQUAL TO GIVEN SIDE A	DRAW ANGLE D-C-E EQUAL TO ANGLE O. PROB. 13.	POINT OFF DIS-TANCE C-F EQUAL TO GIVEN SIDE B	CONNECT D WITH F TO COMPLETE TRIANGLE

Fig. 2-17 To draw a triangle having given one angle and two sides

PROBLEM 15

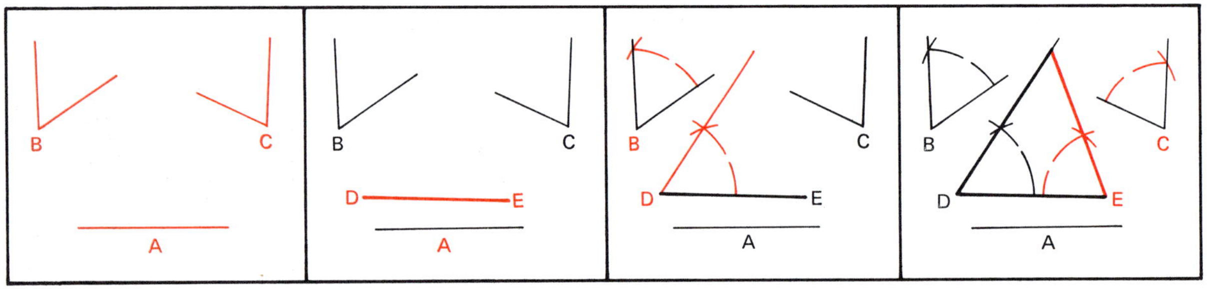

DRAW GIVEN SIDE A AND TWO GIVEN ANGLES B AND C

DRAW LINE D-E EQUAL TO GIVEN SIDE A

AT D DRAW ANGLE EQUAL TO GIVEN ANGLE B PROBLEM 13

AT E DRAW ANGLE EQUAL TO GIVEN ANGLE C TO COMPLETE TRIANGLE

Fig. 2-18 To draw a triangle having given one side and two angles

PROBLEM 16

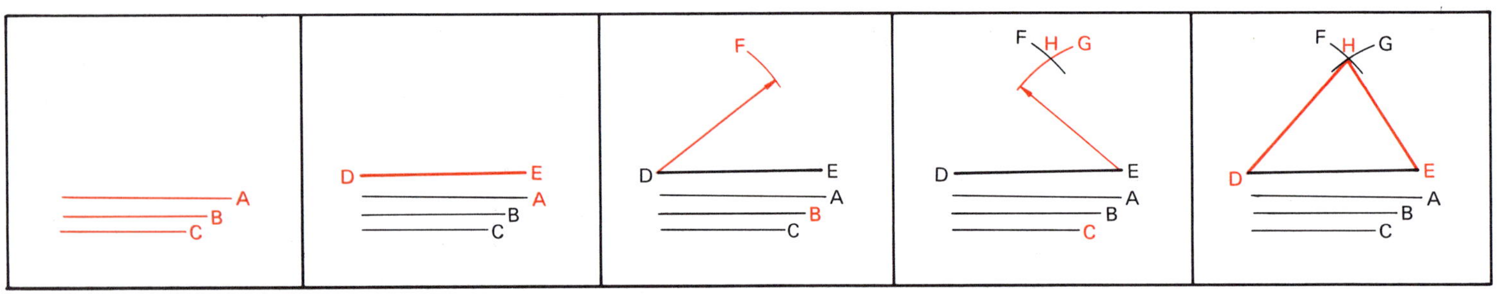

DRAW GIVEN SIDES A, B AND C

DRAW LINE D-E EQUAL TO GIVEN SIDE A

WITH R EQUAL TO GIVEN SIDE B DRAW ARC F

WITH R EQUAL TO GIVEN SIDE C DRAW ARC G TO LOCATE H

CONNECT D-H AND E-H TO COMPLETE THE TRIANGLE

Fig. 2-19 To draw a triangle having given three sides

PROBLEM 17

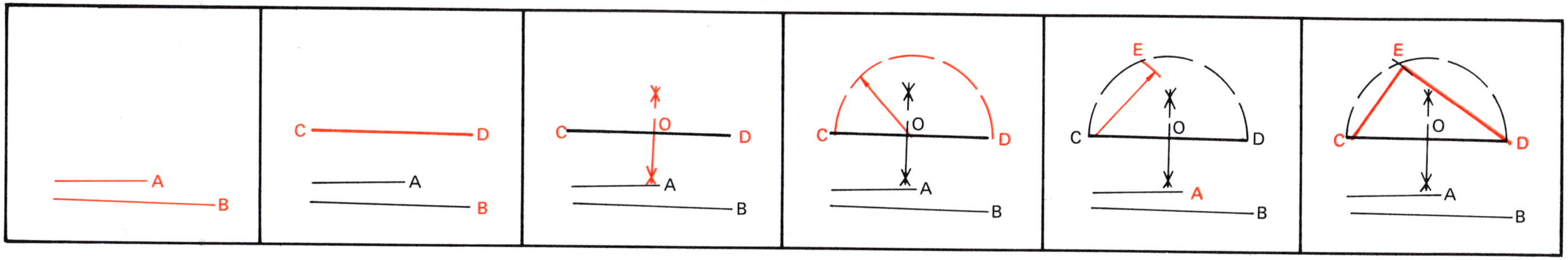

DRAW GIVEN SIDE A AND HYPOTENUSE B

DRAW LINE C-D EQUAL TO GIVEN HYPOTENUSE B

BISECT C-D TO LOCATE CENTER O

DRAW SEMICIRCLE C-D

LAYOFF DIS-TANCE A ON SEMICIRCLE TO LOCATE E

DRAW SIDES C-E AND D-E TO COMPLETE THE TRIANGLE

Fig. 2-20 To draw a right triangle having given one side and the hypotenuse

PROBLEM 18

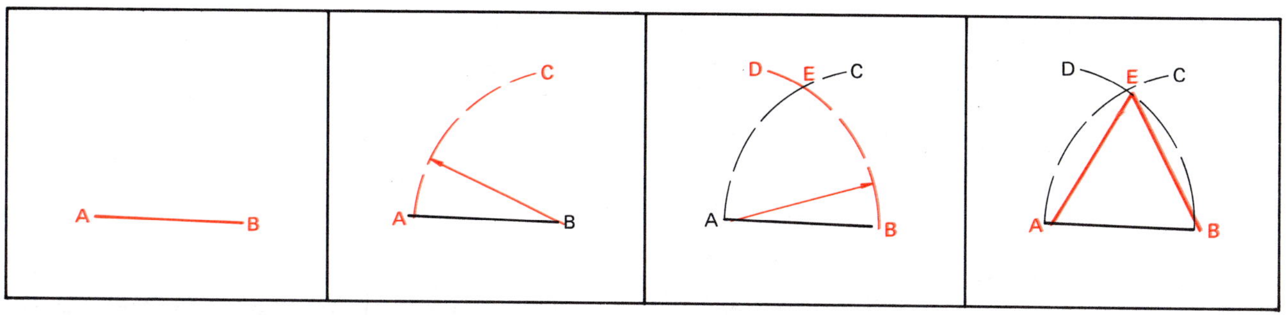

DRAW GIVEN SIDE A-B

DRAW ARC A-C

DRAW ARC B-D TO LOCATE E

DRAW SIDES A-E AND B-E TO COMPLETE THE TRIANGLE

Fig. 2-21 To draw an equilateral triangle having given a side

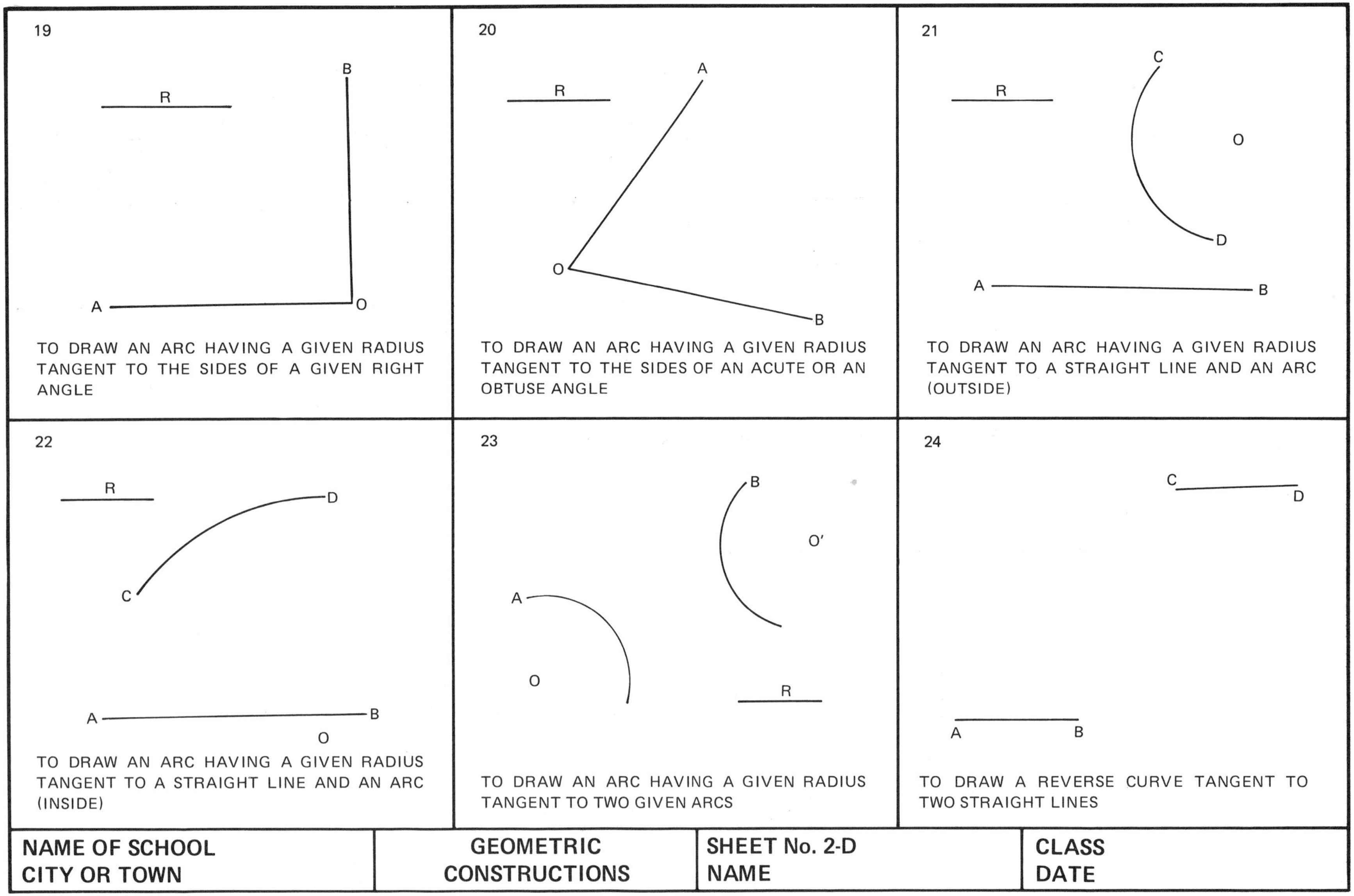
19
R
B
A
O
TO DRAW AN ARC HAVING A GIVEN RADIUS TANGENT TO THE SIDES OF A GIVEN RIGHT ANGLE
20
R
A
O
B
TO DRAW AN ARC HAVING A GIVEN RADIUS TANGENT TO THE SIDES OF AN ACUTE OR AN OBTUSE ANGLE
21
C
R
O
D
A
B
TO DRAW AN ARC HAVING A GIVEN RADIUS TANGENT TO A STRAIGHT LINE AND AN ARC (OUTSIDE)
22
R
D
C
A
B
O
TO DRAW AN ARC HAVING A GIVEN RADIUS TANGENT TO A STRAIGHT LINE AND AN ARC (INSIDE)
23
B
O′
A
O
R
TO DRAW AN ARC HAVING A GIVEN RADIUS TANGENT TO TWO GIVEN ARCS
24
C
D
A
B
TO DRAW A REVERSE CURVE TANGENT TO TWO STRAIGHT LINES
NAME OF SCHOOL
CITY OR TOWN
GEOMETRIC CONSTRUCTIONS
SHEET No. 2-D
NAME
CLASS
DATE

Fig. 2-22

ASSIGNMENT D

The first two problems in figure 2-22 show the construction of an arc tangent to the sides of an angle. The next two constructions show how to draw an arc tangent to a straight line and an arc. The fifth problem gives the construction of an arc tangent to two other arcs. The last problem shows the construction of a reverse curve tangent to two straight lines.

Draw the six geometric constructions specified in figure 2-22.

Procedure

1. Lay out the margin lines and letter the title strip.
2. Divide the sheet into six equal spaces and letter the name of each problem in the proper space.
3. The construction of each problem is shown in figures 2-23 to 2-28. Proceed step-by-step to complete the problems.
4. Check each construction carefully.

TERMS

- A *straight line is tangent to an arc* when the line touches the arc at a single point and does not intersect it.
- *Arcs are tangent to each other* when they touch at a single point and do not intersect.
- An *acute angle* is less than 90 degrees.
- An *obtuse angle* is more than 90 degrees and less than 180 degrees.
- A *reflex angle* is more than 180 degrees.

PROBLEM 19

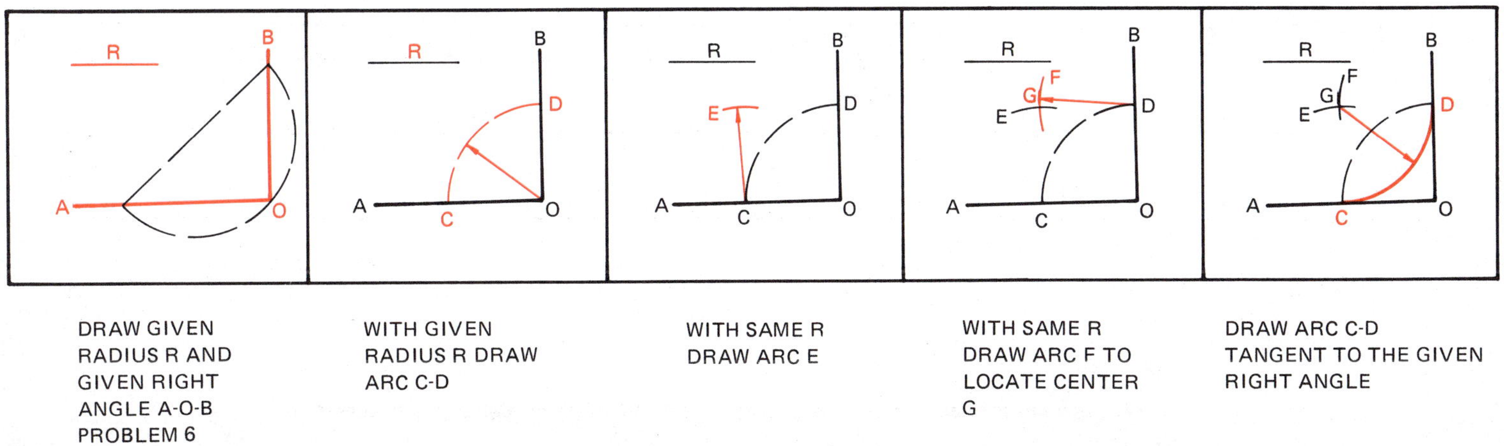

Fig. 2-23 To draw an arc having a given radius tangent to a given right angle

PROBLEM 20

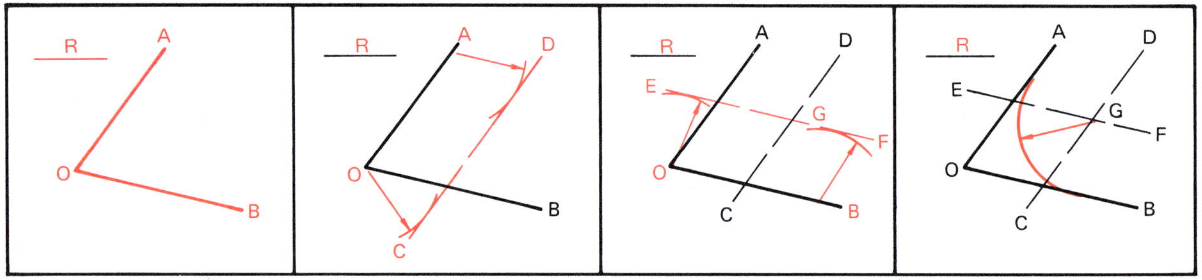

DRAW GIVEN ANGLE A-O-B AND GIVEN RADIUS R

DRAW C-D PARALLEL TO O-A AT GIVEN DISTANCE R PROBLEM 8

DRAW E-F PARALLEL TO O-B AT GIVEN DISTANCE R TO LOCATE CENTER G

WITHIN GIVEN RADIUS R DRAW ARC TANGENT TO THE GIVEN ANGLE

Fig. 2-24 **To draw an arc having a given radius tangent to the sides of an acute or an obtuse angle**

PROBLEM 21

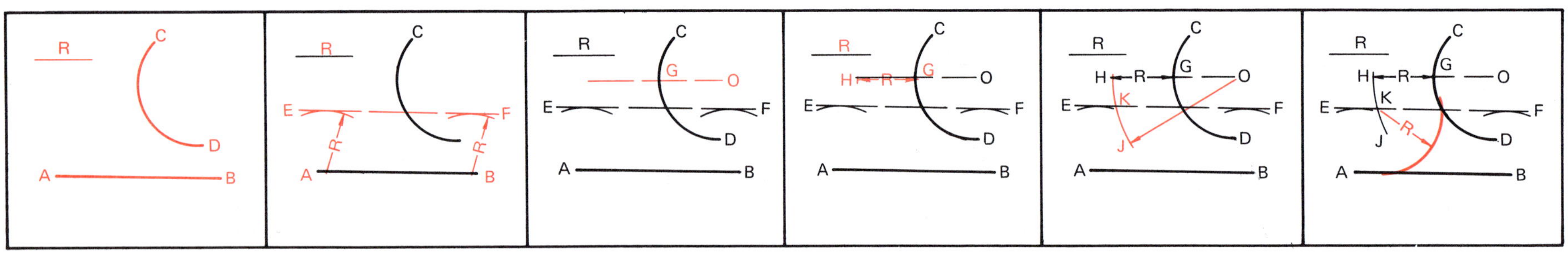

DRAW GIVEN LINE A-B, GIVEN ARC C-D, AND RADIUS R

DRAW LINE E- F PARALLEL TO A-B AT GIVEN DISTANCE R. PROBLEM 8

DRAW LINE O-G

LAYOFF G-H EQUAL TO GIVEN RADIUS R

DRAW ARC J TO LOCATE CENTER K

DRAW ARC TANGENT TO GIVEN ARC AND GIVEN LINE

Fig. 2-25 **To draw an arc having a given radius tangent to a straight line and an arc (outside)**

PROBLEM 22

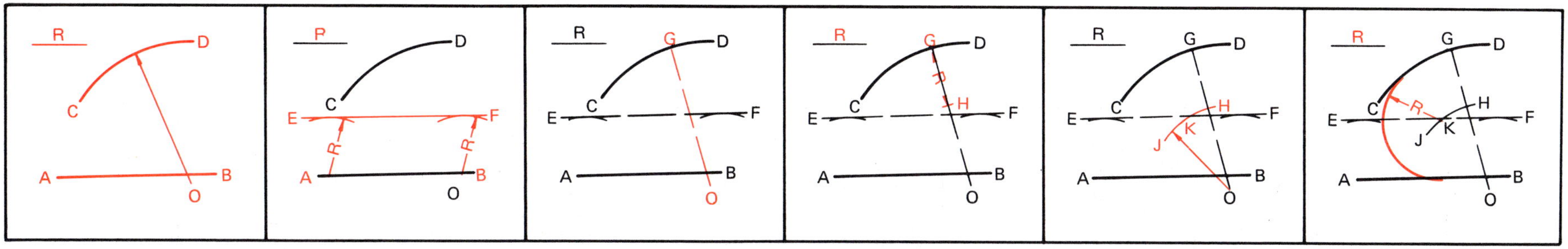

DRAW GIVEN LINE A-B, GIVEN ARC C-D, AND GIVEN RADIUS R

DRAW LINE E-F PARALLEL TO A-B AT GIVEN DISTANCE R PROBLEM 8

DRAW LINE O-G

LAYOFF G-H EQUAL TO GIVEN RADIUS R

DRAW ARC H-J TO LOCATE CENTER K

DRAW ARC TANGENT TO GIVEN ARC AND GIVEN STRAIGHT LINE

Fig. 2-26 To draw an arc having a given radius tangent to a straight line and an arc (inside)

PROBLEM 23

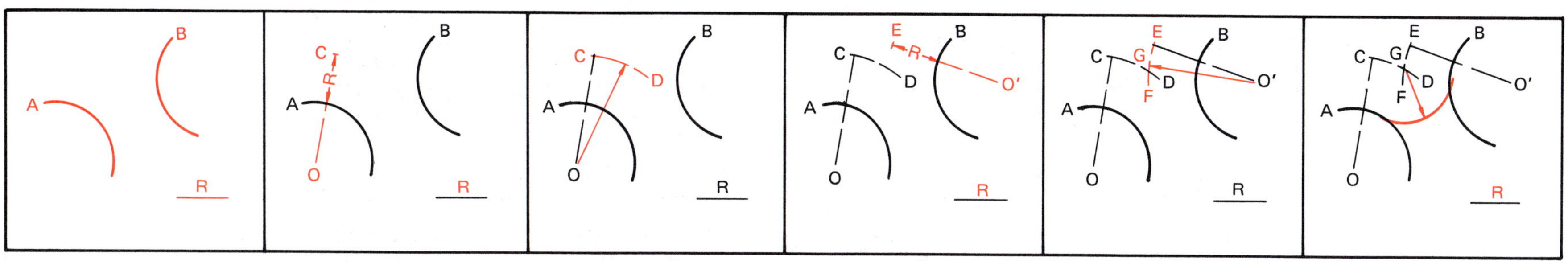

DRAW TWO GIVEN ARCS A AND B AND GIVEN RADIUS R

DRAW LINE FROM O AND LAYOFF DISTANCE R TO LOCATE C

DRAW ARC C-D

DRAW LINE FROM O' AND LAYOFF DISTANCE R TO LOCATE E

DRAW ARC E-F TO LOCATE CENTER G

WITH GIVEN RADIUS R DRAW ARC TANGENT TO TWO GIVEN ARCS

Fig. 2-27 To draw an arc having a given radius tangent to two given arcs

PROBLEM 24

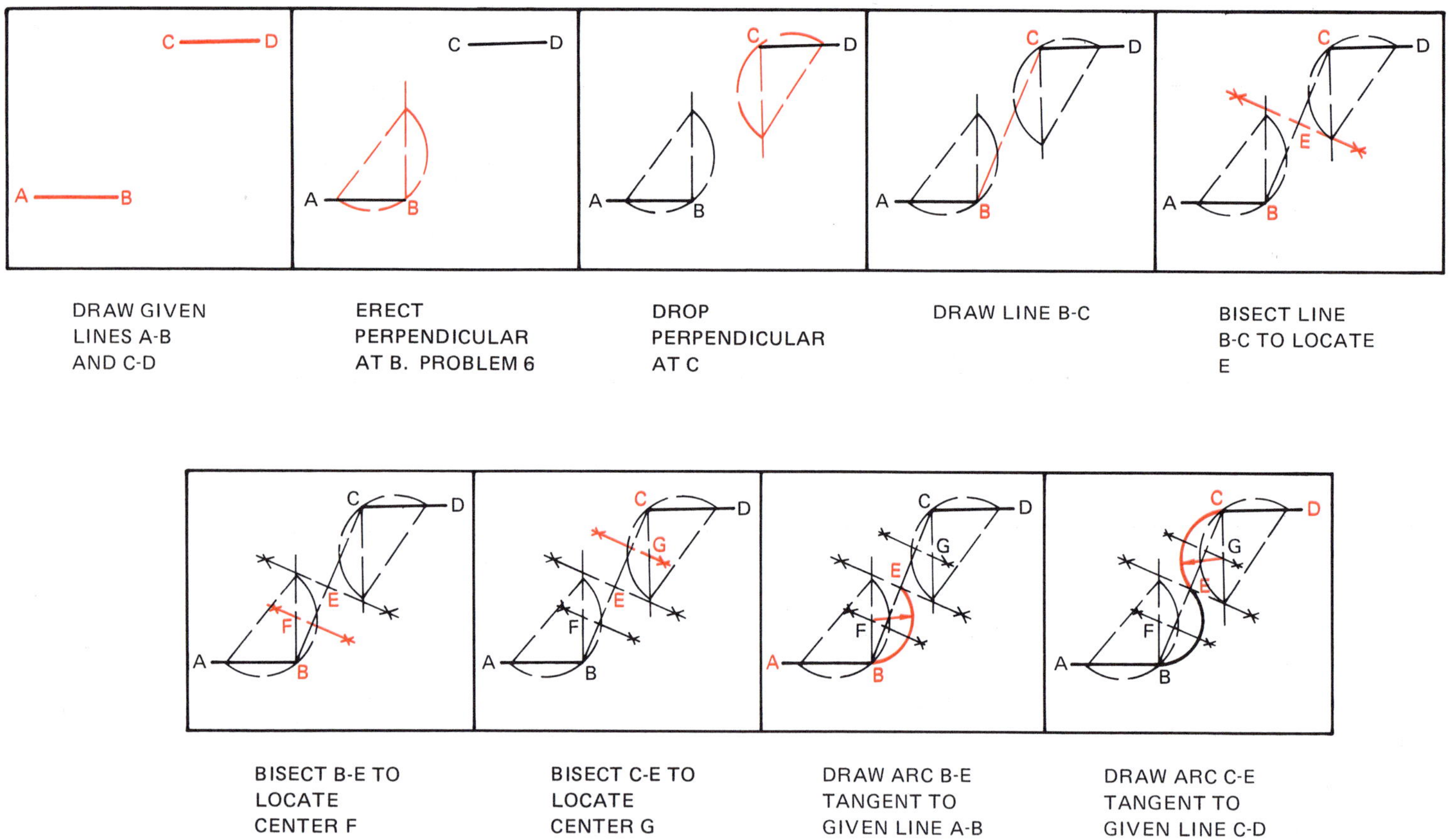

Fig. 2-28 To draw a reverse curve tangent to two straight lines

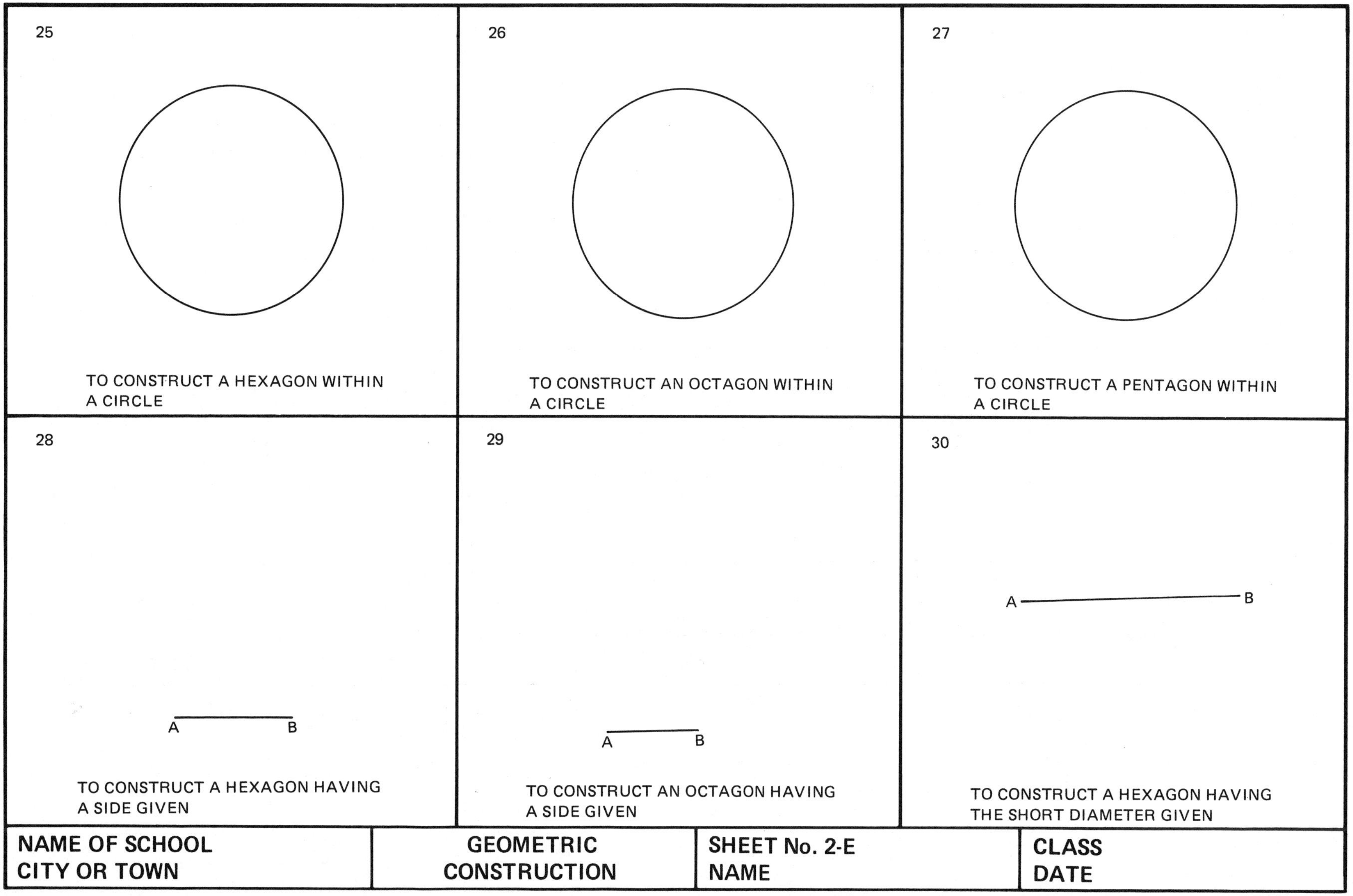
25
TO CONSTRUCT A HEXAGON WITHIN A CIRCLE
26
TO CONSTRUCT AN OCTAGON WITHIN A CIRCLE
27
TO CONSTRUCT A PENTAGON WITHIN A CIRCLE
28
A
B
TO CONSTRUCT A HEXAGON HAVING A SIDE GIVEN
29
A
B
TO CONSTRUCT AN OCTAGON HAVING A SIDE GIVEN
30
A
B
TO CONSTRUCT A HEXAGON HAVING THE SHORT DIAMETER GIVEN
NAME OF SCHOOL
CITY OR TOWN
GEOMETRIC CONSTRUCTION
SHEET No. 2-E
NAME
CLASS
DATE

Fig. 2-29

ASSIGNMENT E

The problems in figure 2-29 show six methods of constructing polygons.

Draw the six geometric constructions specified.

Procedure

1. Lay out the margin lines and letter the title strip.
2. Divide the sheet into six equal spaces and letter the name of each problem in the proper space.
3. The construction of each polygon is described in figures 2-30 to 2-35. Proceed step-by-step to complete the problems.
4. Check each construction carefully by going back over each line to check each step.

TERMS

- A *polygon* is a plane closed figure having many sides and angles.
- A *regular polygon* has equal sides and equal angles.
- A *hexagon* is a regular polygon having six sides.
- An *octagon* is a regular polygon having eight sides.
- A *pentagon* is a regular polygon having five sides.

PROBLEM 25

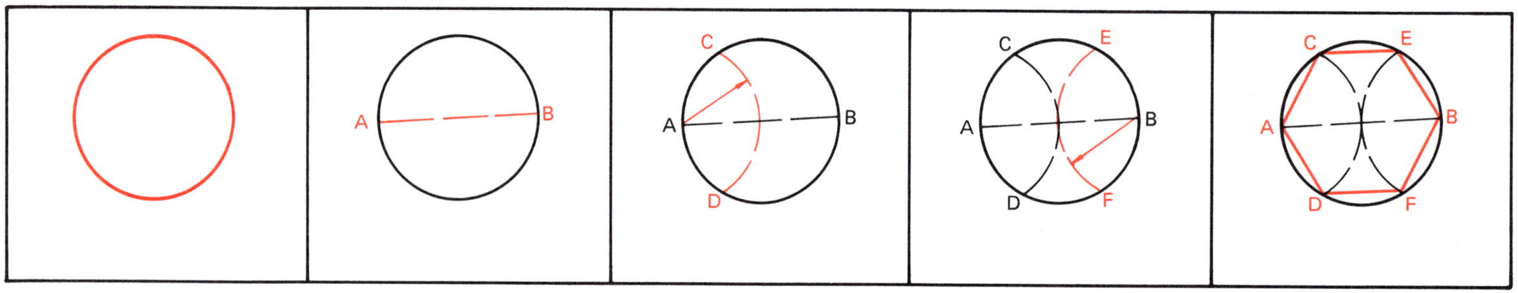

DRAW GIVEN CIRCLE | DRAW DIAMETER A-B | DRAW ARC C-D | DRAW ARC E-F | CONNECT A-C, C-E, E-B, B-F, F-D, AND D-A

Fig. 2-30 To construct a hexagon within a circle

PROBLEM 26

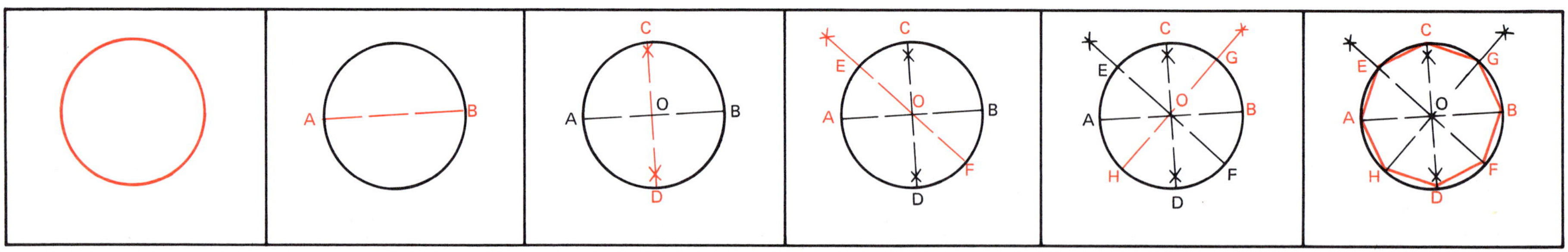

DRAW GIVEN CIRCLE | DRAW DIAMETER A-B | DRAW PERPENDICULAR C-D | BISECT ANGLE A-O-C TO LOCATE E AND F | BISECT ANGLE B-O-C TO LOCATE G AND H | CONNECT A-E, E-C, C-G, G-B, B-F, F-D, D-H, AND H-A

Fig. 2-31 To construct an octagon within a circle

PROBLEM 27

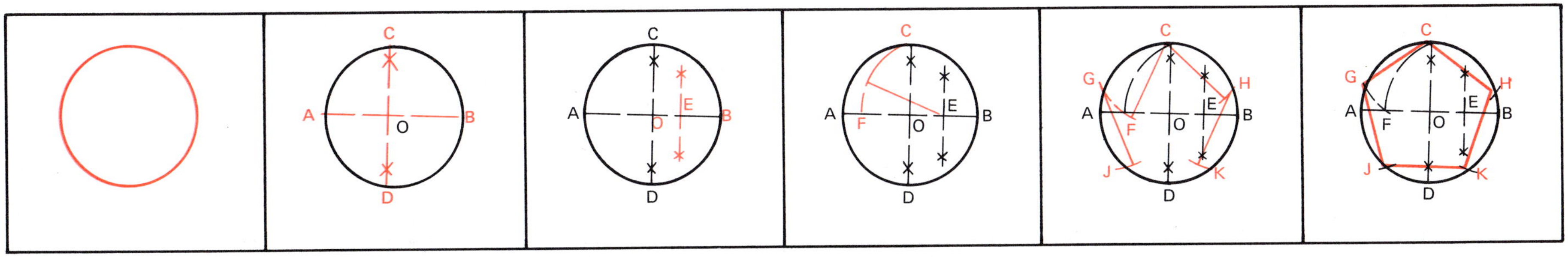

DRAW GIVEN CIRCLE

DRAW DIAMETER A-B AND PERPENDICULAR C-D

BISECT O-B TO LOCATE CENTER E

DRAW ARC C-F

WITH RADIUS C-F DIVIDE THE CIRCLE INTO FIVE EQUAL PARTS

CONNECT C-G, G-J, J-K, K-H, AND H-C

Fig. 2-32 To construct a pentagon within a circle

PROBLEM 28

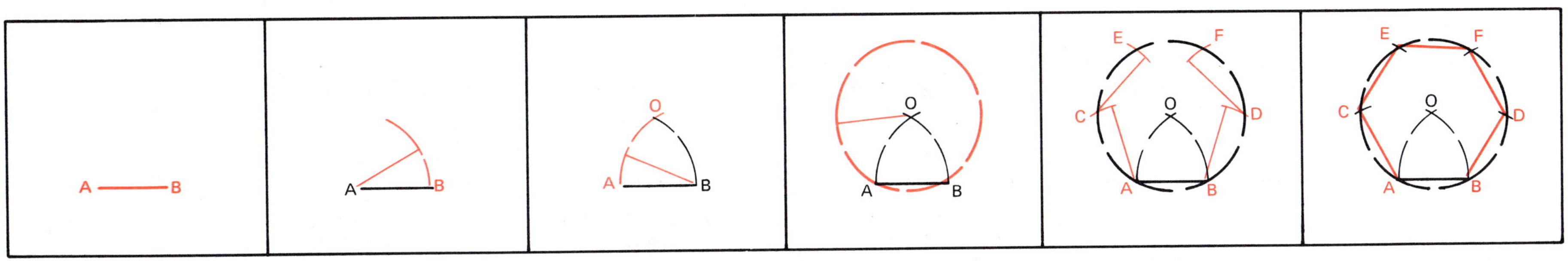

DRAW GIVEN SIDE A-B

DRAW ARC FROM B

DRAW ARC FROM A TO LOCATE CENTER O

DRAW THE CIRCLE

WITH RADIUS A-B DIVIDE THE CIRCLE INTO SIX EQUAL PARTS

CONNECT A-C, C-E, E-F, F-D, AND B-D

Fig. 2-33 To construct a hexagon having given a side

PROBLEM 29

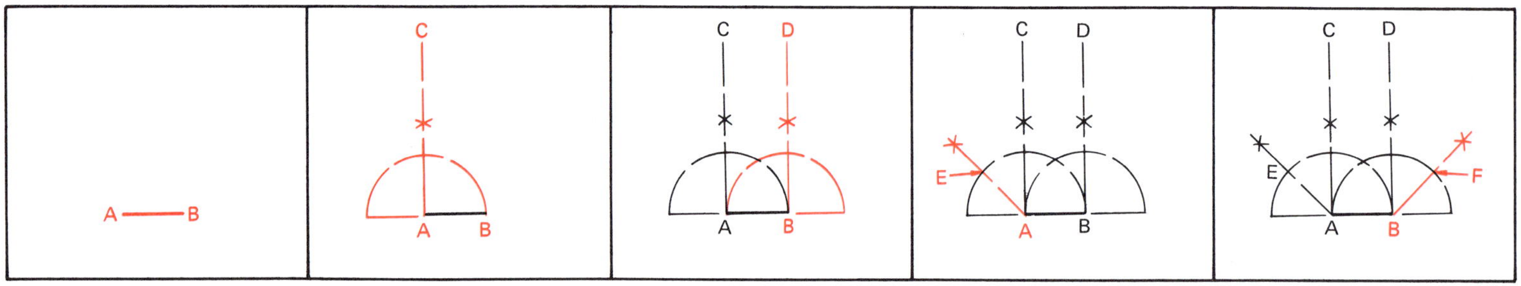

DRAW GIVEN SIDE A-B

EXTEND LINE A-B AND ERECT PERPENDICULAR A-C. PROBLEM 4

ERECT PERPENDICULAR B-D

BISECT RIGHT ANGLE AT A TO LOCATE E. PROBLEM 3

BISECT RIGHT ANGLE AT B TO LOCATE F

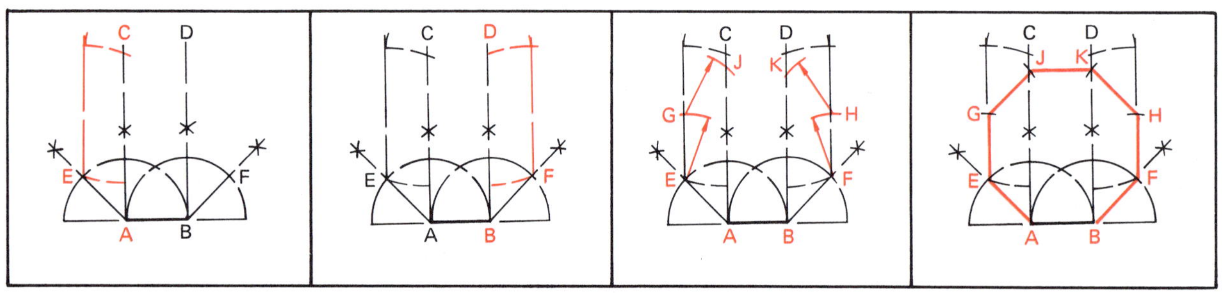

FROM E DRAW LINE PARALLEL TO A-C. PROB. 9

FROM F DRAW LINE PARALLEL TO B-D

WITH RADIUS A-B LAYOFF E-G, G-J, F-H, AND H-K

CONNECT A-E, E-G, G-J, J-K, B-F, F-H, AND H-K

Fig. 2-34 To construct an octagon having given a side

PROBLEM 30

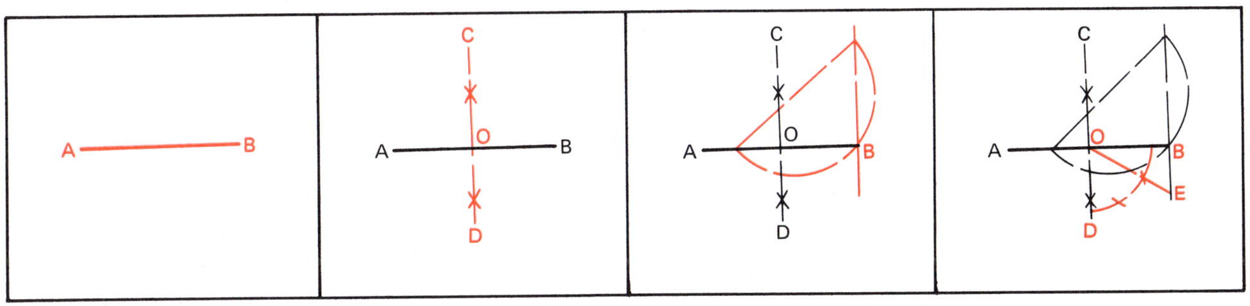

DRAW GIVEN SHORT DIAMETER A-B

DRAW PERPENDICULAR BISECTOR C-D TO LOCATE CENTER O. PROBLEM 1

ERECT A PERPENDICULAR AT B PROBLEM 6

DIVIDE ANGLE B-O-D IN INTO THREE EQUAL ANGLES TO LOCATE CORNER E

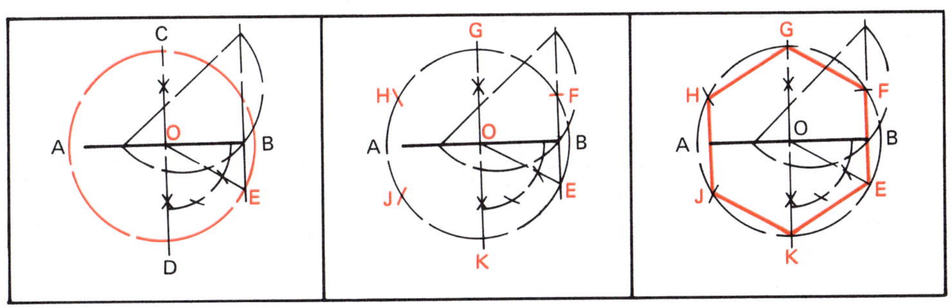

WITH RADIUS O-E DRAW A CIRCLE.

WITH RADIUS O-E LAYOFF E-F, F-G, G-H, H-J, J-K, AND K-E

CONNECT E-F, F-G, G-H, H-J, J-K, AND K-E

Fig. 2-35 To construct a hexagon having given the short diameter

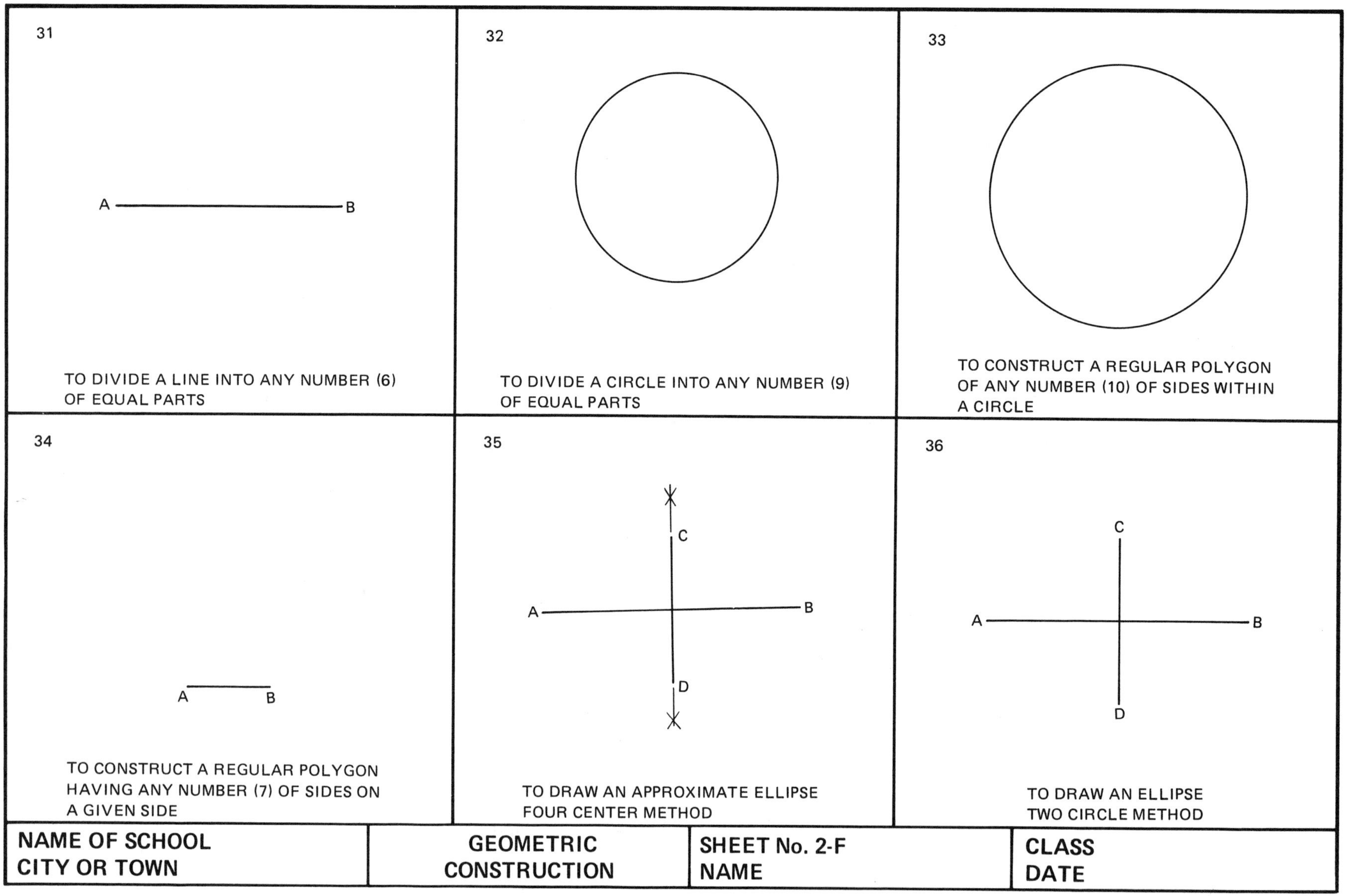
31
A
B
TO DIVIDE A LINE INTO ANY NUMBER (6) OF EQUAL PARTS
32
TO DIVIDE A CIRCLE INTO ANY NUMBER (9) OF EQUAL PARTS
33
TO CONSTRUCT A REGULAR POLYGON OF ANY NUMBER (10) OF SIDES WITHIN A CIRCLE
34
A
B
TO CONSTRUCT A REGULAR POLYGON HAVING ANY NUMBER (7) OF SIDES ON A GIVEN SIDE
35
C
A
B
D
TO DRAW AN APPROXIMATE ELLIPSE FOUR CENTER METHOD
36
C
A
B
D
TO DRAW AN ELLIPSE TWO CIRCLE METHOD
NAME OF SCHOOL
CITY OR TOWN
GEOMETRIC CONSTRUCTION
SHEET No. 2-F
NAME
CLASS
DATE

Fig. 2-36

ASSIGNMENT F

The first two problems in figure 2-36 show the method of dividing a line and a circle into any number of equal parts. These methods are used to construct regular polygons. The last two problems show two methods used to draw an ellipse when the major and minor axes are given.

Draw the six geometric constructions specified.

Procedure

1. Lay out the margin lines and letter the title strip.
2. Divide the sheet into six equal spaces and letter the name of each problem in the proper space.
3. The construction of each problem is described. Proceed step-by-step to complete the problems in figures 2-38 to 2-40.
4. Make side A-B in figure 2-41 approximately 1 inch long and draw it about 1 5/8 inch above the lower margin line.
5. Make the major axis in figure 2-42 approximately 3 1/4 inch long and about 2 1/2 inch above the lower margin line. Draw the minor axis about 1 3/4 inch long.
6. The T square and triangles are to be used in figure 2-43. Draw the major axis 3 1/4 inch long and 2 1/2 inch above the lower margin line. Draw the minor axis 2 inches long.
7. Check each of the six geometric constructions carefully.

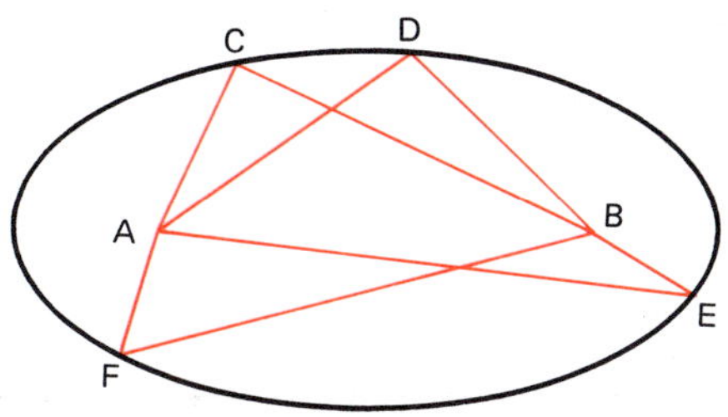

Fig. 2-37 An ellipse

TERMS

- A *decagon* is a regular polygon having ten sides.
- A *nonagon* is a regular polygon having nine sides.
- A *heptagon* is a regular polygon having seven sides.
- An *ellipse* is a plane curve which is the path of a point, the sum of whose distance from two fixed points is constant. This definition is illustrated in figure 2-37. Points A and B are fixed points. Distance A-C plus B-C equals A-D plus B-D, equals A-E plus B-E, equals A-F plus B-F, etc. The length of each pair of lines is constant.

PROBLEM 31

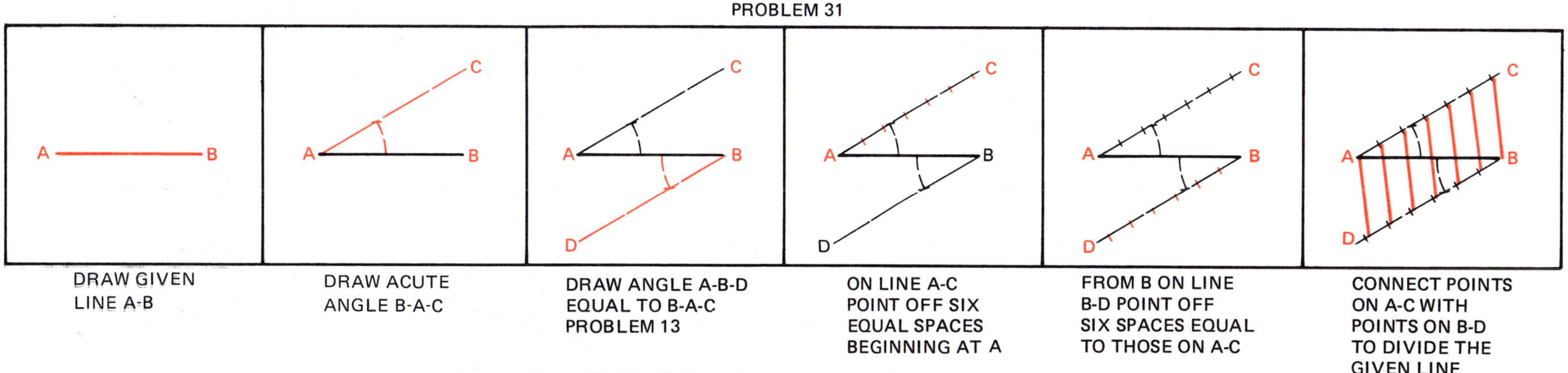

Fig. 2-38 To divide a line into any number of equal parts

PROBLEM 32

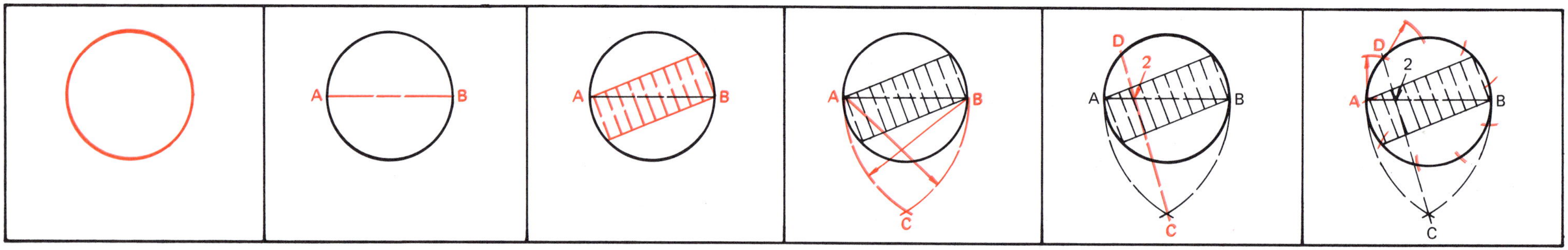

DRAW GIVEN CIRCLE

DRAW DIAMETER A-B

DIVIDE DIAMETER A-B INTO NINE EQUAL PARTS. PROBLEM 31

DRAW ARCS FROM A AND B TO LOCATE C

DRAW A LINE FROM C THRU POINT 2 TO LOCATE D

WITH CHORD A-D DIVIDE THE GIVEN CIRCLE INTO NINE EQUAL PARTS

Fig. 2-39 To divide a circle into any number of equal parts

PROBLEM 33

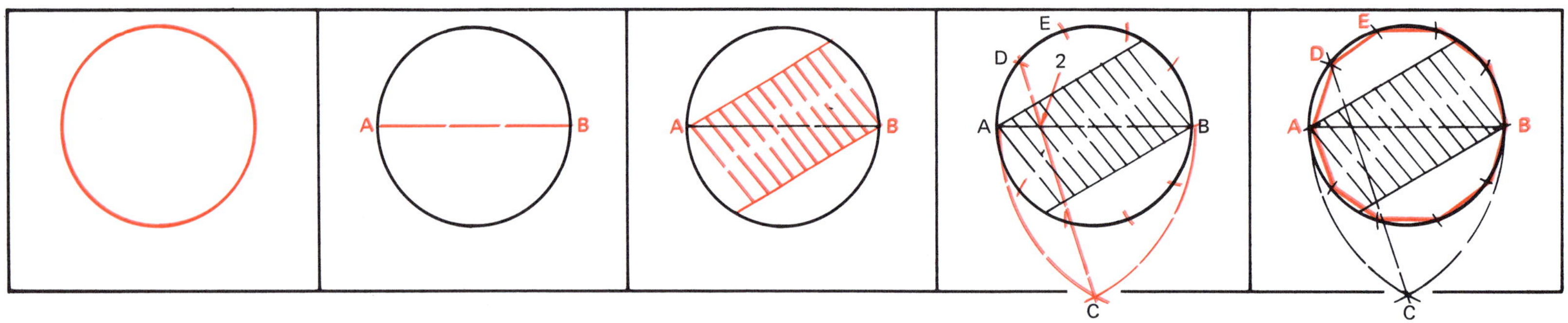

DRAW THE GIVEN CIRCLE

DRAW DIAMETER A-B

DIVIDE DIAMETER A-B INTO TEN EQUAL PARTS PROBLEM 31

DIVIDE THE GIVEN CIRCLE INTO TEN EQUAL PARTS PROBLEM 32

CONNECT A-D, D-E, ETC., TO COMPLETE THE POLYGON

Fig. 2-40 To construct a regular polygon of any number of sides within a circle

PROBLEM 34

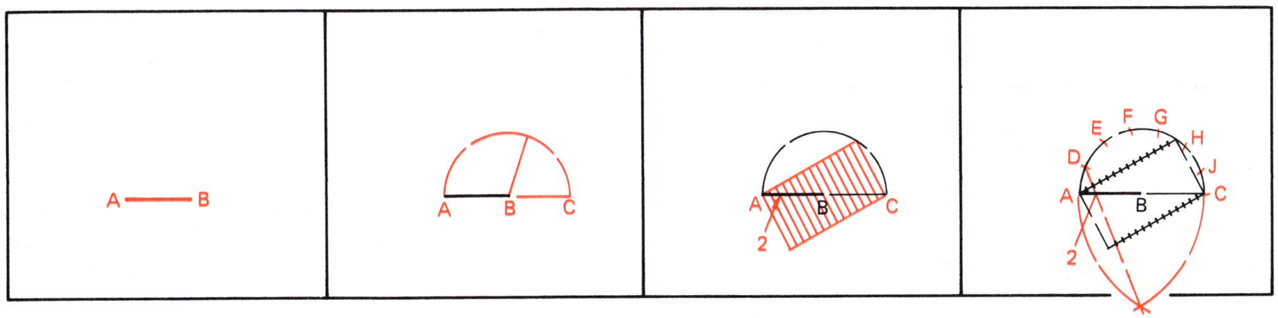

DRAW GIVEN SIDE A-B

EXTEND A-B AND DRAW SEMICIRCLE A-C

DIVIDE A-C INTO FOURTEEN EQUAL PARTS TO LOCATE 2. PROBLEM 31

DIVIDE SEMICIRCLE INTO SEVEN EQUAL PARTS. PROBLEM 32

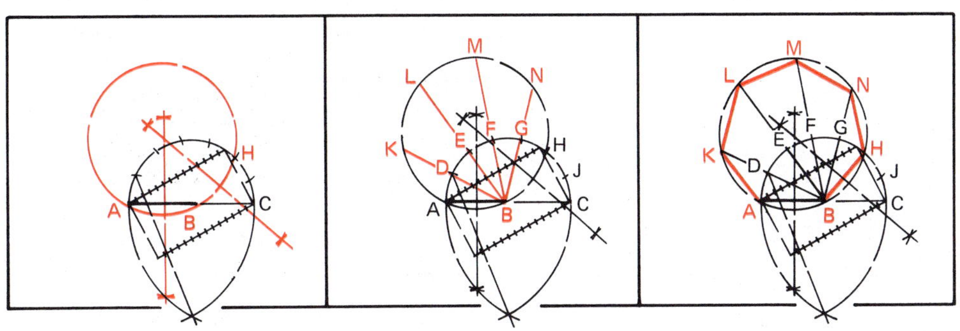

DRAW A CIRCLE THRU POINTS A, B AND H. PROBLEM 10

DRAW LINES B-D, B-E, B-F, AND B-G TO LOCATE K, L, M, AND N

CONNECT A-K, K-L, L-M, M-N, N-H AND H-B TO COMPLETE THE POLYGON

Fig. 2-41 To construct a regular polygon having any number of sides on a given side

PROBLEM 35

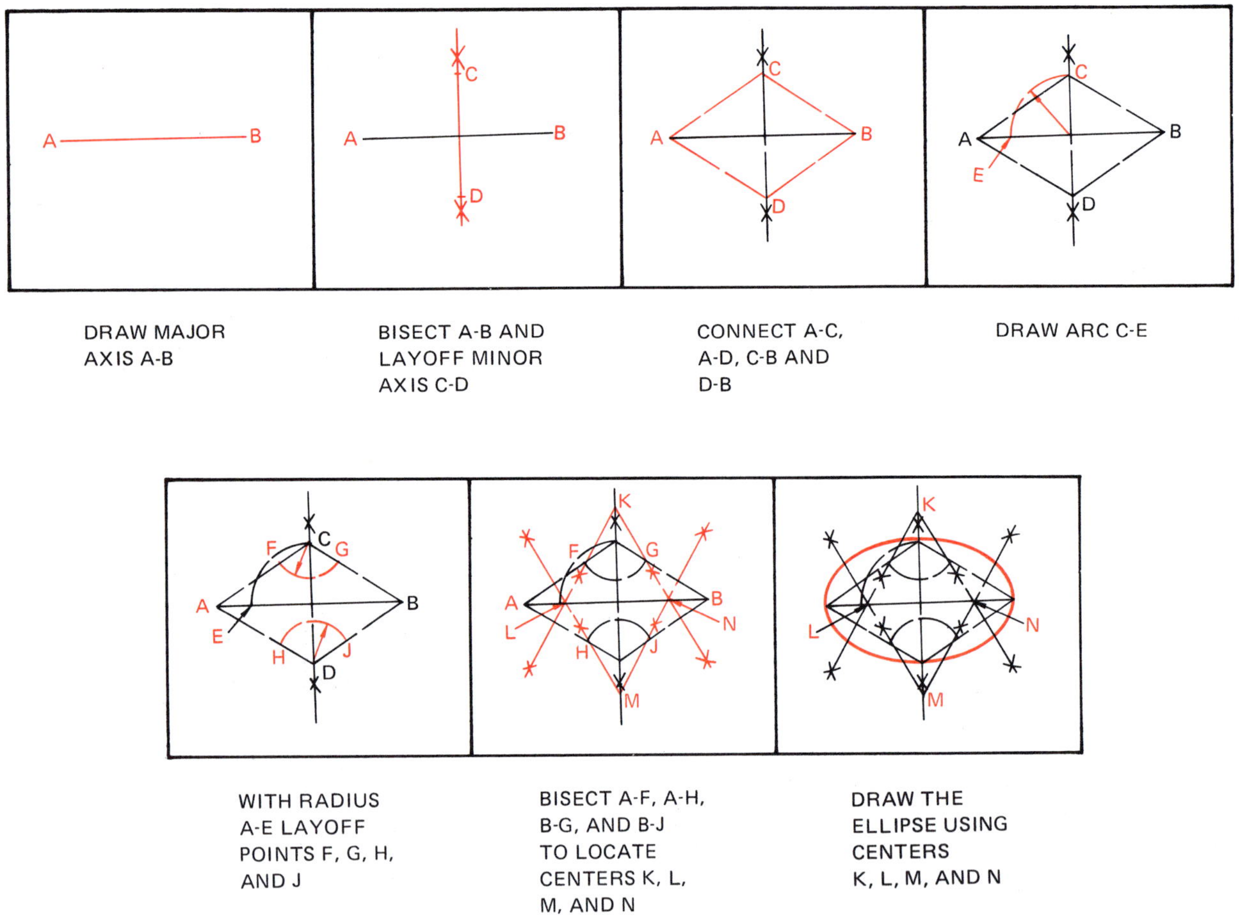

Fig. 2-42 To draw an approximate ellipse. Four center method

PROBLEM 36

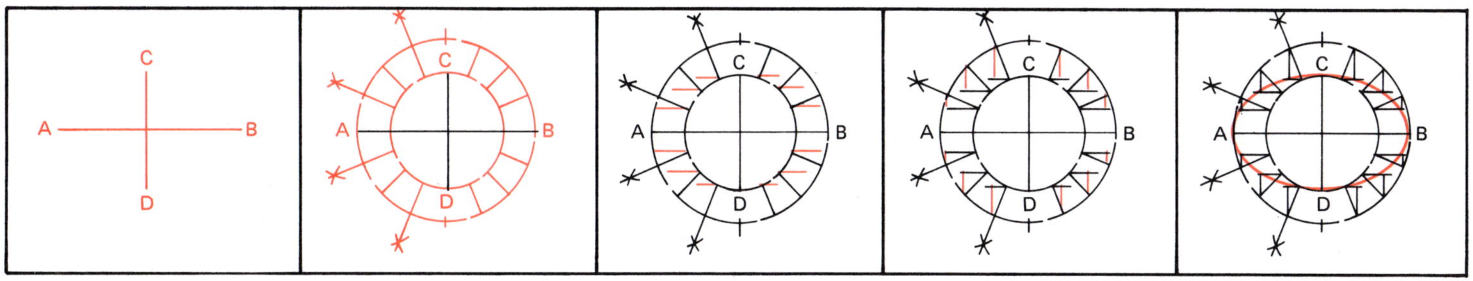

DRAW MAJOR AXIS A-B AND MINOR AXIS C-D	DRAW A CIRCLE THRU A AND B, AND THRU C AND D. DIVIDE CIRCLES INTO 16 EQUAL PARTS	DRAW A HORIZONTAL LINE FROM EACH POINT ON MINOR AXIS CIRCLE	DRAW A VERTICAL LINE FROM EACH POINT ON MAJOR AXIS CIRCLE	CONNECT THE INTERSECTIONS OF THESE LINES TO COMPLETE THE ELLIPSE

Fig. 2-43 To draw an ellipse. Two circle method

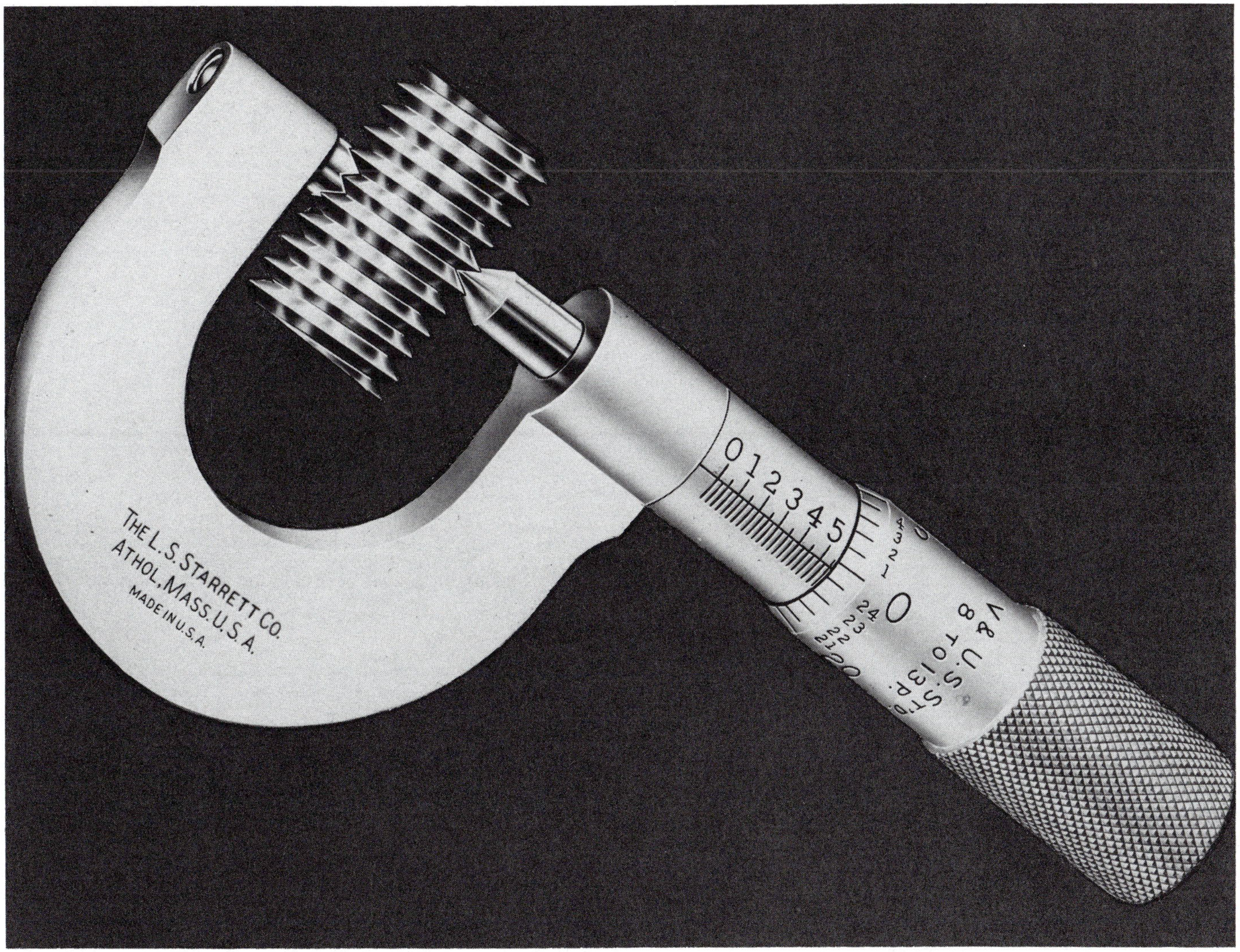

UNIT 3
SHAPE DESCRIPTION

Ideas explained through drawings go back as far as there was need to communicate. Since the industrial revolution, items and the manufacturing processes used to make them have become more complex. This caused the type of drawing needed to explain the ideas to change.

Pictorial drawings, such as the rectangular block in figure 3-1, do not show the actual shape of each surface. A more exact method of drawing objects was necessary as parts became interchangeable and more than one person or company worked on a single product. Technical drawing or drafting was developed. It is the most precise method of communicating complicated ideas visually.

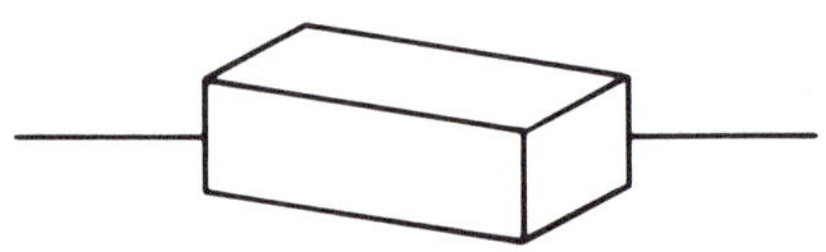

Fig. 3-1 A pictorial drawing

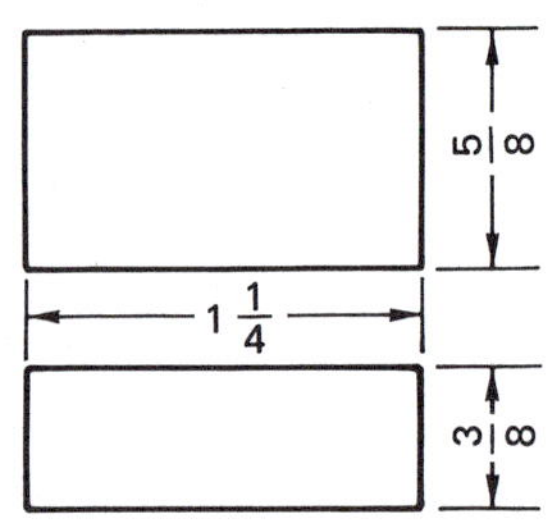

Fig. 3-2 A working drawing

Technical drawings, also called working drawings, accurately describe the form, size, location, materials, finish, construction, and processing details of any object to be manufactured, figure 3-2. All of this is done through lines, symbols, dimensions and notations.

MULTIVIEW DRAWINGS

The most common method of presenting all details of an object is with multiview drawings. A multiview drawing is a number of separate views of an object in relation to each other. The selection, placement, and relationship of each view is determined by

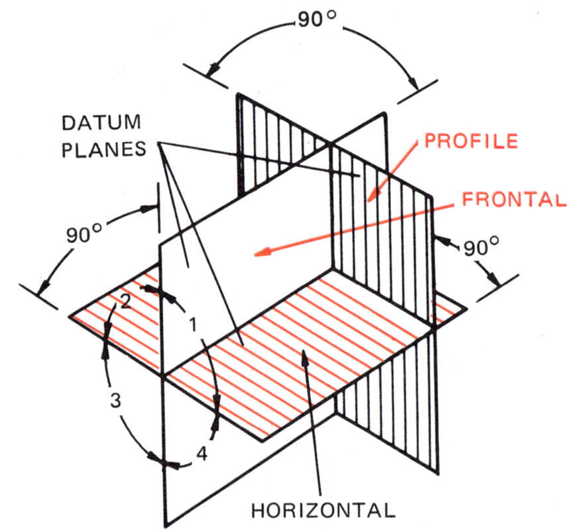

Fig. 3-3 Intersecting planes of projection form quadrants

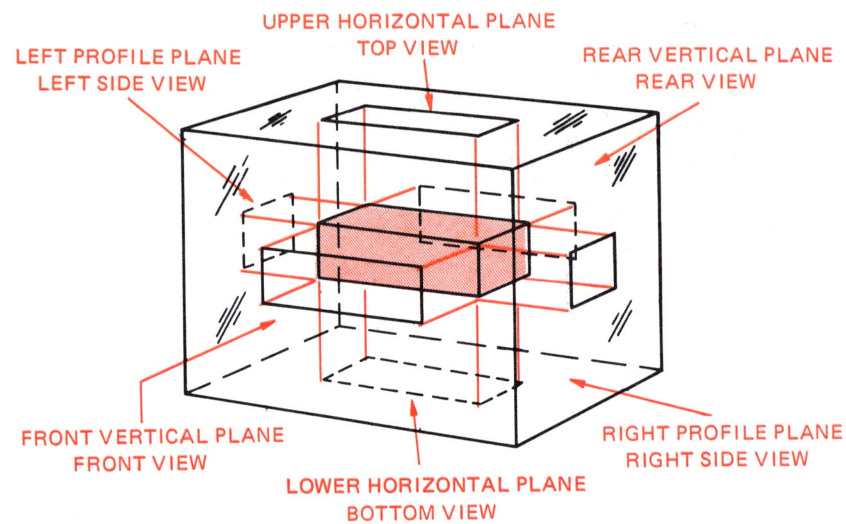

Fig. 3-4 The transparent box

using the theory of orthographic projection. *Orthographic projection* uses three planes intersecting at right angles to each other, figure 3-3. These are the frontal, horizontal, and profile *planes of projection*. They divide space into four quadrants. A *quadrant* is a quarter of a circle. Quadrants are usually referred to as angles in orthographic projection.

It is often helpful to visualize the object as suspended in a glass box within these planes of projection, figure 3-4.

FIRST-ANGLE PROJECTION

First-angle projection is used in countries that have the metric system of measurements. The object is placed in the first quadrant. The viewer is assumed to be in front of the profile plane and above the horizontal plane, figure 3-5. The position of the profile plane is not critical. It can be to the right or left of the object. To the viewer, the three planes of projection are behind the object. When the views are projected to the three planes and the six-sided glass box is unfolded on a flat surface, it appears as in figure 3-6. The front view is above the top view, and the left-side view is to the right of the front view.

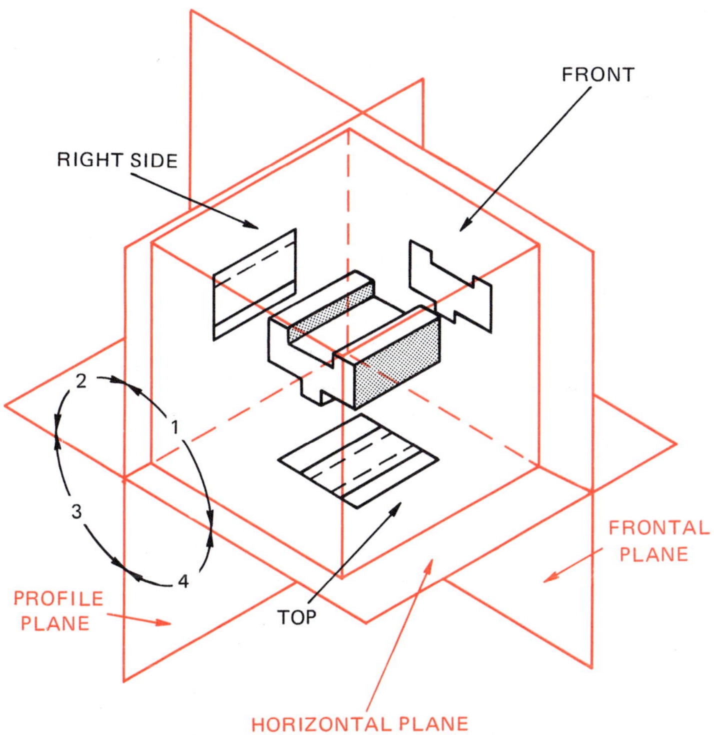

Fig. 3-5 First-angle projection

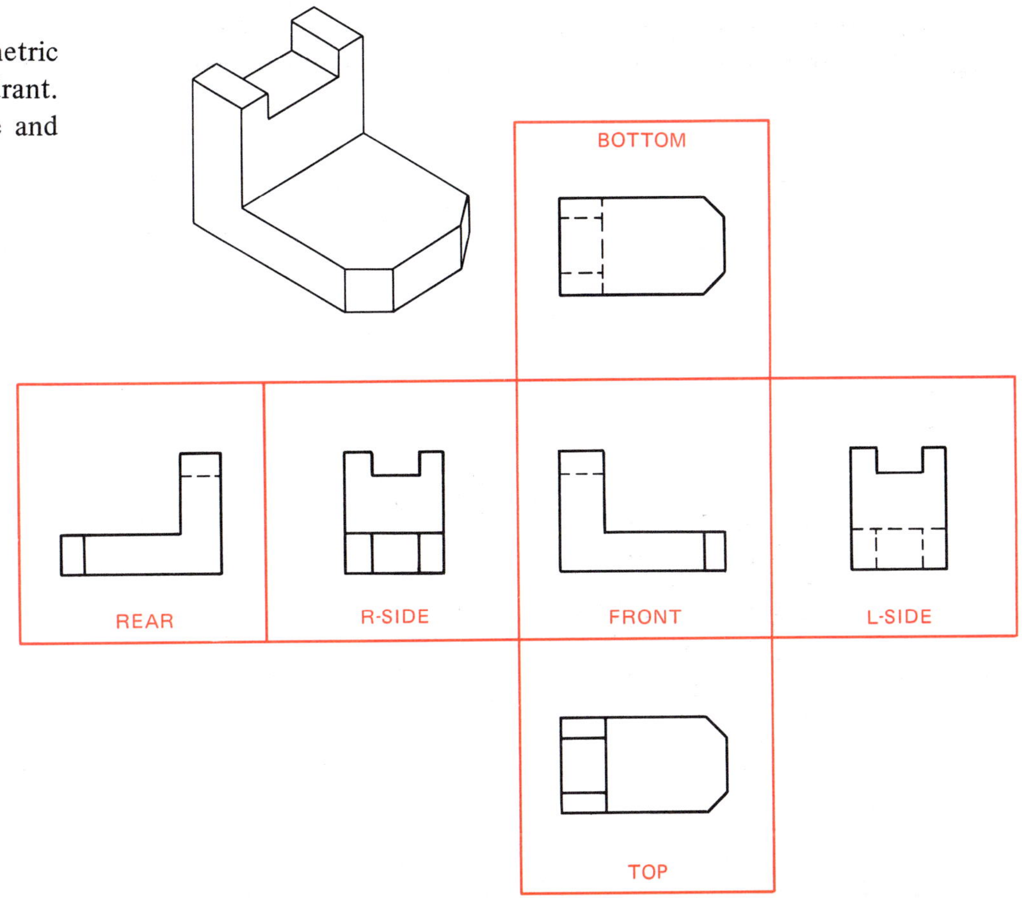

Fig. 3-6 First-angle projection

THIRD-ANGLE PROJECTION

Third-angle projection is used in the United States. The main difference between this approach and first-angle projection is the placement of the object in the third quadrant. This puts the planes of projection between the viewer and the object, figure 3-7. When the six-sided box is unfolded, the top view is above the front view, and the right-side view is to the right of the front view, figure 3-8.

It is possible to draw six principal views of any object. These views are seen by looking at the object from six different points, figure 3-9. Most objects, however, are adequately explained with less drawing. To save drafting time and conserve space, a drafter draws only those views necessary to describe the shape of the object. If two views are sufficient, the front and top views, or the front and right-side views, are preferred. However, if some detail is shown more clearly in the bottom view, it is drawn instead of the top view, figure 3-10. The same is true if some detail is shown

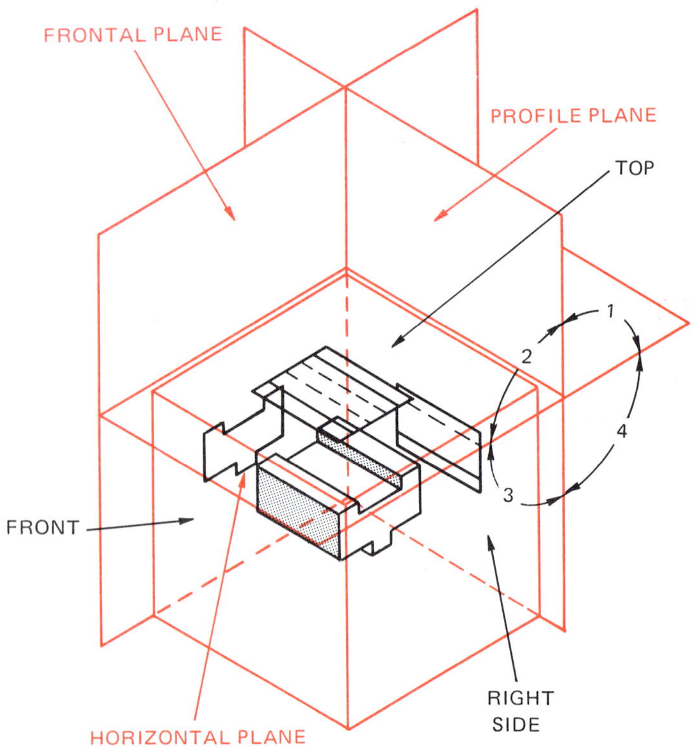

Fig. 3-7 Third-angle projection

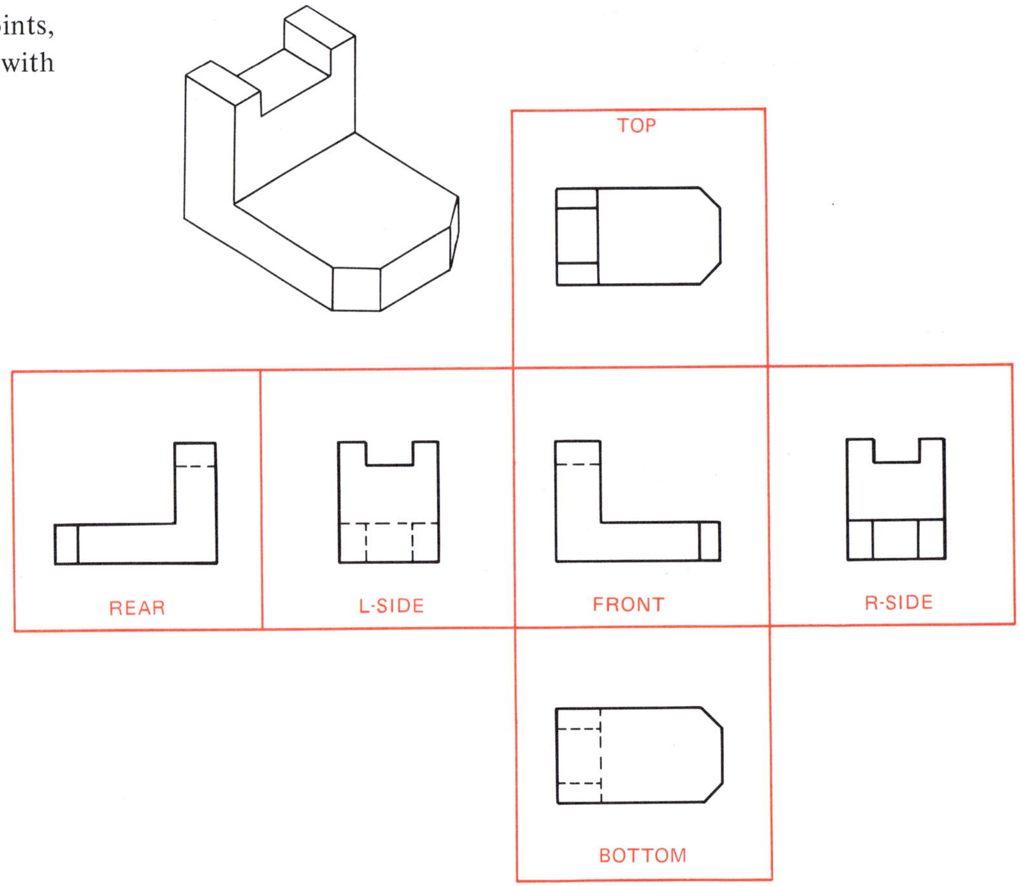

Fig. 3-8 Third-angle projection

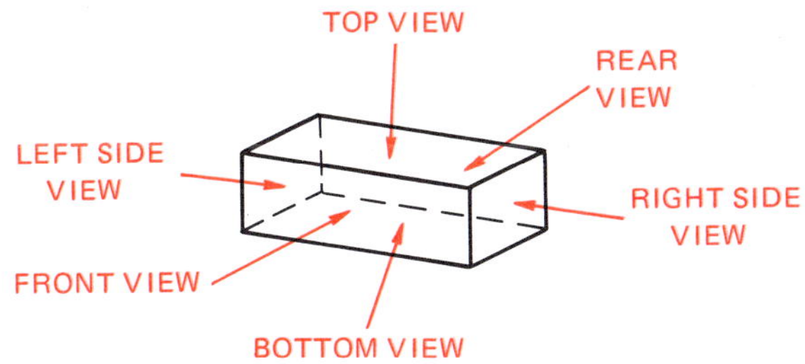

Fig. 3-9 Six principal views

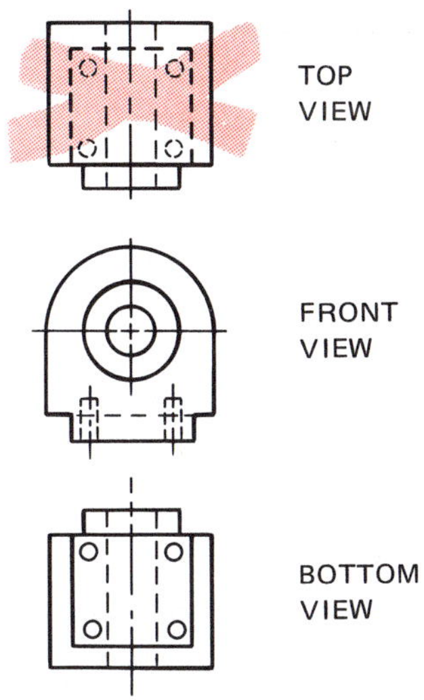

Fig. 3-10 The bottom view preferred

more clearly in the left-side view. It is drawn in preference to the right-side veiw. Some objects such as washers, gaskets, and simple cylindrical parts need only one view.

When three views are required, the front, top, and right-side view are preferred. The left-side view may be substituted for the right-side view as in figure 3-11.

The rule to follow when a choice of views is permitted is to select those views which have the fewest hidden details.

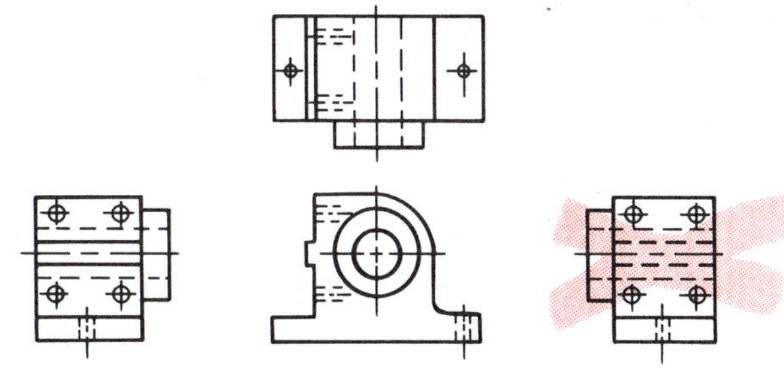

Fig. 3-11 The left-side view preferred

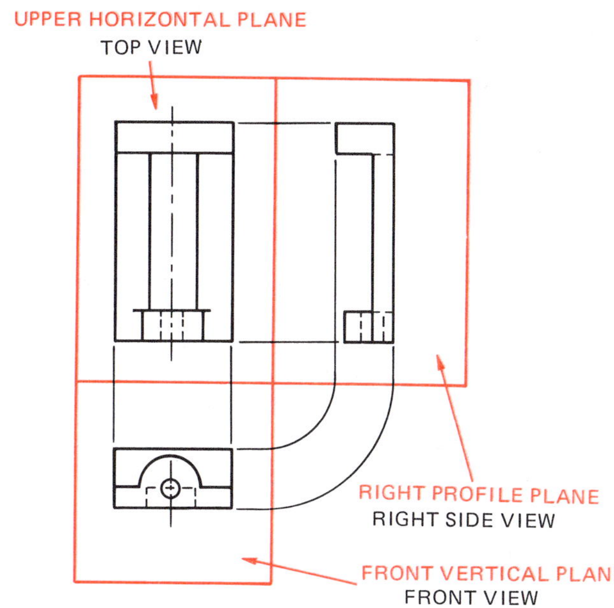

Fig. 3-12 Alternate arrangement of three views

An alternate arrangement of three views is used when the space on a drawing is limited. Then the side view is drawn next to, and in line with, the top view as shown in figure 3-12. In this arrangement the profile plane is hinged to the top horizontal plane.

WORKING DRAWINGS

A *working drawing* shows the correct shape, the exact dimensions, and other information necessary for the manufacture of an object. It consists of only those views that are necessary to explain the shape of an object completely.

The views of a working drawing have the following characteristics:

- The front view is seen from a point directly in front of the object.
- The front view shows the shape, length, and height of the front of the object.
- The top view is seen from a point directly above the object.
- The top view shows the shape, length, and width of the top of the object.
- The top view is drawn directly above and in line with the front view.
- The side views are seen from the point directly to the right or left of the object.
- The side views show the shape, width, and height of the sides.
- The right-side view is drawn at the right of and in line with the front view.
- The left-side view is drawn at the left of and in line with the front view.

UNIT 4
SIZE AND LOCATION DESCRIPTION

While the theory of projection explains the shape of an object, dimensioning describes the exact size and location of each feature of that object, figure 4-1. The dimensioning practices indicated here are based upon those standards used by the American National Standards Institute (ANSI) which is the accepted authority in drawing in the United States.

ALPHABET OF LINES

Before dimensioning, however, a student must understand the *alphabet of lines*, figure 4-2. This alphabet is the basic vocabulary of drafting understood by drafters, engineers, and technicians throughout the world. Each line is precise in form and weight, and consistent no matter which system of dimensioning is used.

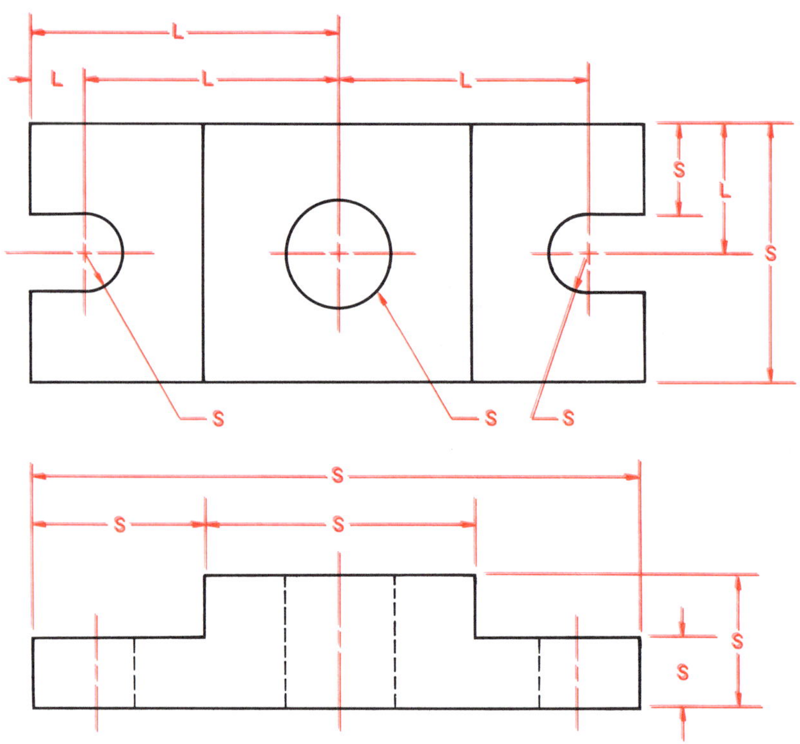

Fig. 4-1 Dimensioning describes the size and location of each feature of an object

Visible Edge Lines and Visible Outlines

The lines on working drawings which represent visible edges and visible outlines of parts are called *visible edge lines* or *visible outlines*. They are heavy unbroken lines because these parts should stand out from the rest of the drawing, figure 4-3.

Hidden or Invisible Edges

Looking at the object from a point directly in front, only those edges on the front of the object are seen. Other edges on the back, or edges which lie directly behind the front edges, cannot be seen. These *hidden edges* may not be drawn in the front view because they are directly beneath the visible edge lines.

The hidden edges on many objects do not always lie behind or beneath visible edges. The position of such hidden edges are shown in the views. The symbol representing hidden edges is a line drawn with a series of short dashes. The dashes are about 1/8 inch long and spaced 1/32 inch apart. They are heavier than extension or dimension lines, figure 4-2, but not as dark or thick as visible outlines.

Hidden edge lines begin and end with a full dash, figure 4-4 (A). If the hidden edge is in line with a visible edge, then an opening separates the two lines as shown at (B). The dashes in the bottom of holes join at the corners (C). A hidden edge that is perpendicular to another is drawn as shown at (D). Arcs drawn with hidden edge lines start and end with a full dash (E).

Extension Lines

Extension lines connect a part of a view with its dimension line. They are light, unbroken lines. They begin about 1/16 inch from the corners of the view or edge they measure and extend 1/8 inch beyond the outside arrowhead, figure 4-3.

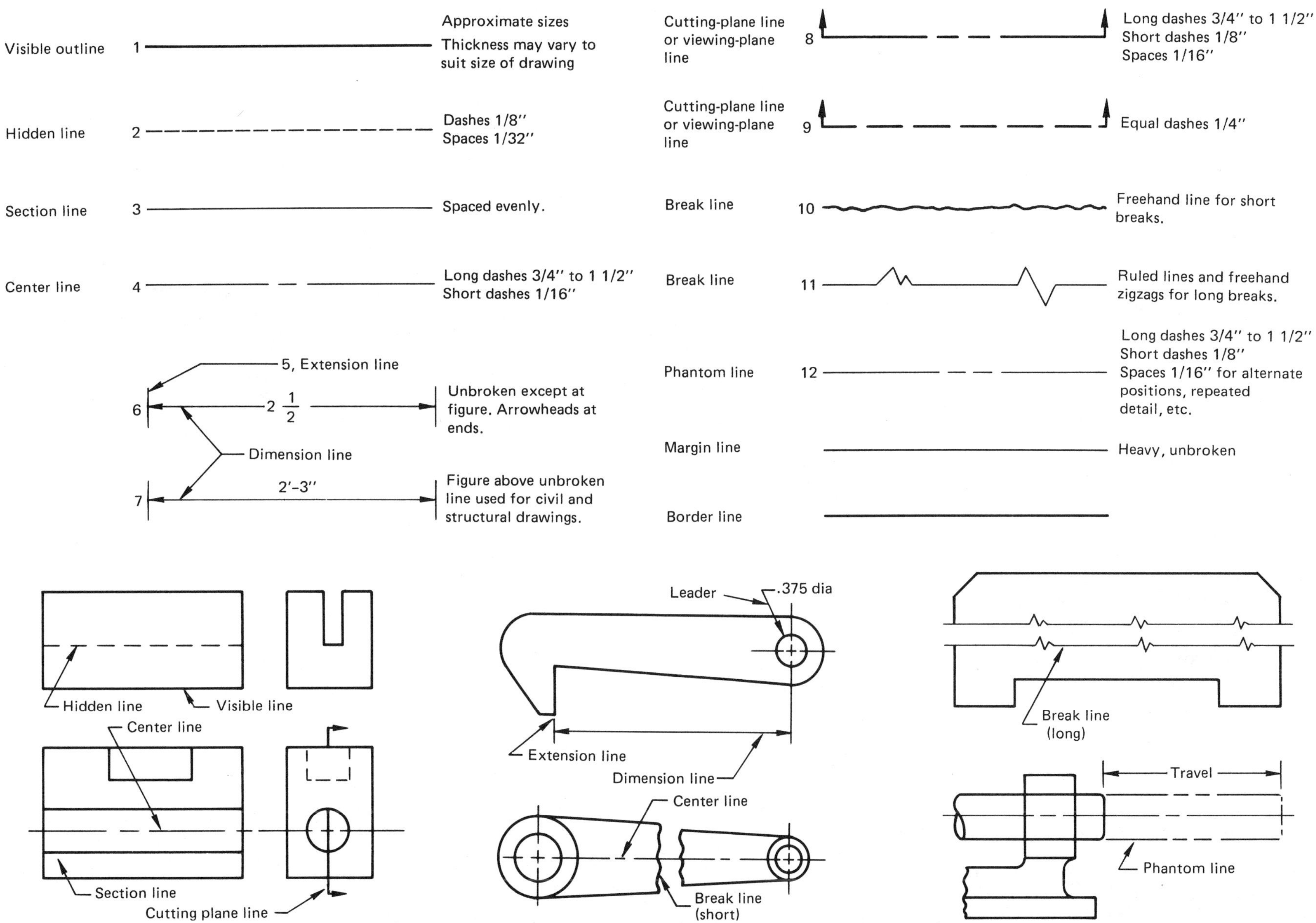

Fig. 4-2 These are the line weights or symbols used on drawings that represent a graphical alphabet recommended by the American National Standards Institute.

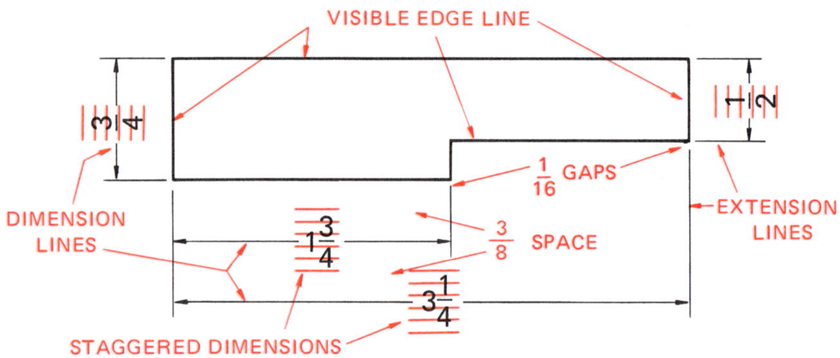

Fig. 4-3 Conventional line symbols

Dimension Lines

Dimension lines are used with numerals and arrowheads to show sizes of features. They are lightweight lines that may be broken to insert numerals. They are often left solid when dual dimensioning or tolerancing, or on architectural and structural drawings, figure 4-5. Broken and unbroken dimension lines are not mixed on a drawing. The first dimension line is spaced 3/8 inch from the view. When several parallel dimension lines are drawn, they are about 1/4 inch apart.

Crossing of dimension or extension lines is reduced by placing the shortest dimensions nearest the view and working out to the longest, figure 4-6. The dimensions are staggered so they are not crowded, figure 4-7.

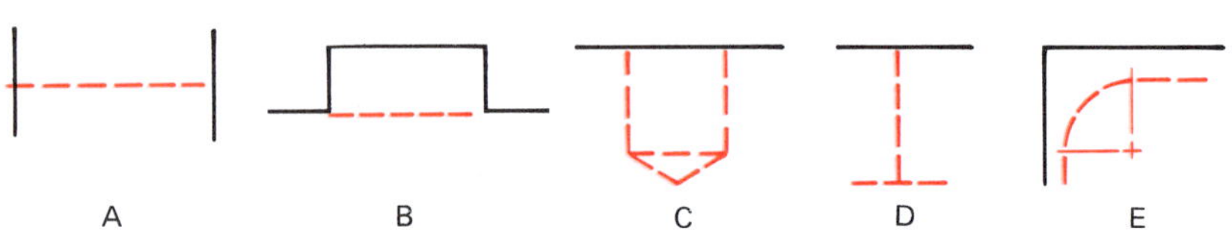

Fig. 4-4 Drawing hidden edge lines

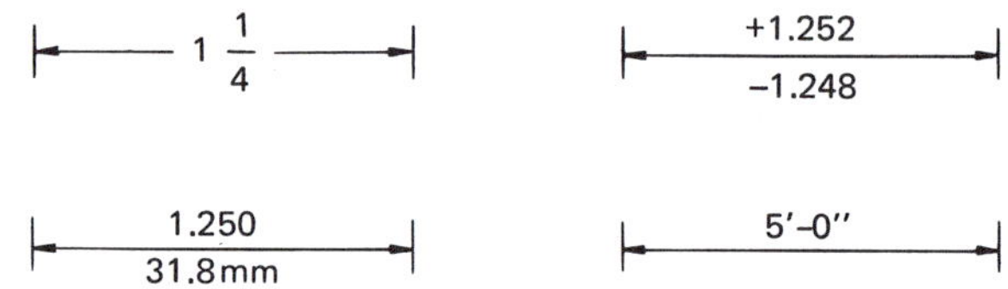

Fig. 4-5 Dimension lines may be broken or solid lines

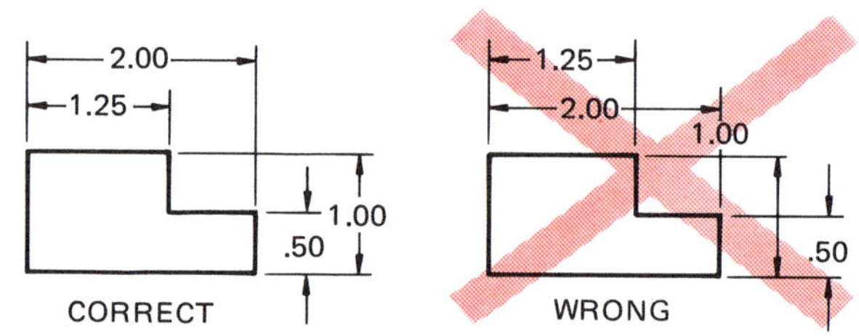

Fig. 4-6 Larger dimensions are placed outside shorter dimensions

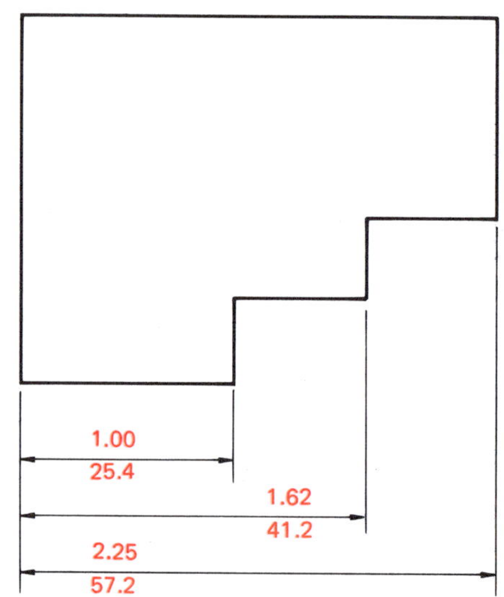

Fig. 4-7 Dimensions are staggered to avoid overcrowding

Arrowheads

Arrowheads are placed on the ends of the dimension lines. They are about 1/8 inch to 3/16 inch long. All arrowheads on a drawing are of uniform length and 1/3 as wide as they are long. Arrowheads are made with two freehand strokes from the barbs to the point. A third stroke is used to fill in, figure 4-8.

Leaders

Special notes sometimes accompany dimensions to specify information on shop processes or placement. They are attached to the view with thin, straight lines called *leaders,* figure 4-2. The leader replaces both extension and dimension lines and ends in one arrowhead. The point of the arrowhead touches the part of the drawing to which the note applies. The note end of the leader becomes a horizontal line about 1/8 inch long. The preferred angle for the leader is 45 or 60 degrees. When drawn to a circle or arc, the leader is drawn in line with the center of that arc or circle, figure 4-9.

Fig. 4-8 Arrowheads

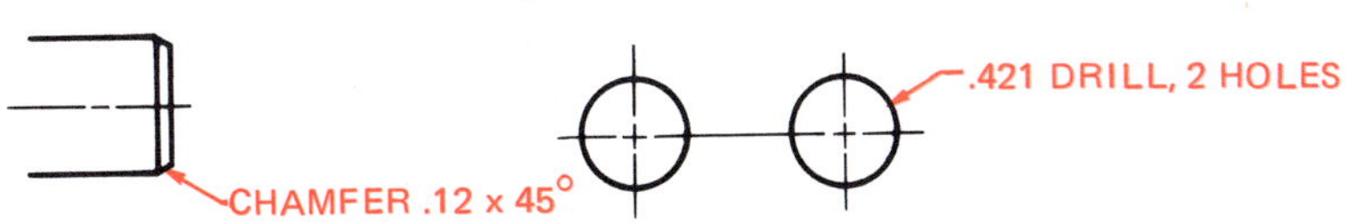

Fig. 4-9 Notes and leaders on drawings

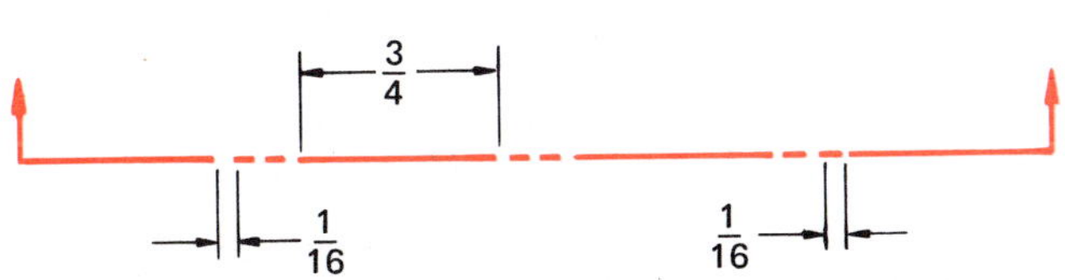

Fig. 4-10 Cutting-plane line

Cutting-Plane Line

The *cutting-plane line,* figure 4-10, is assumed to pass through the object, figure 4-11. Arrowheads on the ends of the cutting-plane line show the direction in which the section is viewed. Letters used with cutting-plane lines identify the cutting plane and the corresponding sectional view. The cutting-plane line, the letters, and the arrowheads are omitted from simple symmetrical views when it is evident that the cutting plane passes through the center of the part.

Centerlines

Centerlines are drawn to show the center of a circle. They are also used to show the axis of spheres, cylinders and cones. Centerlines are frequently drawn through the center of symmetrical objects. Centerlines are light lines drawn with a series of alternating long and short dashes. The long dashes are approximately 3/4″ long; the short dashes are 1/16″ long. The openings between the dashes are 1/16″ long.

Sectioning

Section lining or crosshatching is used on surfaces of an object that touch the cutting plane in sectional views. Some common symbols indicating various materials are shown in figure 4-12.

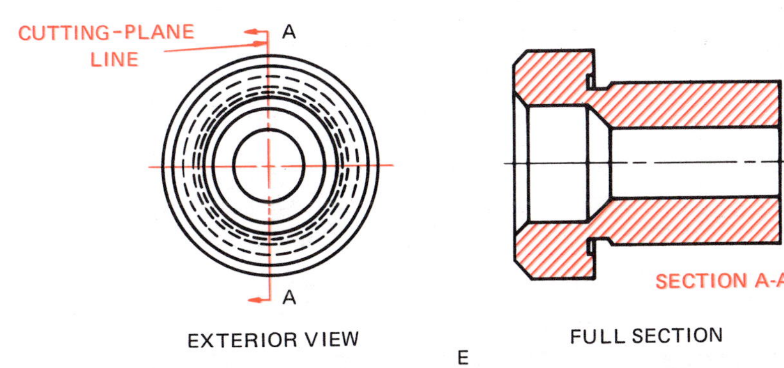

Fig. 4-11 A section view

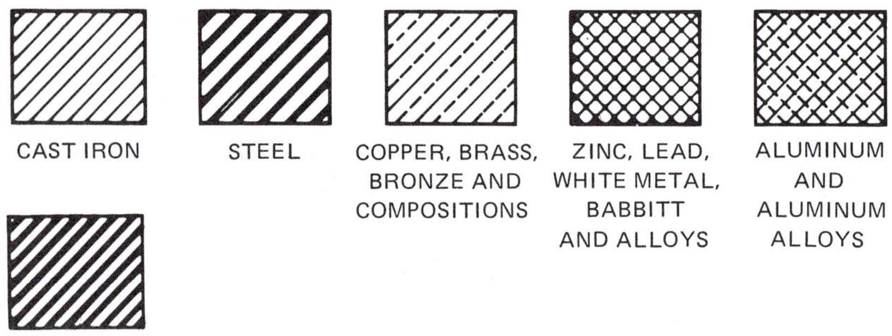

Fig. 4-12 Symbols for section lining

Section lines are uniformly spaced by eye without mechanical aid. For small parts, the lines are 1/32 inch apart; on large parts, they are 3/16 inch apart. For the sectional views in this series of drawings the lines are spaced approximately 3/32 inch apart.

The symbol for cast iron is commonly used on detail drawings to indicate all materials except parts made of wood. The materials, processes, and protective treatments are also shown on the drawing or parts list. This practice saves drawing time. Exceptions to this practice include assembly drawings and details of parts where indicating the type of material is important, figure 4-13.

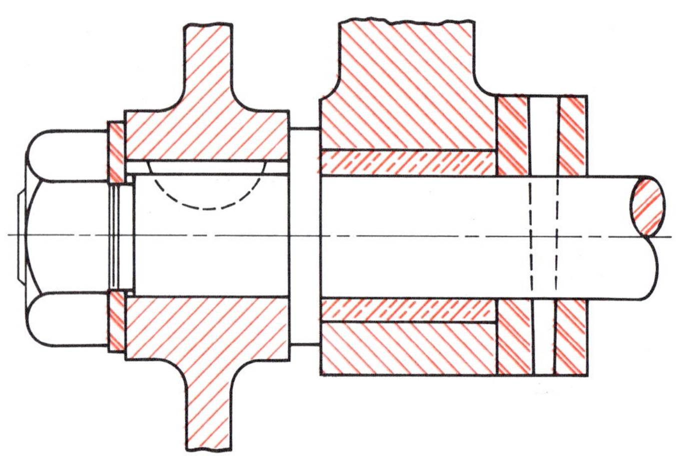

Fig. 4-13 Assembly drawings use material symbols to indicate different parts rather than materials.

DIMENSION NUMBERS AND NOTES

There are two systems of dimensioning accepted by industry. The *aligned system,* in which all dimension numbers are placed parallel to their dimension line, is the original method. In this system the numbers are read from the bottom or the right side of the drawing, figure 4-14. The *unidirectional system* is becoming more popular because of the ease in lettering and reading all numbers from the bottom regardless of the dimension line angle, figure 4-15. Either system of placing dimensions can be used with any of the measuring systems selected.

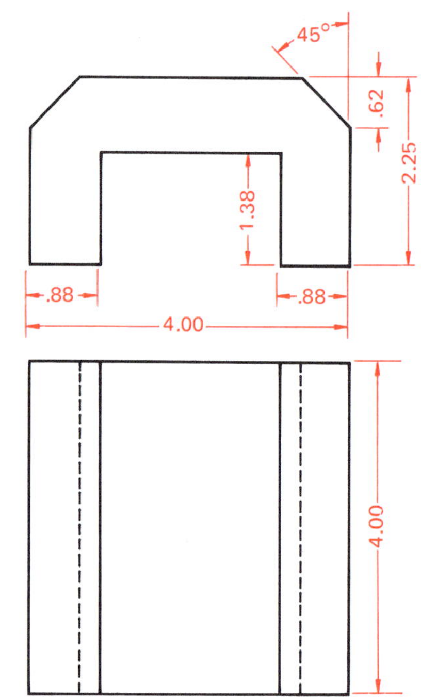

Fig. 4-14 Aligned dimensioning was the only acceptable system for many years.

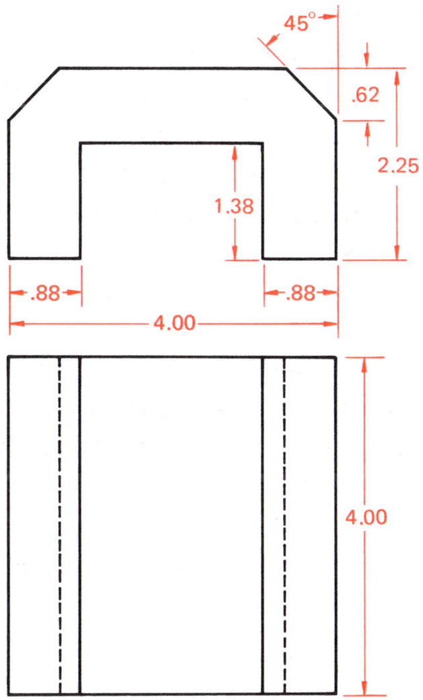

Fig. 4-15 Unidirectional dimensioning is more practical on very large drawings such as those done in the aerospace industries.

MEASURING SYSTEMS

Drawings may be dimensioned with the English system, metric system, or dual-dimensioning system. The choice depends upon the use of the final drawing.

English System

Both the fractional and decimal units of dimensioning are based on the English inch and foot linear measurements. The United States is the only major nation still using these units. They are gradually being phased out.

Fractions are only used on drawings of parts and products where close tolerances are not critical, and in the architectural and structural fields. However, common fractions still designate many parts of objects. A thread notation, 1 1/8 - 12 UNF - 2 for example, continues to be designated by its common fraction even though the rest of the drawing is dimensioned in decimals. When dimensioning in fractions, the whole number is usually 1/8 inch high with the fraction 1/4 inch. Each numeral in the fraction is slightly smaller than 1/8 inch so there is space between numbers for the fraction bar, figure 4-16. When all dimensions on a drawing are in inches, the standard practice is to omit the (″) symbol. The symbols (′) for feet and (″) for inches are used when any dimension on a drawing exceeds 72 inches. The pattern must be consistent as in figure 4-17.

Standard nominal sizes of materials, or features designated by fractions, are used in fraction form when the rest of the dimensions are given in decimal inches.

Decimal dimensioning, figure 4-18, is usually done to the hundredth (two-place decimals) increment of an inch and rounded to even numbers, figure 4-19, where tolerances of ±.01 are allowed. Some exceptions to this practice are when using ordinate dimensioning, figure 4-20; when dimensioning from common fractions,

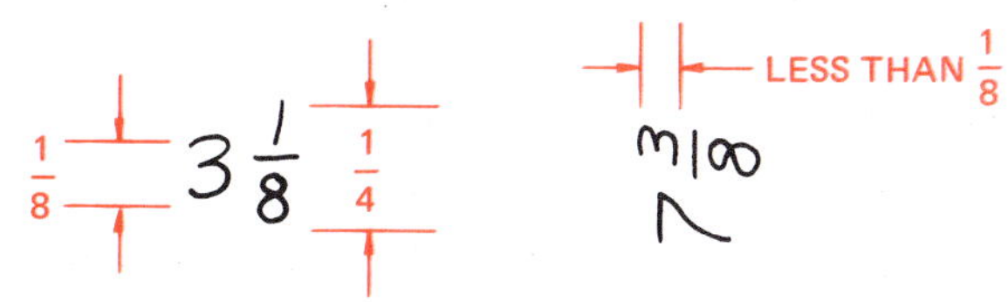

Fig. 4-16 The fraction bar is either horizontal or vertical-never slanted

$10' - 8\frac{1}{2}''$ $0' - 9''$ $3' - 0''$

Fig. 4-17 When measuring with feet and inches, both units must be indicated, whether they can be measured there or not, to keep the dimensioning consistent .

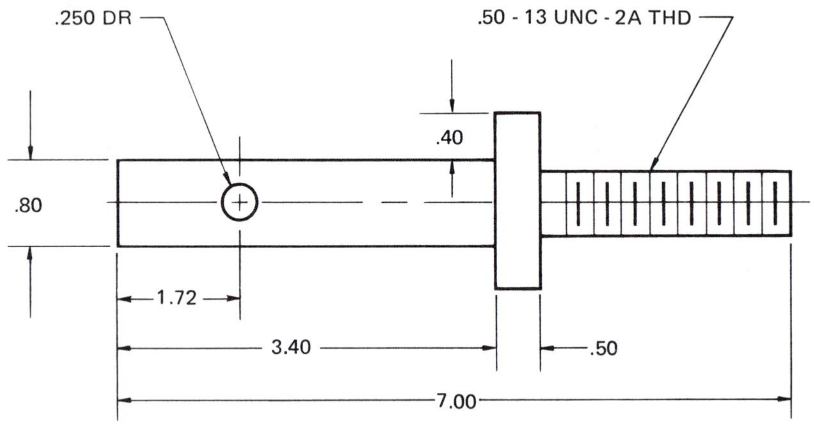

Fig. 4-18 Applying the rules of decimal dimensioning to a drawing

such as 1/4 inch = .25 or 3/4 inch = .75; or when it is necessary to use three-place decimals for tolerance limits less than ±.010.

Decimal dimensions are the same height as the lettering on a drawing. Zeros to the left of the decimal point are omitted. The decimal point must be definite and large enough to reproduce when drawings are reduced in size. Fractions are often converted to decimals by using a chart similar to the one in the Appendix.

Metric System

In 1960, the International Bureau of Weights and Measures located in France established Le Systeme International D'Unites (The International System of Measurements). It is abbreviated SI. This SI metric system revised and simplified the existing metric system. It still uses the metre as its basic linear measure, however. Internationally, most scientific and technical work is done in this measuring system. The United States is gradually converting to metrics because of its dependence on international exchange.

The metric units are based on multiples of ten. Although English-speaking countries use a period for the decimal point in metric dimensioning, most countries use a comma. Spaces, instead of commas, group numerals into thousands (21 734 186.25). Whenever a number is less than one, a zero is placed to the left of the decimal point. Avoid using the centimetre (cm) on drawings. The millimetre is used on most drawings where dimensions do not exceed 10 000 millimetres. The SI units are not mixed with units from another system of measurement. The total tolerance applied to an inch dimension is the basis for the accuracy in rounding off dimensions and tolerances converted to millimetres.

5/8 = .625 = .62

41/64 = .6406 = .64

21/32 = .6562 = .66

Fig. 4-19 Even numbers are preferred when rounding decimals to hundredths.

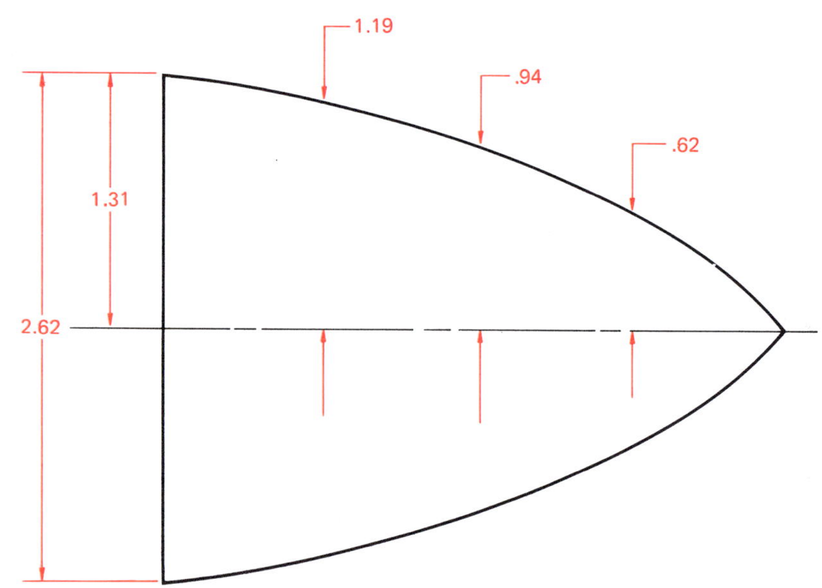

Fig. 4-20 Odd hundredth dimensions may be used on ordinate dimensioning of curves.

Dual Dimensioning

During the period of conversion to the metric system in the United States, both the English inch and metric (SI) dimensions are often stated on the same drawing.

When the drawing is primarily intended for use in the United States, the decimal-inch dimension appears above the dimension line or to the left. The millimetre value is placed below the line or to the right of the inch unit. The millimetre value can also be put in brackets. The reverse is true in countries using SI practices, figure 4-21. Another method is to dimension the drawing in millimetres, place the word "METRIC" on the drawing at an angle, and provide a chart on the drawing converting the metric units to English units.

Because the SI method uses first-angle projection and the U.S. method uses third-angle projection, a symbol for third-angle projection, figure 4-22, is included on all dual-dimensionirg drawings to avoid confusion in international interpretation.

GENERAL RULES FOR DIMENSIONING

- Select the system of measurement best suited to the purpose of the drawing and the degree of accuracy required.
- Use either the aligned or the unidirectional system of placing the dimensions on the drawing.
- Plan adequate space for the drawing and dimensions.
- Dimension the object so that the engineering or design concept cannot be misinterpreted.
- Dimension size, form, and location completely so the object is produced without measuring the drawing or assuming distances.
- Show dimensions between all points or features that have a specific functional relationship to one another.
- Make sure all line symbols are accurate and contrast noticeably.
- Dimension each feature on the view where its most characteristic contour appears.
- Avoid duplicating of dimensions. Omit one dimension in a chain, except as a reference only, to avoid locating a feature from more than one point.
- Place dimensions between the views to which they relate and outside the views wherever possible.
- Avoid crossing dimension lines with extension lines or leaders.

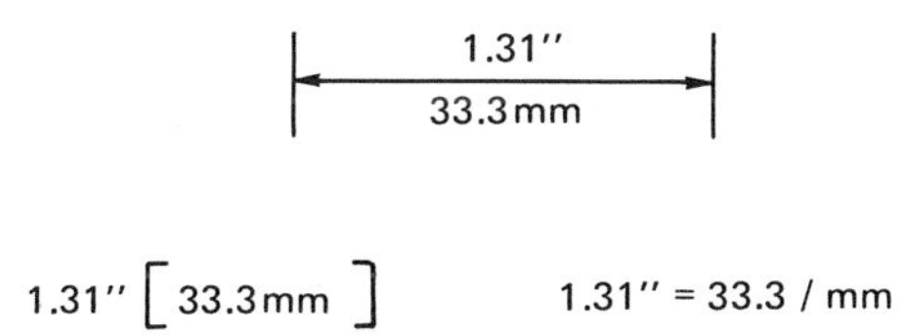

Fig. 4-21 United States practice for dual dimensioning. SI countries reverse the positions.

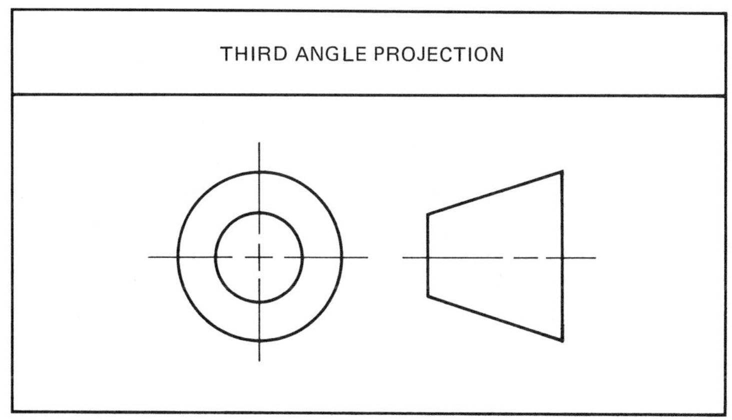

Fig. 4-22 Standard symbol to indicate United States projection practice that differs from SI practice.

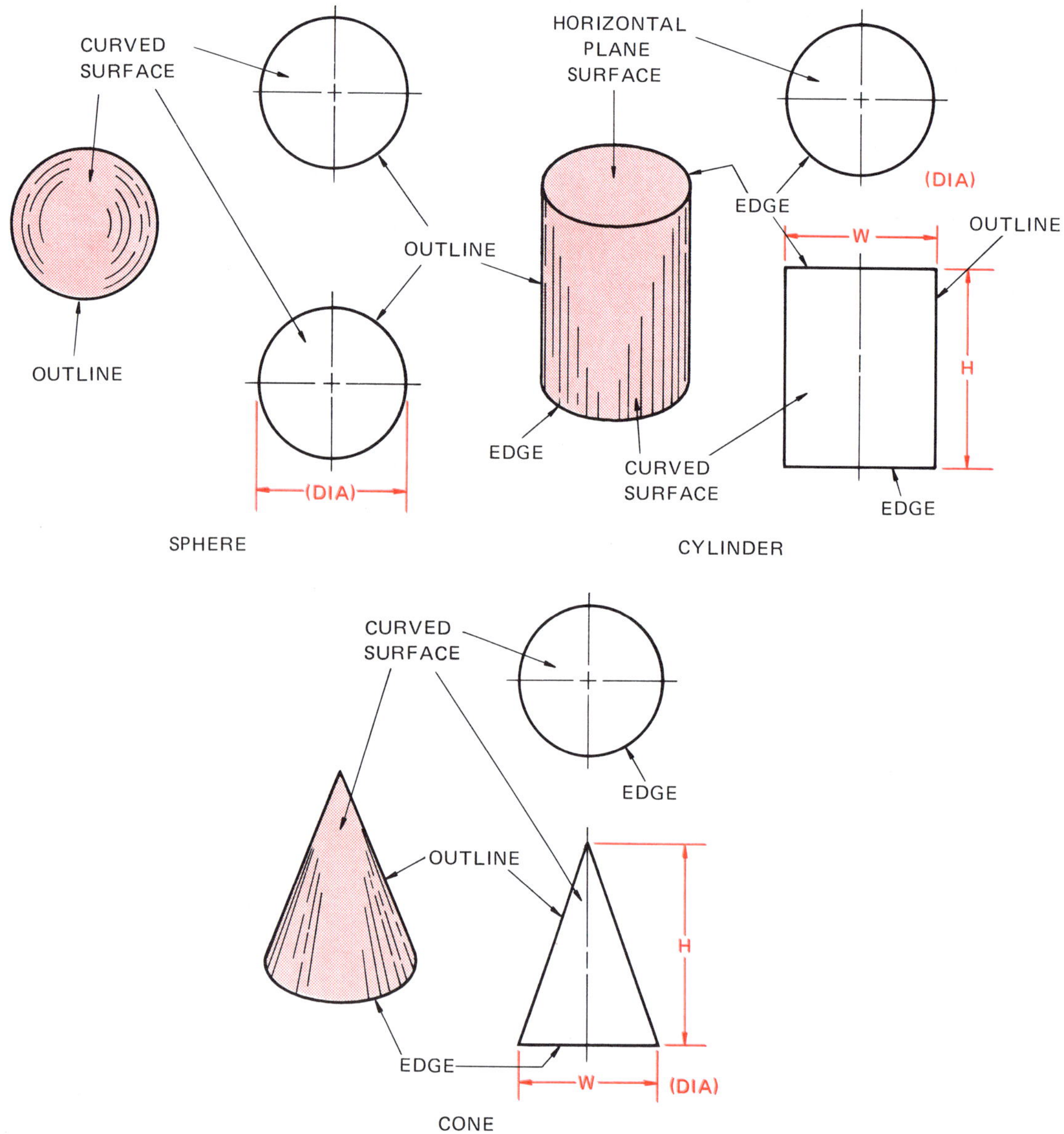

Fig. 4-23 Cylindrical shapes. The common geometric forms having curved surfaces are the sphere, cylinder, and cone.

Cylinders, Cones, Pyramids, and Spheres

All objects are compositions of basic geometric shapes: the prism cylinder, cone, pyramid, and sphere. They all have length, width, and height. In some cases, the diameter of an object takes the place of two or even three of the basic dimensions. Figure 4-23 indicates the location of these dimensions on objects with curved surfaces.

The true form of curved surfaces cannot be shown on a two-dimensional surface. Only the outlines of the curved surfaces are shown. The upper and lower surfaces of the cylinder are horizontal surfaces that are represented as horizontal lines. In a two-view drawing of a cylinder, one view appears as a rectangle. A center line in cylindrical shapes, such as holes, indicates the objects are round instead of rectangular, figure 4-24. Round holes are dimensioned on the view where they appear as circles, figure 4-25. Small holes are usually dimensioned with a leader. Larger holes use diagonal line across the diameter. Holes that require special

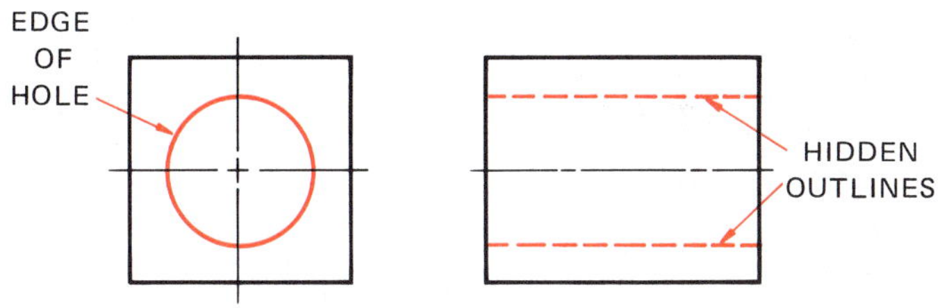

Fig. 4-24 A hole can be a cylinder

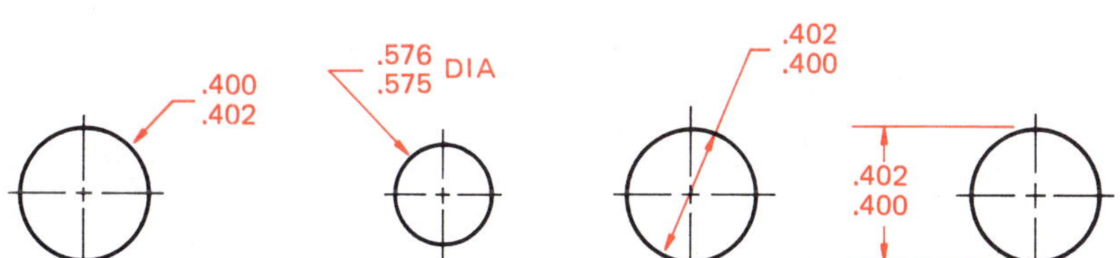

Fig. 4-25 Dimensioning holes in the circular view

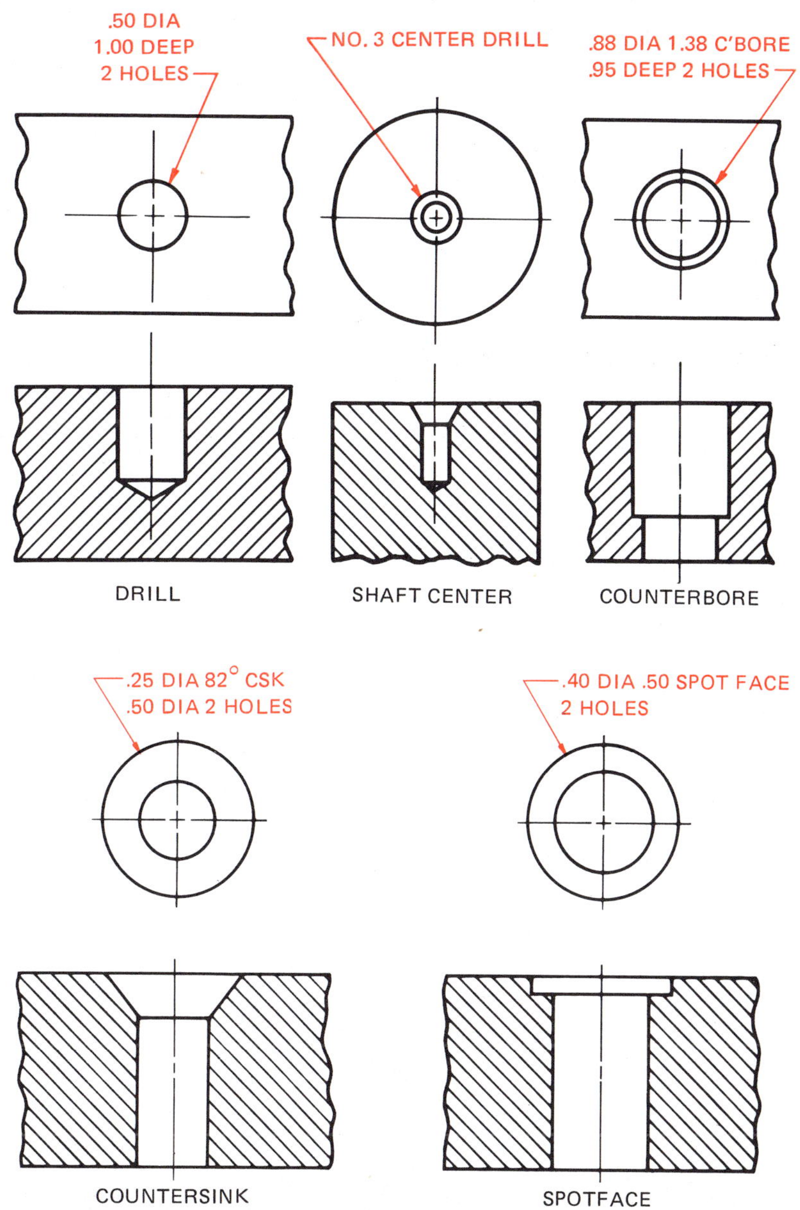

Fig. 4-26 Dimensioning special holes in circular and section views

machining processes or limitation of depth need a note to specify the conditions as well as the size, figure 4-26. In single-view drawings, if the shape of a cylinder is not obvious, the diameter is abbreviated (DIA). The diameter symbol (∅) is used instead of (DIA) in SI dimensioning practice. Where a circular view is shown, the word diameter is not necessary.

Where the diameters of a number of concentric circles are indicated, the dimensions are usually placed on the longitudinal view.

Circular Arcs and Curved Surfaces

Circular arcs are frequently dimensioned with a combination radius and leader. A small cross indicates the center from which the radius dimension is taken. The different methods of indicating that dimension are shown in figure 4-27.

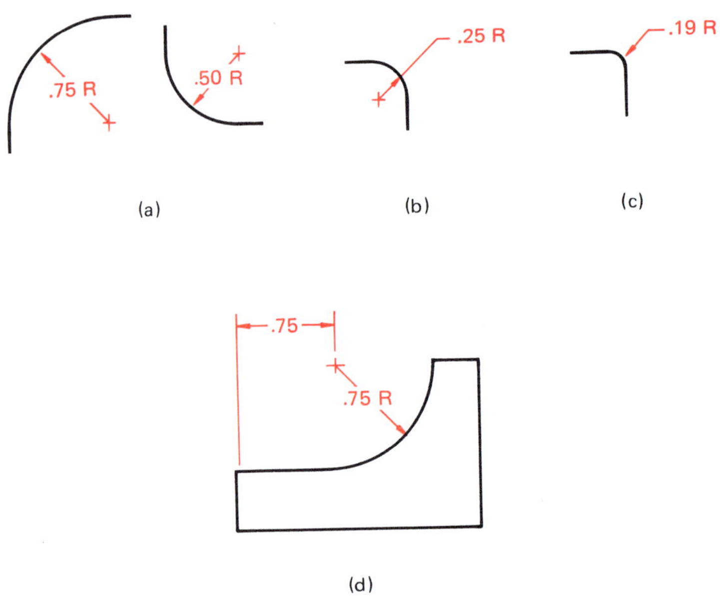

Fig. 4-27 Dimensioning arcs

Fillets and rounds are dimensioned with the radius value and a leader, or with a general note that indicates the size of all rounds and fillets for the drawing, figure 4-28. Linear distances are dimensioned along the curve with a radius dimension, or across the chord of the arc as demonstrated in figure 4-29. Symmetrical

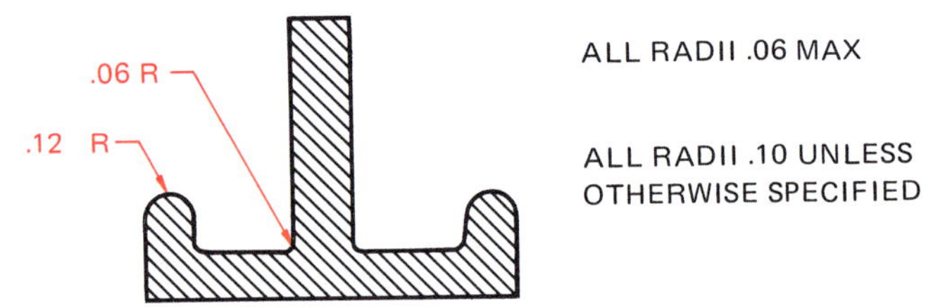

Fig. 4-28 Dimensioning rounds and fillets

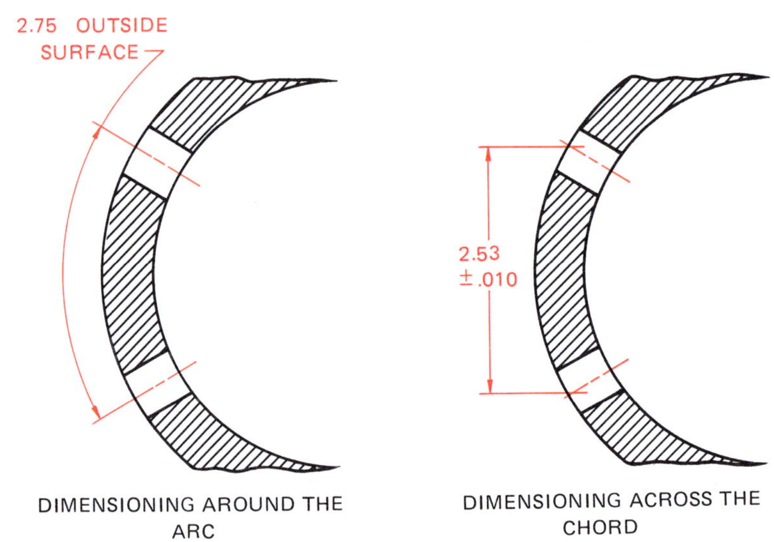

Fig. 4-29 Dimensioning around the arc and across the chord

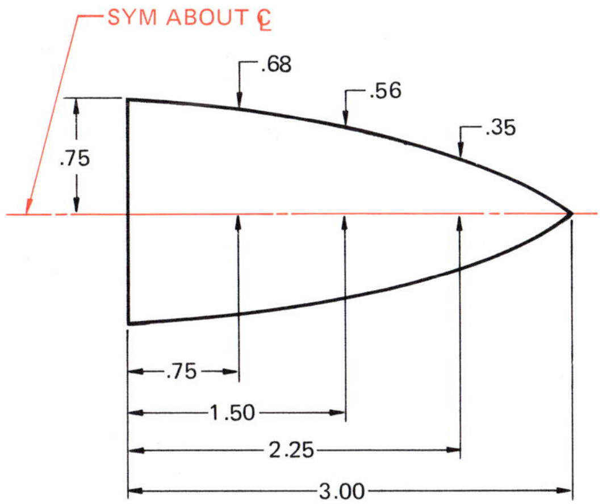

Fig. 4-30 Dimensioning with a symmetrical center line symbol

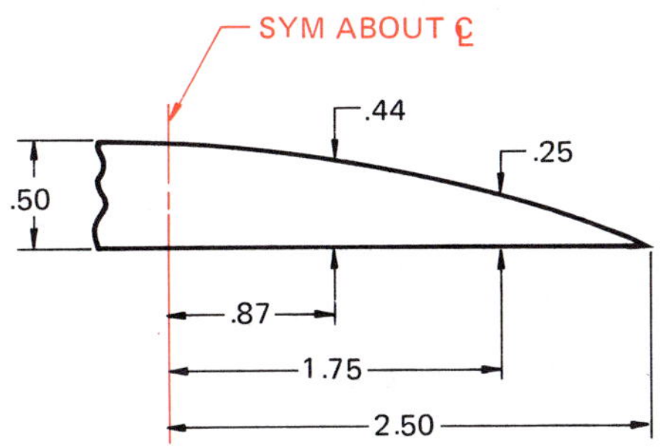

Fig. 4-31 Dimensioning a symmetrical part when only one side is drawn

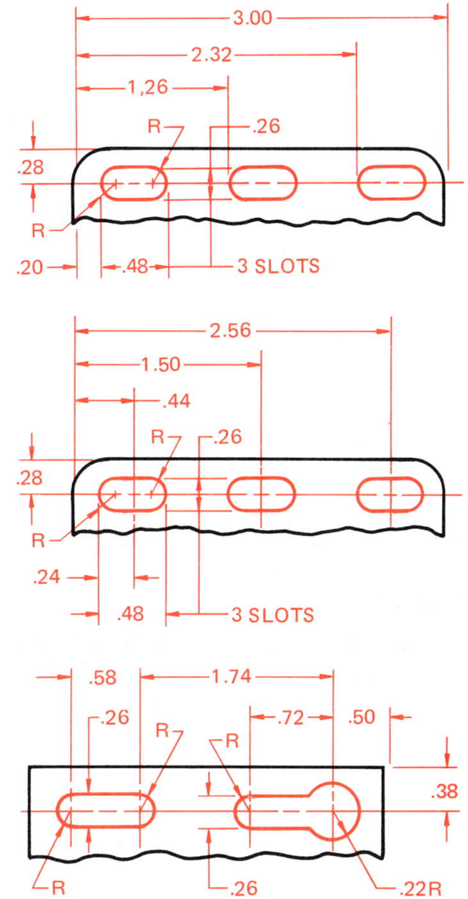

Fig. 4-32 Standard methods for dimensioning slotted holes

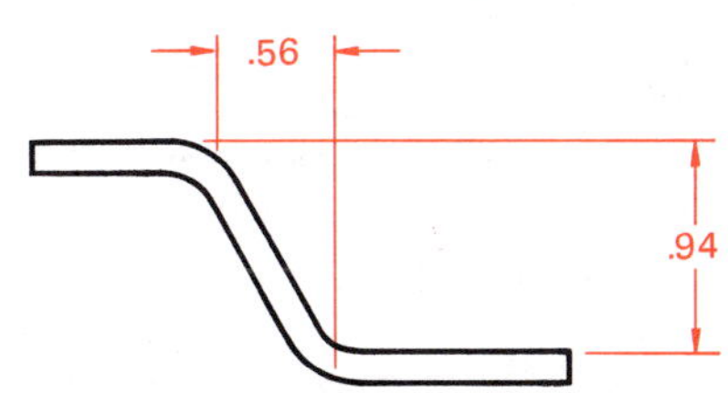

Fig. 4-33 Dimensioning offsets from points of intersection

curves are dimensioned correctly in either of the methods indicated in figures 4-30 and 4-31.

Slotted holes are position dimensioned by center lines and size dimensioned as indicated in figure 4-32.

Offsets on a part are dimensioned from the points of intersection of the tangent arcs along one side of the part, figure 4-33.

Prisms and Angles

Prisms are constructed of planes that meet to form definite edges which can be indicated on a drawing sheet. Figure 4-34 indicates the recommended methods of dimensioning prisms.

Dimensioning Inclined Surfaces

The slope of an inclined line is dimensioned from a horizontal or vertical line, or from another inclined line, by an angle dimension, figure 4-35. To dimension an angle, the needle point of the compass is placed on the vertex of the angle. An arc is drawn from one side of the angle to the other leaving a space in the center for the numerals. Arrowheads are placed on the ends of the arc. In restricted places and for small angles, the arrowheads are reversed. Except for large angles, the numerals are placed so as to read from the horizontal position. Angular dimensions are specified in degrees (°), minutes (′), and seconds (″).

Chamfer Dimensioning

Chamfers of 45 degrees are commonly dimensioned by a note, figure 4-36. All chamfers using other angles are dimensioned by identifying the angle and the distance along the length of the part, figure 4-37.

NOTES

Notes on drawings provide additional information, such as thread notations, processes, or general information.

Leaders connect local notes, figure 4-38, to the detail. General notes, such as "finish all over" or "rounds and fillets," apply to the entire drawing. Common locations for general notes are above the title strip on the right side of the sheet, or to the left of the title block. Tolerances, material specifications, and heat treatment notations are often included in the title block.

Notes on drawings are always made in capital letters. The lettering is the same size as the dimensions on the drawing, usually 1/8 inch. Two or more lines of lettering in the same note are spaced one-half to one letter height apart. Two separate notes are spaced at least 1/4 inch apart.

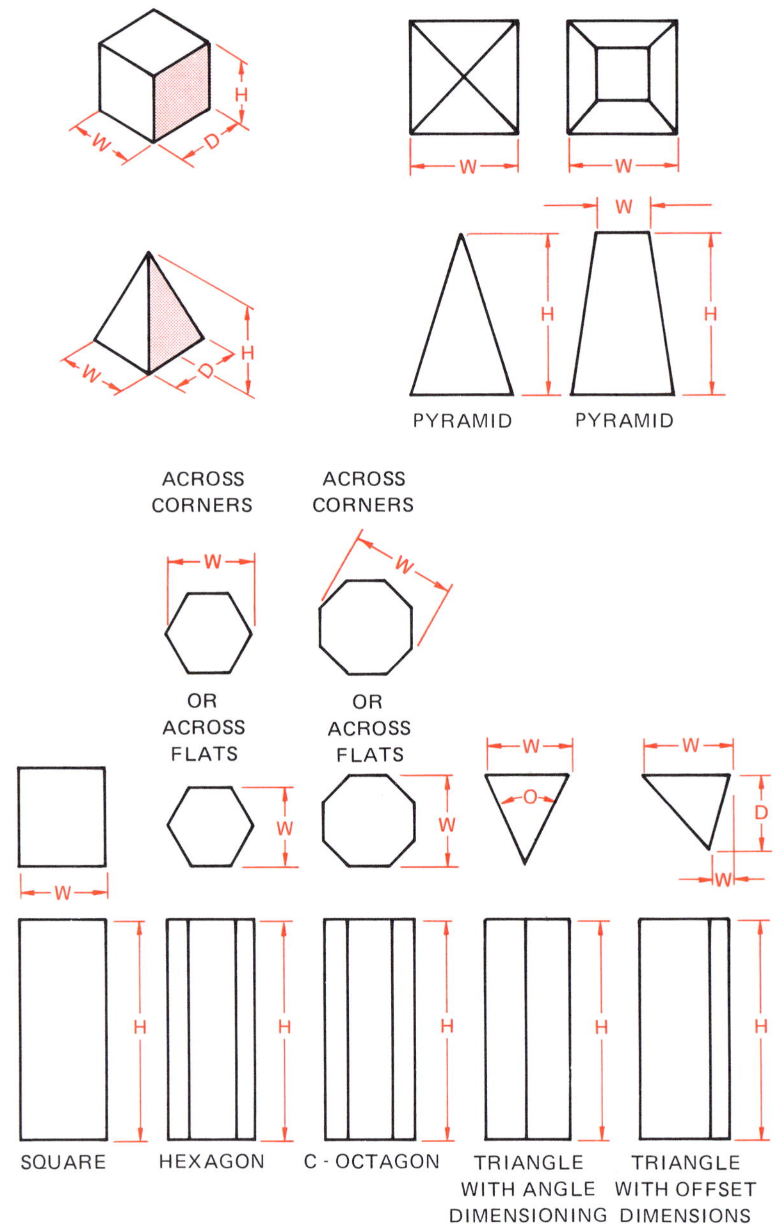

Fig. 4-34 Examples of prism dimensioning

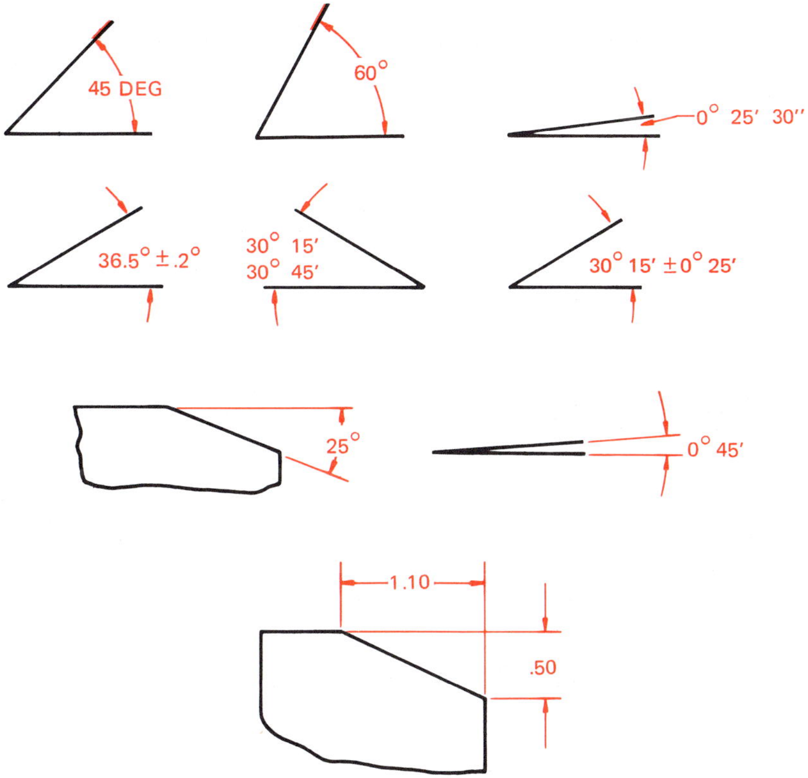

Fig. 4-35 Dimensioning angles

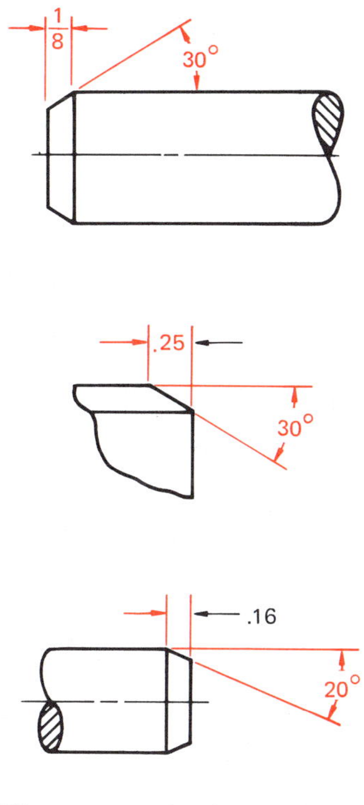

Fig. 4-37 Different methods of dimensioning a chamfer

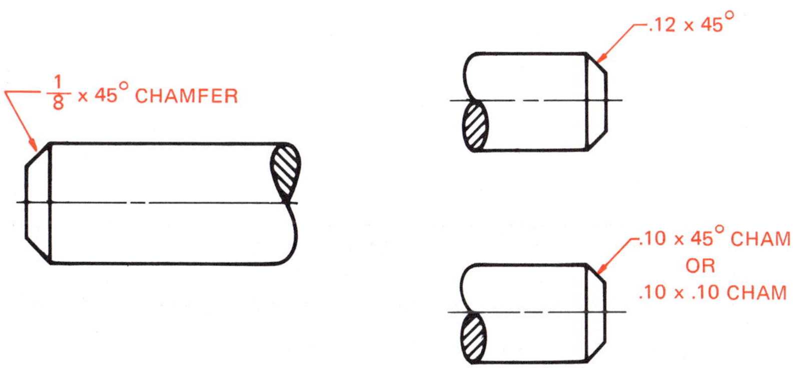

Fig. 4-36 Dimensioning chamfers

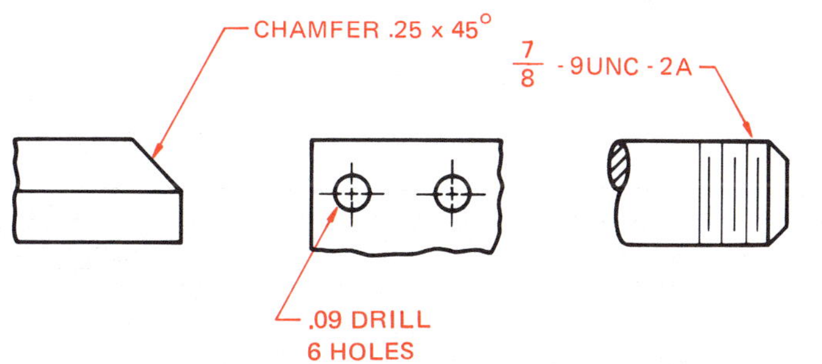

Fig. 4-38 Local notes with leaders are used to describe size and form.

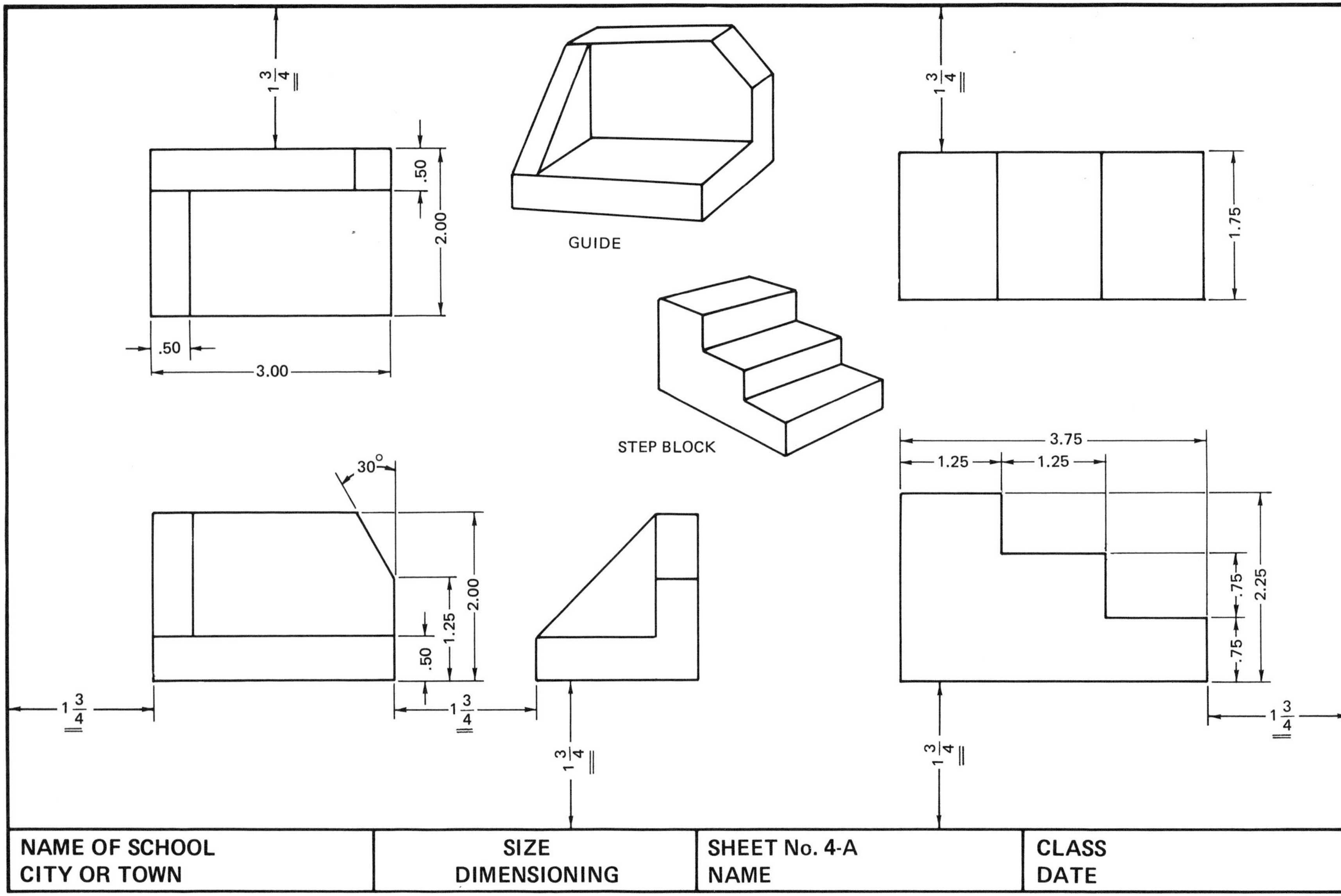
GUIDE
STEP BLOCK
.50
2.00
.50
3.00
30°
2.00
1.25
.50
1.75
3.75
1.25
1.25
.75
.75
2.25
1 3/4
NAME OF SCHOOL
CITY OR TOWN
SIZE
DIMENSIONING
SHEET No. 4-A
NAME
CLASS
DATE

Fig. 4-39

All notes should be short and precise. They may use standard abbreviations and symbols. Abbreviations do not need periods unless they could be misinterpreted.

ASSIGNMENT A

Draw the guide and step block as shown in figure 4-39.

Procedure

1. Lay out the margin lines on the drawing sheet and letter the title strip.
2. Draw the front view of the guide. Locate it 1-3/4 inch from the left margin and 1-3/4 inch above the title strip.
3. Draw the top view of the guide 1-3/4 inch below the top margin line.
4. Draw the right side view 1-3/4 inch from the front view.
5. Dimension the views as indicated. Add the metric equivalent below the dimension line for dual dimensioning.
6. Draw the front view of the step block 1-3/4 inch from the right margin line.
7. Draw the top view 1-3/4 inch below the top margin line.
8. Dimension the views as indicated. Add the metric equivalent below the dimension line for dual dimensioning.
9. Underlined dimensions are for information on the placement of the views and should not appear on the finished drawing.
10. Check the drawing for accuracy.

LOCATION DIMENSIONING

When the position of any feature in relation to another is important to the manufacture and function of a part, it is dimensioned by location or position. The design concept and the tolerances allowed must be understood by the drafter in order to know where location dimensions are necessary and what to establish as a reference for exact measuring.

Point to Point Cumulative Dimensioning

Locations are dimensioned from point to point on simple objects, figure 4-40. Figure 4-41 shows the acccumulated margin of error allowed when maximum and minimum tolerances are applied.

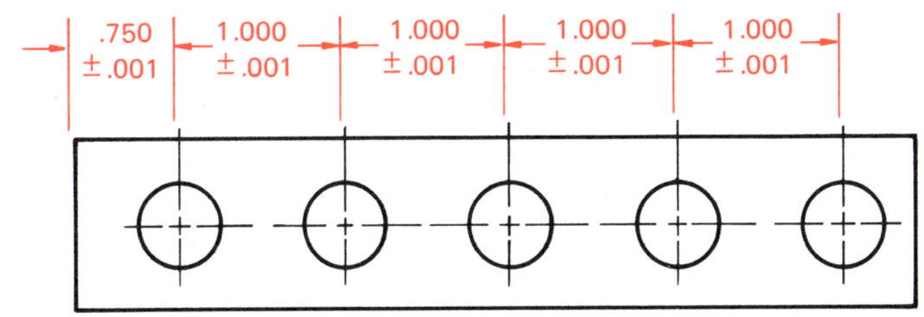

Fig. 4-40 Point-to-point, chain, or cumulative dimensioning

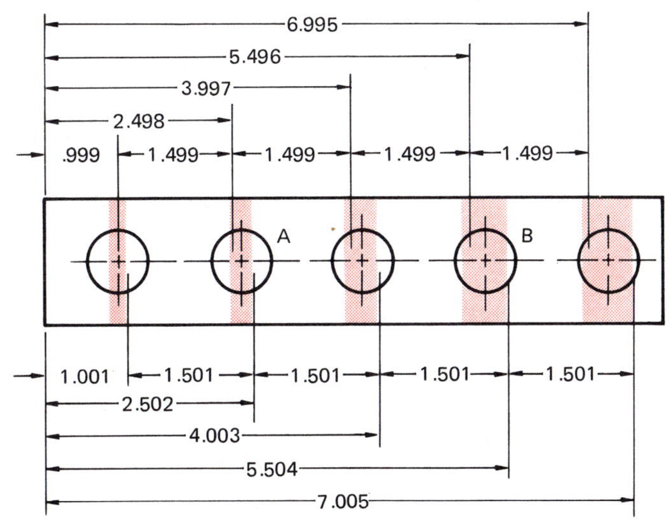

Fig. 4-41 Accumulated tolerance

Coordinate Dimensioning

An object that contains features mating with another part is dimensioned using coordinate dimensioning to datums to avoid the accumulation of tolerances. The coordinate dimensioning method, with datums to control precision mating of interchangeable parts, is shown in figure 4-42.

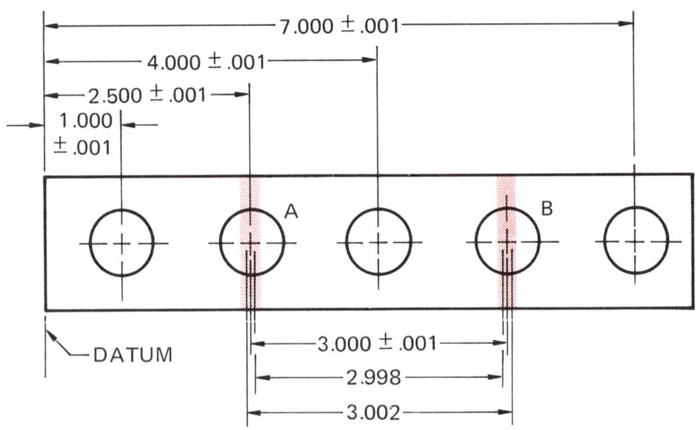

Fig. 4-42 Coordinate dimensioning to a datum. Accumulation of tolerance is less.

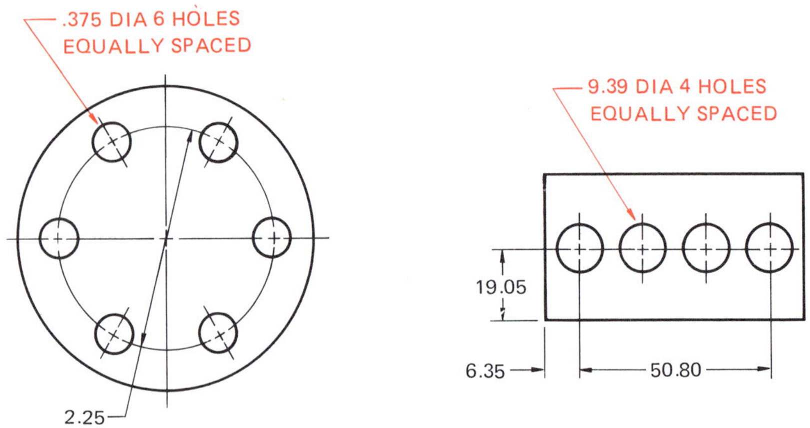

Fig. 4-43 Locating equally spaced holes with a note

A series of equally spaced holes are located by indicating the relationship of the common center line or distance to another feature or plane of the part. A note can also be used, figure 4-43.

Rectangular coordinate dimensioning locates features that lie in a rectangular pattern. This type of dimensioning is necessary on drawings of parts to be made on numerical control machines,

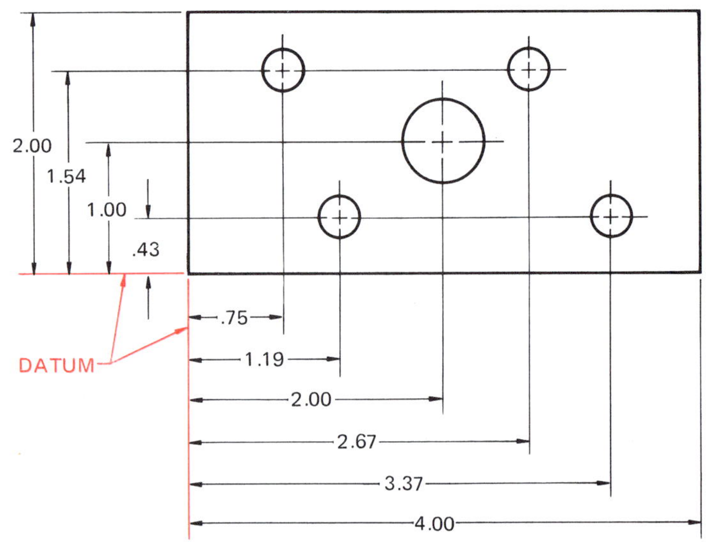

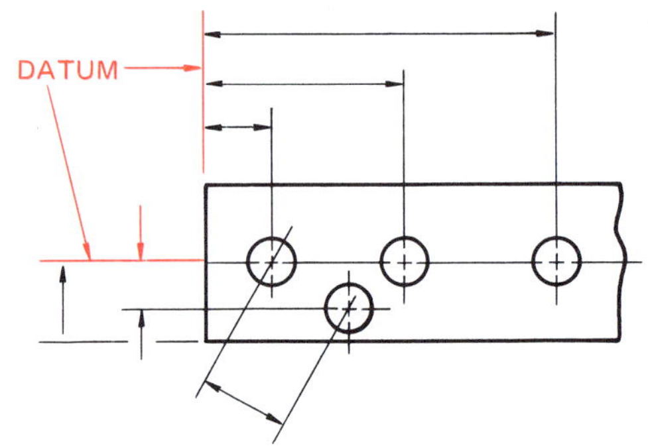

Fig. 4-44 Coordinate dimensions

figure 4-44 and figure 4-45. Noncircular curves are also dimensioned by locating coordinates from datum lines, figure 4-46.

Polar coordinate dimensioning is used when holes or other features lie in a circular pattern that are given a radial dimension. A radial dimension is given from the center and an angular dimension from a datum plane, figure 4-47.

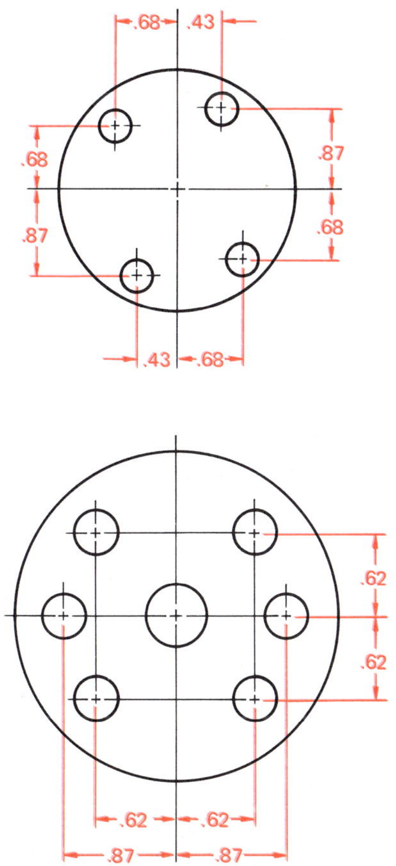

Fig. 4-45 Round holes dimensioned by rectangular coordinates

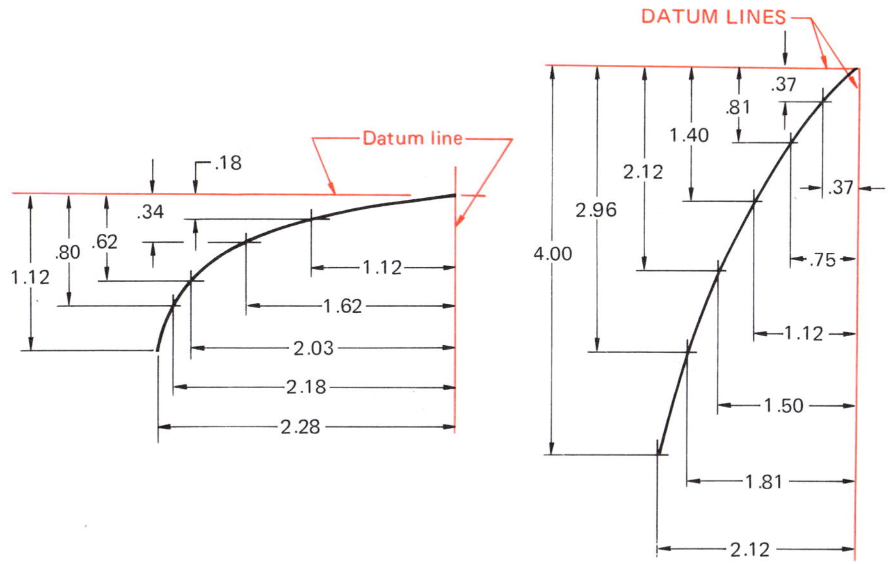

Fig. 4-46 Curves dimensioned by locating coordinates from datum lines

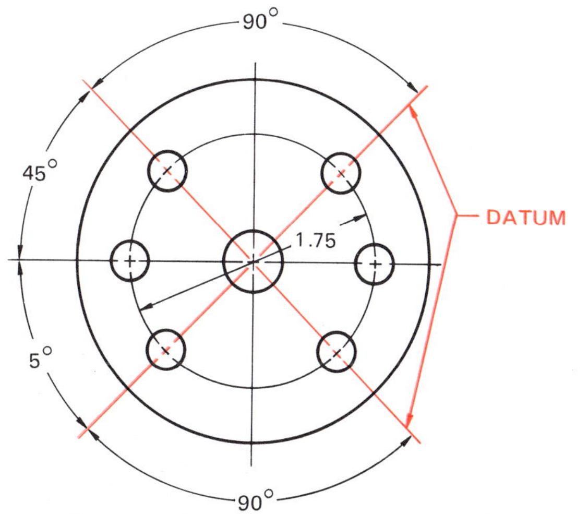

Fig. 4-47 Use polar coordinates for features in a circular pattern.

Dimensioning to Rounded Ends

Parts with rounded ends are dimensioned in several ways with emphasis on either the size or the location. If overall length and width are not critical, dimension as in figure 4-48. If the overall length and width are important, dimension as in figure 4-49. If overall dimensions and center-to-center placement are important, dimension as in figure 4-50. Other curved surfaces are dimensioned as in figures 4-51, 4-52, and 4-53.

Overdimensioning

Any dimension that is not required in the manufacture and assembly of a product should not be placed on the drawing. If proper dimensions are given for overall size, and the location and size of each of its parts is related to a datum surface or basic feature, there is no need for reference or extra dimensions. Examples of common errors in overdimensioning are shown in figure 4-54A.

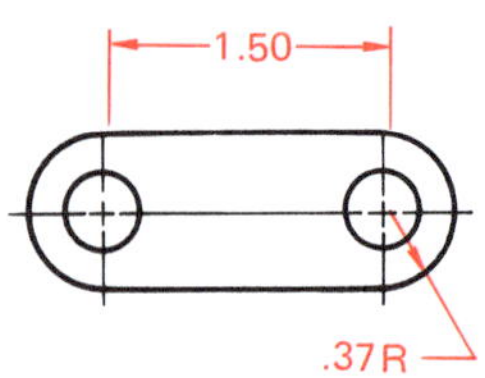

Fig. 4-48

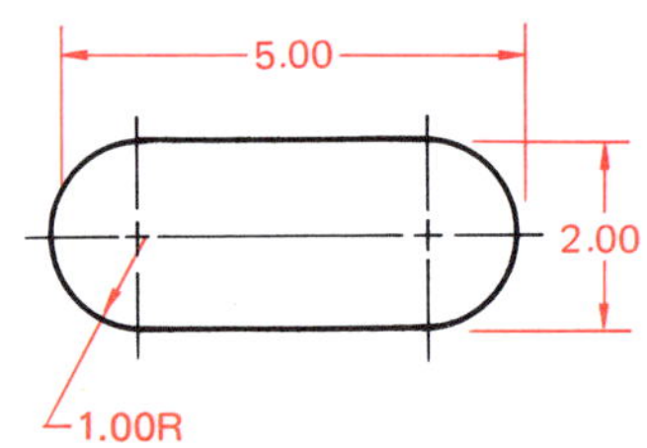

Fig. 4-49

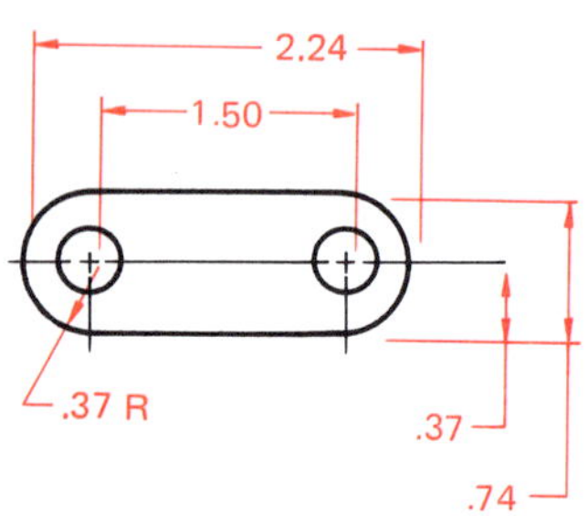

Fig. 4-50

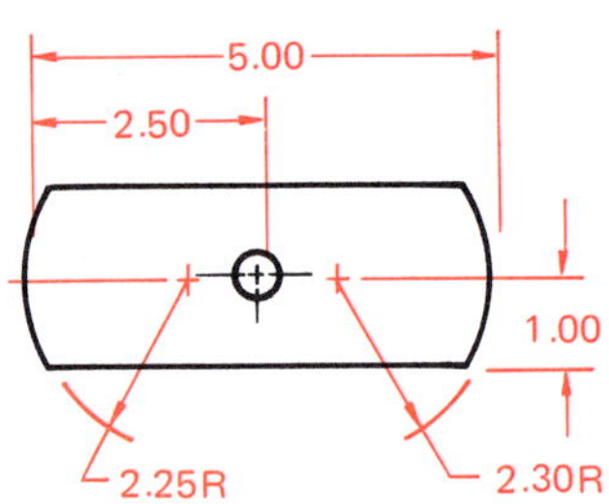

Fig. 4-51

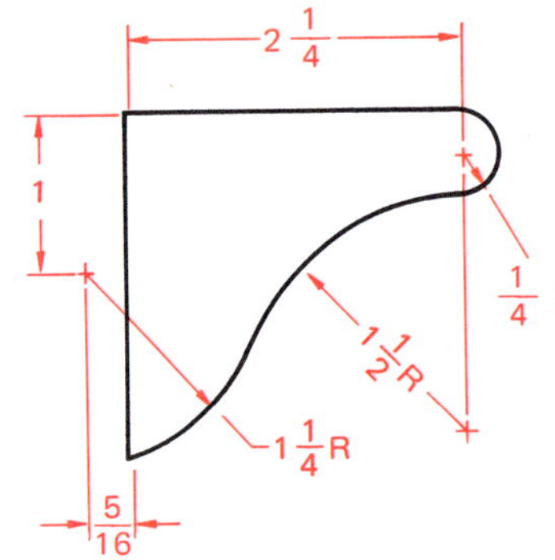

Fig. 4-52

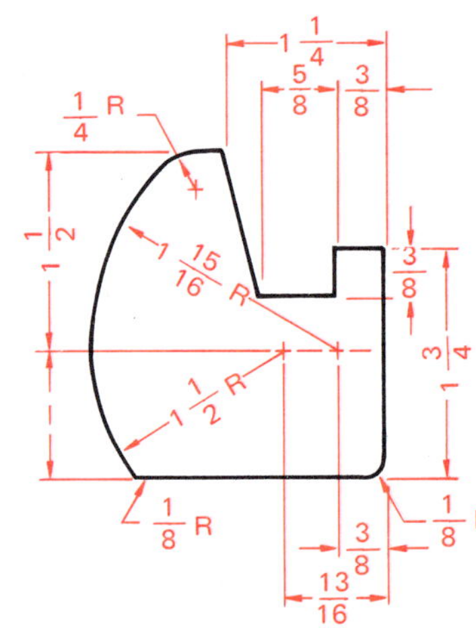

Fig. 4-53

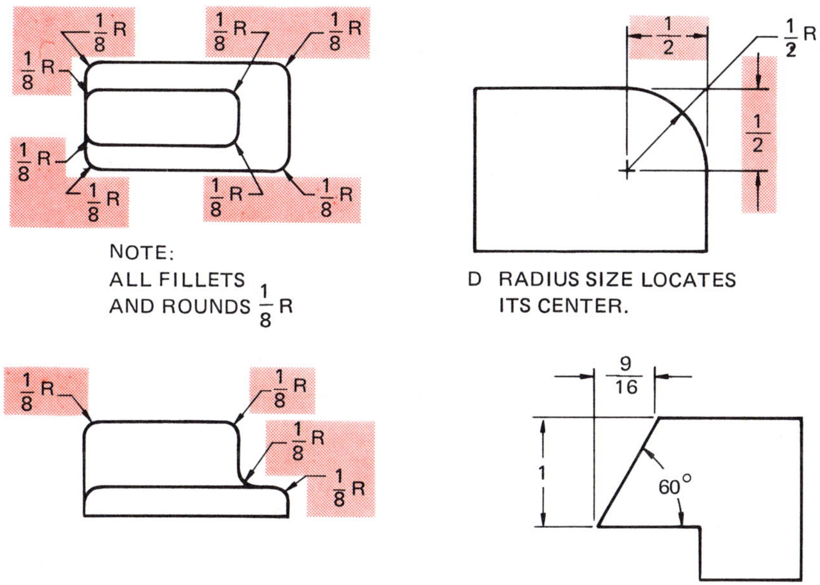

H DO NOT REPEAT DIMENSIONS GIVEN IN NOTES.

K OMIT THE ANGLE OR ONE DIMENSION.

Fig. 4-54A. Examples of common errors committed by giving unneeded dimensions.

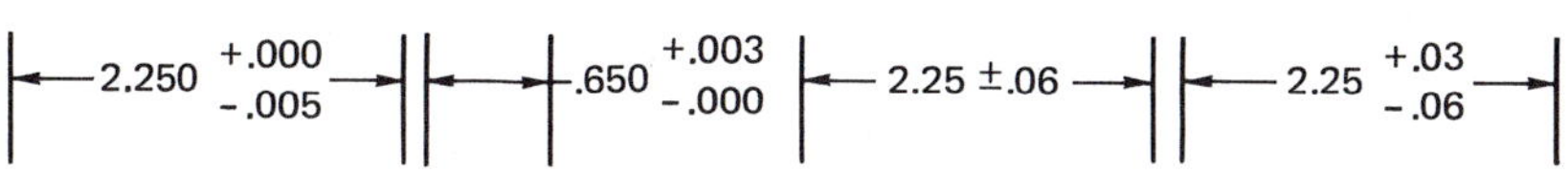

(A) UNILATERAL TOLERANCING (B) BILATERAL TOLERANCING

Fig. 4-54B. Plus and minus tolerancing.

TOLERANCE OR PRECISION DIMENSIONING

Before interchangeable parts were made, the responsibility for accuracy depended upon the skill of the worker. Today, the engineer, designer, and drafter are responsible for thinking through every detail of a product. Careful engineering and precise dimensioning are the basis of every product manufactured. When parts are manufactured anywhere in the world, only a system of precision dimensioning and correct interpretation of these values can make mass production of interchangeable parts a reality.

It is impossible to use fractional values. Dimensions must be given to a much finer degree. Also, some allowance for error without interfering with function must be allowed. The allowable difference over or under the specified size is called the *tolerance,* figure 4-54B.

Precision instruments measure these values. They are expressed as decimals of inches or decimals of millimetres. Some industries that manufacture parts in large quantity for interchangeable assembly or that follow SI practices want only an *end product drawing.* This states the size, location, and tolerance dimension with no information about the manufacturing method used. The process engineering department selects which method of generating a hole, for example, will meet the tolerance specifications and still be economical. The end product drawing does not specify drill, ream, core, or bore. It simply gives the diameter of the hole and the tolerance limitations.

Other industries prefer preliminary operation notations on the drawings. A *process drawing* includes information about the process to be used and the dimension for each operation.

The type of drawing selected depends upon its purpose and the industry for which it is intended.

A few definitions of terms used in drafting must be explained. They are found in the geometric tolerancing and dimensioning system used by SI nations.

A plane surface, cylindrical surface, or geometric shape that is part of an object is described as a *feature.* It is the size definition and location of various features that the rules of dimensioning are all about.

Tolerancing recognizes how impossible it is to achieve exact sizes through machining and establishing a range of allowable variation. The two types of tolerances are positional and form. *Positioning tolerancing* controls a feature's location in relation to a reference point. *Form tolerancing* controls the geometric shape of those features.

A *basic dimension* is an exact value describing size, shape, or location of a feature. It is not toleranced. It is used as a base from which to dimension other features or establish tolerances. Basic dimensions are identified by the word BASIC, the BSC abbreviation, or by enclosing the dimension in a rectangular frame, figure 4-55.

True position is theoretically the exact location of a feature established by basic dimensions.

Datums are usually planes or points on a part from which other features are located, figure 4-56. *Datum dimensioning* is sometimes called base line dimensioning, figure 4-57. This system avoids accumulated error by adding one tolerance to another during cumulative dimensioning from feature to feature. Where the distance between two features must be controlled for mating parts without small tolerance limitations, datum dimensioning is used.

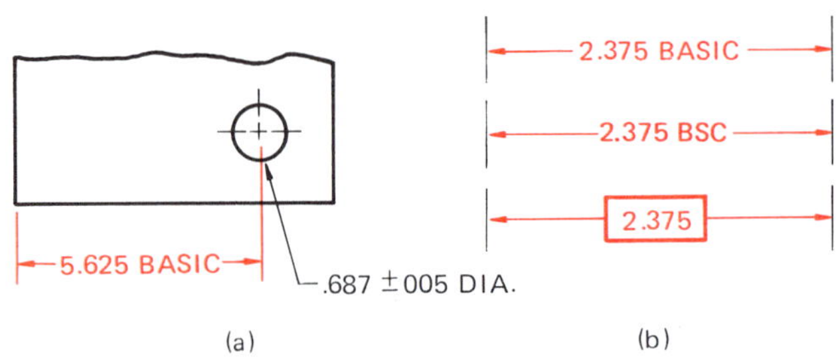

Fig. 4-55 Indicating BASIC dimensions

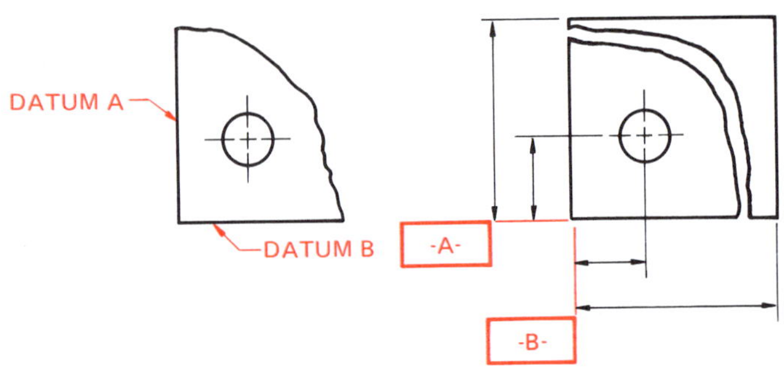

Fig. 4-56 Identifying datums

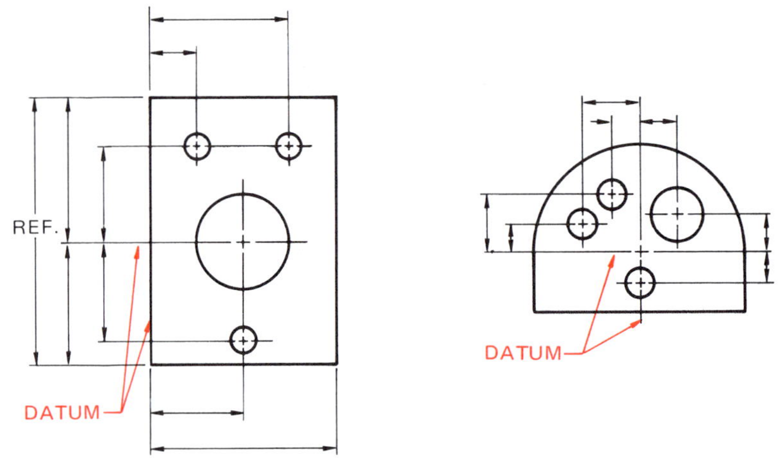

Fig. 4-57 A center line or edge can be used as a datum.

Maximum material condition (MMC) recognizes each individual feature of an object as it contains the greatest amount of material. That means the holes are their smallest, and external features are their largest, figure 4-58.

Regardless of feature size (RFS) indicates that the location of a feature, or its position in relation to a datum, basic dimension, or other feature, is more important than its size. It can be toleranced in a different way than any form tolerance applied to the object.

Many mass-produced parts, such as fasteners, steel bars and rods, lumber, and tools, are not dimensioned to exact size. Instead, they are referred to by more common dimensions called *nominal sizes.* For example a nominal 2 by 4 inch piece of lumber actually measures 1 1/2 by 3 1/2 inches. These nominal sizes are not changed to exact sizes on the drawing, but are placed in the note section with the manufacturer's sizes indicated.

Allowance is the difference in the dimension of mating parts providing for minimum clearance or maximum interference, figure 4-59.

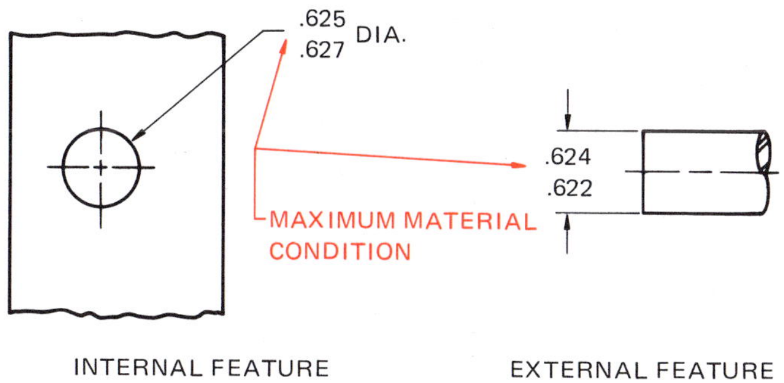

Fig. 4-58 Maximum material condition

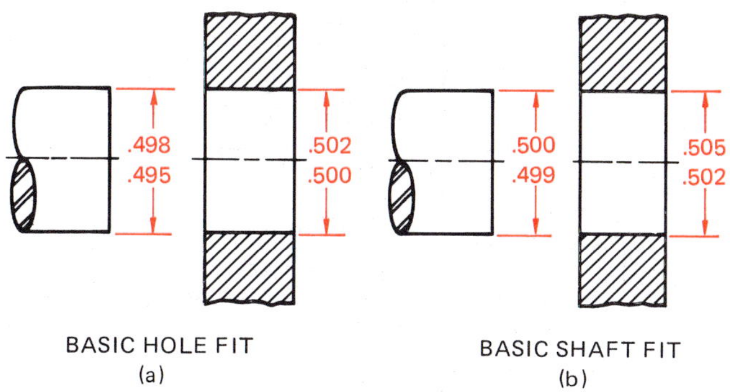

Fig. 4-59 Basic hole and basic shaft systems

Design size is the result of applying the tolerance limit to the allowance provided for, figure 4-60.

Size limits identify the maximum size and the minimum size of any part. In tolerancing a dimension by limit dimensioning, only the maximum and minimum sizes are given. The limits are arranged by one of the methods shown in figure 4-61.

Reference dimensions do not govern production operations and are not toleranced. They are a result of values already placed on the drawing, or a related drawing, and are used for information

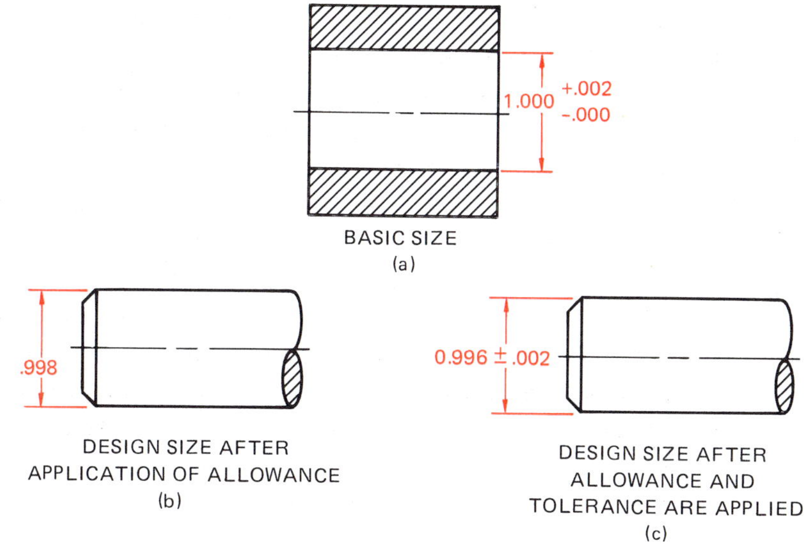

Fig. 4-60 Basic size, allowance, and design size

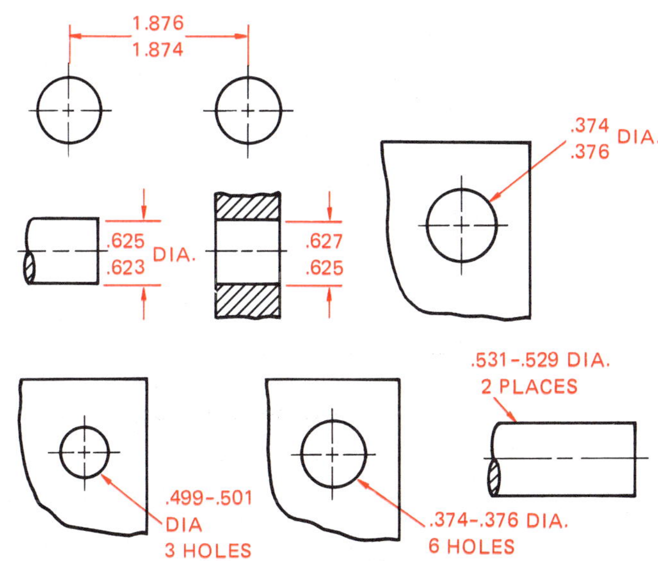

Fig. 4-61 Limit dimensioning

only. A reference dimension is enclosed in parentheses, or by the abbreviation, REF to the right or under the dimension, figure 4-62.

A system of symbols has been worked out to help the drafter convey the design intent and the engineering controls on sizes, shapes, and relationships of parts. The International System of Units and its United States counterpart, the American National Standards Institute (ANSI), both use a set of symbols to control different characteristics of the features and their relationships, figure 4-63. These symbols are increasingly being used in industrial drafting departments, particularly those that depend on mass-produced interchangeable parts from other companies and other countries. These symbols are given as reference information only and in selected detail. The more common symbols and their application are shown in figures 4-64 to 4-72.

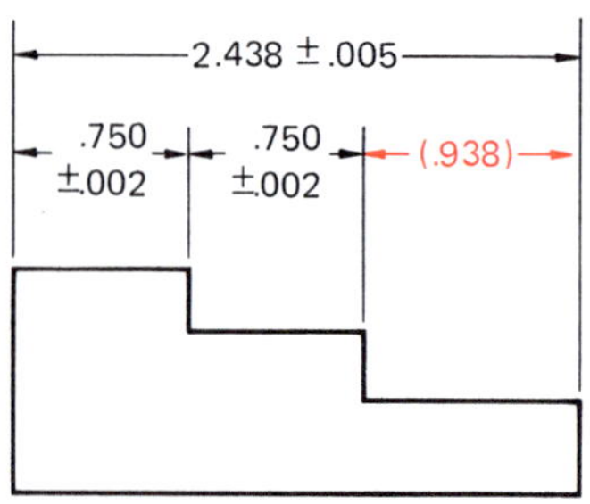

Fig. 4-62 Reference dimensions

AMERICAN NATIONAL STANDARD
DIMENSIONING AND TOLERANCING

ANSI Y 14.5–1973

		CHARACTERISTIC	SYMBOL	NOTES
INDIVIDUAL FEATURES	FORM TOLERANCES	STRAIGHTNESS	⏤	1
		FLATNESS	⏥	1
		ROUNDNESS (CIRCULARITY)	○	
		CYLINDRICITY	⌭	
INDIVIDUAL OR RELATED FEATURES		PROFILE OF A LINE	⌒	2
		PROFILE OF A SURFACE	⌓	2
RELATED FEATURES		ANGULARITY	∠	
		PERPENDICULARITY (SQUARENESS)	⊥	
		PARALLELISM	//	3
	LOCATION TOLERANCES	POSITION	⌖	
		CONCENTRICITY	◎	3, 7
		SYMMETRY	⌯	5
	RUNOUT TOLERANCES	CIRCULAR	↗	4
		TOTAL	↗	4, 6

NOTE: 1) The symbol ∼ formerly denoted flatness.

The symbol ⌒ or — formerly denoted flatness and straightness.

2) Considered "related" features where datums are specified.

3) The symbol || and ⦿ formerly denoted parallelism and concentricity, respectively.

4) The symbol ↗ without the qualifier "CIRCULAR" formerly denoted total runout.

5) Where symmetry applies, it is preferred that the position symbol be used.

6) "TOTAL" must be specified under the feature control symbol.

7) Consider the use of position or runout.

Where existing drawings using the above former symbols are continued in use, each former symbol denotes that geometric characteristic which is applicable to the specific type of feature shown.

Fig. 4-63 Geometric characteristics symbols

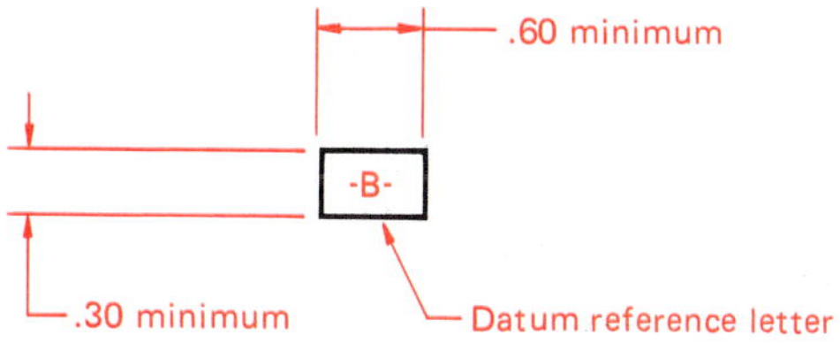

Fig. 4-64 Datum identifying symbol

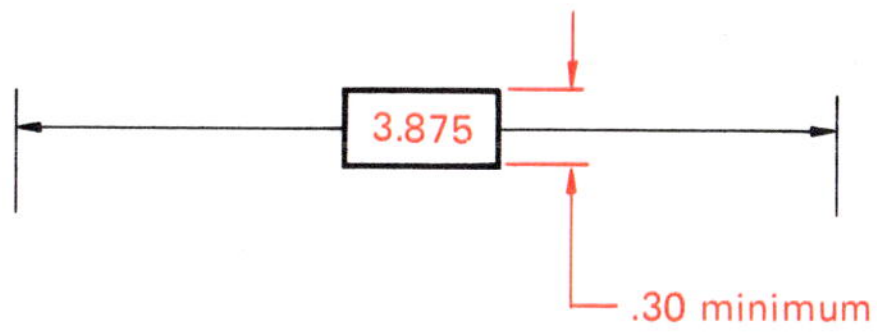

Fig. 4-65 Basic dimension symbol

TERM	ABBREVIATION	SYMBOL
Maximum Material Condition	MMC	Ⓜ
Regardless of Feature Size	RFS	Ⓢ
Diameter	DIA	⌀
Projected Tolerance Zone	TOL ZONE PROJ	Ⓟ
Reference	REF	(1.250)
Basic	BSC	3.875 (boxed)

Fig. 4-66 Other symbols

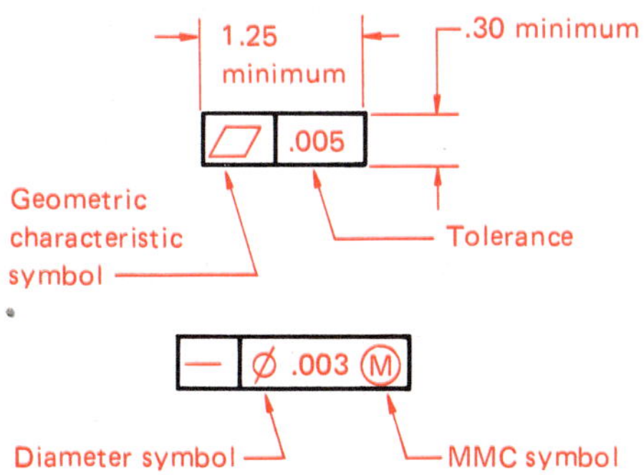

Fig. 4-67 Feature control symbols

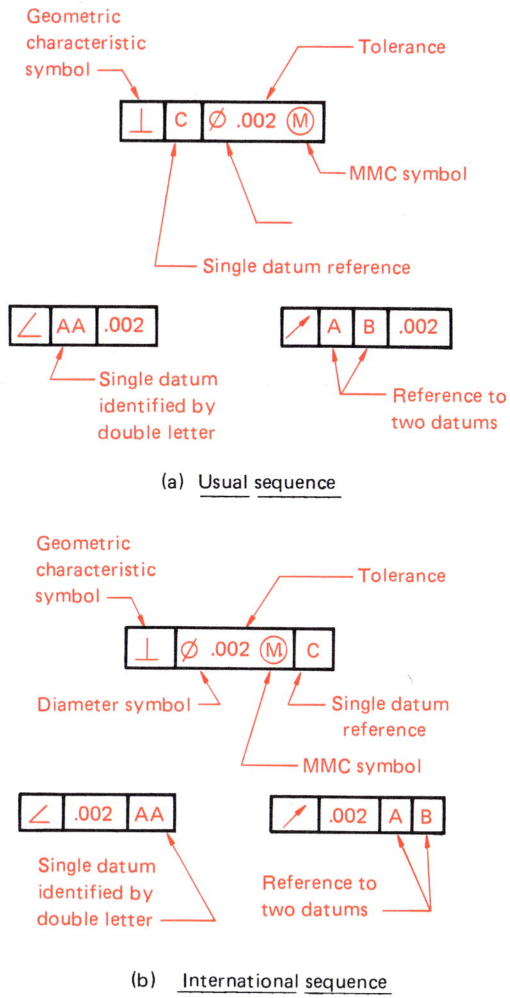

Fig. 4-68 Feature control symbols incorporating datum references

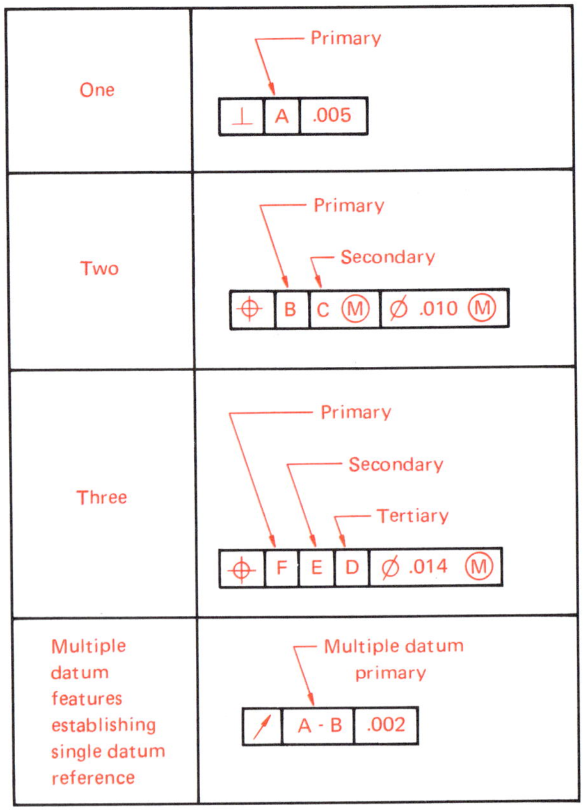

Fig. 4-69 Order of precedence of datum references

AMERICAN NATIONAL STANDARD
DIMENSIONING AND TOLERANCING

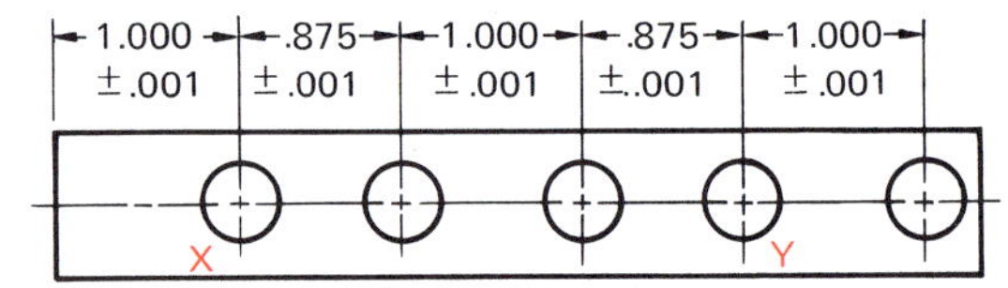

(a) Chain Dimensioning - Greatest tolerance accumulation between X and Y

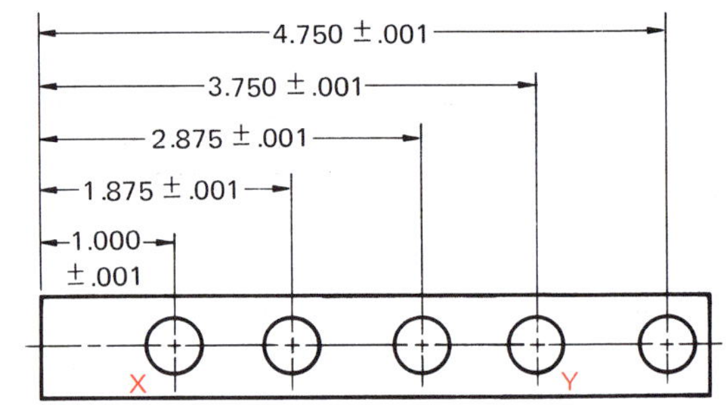

(b) Datum Dimensioning - Lesser Tolerance Accumulation between X and Y

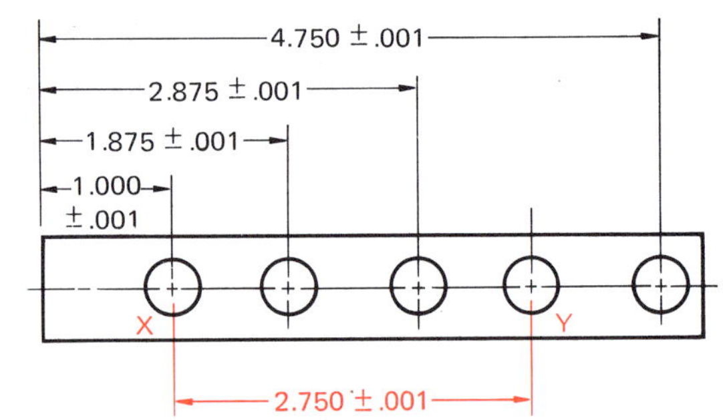

(c) Direct Dimensioning - Least Tolerance Accumulation between X and Y

Fig. 4-70 Dimensioning comparison

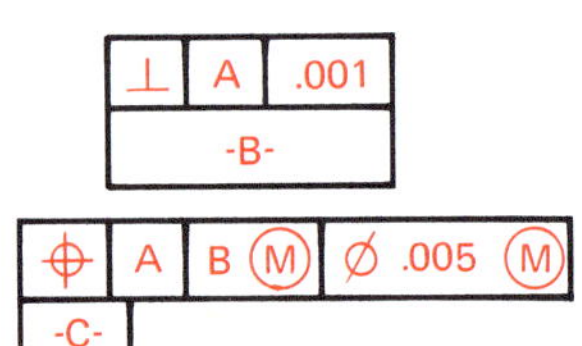

Fig. 4-71 Combined feature control and datum identifying symbol

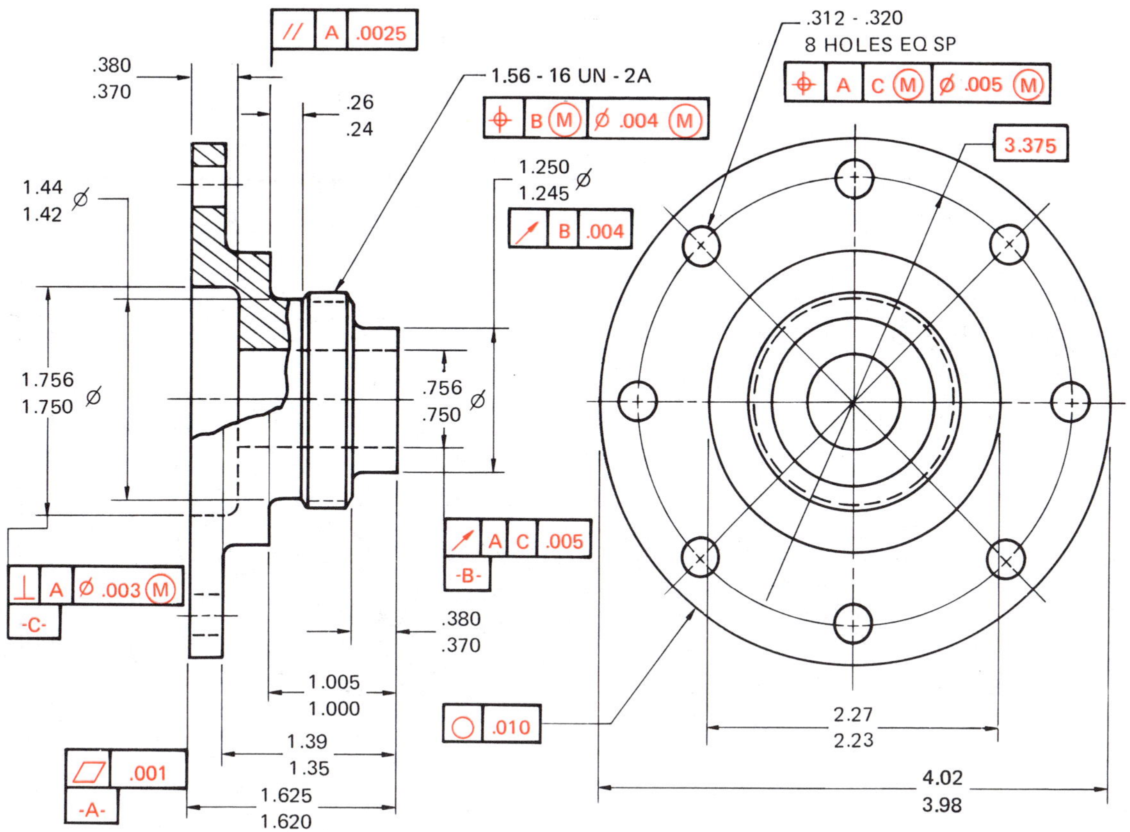

Fig. 4-72 Applications of geometric characteristics symbols

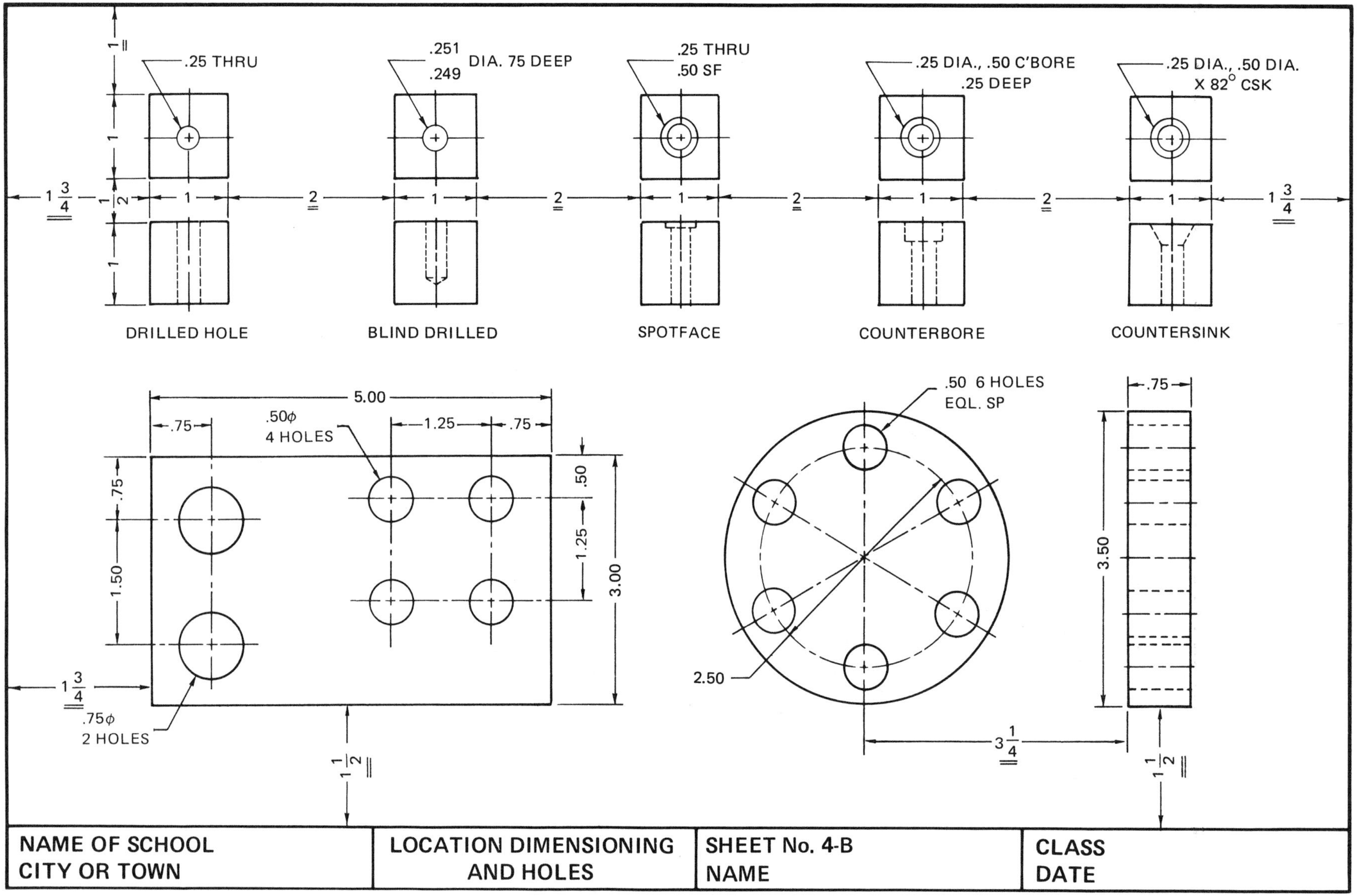
.25 THRU
.251
.249
DIA. 75 DEEP
.25 THRU
.50 SF
.25 DIA., .50 C'BORE
.25 DEEP
.25 DIA., .50 DIA.
X 82° CSK
DRILLED HOLE
BLIND DRILLED
SPOTFACE
COUNTERBORE
COUNTERSINK
5.00
.75
.50φ
4 HOLES
1.25
.75
.50
1.25
3.00
1.50
.75φ
2 HOLES
.50 6 HOLES
EQL. SP
2.50
3.50
NAME OF SCHOOL
CITY OR TOWN
LOCATION DIMENSIONING
AND HOLES
SHEET No. 4-B
NAME
CLASS
DATE

Fig. 4-73

ASSIGNMENT B

Draw the location dimensioning exercise shown in figure 4-73.

Procedure

1. Lay out the margin lines on the drawing sheet and letter the title strip.
2. Using light construction lines, locate the squares for different holes in the position indicated by the underlined dimensions.
3. Draw each type of hole as indicated and dimension as suggested.
4. Locate the objects for dimensioning as indicated by the underlined dimensions.
5. Draw both objects using the correct line symbols.
6. Dimension as shown and add the metric equivalents under each dimension line to make it dual dimensioning.
7. Omit all underlined dimensions from the sheet.
8. Check the drawing for accuracy.

REVIEW QUESTIONS

1. Define the following terms.
   - A. Visible outline
   - B. Dimension line
   - C. Extension line
   - D. Leader
   - E. Section line
   - F. Cutting plane line
   - G. Datum surface
   - H. Datum dimensioning
   - I. Basic feature
   - J. Tolerance
   - K. Allowance
   - L. Form tolerancing
   - M. Reference dimension
   - N. Feature control
   - O. Nominal size
   - P. Design size
   - Q. Size limits
   - R. Maximum material condition (MMC)
   - S. Regardless of feature size (RFS)
   - T. Millimetre
2. What is the difference between aligned and unidirectional dimensioning?
3. When is it permissible to use fractions in dimensioning?
4. What units of measurement are used only in the United States?
5. What units of measurement are used in thread notations?
6. What is an accumulation of tolerance?
7. When is it necessary to use rectangular coordinate dimensioning to locate features?
8. What is the standard symbol to indicate U.S. projection practice?
9. What is the difference between end product drawing and process drawing?
10. What does the abbreviation, SI, represent?

UNIT 5 TWO-VIEW DRAWINGS

Before starting a working drawing of a part, study it to determine the number of surfaces, edges, and corners to show in each view.

The block in the pictorial drawing, figure 5-1, shows that the front is a plane surface. Four corners are formed where the horizontal and vertical edges meet.

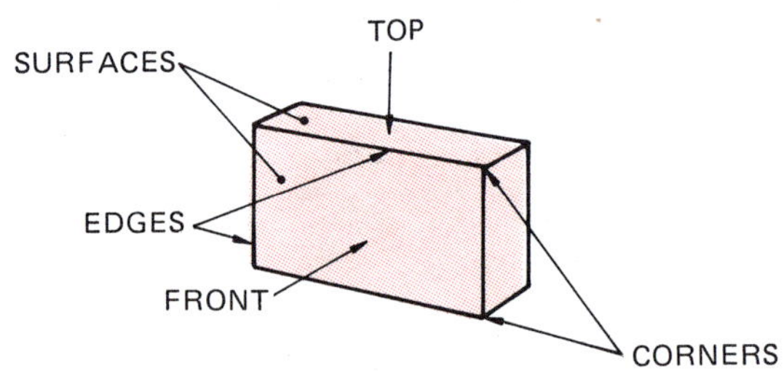

Fig. 5-1 A pictorial drawing

In the front view of the block, figure 5-2, four lines are drawn to represent the four edges. The front view shows the shape, the length, and the height of the front of the block.

The top of the block is also a plane surface bounded by four edges and having four corners. In the top view, figure 5-2, lines are drawn to represent these four edges. They show the shape, the length, and the width of the top.

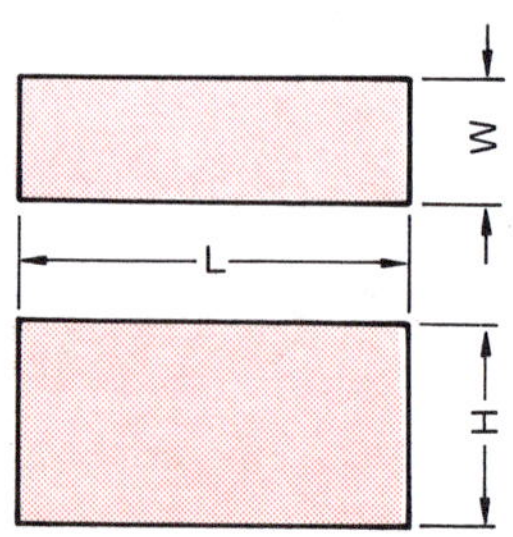

Fig. 5-2 A two-view working drawing

A study of the block in the pictorial drawing, figure 5-3, shows that the front is a plane surface bounded by six edges and having six corners. In the front view, figure 5-4, the six lines represent these six edges. They show the shape of the front, the length of the top and bottom of the block, and its height at the left and right sides.

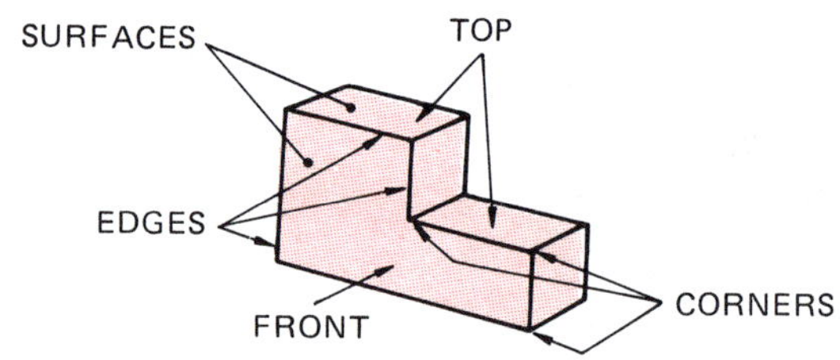

Fig. 5-3 A pictorial drawing

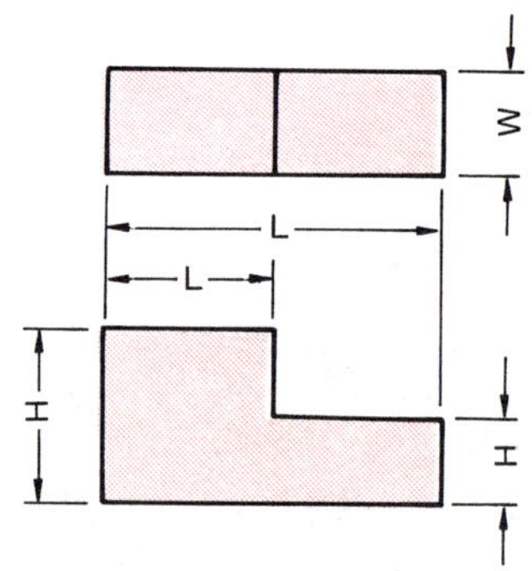

Fig. 5-4 A two-view working drawing

Looking at the block in figure 5-3 from a point directly above, there are two top surfaces, seven edges, and six corners. In the top view, figure 5-4, lines are drawn to represent the seven edges. One edge and two corners of the lower surface are directly beneath those on the upper surface. Therefore, they cannot be shown in the top view. The top view shows the shape, the length, and the width of the top surfaces.

This comparison and analysis of the two blocks show that the number of surfaces, edges, and corners vary with objects of different shapes.

ASSIGNMENT A

Make two working drawings. Draw a front and top view of the step block shown in figure 5-5. Draw a front and top view of the double step block shown in figure 5-6.

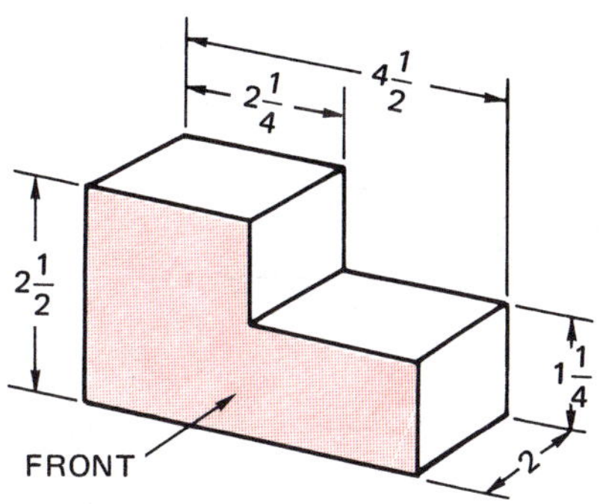

Fig. 5-5 Step block

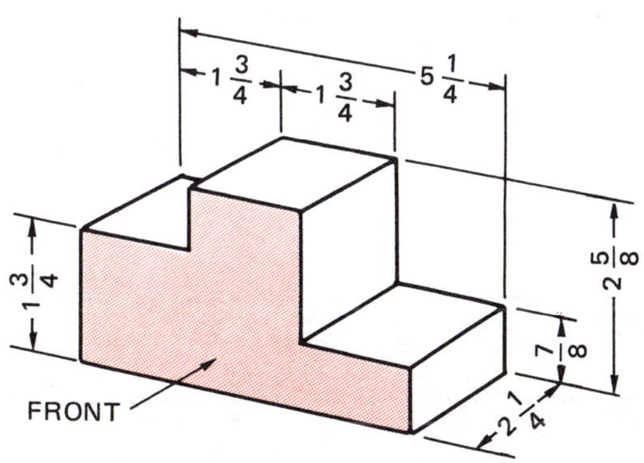

Fig. 5-6 Double step block

Procedure

A. Step Block

1. Lay out the sheet and letter the title strip.
2. Look at the front of the step block. Determine the number of surfaces, edges, and corners which are seen from a point directly in front. Draw the front view, locating it 1 3/4 inch from the left margin line and 1 5/8 inch above the title strip line.
3. Draw very light construction lines up from each upper corner of the front view. These construction lines are important because they show the location of the vertical visible edge lines in the top view.
4. Look at the top of the step block. Determine the number of surfaces, edges, and corners which are seen from a point directly above it. Draw the top view 1 5/8 inch below the top margin line.
5. Dimension the drawing as shown in figure 5-7.

B. Double Step Block

6. Look at the front of the double step block. Determine the number of surfaces, edges, and corners which are seen from a point directly in front. Draw the front view, locating it 2 1/4 inch from the right margin line and 1 5/8 inch above the title strip.
7. Draw very light construction lines up from each upper corner of the front view. These lines show the location of the vertical visible edge lines in the top view.
8. Look at the top of the double step block. Determine the number of surfaces, edges, and corners which are seen from a point directly above it. Draw the top view 1 1/2 inch below the top margin line.
9. Draw the extension and dimension lines, locating them as illustrated in figure 5-7.

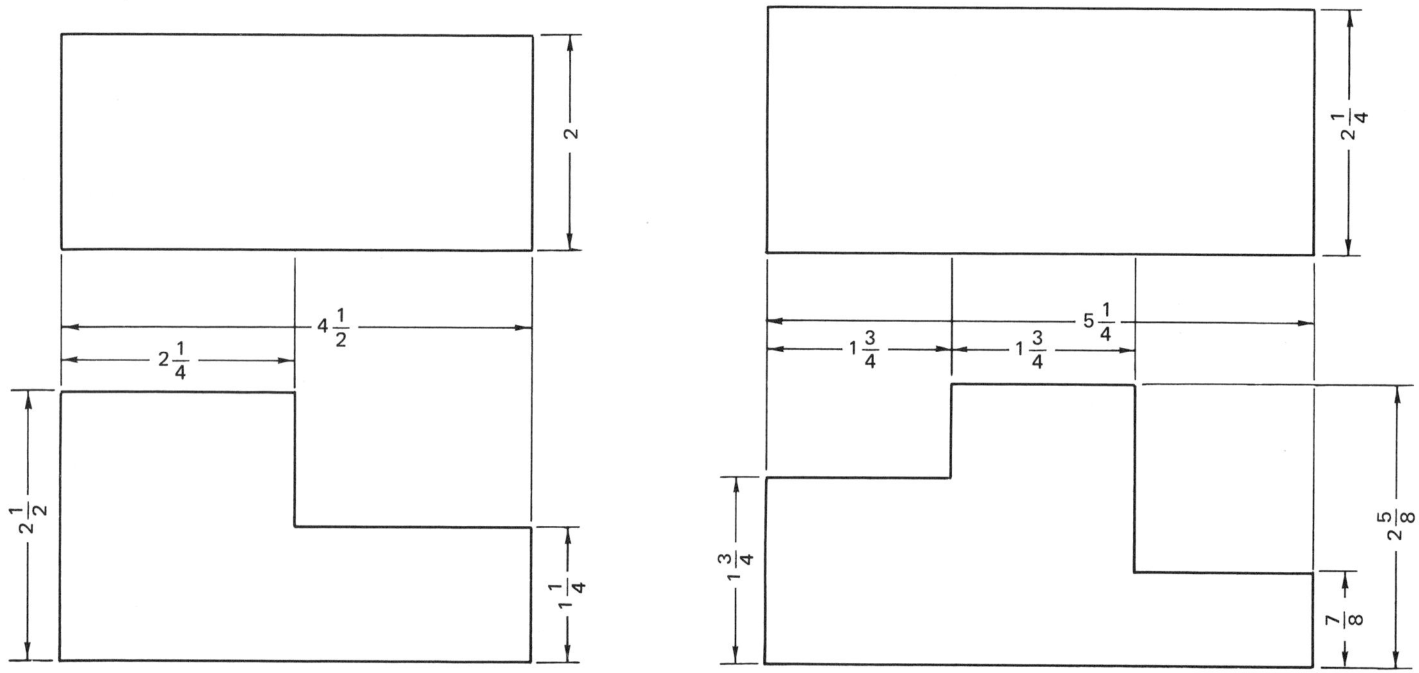

NAME OF SCHOOL CITY OR TOWN	STEP BLOCK DOUBLE STEP BLOCK	SHEET No. 5-A NAME	CLASS DATE

Fig. 5-7

10. Check the drawing for errors and darken the lines to the correct line weight before cleaning the drawing.

REVIEW QUESTIONS

1. How many surfaces are shown in the front view of the step block? The double step block?
2. How many corners are shown in the front view of the step block? The double step block?
3. How many edges are shown in the front view of the step block? The double step block?
4. How many surfaces are shown in the top view of the step block? The double step block?
5. How many corners are shown in the top view of the step block? The double step block?
6. How many edges are shown in the top view of the step block? The double step block?

DRAWING WITH VISIBLE LINES

Toolmakers use the solid steel square to check right angles. A working drawing of the solid steel square illustrated in figure 5-11 has two small fractional dimensions, 1/4 inch and 1/8 inch. The dimension lines for these fractions are drawn outside of the extension lines, and the arrowheads are reversed to point inward, figure 5-8.

This arrangement permits the numerators and denominators to be made 1/8 inch high without crowding. The length of the arrowheads can be the same as other arrowheads on the drawing.

Fig. 5-8 Dimension lines with reversed arrowheads

MULTIPLE STEP BLOCK

Step blocks use strap clamps to secure the work to the table of a machine. The strap clamp must be in a level position with one end resting on the work and the other end on the step block, figure 5-9. The work is then securely clamped by tightening a nut on the end of a T bolt.

The multiple step block, figure 5-10, supports the strap clamp in a level position with work ranging from 1 1/2 inch to 3 1/2 inch high.

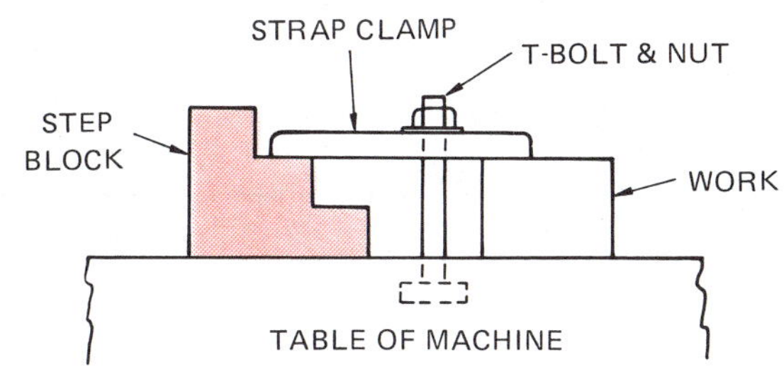

Fig. 5-9 The step block

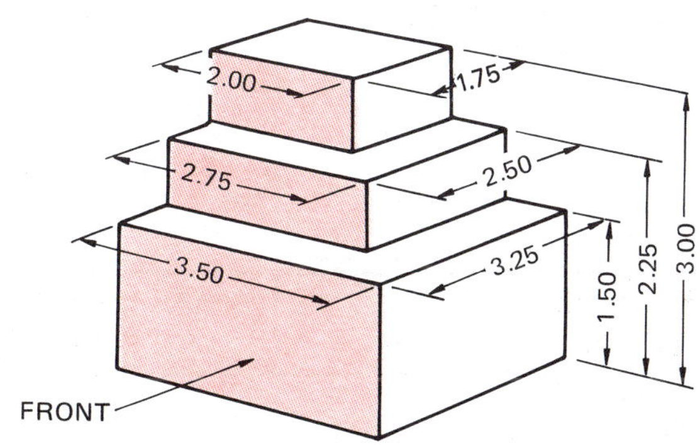

Fig. 5-10 Multiple step block

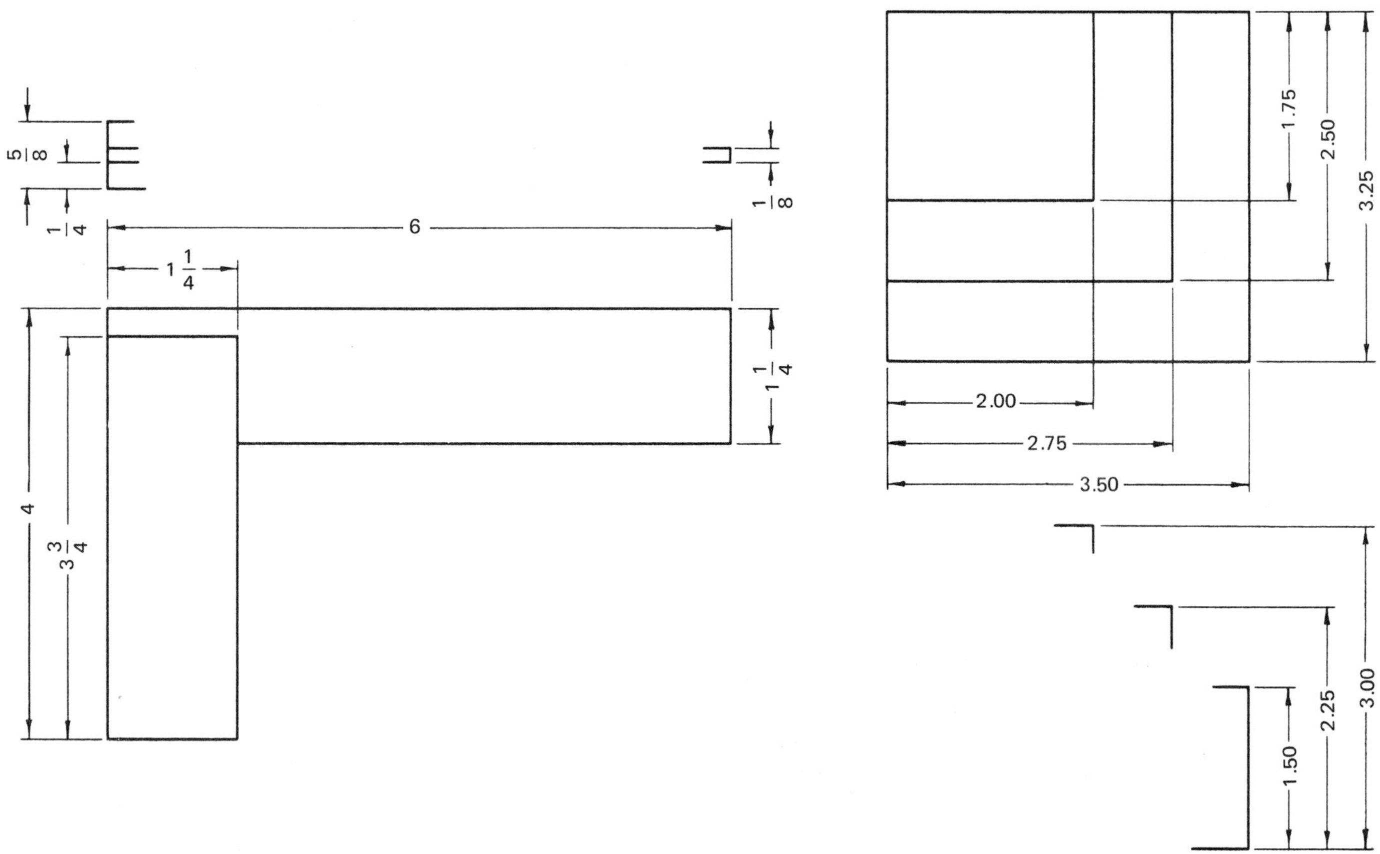

NAME OF SCHOOL	SOLID STEEL SQUARE	SHEET No. 5-B	CLASS
CITY OR TOWN	MULTIPLE STEP BLOCK	NAME	DATE

Fig. 5-11

ASSIGNMENT B

When a working drawing is made, the view which best describes the shape of the object is drawn first.

1. Draw two views of the solid steel square in figure 5-12. The front view describes the shape better than the top view, so the front view is drawn first.
2. Draw two views of the multiple step block shown in figure 5-10. In this drawing, the top view is drawn first, then the front view.

Procedure

A. Solid steel square

1. Lay out the drawing sheet and letter the title strip.
2. Look at the front of the solid steel square to determine the number of surfaces, edges, and corners which are seen from a point directly in front. Draw the front view. Locate this view 2 1/4 inch from the left margin line and 2 inches above the title strip.
3. Draw very light construction lines up from each upper corner of the front view to show the location of the vertical visible edge lines in the top view.
4. Look at the top of the solid steel square. Determine the number of surfaces, edges, and corners which are seen from a point directly above it. Draw the top view 2 inches below the top margin line.
5. Draw the extension and dimension lines.
6. Fill in the dimensions and form the arrowheads carefully.

B. Multiple step block

1. Draw the top view of the multiple step block first. Look at the pictorial drawing to determine the number of surfaces, edges, and corners which are seen from a point directly above it. Locate the view 1 inch below the top margin line and 2 1/2 inches from the right margin line.
2. Draw very light construction lines down from each lower corner of the top view to show the location of the vertical visible edge lines in the front view.
3. Look at the front of the multiple step block, figure 5-10. Determine the number of surfaces, edges, and corners which are seen from a point directly in front. Draw the front view 1 inch above the title strip.
4. Draw the extension and dimension lines.
5. Fill in the dimensions and form the arrowheads carefully.
6. Check the drawing for accuracy and make the necessary corrections. Clean the drawing.

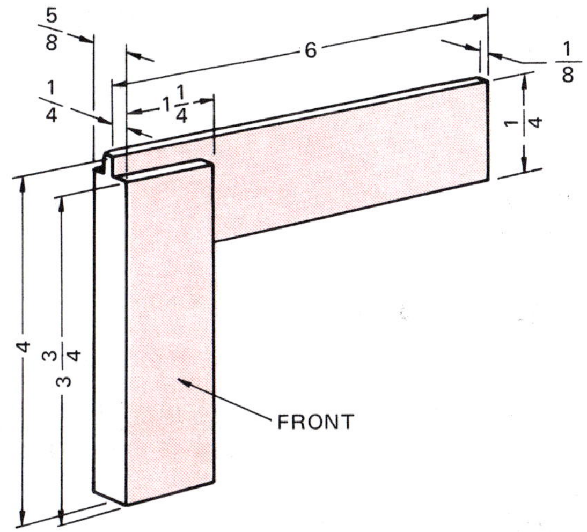

Fig. 5-12 Solid steel square

REVIEW QUESTIONS

1. How many surfaces are shown in the front view of the solid steel square? How many edges? How many corners?

2. How many surfaces are shown in the top view of the solid steel square? How many edges? How many corners?
3. How many surfaces are shown in the top view of the multiple step block? How many edges? How many corners?
4. How many surfaces are shown in the front view of the multiple steel block? How many edges? How many corners?

DRAWINGS WITH HIDDEN LINES

The instructions for the drawings just completed required studying the pictorial drawing from certain points of view to determine which surfaces, edges, and corners were visible. In the various views of these drawings, the visible edges are represented by visible edge lines.

ASSIGNMENT C

Make two working drawings. The first drawing has a front view and a top view of the box parallel shown in the pictorial drawing, figure 5-13. The second drawing has a front view and a top view of the jig base shown in the pictorial drawing, figure 5-14.

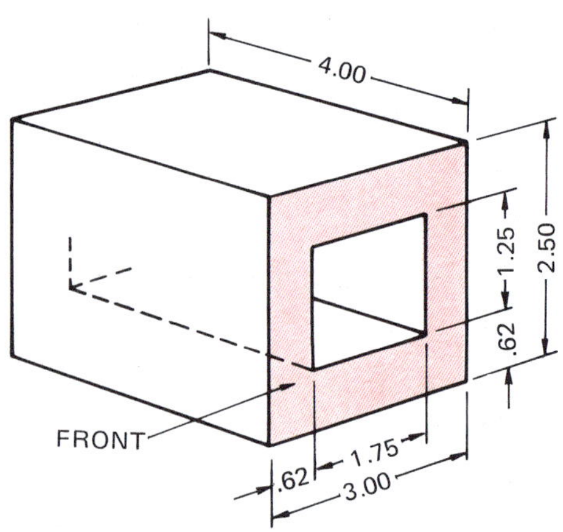

Fig. 5-13 Box parallel

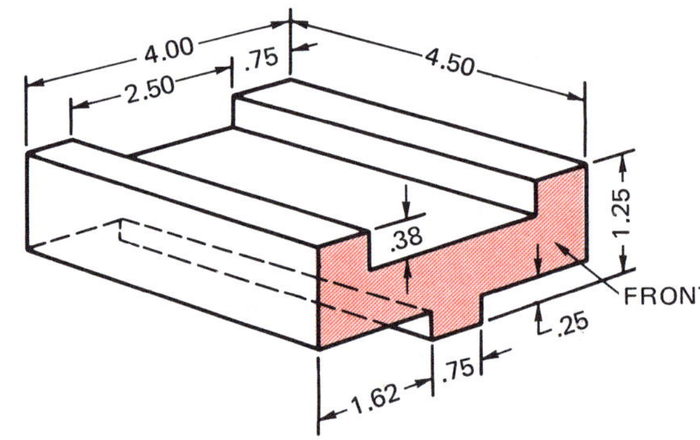

Fig. 5-14 Jig base

The front view of each object describes the shape better than the top view, so it is drawn first.

Procedure

A. Box parallel

1. Lay out the margin lines on the drawing sheet.
2. Letter the title strip.
3. Determine the number of surfaces, edges, and corners on the front of the box parallel. Draw the front view. Locate it 2 3/4 inches from the left margin line and 1 inch above the title strip, figure 5-15.
4. Draw very light construction lines up from each upper corner of the front view. These lines show the location of the vertical visible edge lines and the hidden lines in the top view.
5. Look at the top of the box parallel. Determine the number of visible and hidden edges. Draw the top view 1 inch below the top margin line. Draw the lines for the hidden edges.
6. Locate and draw the extension and dimension lines.
7. Fill in the dimensions and form the arrowheads carefully.

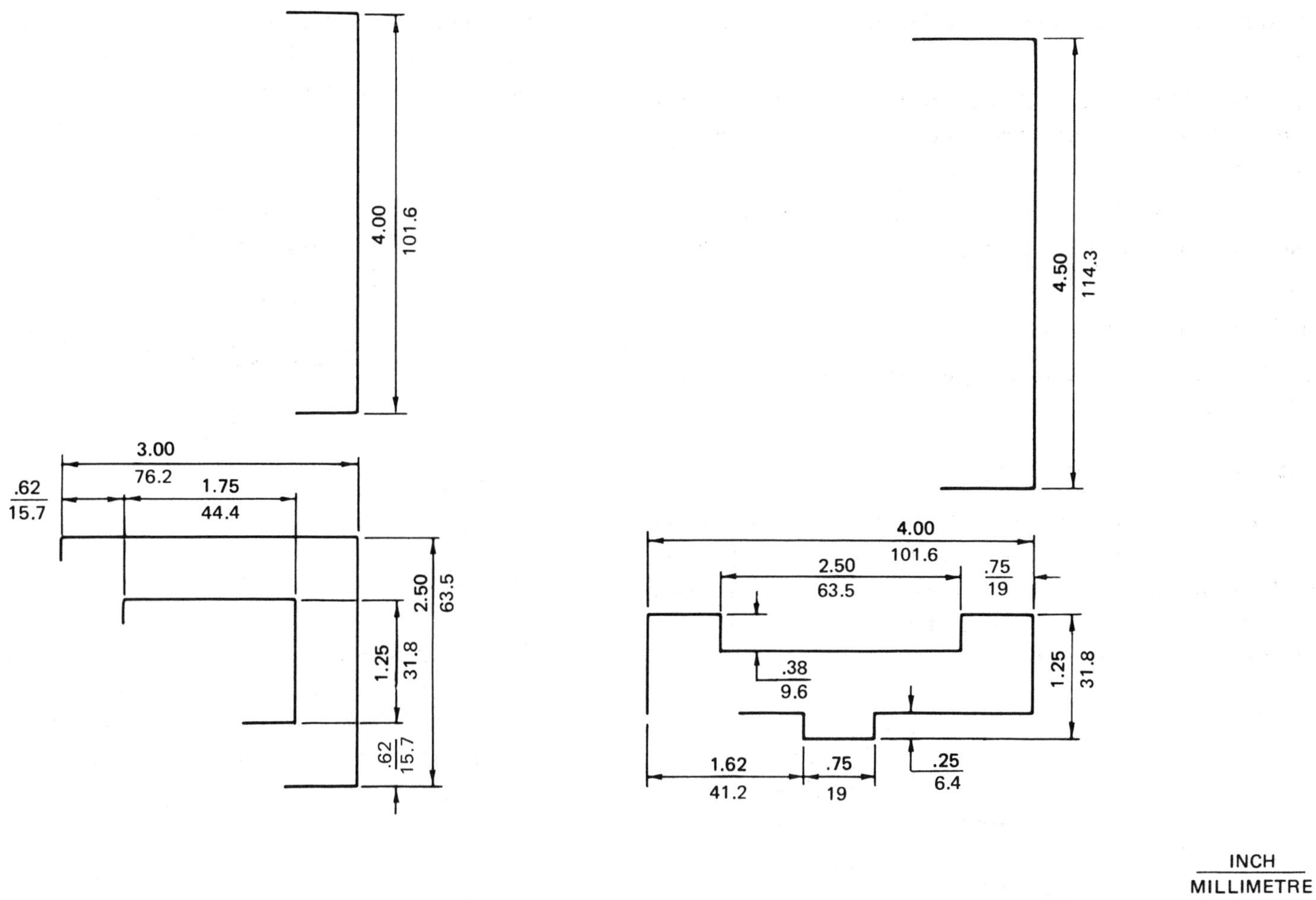

NAME OF SCHOOL CITY OR TOWN	BOX PARALLEL JIG BASE	SHEET No. 5-C NAME	CLASS DATE

Fig. 5-15

B. Jig Base

1. Draw the front view of the jig base locating it 3 1/4 inch from the right margin line and 1 1/2 inch above the title strip.
2. Draw very light construction lines up from each upper corner of the front view. These lines locate the vertical visible edge lines and the hidden lines in the top view.
3. Look at the top of the jig base. Determine the number of visible and hidden edges. Draw the top view 1 1/4 inch below the top margin line.
4. Dual-dimension the drawing.
5. Check the drawing for accuracy and correct any errors. Clean the drawing.

DRAWING THE RIGHT-SIDE VIEW

The objects in the next assignment require that the front and side views be drawn so that the working drawings fit the space on the sheet. In addition, the side view of the bench hook, figure 5-16, is needed to describe its shape. The right-side view of each object is drawn to the right of the front views.

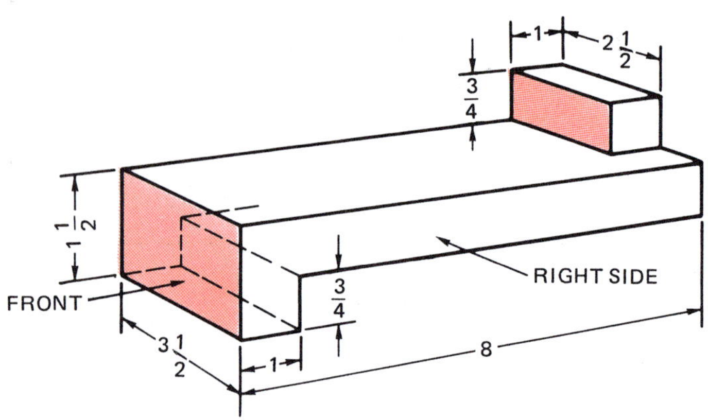

Fig. 5-16 Bench hook

ASSIGNMENT D

Make two working drawings. The first drawing has a front view and a right-side view of the miter box shown in the pictorial drawing figure 5-16. The view which best describes the shape of the object is drawn first.

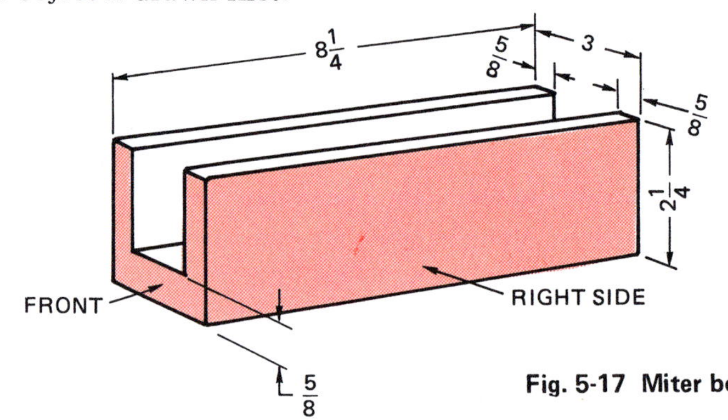

Fig. 5-17 Miter box

Procedure

A. Miter box

1. Lay out the margin lines on the drawing sheet and letter the title strip.
2. The front view of the miter box shows the shape better than the side view, so it is drawn first. Draw the front view, locating it 2 inches below the top margin line and 1 3/4 inches from the left margin line, figure 5-18.
3. Draw very light construction lines to the right from each corner of the front view. These lines locate the horizontal visible and invisible edge lines in the side view.
4. Draw the right-side view, locating it 1 3/4 inches from the right margin line. Draw the line for the hidden edge.
5. Draw extension and dimension lines. Locate them as shown on the drawing.
6. Fill in the dimensions and form the arrowheads carefully.
7. Check the drawing for accuracy.

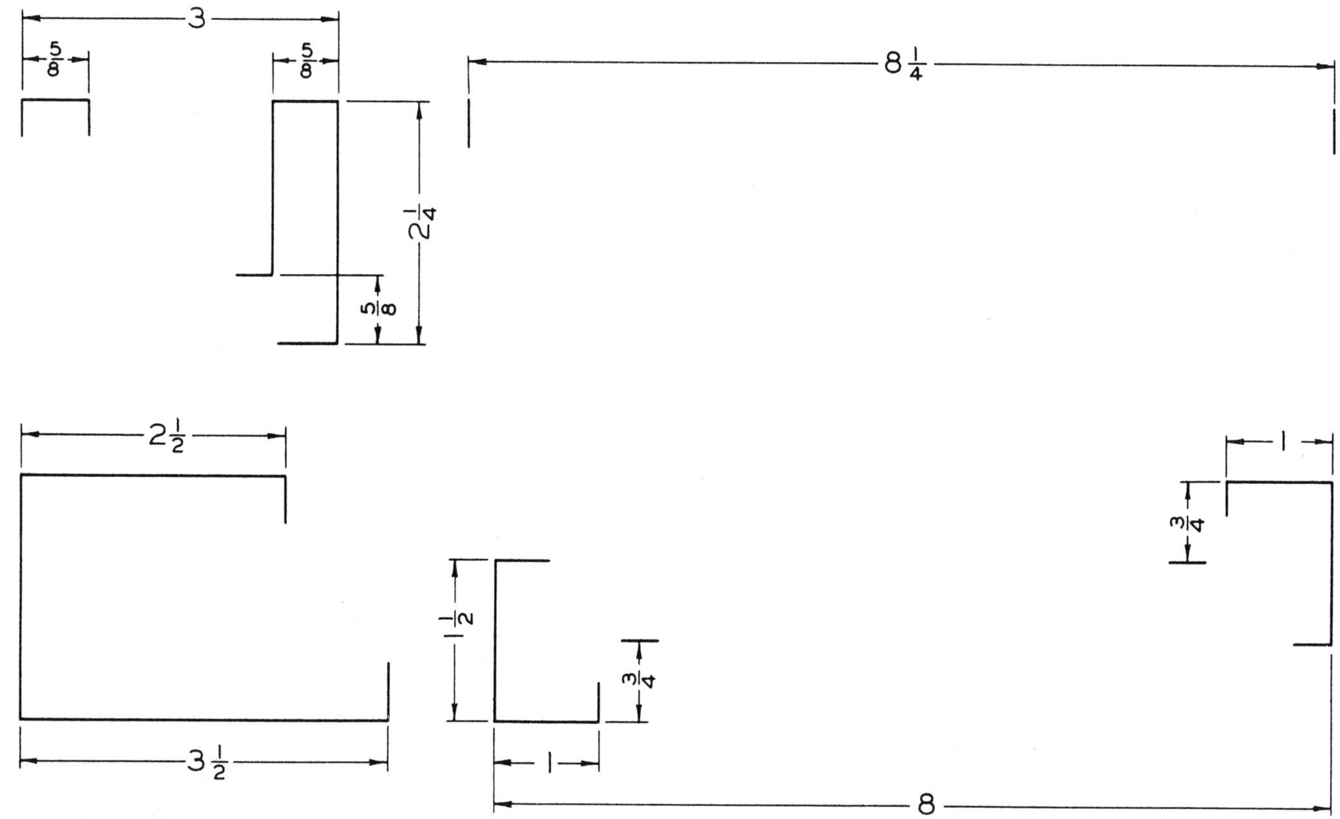

NAME OF SCHOOL CITY OR TOWN	MITER BOX BENCH HOOK	SHEET No. 5-D NAME	CLASS DATE

Fig. 5-18

B. Bench hook

1. The side view of the bench hook is drawn first because it describes the shape better than the front view. Draw the right-side view 1 3/4 inches from the right margin line and 2 inches above the title strip.
2. Draw very light construction lines to the left from each corner of the side view. These lines locate the horizontal visible and invisible edge lines in the front view.
3. Draw the front view 1 3/4 inches from the left margin line.
4. Dimension the views.
5. Check the completed drawing for accuracy.

REVIEW QUESTIONS

1. How many edges are in the front view of the miter box? In the side view?
2. How many edges are in the side view of the bench hook?
3. How many surfaces are shown in the front view of the bench hook? How many edges?
4. When several views are to be drawn, which view is drawn first?

DRAWING INCLINED SURFACES

In the example of the rectangular block suspended within a transparent box, each surface of the block is parallel with one of six transparent planes of the box. Therefore, the projected surfaces are their correct shapes, and the edges are their true lengths.

When a surface is not parallel to a plane of projection, it is said to be inclined to that plane. An edge not parallel to a plane is also inclined to it.

The length of an inclined edge when projected onto the plane to which it is inclined is shorter than its real length. An inclined surface when projected onto the plane to which it is inclined is also shorter than its real length. Therefore, inclined edges cannot be drawn their true lengths on the planes to which they are inclined. Inclined surfaces cannot be shown in their true shapes on the planes to which they are inclined.

ASSIGNMENT E

Draw a front view and a top view of the V-block shown in the pictorial drawing, figure 5-19, and a top view and a front view of the guide shown in figure 5-20.

Procedure

A. V-block

1. Lay out the margin lines on the drawing sheet.

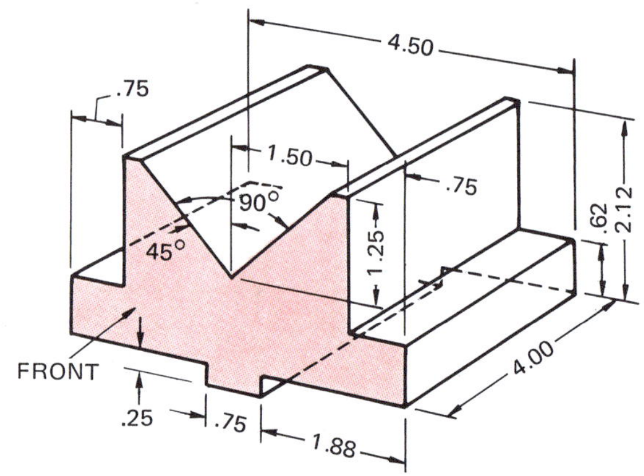

Fig. 5-19 V-block

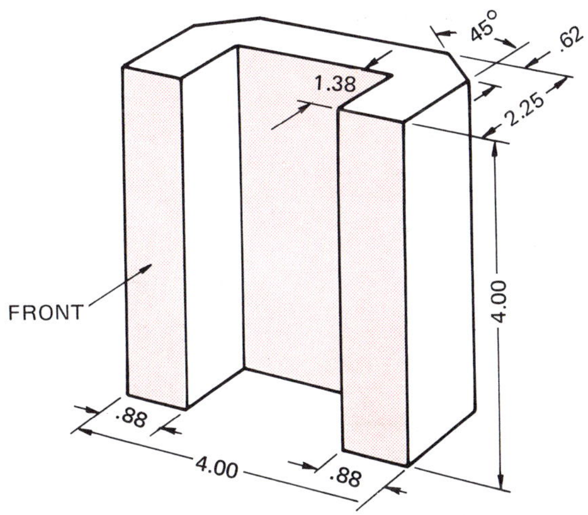

Fig. 5-20 Guide

2. Letter the title strip.
3. Draw the front view of the V-block, locating it 1 1/4 inches above the title strip and 2 inches from the left margin line.
4. Draw a very light construction line up from each corner of the front view.
5. Draw the top view 1 inch below the top margin line. Be sure to draw the hidden lines.
6. Dual-dimension the view as shown in figure 5-21.

B. Guide

1. Draw the top view of the guide locating it 1 3/8 inches below the top margin line and 2 3/4 inches from the right margin line.
2. Draw very light construction lines down from each corner.
3. Draw the front view 1 inch above the title strip. Be sure to draw the hidden lines.
4. Dual-dimension the views.
5. Check the drawing for accuracy and correct any errors.
6. Clean with an eraser.

REVIEW QUESTIONS

1. Indicate the inclined edges in the drawing.
2. Show the inclined surfaces in the drawings.
3. Show where the foreshortened edges are.

DRAWING THE LEFT-SIDE VIEW

When the two parts of the adjustable parallels shown in figure 5-23, page 110, are placed together, the upper and lower surfaces are parallel. The height between these parallel surfaces is adjusted by sliding the upper part to the right or left. Each part of the adjustable parallels has three inclined surfaces.

A study of these two parts shows that a front view and a side view is required to adequately describe their shapes. In deciding which view to draw, the rule is to select the view which has the fewest hidden details.

In the upper part of the parallel, the right-side view contains many hidden lines while the left-side view has none. Therefore, the left-side view is selected. The left-side view of the lower parallel contains many hidden lines while the right-side view has none. Therefore, the right-side view of this part is used.

ASSIGNMENT F

Draw the front and left-side views of the upper and lower parallel, figure 5-23.

Procedure

The rule of drawing the view first which best describes the shape of the object is not used here because there is no view which "best" describes the shape. The front and side views are both necessary to adequately describe the shape of these parallels.

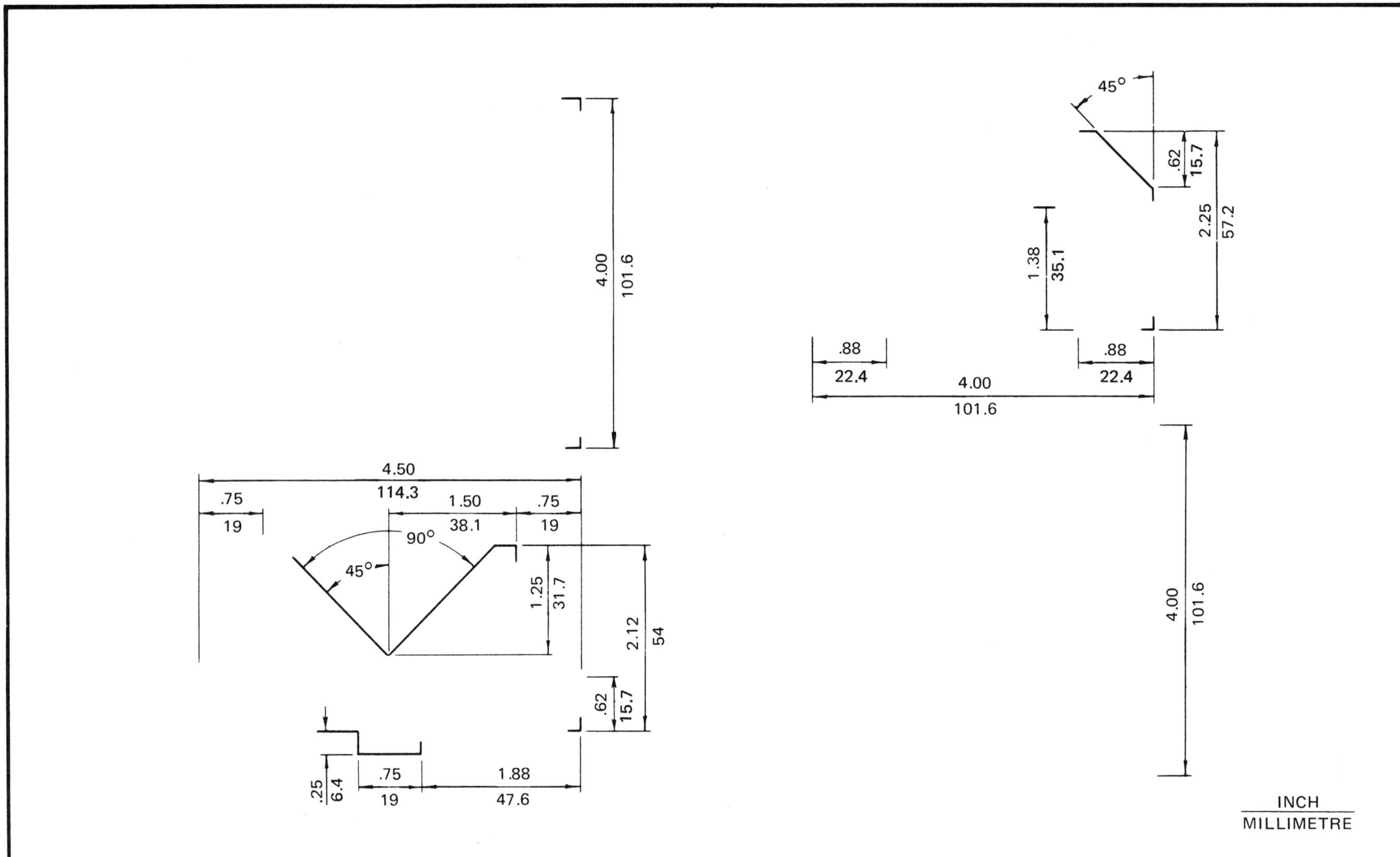

NAME OF SCHOOL CITY OR TOWN	V BLOCK GUIDE	SHEET No. 5-E NAME	CLASS DATE

Fig. 5-21

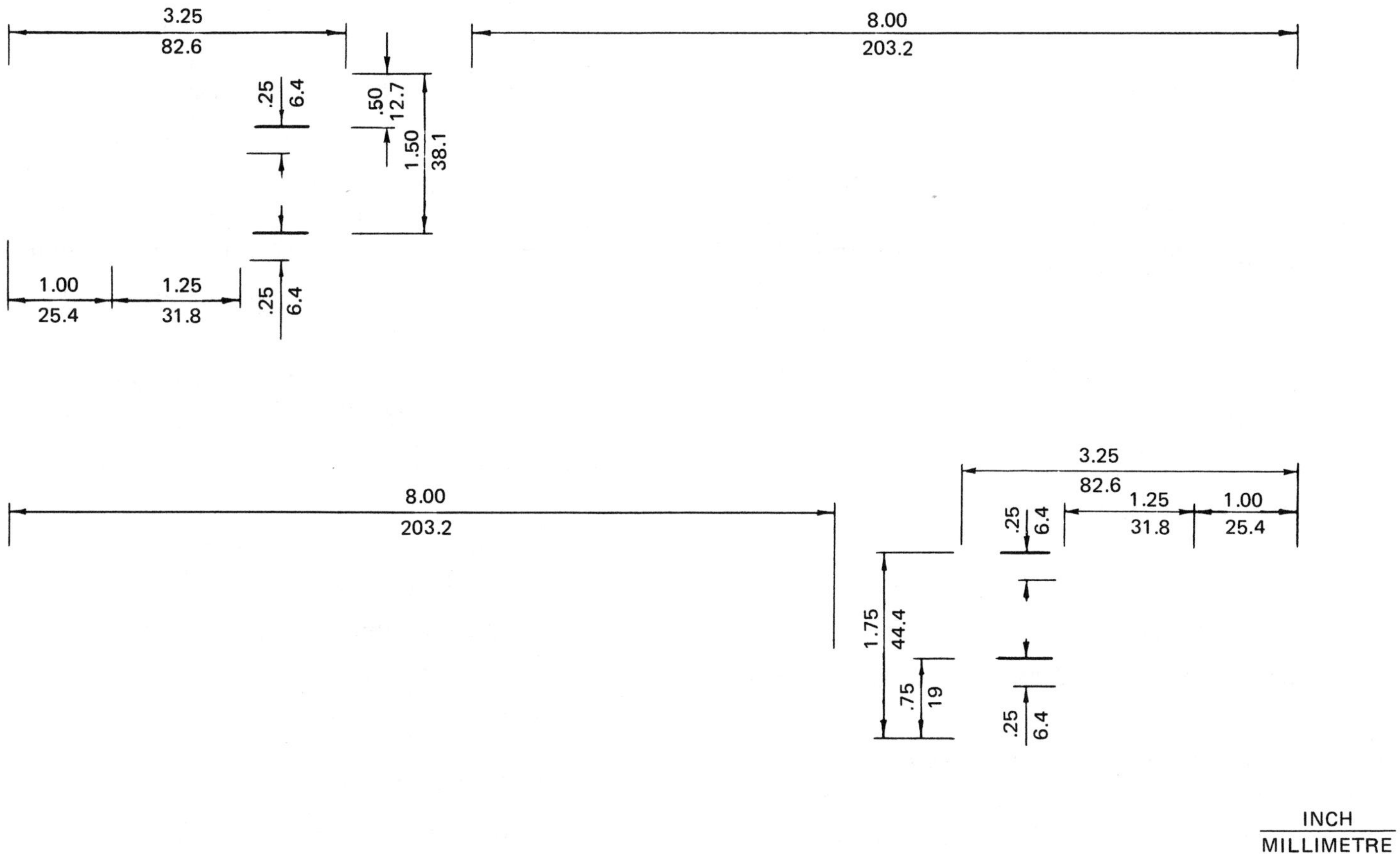

NAME OF SCHOOL CITY OR TOWN	ADJUSTABLE PARALLELS	SHEET No. 5-F NAME	CLASS DATE

Fig. 5-22

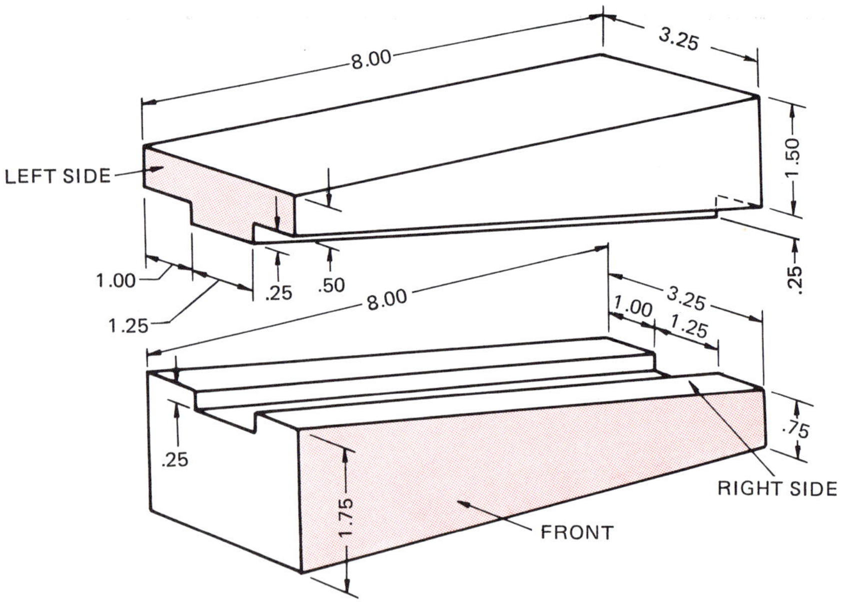

Fig. 5-23 Adjustable parallels

A. Upper part of adjustable parallels

1. Lay out the margin line on the drawing sheet and letter the title strip.
2. Start with a partial left-side view. Draw only the surface on the left side of the upper parallel. Draw this partial view 1 3/4 inches below the top margin line and 1 3/4 inches from the left margin line, figure 5-22.
3. Draw construction lines to the right from each corner of the left-side view.
4. Draw the complete front view. Locate it 1 3/4 inches from the right margin line.
5. From the two lower corners of front view, draw construction lines to the left.
6. Complete the left-side view by adding lines to enclose the inclined surfaces.
7. Dual-dimension the views.
8. Check the drawing. Correct any errors.

B. Lower part of adjustable parallels

1. Start the drawing of the lower part with a partial right-side view. Draw only that surface which is on the right side. Locate this partial view 1 3/4 inches from the right margin line and 1 3/4 inches above the lower margin line.
2. From each corner of the right-side view, draw a construction line to the left.
3. Draw the complete front view. Locate it 1 3/4 inches from the left margin line.
4. Draw construction lines to the right from the two upper corners of the front view.
5. Complete the right-side view by drawing lines which enclose the inclined surfaces.
6. Dual-dimension the views. Check the drawing and correct any errors.
7. Clean the drawing.

REVIEW QUESTIONS

1. Indicate the inclined edges in the views of the drawing.
2. Show the inclined surfaces on the drawing.
3. Show where the foreshortened edges are.

DRAWING CURVED SURFACES

Up to this point, the objects drawn contained horizontal, vertical or inclined surfaces. These surfaces are classified as plane (flat) surfaces. The next group of objects have some curved surfaces.

ASSIGNMENT G

1. Draw a top view and a front view of the adapter shank, figure 5-24.
2. Draw a top view and a front view of the guide post bushing, figure 5-25.

Procedure

A. Adapter shank

1. Lay out the margin lines on the drawing sheet.
2. Letter the title strip.
3. Draw the horizontal center line for the top view 2 1/2 inches below the top margin line, figure 5-26.
4. Draw the vertical center line 4 1/2 inches from the left margin line.
5. The two circles in the top view represent the edges enclosing the two upper surfaces of the adapter shank.

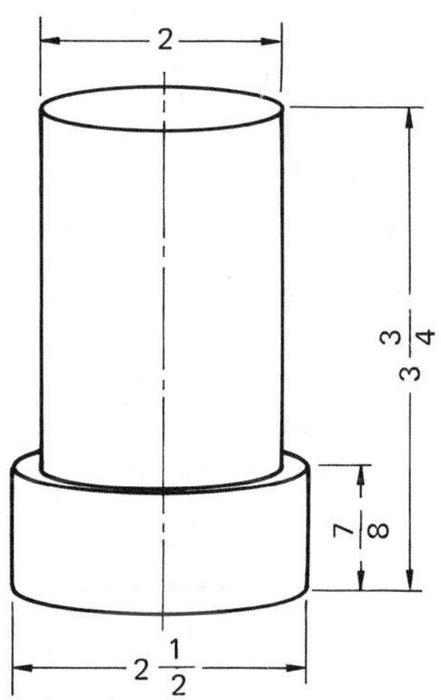

Fig. 5-24 Adapter shank

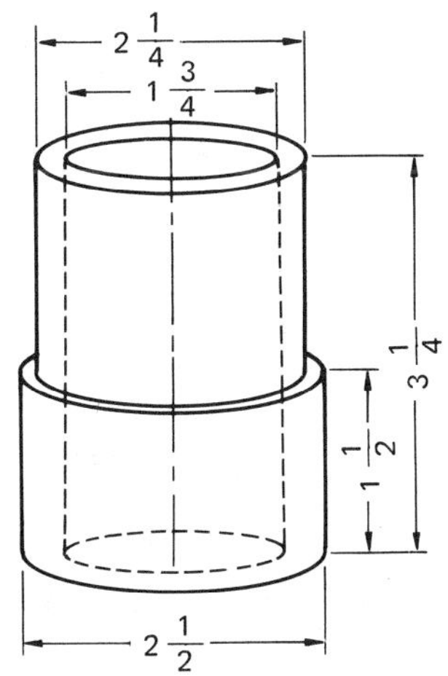

Fig. 5-25 Guide post bushing

6. Draw construction lines down from the sides of the circles. These construction lines locate the outlines of the curved surfaces in the front view.
7. Draw the front view 1 1/2 inches above the title strip. Draw the center line.
8. Dimension the views. Note: When dimensioning cylindrical shapes, it is customary to place all dimensions in one view.
9. Check the drawings and correct errors.

B. Guide post bushing

1. Draw the horizontal center line for the top view of the guide post bushing 2 1/2 inches below the top margin line.
2. Draw the vertical center line 5 1/4 inches from the right margin line.
3. Draw the top view.
4. Draw the construction lines down from the sides of the circles. These construction lines locate the outlines of the curved surfaces in the front view.

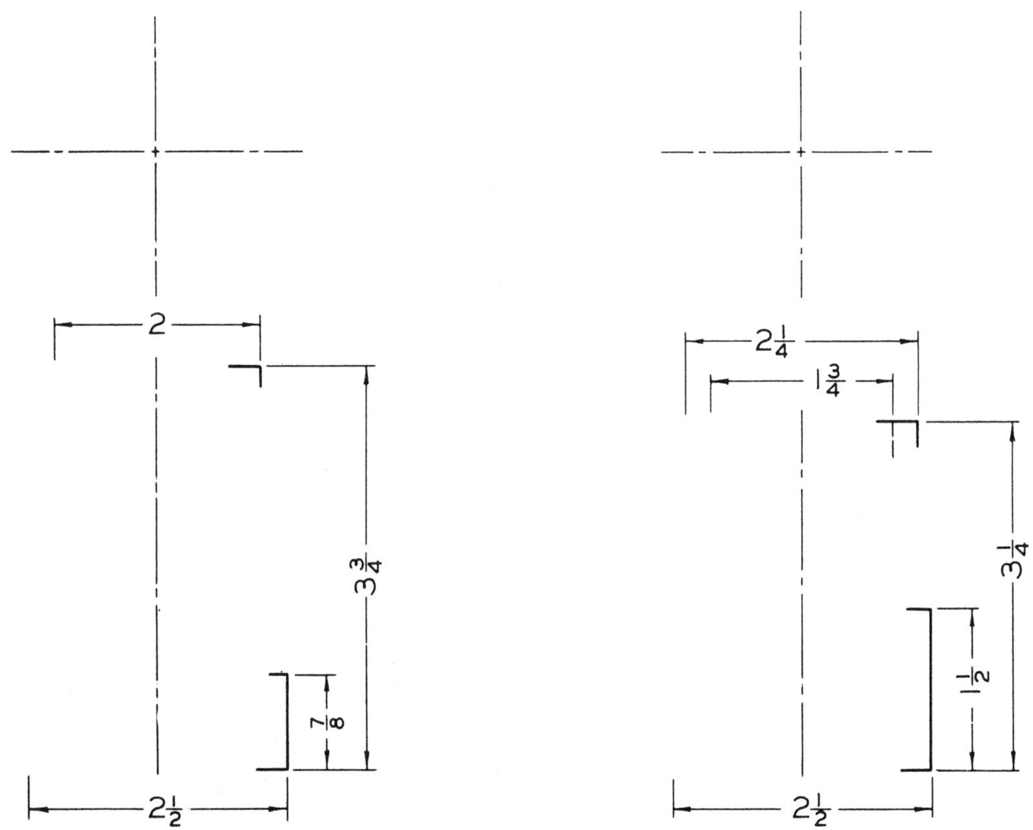

NAME OF SCHOOL CITY OR TOWN	ADAPTER SHANK GUIDE POST BUSHING	SHEET No. 5-G NAME	CLASS DATE

Fig. 5-26

5. Draw the front view 1 1/2 inches above the title strip. Be sure to draw the hidden outlines of the hole and the center line.
6. Dimension the front view.
7. Check the drawing and correct any errors.
8. Clean the drawing.

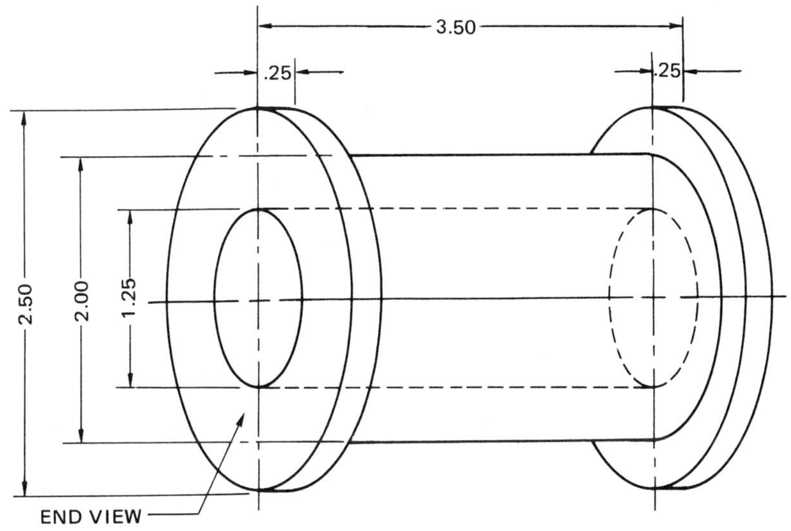

Fig. 5-27 Flanged bushing

TERMS

1. A *circle* is a closed curved line on a plane where every part is an equal distance from a point within, called the center.
2. The *diameter* is a straight line through the center of the circle and connecting 2 points on the circle.
3. The *radius* is a straight line or distance from the center to the circle.
4. The *circumference* is the distance around a circle.
5. An *arc* is a part of a circle.

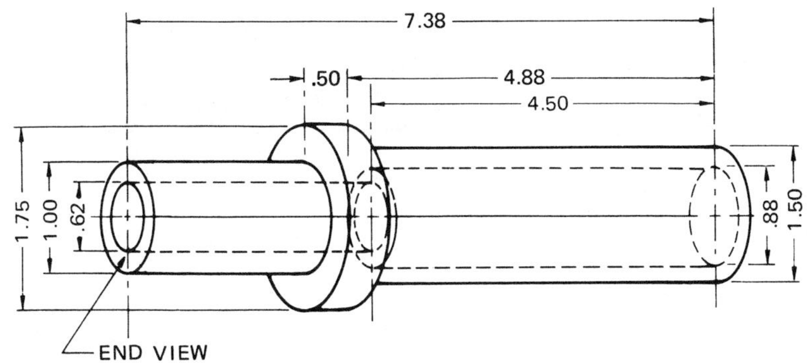

Fig. 5-28 Hollow shaft

DRAWING HOLES

When two views of long cylindrical parts, such as bushings, shafts, punches, and dowel pins, are drawn horizontally on a sheet, the view showing the outside of the cylindrical shapes is usually called the front view. The view showing the circles is called the end view instead of the side view.

The next objects to be drawn have a number of parts in the form of a cylinder. For example, the flanged bushing in figure 5-27 includes four cylindrical parts: the hole, two flanges, and the body. The hollow shaft in figure 5-28 includes five cylindrical parts: the left end, the center, the right end, and two holes. Each of these parts is a different diameter. Circles are drawn in the left-end view to show the shape of these cylindrical parts. Some of these circles are invisible and are drawn with hidden lines.

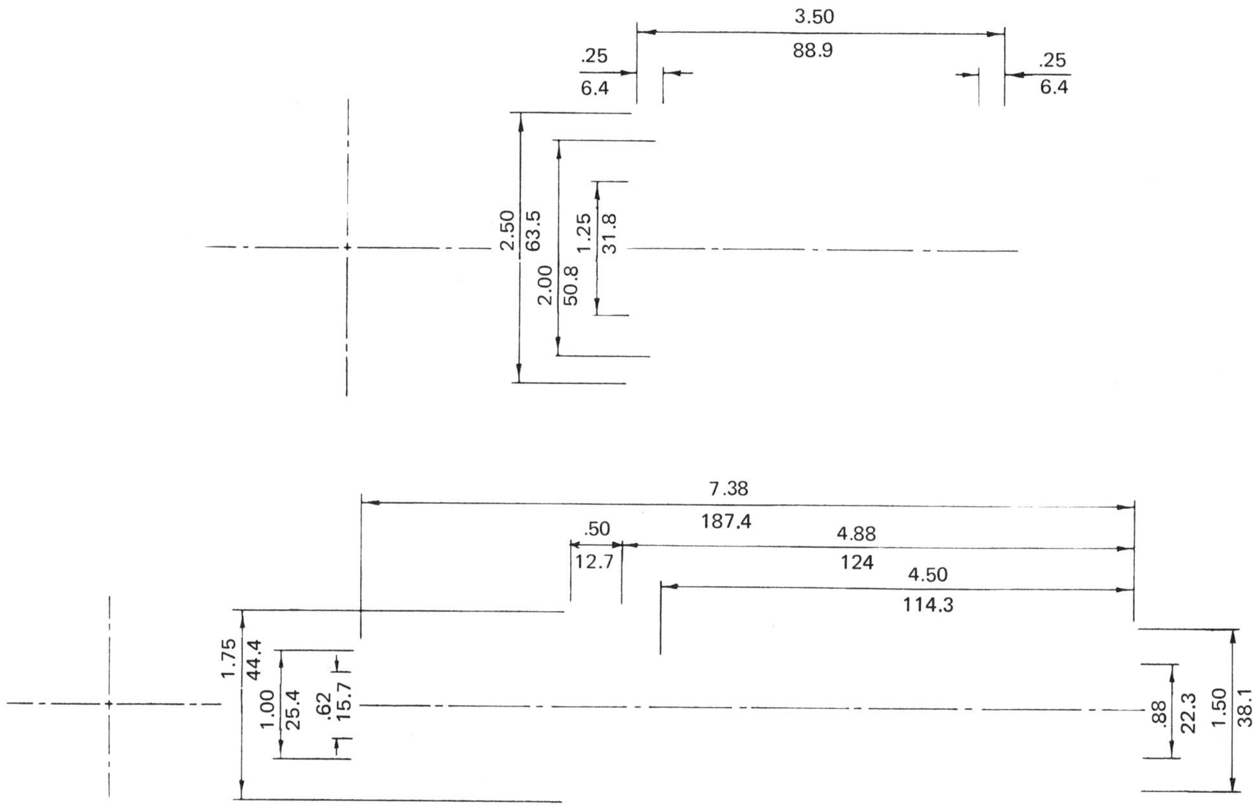

INCH
MILLIMETRE

NAME OF SCHOOL CITY OR TOWN	FLANGED BUSHINGS HOLLOW SHAFT	SHEET No. 5-H NAME	CLASS DATE

Fig. 5-29

ASSIGNMENT H

1. Draw an end view and a front view of the flanged bushing, figure 5-27.
2. Draw an end view and a front view of the hollow shaft, figure 5-28.

Procedure

A. Flanged bushing

1. Lay out the margin lines on the drawing sheet and letter the title strip.
2. Draw the horizontal center line for the end view of the flanged bushing 3 inches below the top margin line, figure 5-29.
3. Draw the vertical center line 5 1/2 inches from the left margin line.
4. Draw the circles in the end view. Be sure to draw the hidden circle.
5. Draw construction lines to the right from the circles. Draw the front view 4 1/4 inches from the right margin line.
6. Dual-dimension the front view. Check the accuracy of all lines on the drawing.

B. Hollow shaft

1. Draw the horizontal center line for the end view of the hollow shaft 2 1/2 inches above the title strip.
2. Draw the vertical center line 3 1/4 inches from the left margin line.
3. Draw the circles in the end view.
4. Draw construction lines to the right from the circles. Draw the front view 3 inches from the right margin line.
5. Dual-dimension the front view. Check for accuracy in representing and dimensioning the object.

REVIEW QUESTIONS

1. What is (a) a circle? (b) a radius? (c) a diameter? (d) a circumference? (e) an arc?
2. How is a center line drawn?
3. For what purpose are center lines used?
4. Indicate the curved surfaces in the front views of the drawing.

DRAWING CONICAL SURFACES

The next objects to be drawn are considered to be made up of separate parts. Some are in the form of a cylinder, others are a part of a cone.

The washer illustrated in figure 5-30 has a beveled or sloping upper surface. If this beveled surface were continued to a point, it would form a cone. Therefore, the upper part of the washer within the cone is called a conical shape.

The collet, figure 5-31, has two surfaces which, if continued to a point, would form cones. The parts of the collet within these cones are conical shapes. Other parts of the washer and collet are cylindrical in shape.

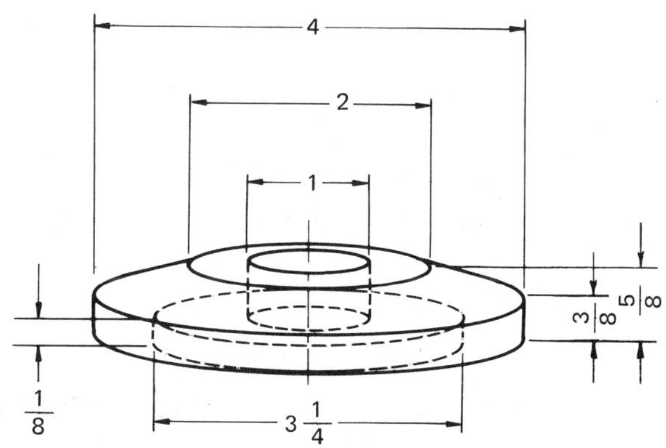

Fig. 5-30 Washer

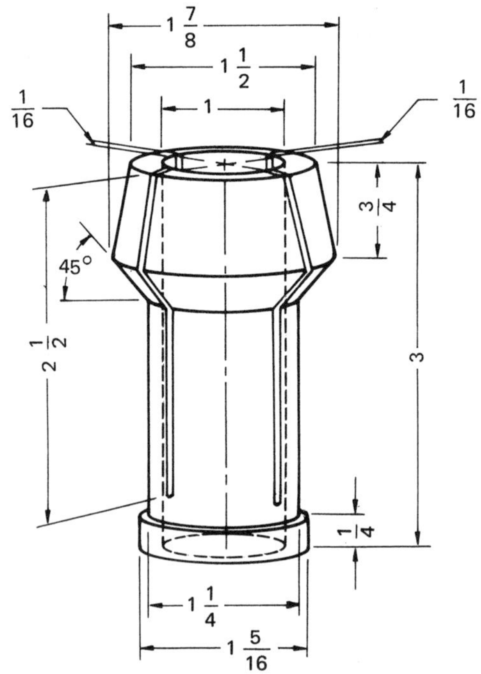

Fig. 5-31 Collet

Conical surfaces are dimensioned by two methods. One method, figure 5-31, shows the angle between the outline of the conical surface and a horizontal line. The other method gives the height of the conical surface and the diameter at each end.

ASSIGNMENT I

1. Draw a top view and a front view of the washer, figure 5-30.
2. Draw a top view and a front view of the collet, figure 5-31.

Procedure

A. Washer

1. Lay out the margin lines on the drawing sheet and letter the title strip.
2. Draw the horizontal center line for the top view of the washer 3 1/2 inches below the top margin line, figure 5-32.
3. Draw the vertical center line 4 3/4 inches from the left margin line.
4. Draw the top view.
5. Draw construction lines down from the circles. Draw the front view 2 inches above the title strip.
6. Dimension the front view. Check the drawing for accuracy.

B. Collet

1. Draw the horizontal center line for the top view of the collet 2 1/4 inches below the top margin line.
2. Draw the vertical center line 4 3/4 inches from the right margin line.
3. Draw the top view. Two of the circles are cut into four parts by the 1/16 inch wide slots.
4. Draw construction lines down from the circles. Draw the front view 2 inches above the title strip.
5. Dimension the views.
6. Check your drawing and correct any errors. Clean the drawing.

REVIEW QUESTIONS

1. Indicate the cylindrical surfaces in the front views of the drawing.
2. Indicate the conical curved surfaces.
3. Explain the two methods of dimensioning conical surfaces.

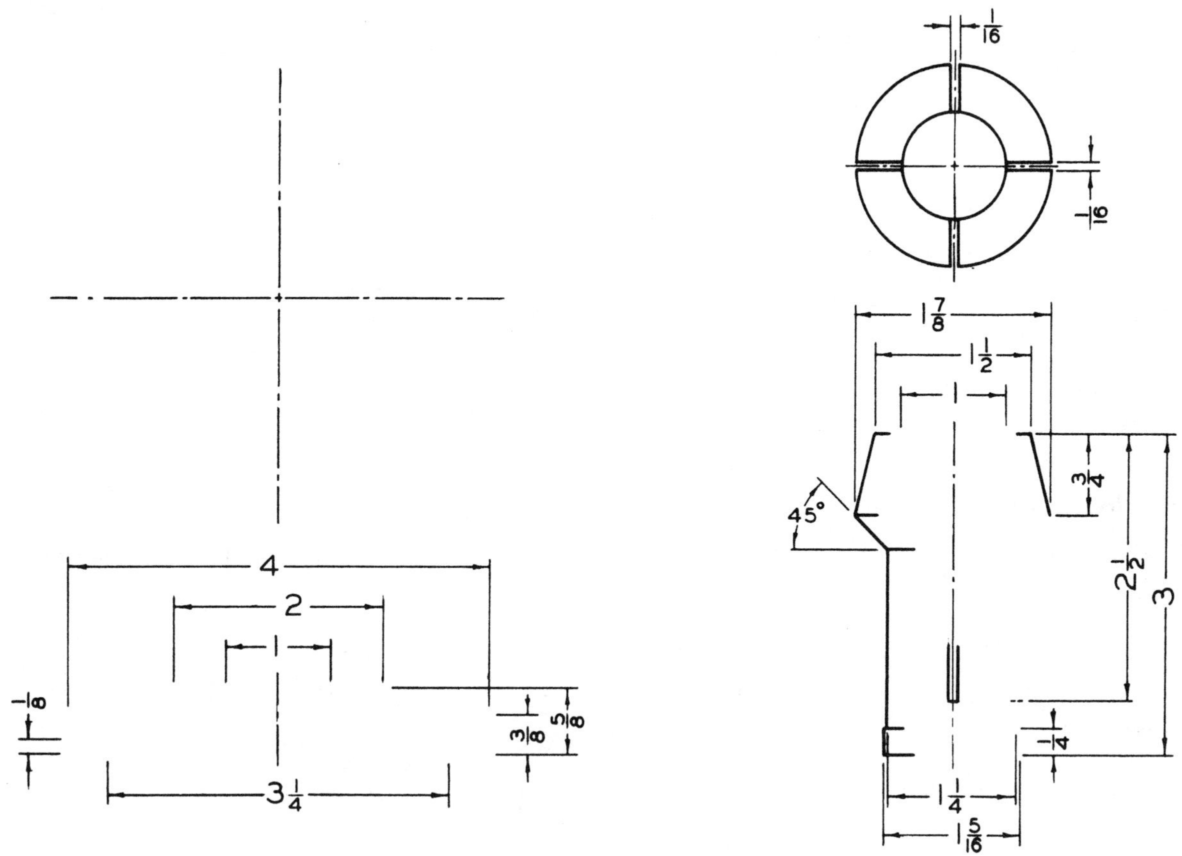

NAME OF SCHOOL CITY OR TOWN	WASHER COLLET	SHEET No. 5-I NAME	CLASS DATE

Fig. 5-32

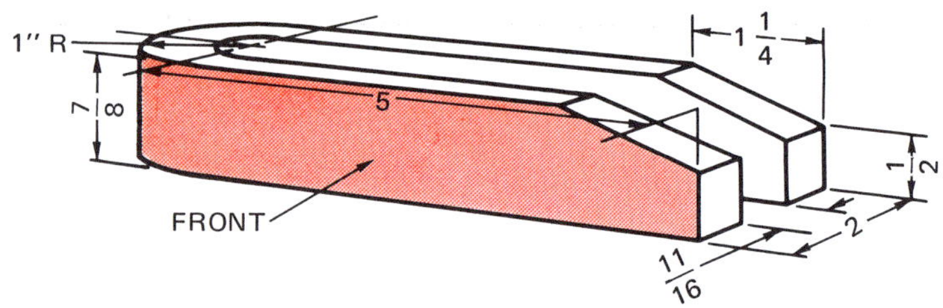

Fig. 5-33 U-strap clamp

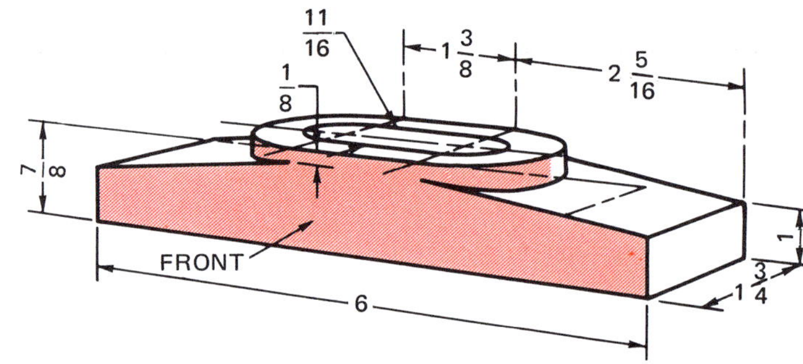

Fig. 5-34 Plain strap clamp

DRAWING TANGENT POINTS

The clamps illustrated in figures 5-33 and 5-34 have some horizontal, vertical, and inclined surfaces. Other surfaces are curved like those on a cylinder. The curved surfaces merge or blend into the plane surfaces.

In the top view, the curved surfaces are represented by arcs, and the plane surfaces are represented by straight lines. The points in the top view, where arcs meet the straight lines, are called the *points of tangency* or *tangent points.*

To find the tangent point, a line is drawn from the center of the arc at a right angle (90°) to the straight line. Figure 5-35 shows how tangent points are found.

ASSIGNMENT J

1. Draw a top and front view of the U-strap clamp, figure 5-33.
2. Draw a top and front view of the plain strap clamp, figure 5-34.

Procedure

A. U-strap clamp

1. Lay out the margin lines on the sheet and letter the title strip.
2. Draw the horizontal center line for the top view 3 1/2 inches below the top margin line, figure 5-36.
3. Draw the vertical center line 2 1/4 inches from the left margin line.

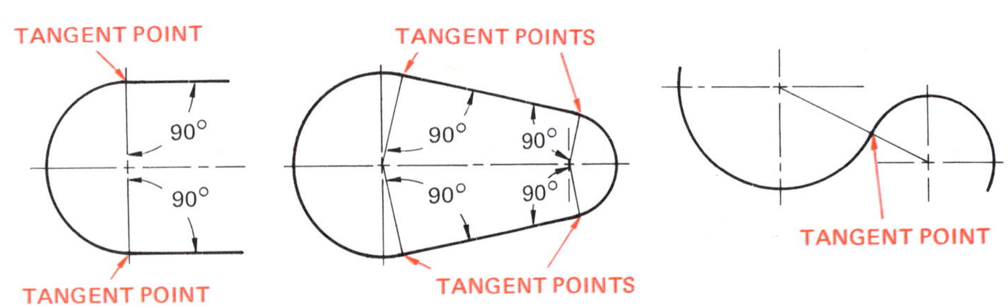

Fig. 5-35 Locating tangent points

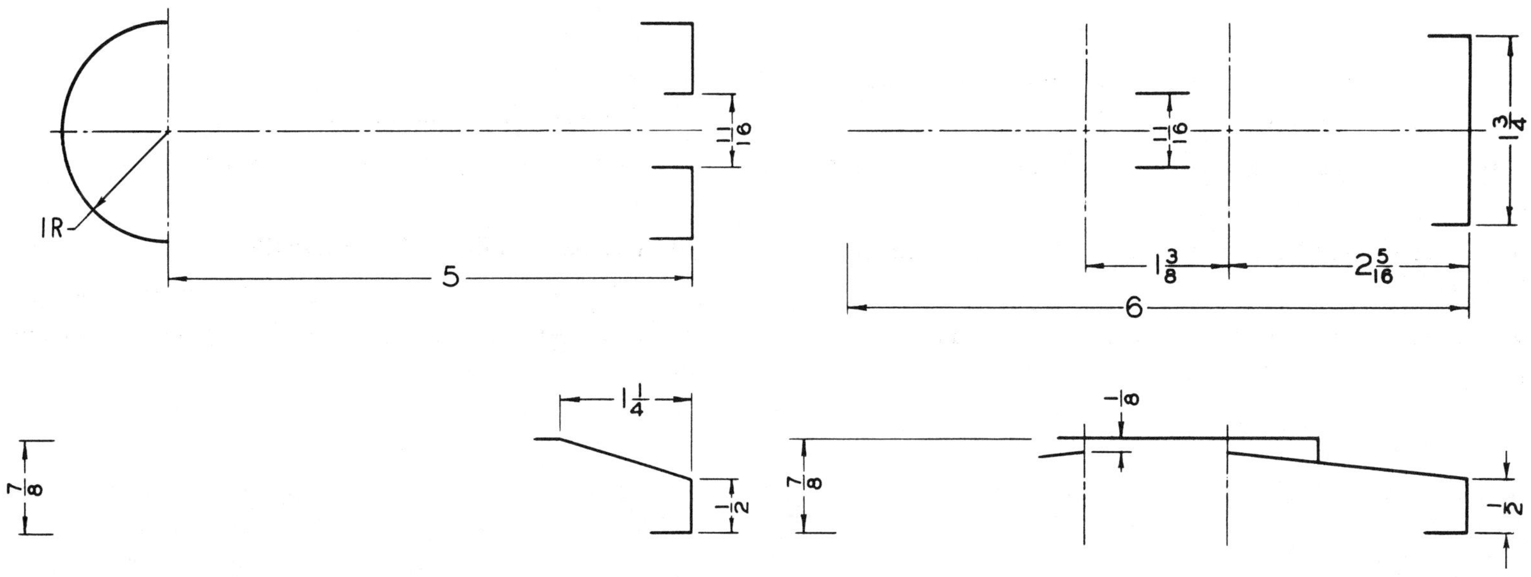

NAME OF SCHOOL CITY OR TOWN	"U" STRAP CLAMP PLAIN STRAP CLAMP	SHEET No. 5-J NAME	CLASS DATE

Fig. 5-36

4. Draw the arcs in the top view first before completing the rest of the view.
5. Draw construction lines down from the arcs and from the corners. Draw the front view 2 1/2 inches above the title block.
6. Dimension the views. Check the accuracy of all lines and dimensions on the drawing.

B. Plain strap clamp

1. Draw the horizontal center line for the top view 3 1/2 inches below the top margin line.
2. Draw the first vertical center line 3 9/16 inches from the right margin line.
3. Draw the second vertical center line 1 3/8 inches to the left of the first.
4. Draw the arcs in the top view first before completing the rest of the view.
5. Draw construction lines down from the arcs, the corners, and the tangent points. Draw the front view 2 1/2 inches above the lower margin line. Two lines in the front view are terminated at the tangent points.
6. Dimension the views. Check the drawing for accuracy.

REVIEW QUESTIONS

1. Indicate the tangent points in the top and front views of the drawing.
2. Indicate the plane surfaces.
3. Indicate the cylindrical curved surfaces.

DRAWING NOTES AND LEADERS

The crank, figure 5-37, and the stuffing box gland, figure 5-38, have some inclined plane surfaces which are tangent to curved surfaces.

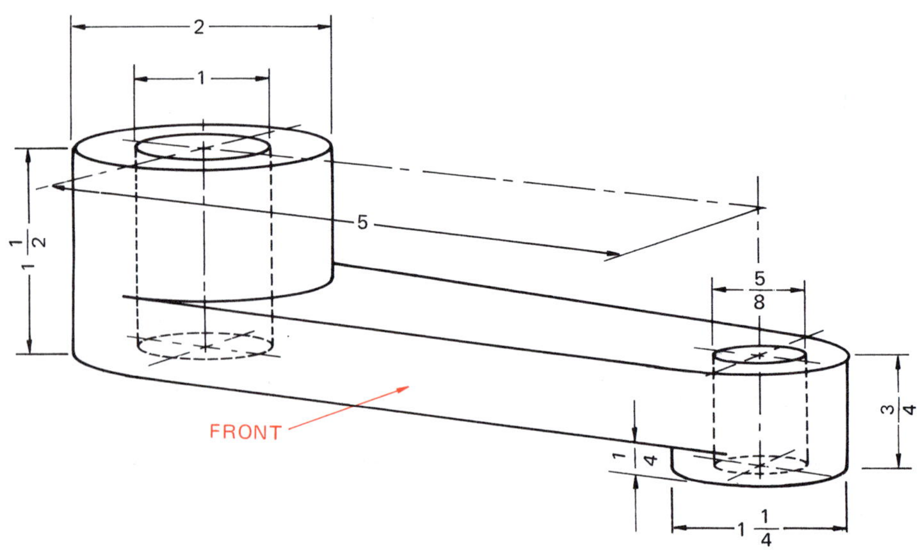

Fig. 5-37 Crank

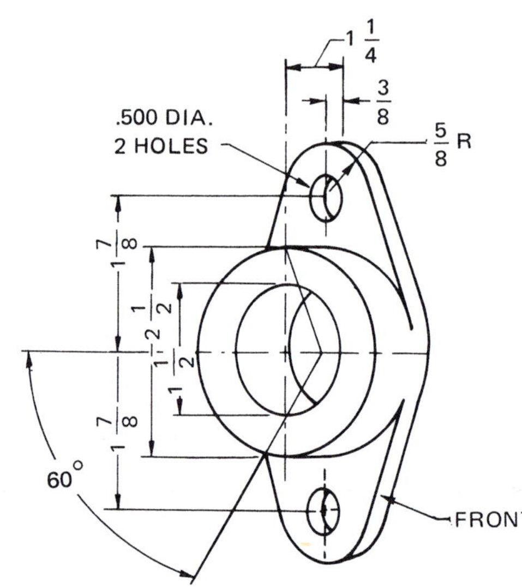

Fig. 5-38 Stuffing box gland

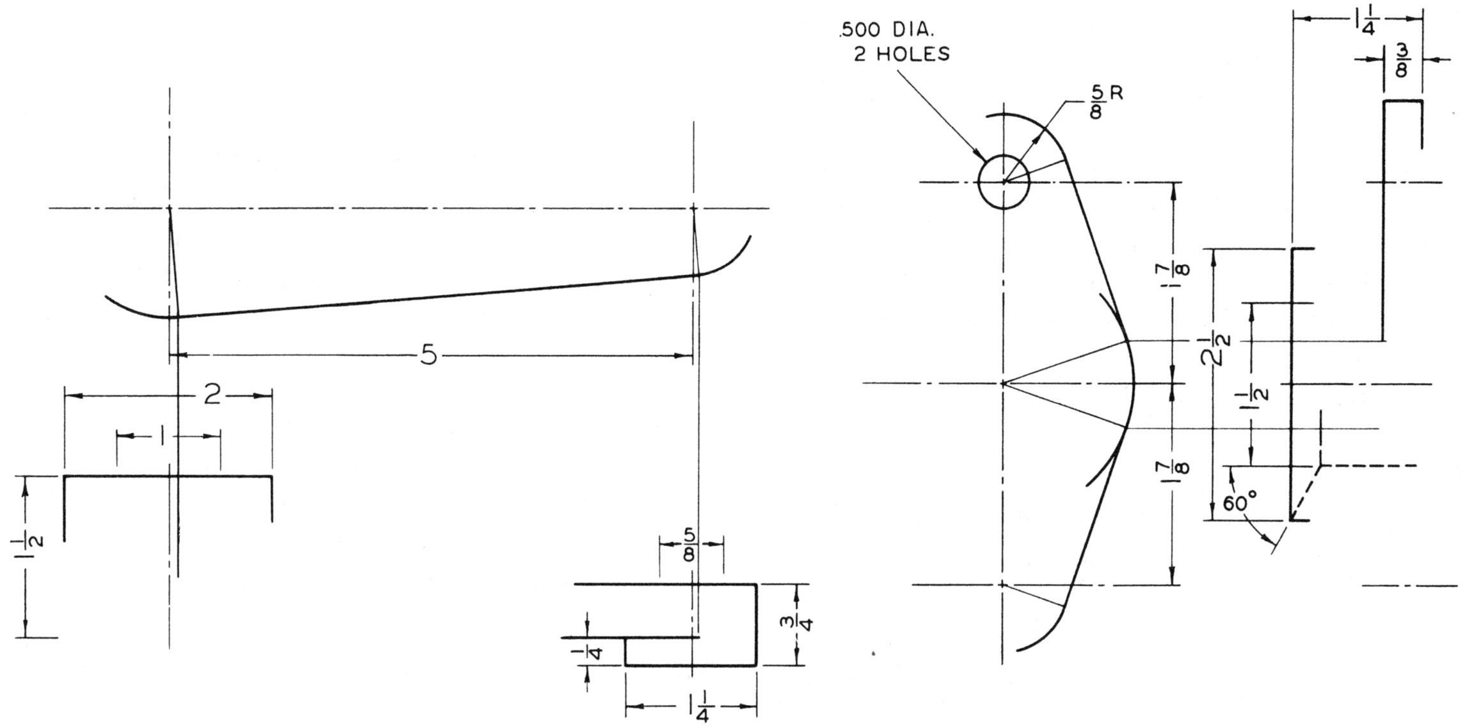

NAME OF SCHOOL CITY OR TOWN	CRANK STUFFING BOX GLAND	SHEET No. 5-K NAME	CLASS DATE

Fig. 5-39

The tangent points are found by the method previously illustrated. The tangent points are always marked in the views of a drawing. Then, if the drawing is inked, the arcs and straight lines are drawn to these points without overlapping. The tangent points are marked in the front views of this drawing because some of the straight lines terminate at the tangent points.

The stuffing box gland has an internal curved surface which, if continued to a point, forms a cone. The outlines of this conical curved surface are invisible in the front view.

Two of the holes in the stuffing box gland are drilled holes. These holes are dimensioned by a note that specifies the size of the drill and the number of holes to be drilled. The note is connected with the holes by a leader. A leader is a fine line with an arrowhead at one end. The leader is drawn in line with the center of the circle. The arrowhead touches the circle.

ASSIGNMENT K

1. Draw a top view and a front view of the crank, figure 5-37.
2. Draw a left-side view and a front view of the stuffing box gland, figure 5-38.

Procedure

A. Crank

1. Lay out the margin lines on the sheet and letter the title strip.
2. Draw the horizontal center line for the top view 3 inches below the top margin line.
3. Draw the first vertical center line 2 1/4 inches from the left margin line.
4. Draw the second vertical center line 5 inches to the right of the first vertical line.
5. Draw the circles. Complete the top view. Locate the tangent points.
6. Draw construction lines down from the circles and from the tangent points. Draw the front view 2 inches above the title strip. The two lines in the front view terminate at the tangent points.
7. Dimension the views as shown in figure 5-39.

B. Stuffing box gland

1. Draw the horizontal center line for the left-side view 4 5/8 inches below the top margin line.
2. Draw the vertical center line 5 1/4 inches from the right margin line.
3. Draw the horizontal center lines for the two 1/2 inch holes.
4. Draw the circles and arcs. Complete the left-side view. Locate the tangent points.
5. Draw construction lines to the right from the circles, the arcs, and the tangent points. Draw the front view 1 1/4 inches from the right margin line. The two lines in this view terminate at the tangent points.
6. Dimension the views as shown in figure 5-39.

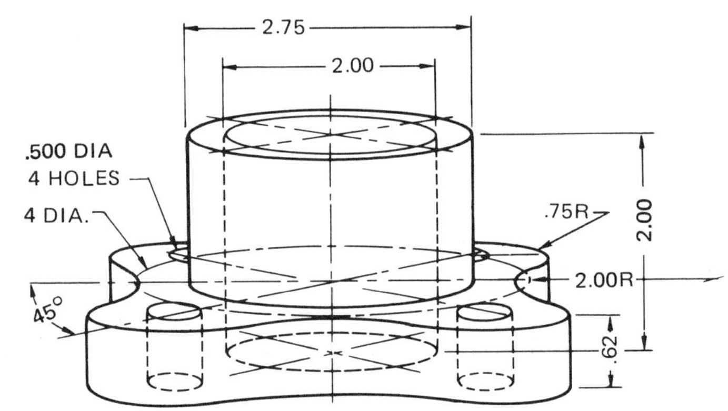

Fig. 5-40 End bearing

DRAWING FILLETS AND ROUNDS

Both the end bearing, figure 5-40, and the fixture, figure 5-41, use different methods of locating the centers of tangent arcs. These objects have some curved surfaces which are tangent to other curved surfaces. The tangent curved surfaces are represented by tangent arcs in the top view.

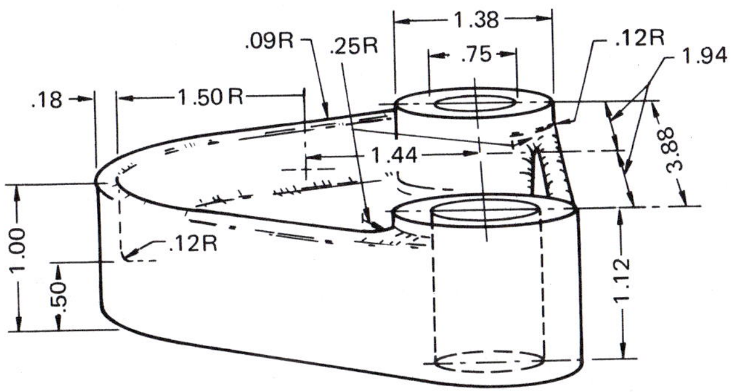

Fig. 5-41 Fixture head

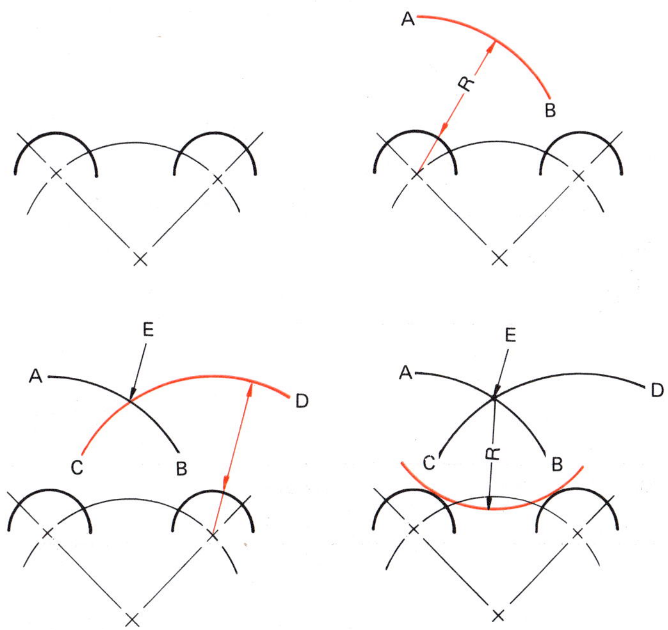

Fig. 5-42 Arc tangent to two given arcs

The centers of the four .75 inch radius arcs in figure 5-40, are located in the center of the .500 inch drilled holes. The location of the centers of the 2-inch radius tangent arcs is not specified on the drawing. These center points must be carefully located.

The fixture head has some fillets and rounds. *Fillets* and *rounds* are additional metal left on a part to make it stronger. One method of locating the center of tangent arcs is illustrated in figure 5-42.

The fixture head also has some fillet runouts where the .12 inch radius fillets go partly around the 1.38 inch diameter cylindrical shaped parts. These fillets run out or disappear at the tangent points. Figure 5-43 shows how the runouts are located and drawn in the front view. The radius of the runouts is the same as the radius of the fillets, but only one-eighth of a circle is drawn.

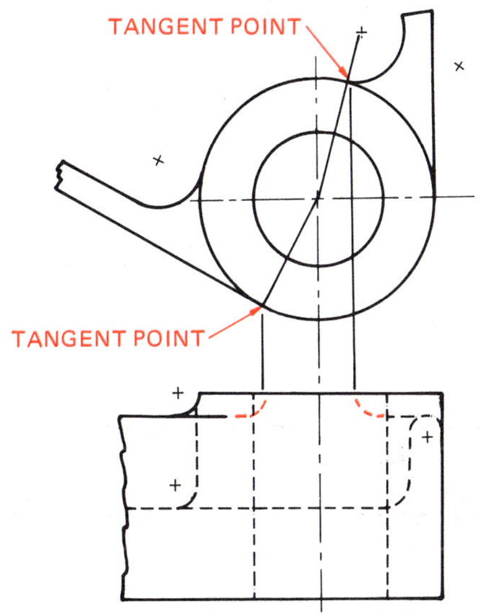

Fig. 5-43 Runout of fillets

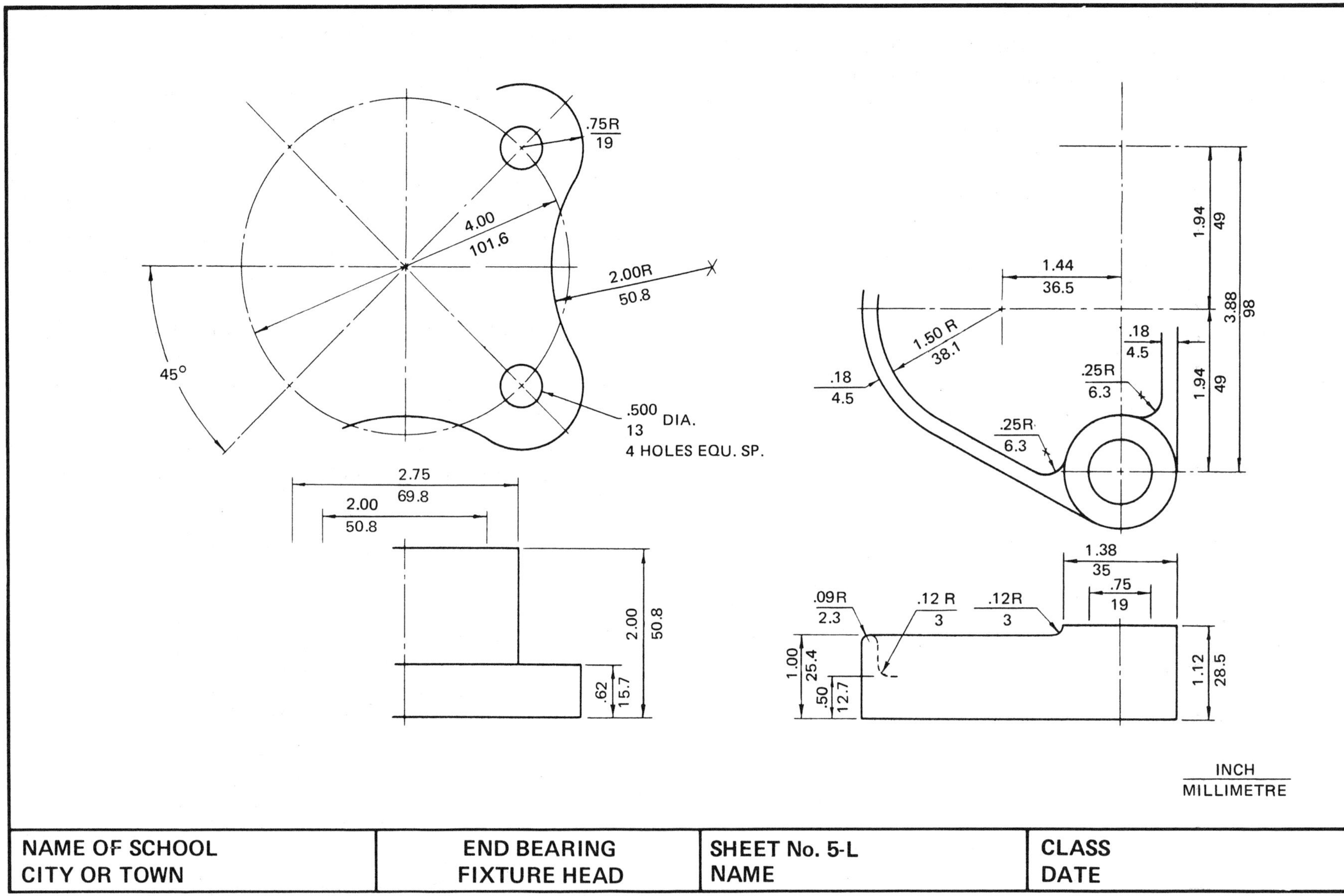
.75R
19
4.00
101.6
2.00R
50.8
45°
.500
13
DIA.
4 HOLES EQU. SP.
2.75
69.8
2.00
50.8
2.00
50.8
.62
15.7
1.44
36.5
1.94
49
3.88
98
1.50 R
38.1
.18
4.5
.18
4.5
.25R
6.3
.25R
6.3
1.94
49
1.38
35
.75
19
.09R
2.3
.12 R
3
.12R
3
1.00
25.4
.50
12.7
1.12
28.5
INCH
MILLIMETRE
NAME OF SCHOOL
CITY OR TOWN
END BEARING
FIXTURE HEAD
SHEET No. 5-L
NAME
CLASS
DATE

Fig. 5-44

ASSIGNMENT L

1. Draw a top view and a front view of the end bearing, figure 5-40.
2. Draw a top view and a front view of the fixture head, figure 5-41.
3. Dimension the drawing as in figure 5-44.

Procedure

A. End bearing

1. Lay out the margin lines on the drawing sheet.
2. Letter the title strip.
3. Draw the horizontal center line for the top view of the end bearing 3 1/4 inches below the top margin line.
4. Draw the vertical center line 4 inches from the left margin line.
5. Draw the two center lines which slope 45 degrees.
6. Draw the 4-inch diameter center line circle.
7. Draw the four .75 inch radius arcs.
8. Locate the centers of the 2-inch radius arcs using the method shown in figure 5-42. Draw the tangent arcs.
9. Locate and mark the eight tangent points. If the drawing is inked, the tangent arcs meet at these points.
10. Complete the top view.
11. Draw the front view 1 1/8-inch above the title strip. Be sure to draw the outlines of all of the curved surfaces and the outlines of the holes.
12. Dual-dimension the views.
13. Check every view, line, and dimension for accuracy.

B. Fixture head

1. Draw the horizontal center line for the top view of the fixture head 3 1/2 inches below the top margin line.
2. Draw the vertical center line 3 1/4 inches from the right margin line.
3. Locate and draw the other center lines.
4. Draw the circles and the outline of the top view.
5. Locate the centers of the .25 inch radius fillets by the method shown in figure 5-43. Draw the tangent arcs.
6. Mark the sixteen tangent points in the top view.
7. Draw the front view 1 1/8 inch above the title strip. Locate the centers for the .12 inch fillets.
8. Draw the runouts of the fillets.
9. Dual-dimension the views.
10. Check the drawing of the fixture head for errors.
11. Clean the drawing.

REVIEW QUESTIONS

1. What is a fillet? For what purposes are fillets used?
2. What is a round?
3. Indicate the fillets in the different views of the drawing.
4. Indicate the fillet runouts on the drawing.
5. Indicate the rounds.

UNIT 6 THREE-VIEW DRAWINGS

Before proceeding with the next group of drawings, the theory of projecting views onto the planes of projection is shown in figure 6-1. The top, front, and bottom views are in line vertically and are the same length. Vertical construction lines are drawn to keep these views in line on working drawings.

The front view, both side views, and the rear view are in line horizontally and are the same height. Horizontal construction lines are drawn to keep these views in line on a working drawing.

The top view, both side views, and the bottom view are the same width but are not in line. To make these views the same width on working drawings, construction lines are drawn using one of the methods illustrated in figure 6-2. The method shown at B is preferred.

The use of construction lines is always helpful whenever a three-view drawing is made. In figure 6-3, two views of a triangular pyramid are given. The third view is found with the aid of construction lines drawn from the two given views. The triangular pyramid has four points. A number is placed at each point in the

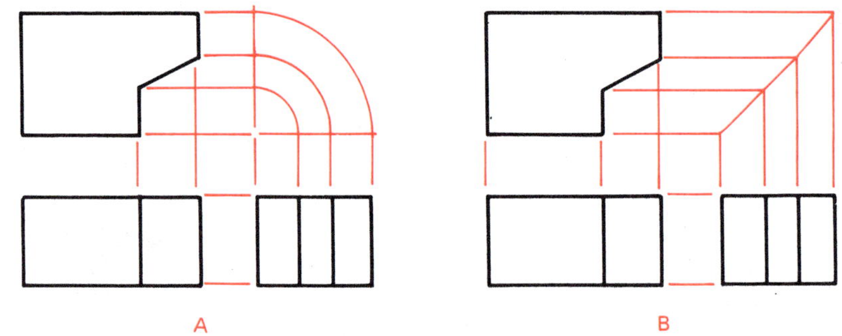

Fig. 6-2 Drawing construction lines

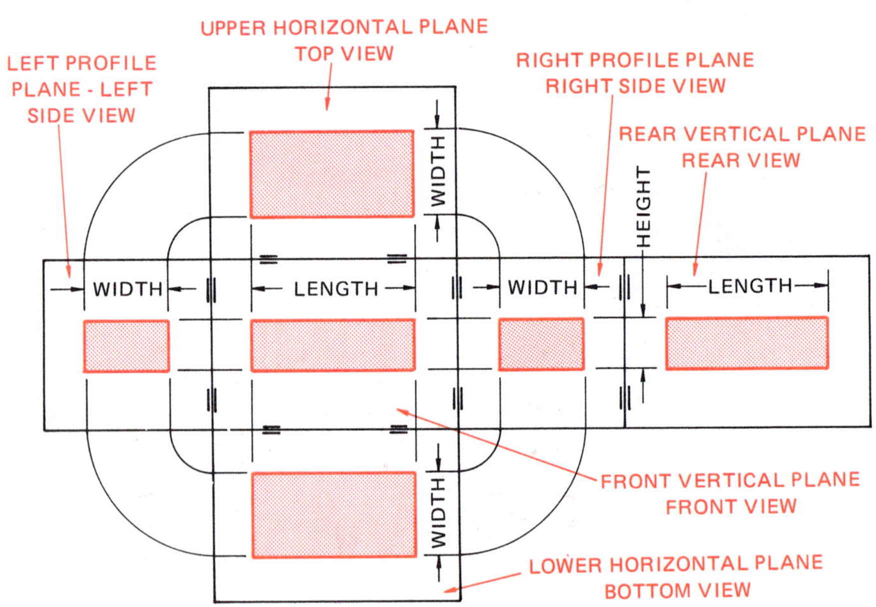

Fig. 6-1 The position of the six planes on a flat surface

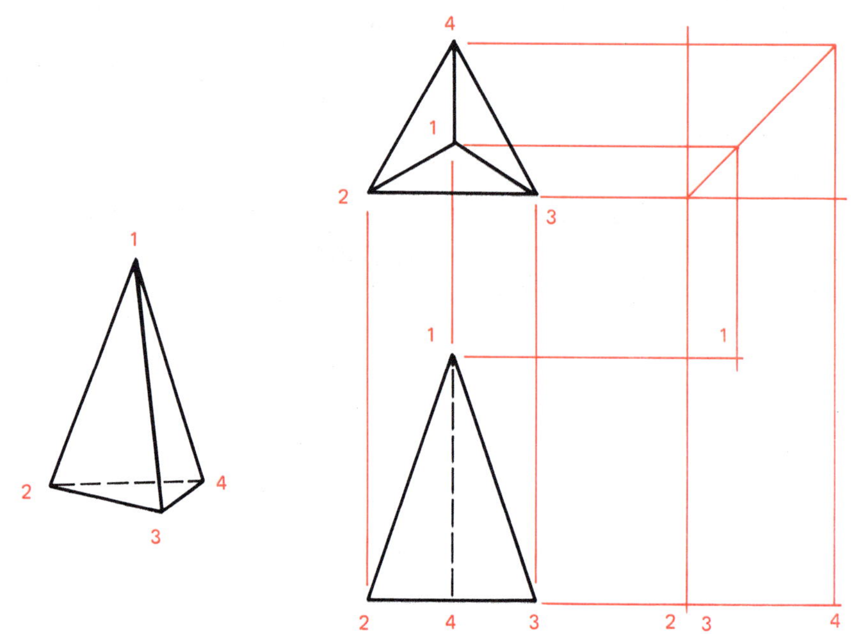

Fig. 6-3 Locating the side view

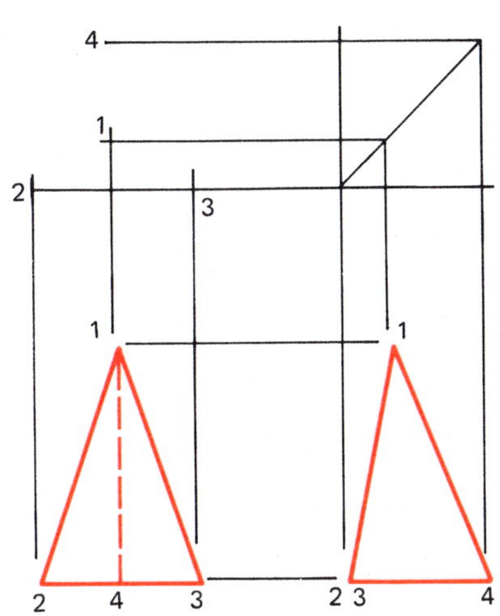

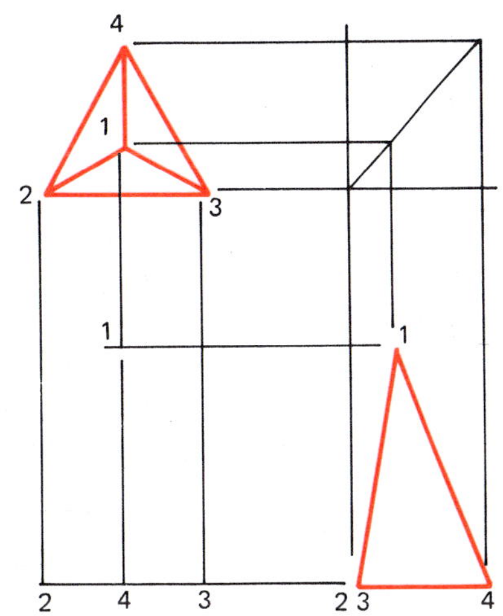

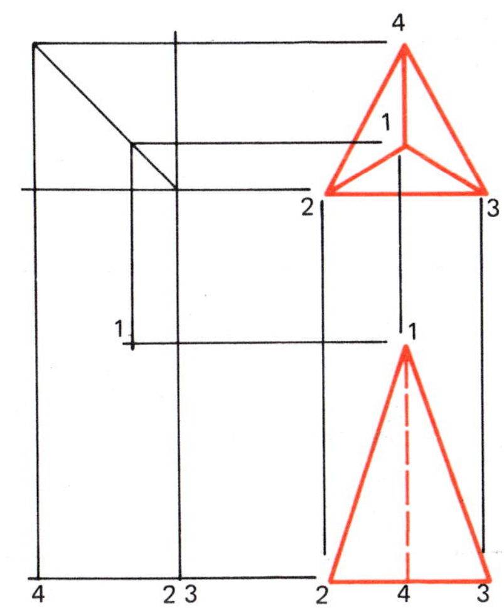

top view. The same numbers are placed directly below at corresponding points in the front view. The points in the side view are located at the intersection of construction lines drawn from the two given views. For example, the line from point 1 in the front view, and the line from point 1 in the top view, intersect to locate point 1 in the side view. The other points are located the same way. The points are connected with lines to complete the side view.

Figure 6-4 shows how construction lines and numbers are used to find the top view, front view, or left-side view when the other two views are given.

In the following drawings, construction lines and numbers are used to find the missing views.

REVIEW QUESTIONS

1. Which views are in line vertically?
2. Which views are in line horizontally?
3. Which views are the same length?
4. Which views are the same height?
5. Which views are the same width?

MISSING SIDE VIEWS

Two views of the tongued block, figure 6-6, are shown in figure 6-5. The tongued block has sixteen corners. A number is placed in the front view for each of these corners. The same numbers are placed at the corresponding corners in the top view. The numbers for the corners of the side view are located by the method previously illustrated in figure 6-3.

Two views of the toolmaker's knee, figure 6-7, are also given in figure 6-5. The missing side view is located by drawing construction lines from the two given views.

BREAK ALL SHARP EDGES

The shop note, *break all sharp edges* means that all of the sharp edges on the object must be rounded slightly. This is done freehand or with a compass set at a 1/16-inch radius.

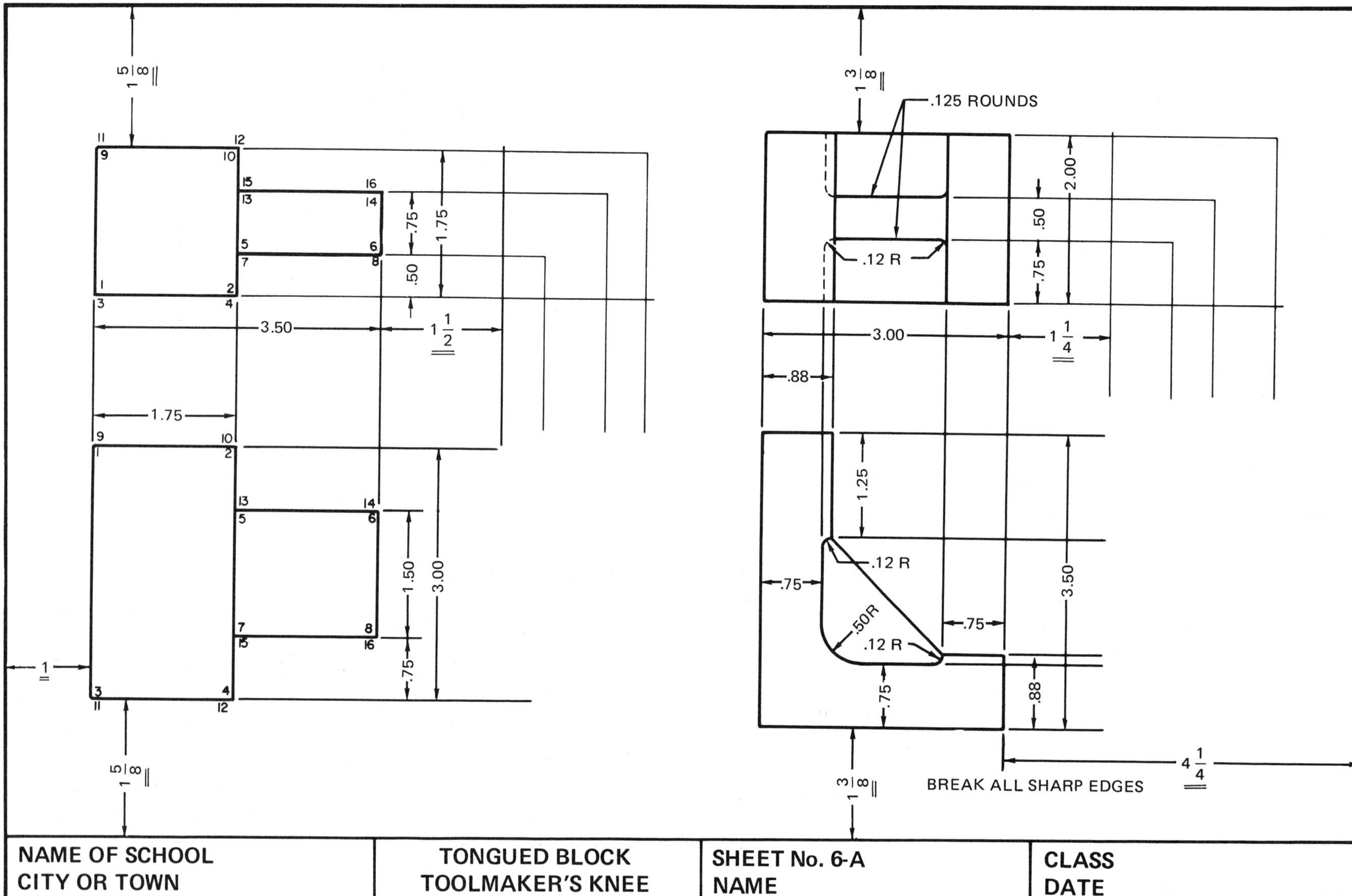
.125 ROUNDS
.12 R
.50R
BREAK ALL SHARP EDGES
3.50
1.75
3.00
.88
.75
.50
2.00
1.25
1.50
3.50
.88
NAME OF SCHOOL
CITY OR TOWN
TONGUED BLOCK
TOOLMAKER'S KNEE
SHEET No. 6-A
NAME
CLASS
DATE

Fig. 6-5

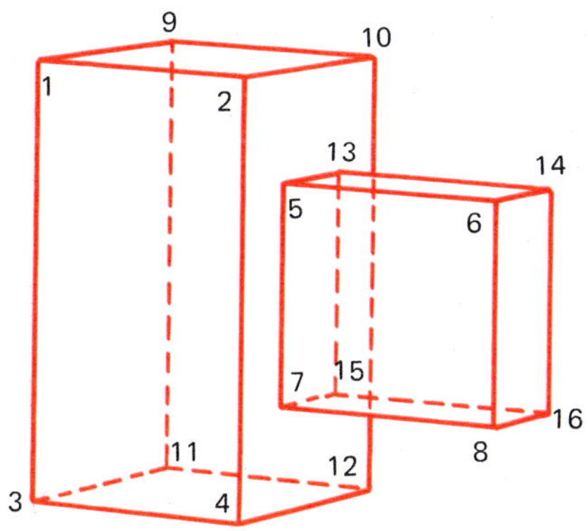

Fig. 6-6 Tongued block

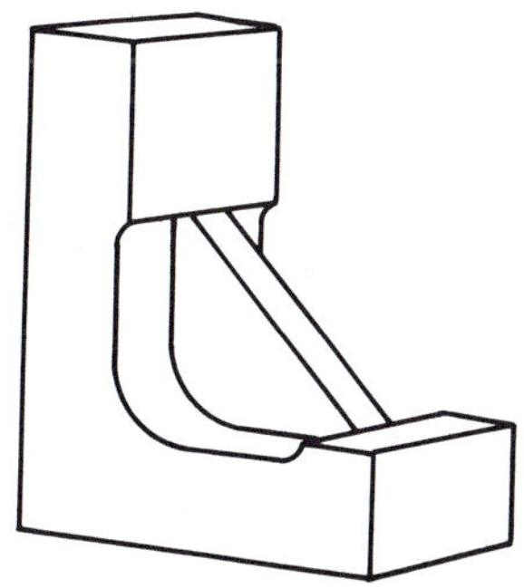

Fig. 6-7 Toolmaker's knee

ASSIGNMENT A

1. Draw three views of the tongued block.
2. Draw three views of the toolmaker's knee.

NOTE: The dimensions underlined on the drawing locate the views on the sheet. They are not to be placed on your drawing.

Procedure

A. Tongued block

1. Lay out the margin lines on the drawing sheet and letter the title strip.
2. Draw the front view of the tongued block.
3. Draw light construction lines up from the corners of the front view. Draw the top view.
4. Number the corners of the front view. Two numbers appear at each corner of the front view because the corners on the back of the tongued block are directly behind the corners on the front.
5. Place the same numbers on each of the corresponding corners in the top view.
6. Draw a horizontal line to the right from the lower edge of the top view. Draw a vertical line 1 1/2 inches to the right of the top view. These two lines form a right angle.
7. From each corner of the front view, draw a horizontal construction line to the right.
8. From each corner of the top view, draw a horizontal construction line to the right as far as the side of the right angle. Place the needle point of the compass at the vertex of the right angle and extend the construction line through an arc of 90 degrees. Continue the construction lines vertically until they intersect the horizontal construction lines.
9. Using the method illustrated in figure 6-2, locate the numbers for the corners of the side view. Draw the side view.
10. Dimension the views in decimal inches to two places, omitting the underlined dimensions. Check the drawing for placement and completeness of views, dimensions, and line weights.

B. Toolmaker's knee

1. Draw the front view of the toolmaker's knee.
2. Draw the top view.
3. Draw the right angle. Draw construction lines from the front and top views. Complete the side view. Locate the side view without using numbers.
4. Dimension the views in decimal inches to two places, omitting the underlined dimensions. Check both drawings and correct errors.
5. Clean the drawing.

REVIEW QUESTIONS

1. What is meant by the shop note "break all sharp edges"?
2. Explain how the numbers are located in the side view of the tongued block.

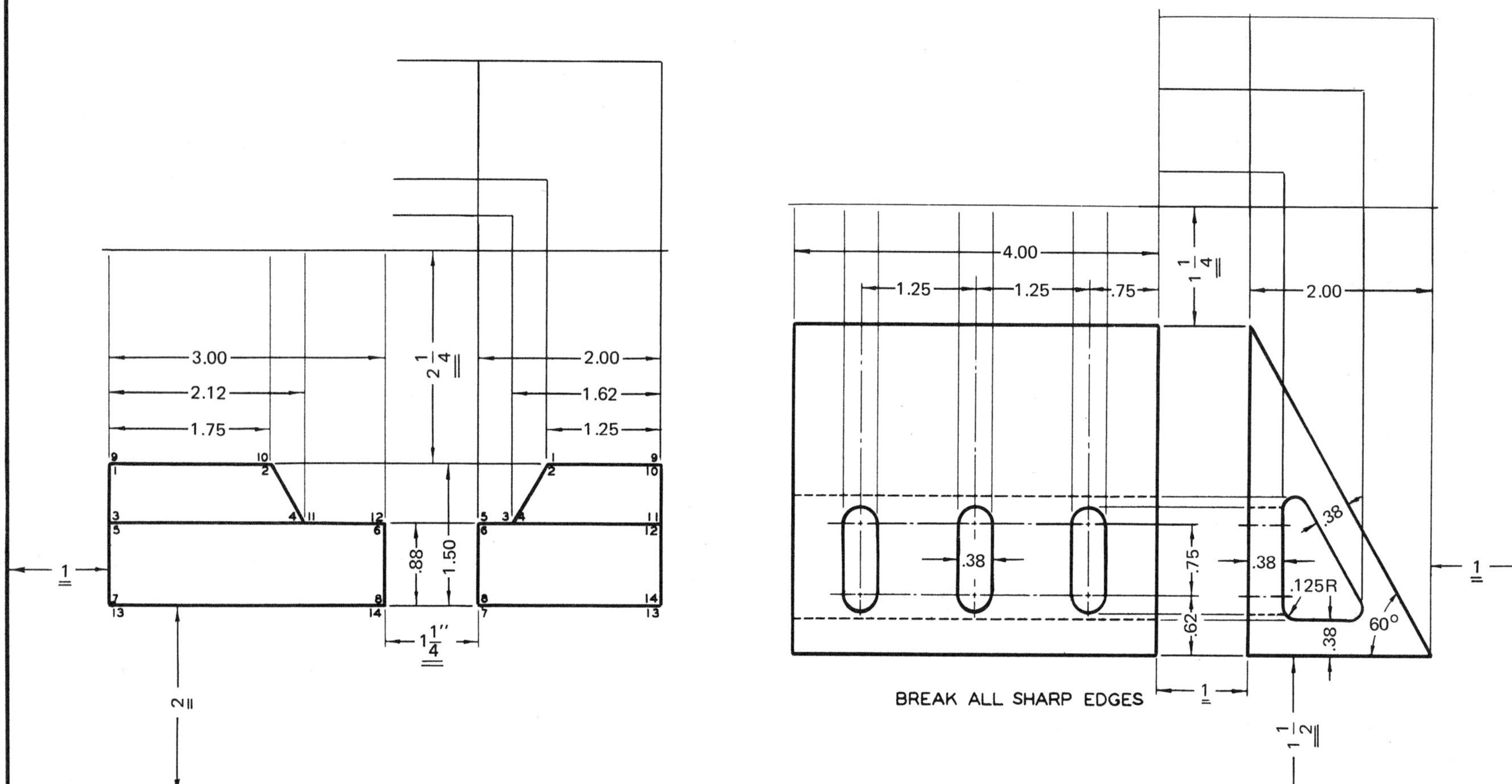

NAME OF SCHOOL CITY OR TOWN	BLOCK ANGLE ATTACHMENT	SHEET No. 6-B NAME	CLASS DATE

Fig. 6-8

MISSING TOP VIEWS

The top front view and the side view of a block are shown in figure 6-8. A pictorial drawing of the block is shown in figure 6-9. This block has fourteen corners. A number is placed in the front view for each of these corners. The same numbers are placed at the corresponding corners in the side view. The numbers for the corners of the missing top view are found by the method illustrated in figure 6-4.

The front view and the side view of the angle attachment, figure 6-10, are also given in figure 6-8. Locate the missing top view by drawing construction lines from the front and the side view.

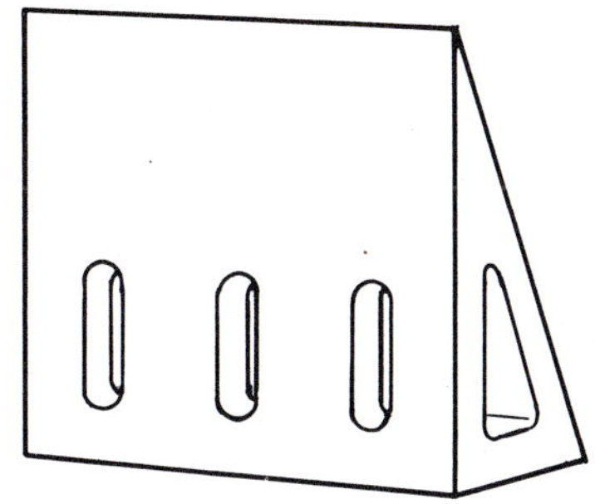

Fig. 6-10 Angle attachment

ASSIGNMENT B

1. Draw a three-view working drawing of the block, figure 6-9.
2. Make a three-view working drawing of the angle attachment, figure 6-10.

Procedure

A. Block

1. Lay out the margin lines on the drawing sheet and letter the title strip.
2. Draw the front view of the block.

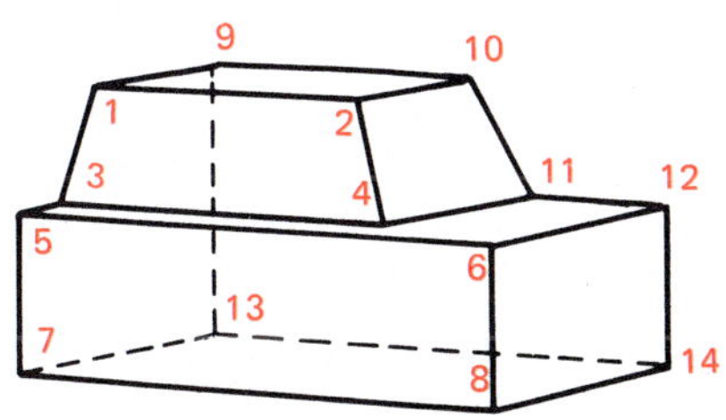

Fig. 6-9 Block

3. Draw the side view.
4. Place a number in the front view for each corner of the block.
5. Place the same numbers at the corresponding corners in the side view.
6. Draw a horizontal line 2 1/4 inches above the front and side views.
7. Draw a vertical line up from the left side of the side view. These two lines form a right angle above the side view.
8. Draw a construction line up from each corner of the front view.
9. From each corner of the side view draw a vertical line up to the side of the right angle. Continue these lines through an arc of 90 degrees. Extend the lines horizontally to the left until they intersect the vertical construction lines.
10. Locate the numbers for the corners of the top view. Draw the top view.
11. Dual-dimension the views omitting the underlined dimensions.
12. Check the drawing for accuracy of projection and dimensioning. Correct any errors.

B. Angle attachment

1. Draw the side view of the angle attachment.
2. Draw horizontal construction lines to the left. Draw the front view.
3. From the holes in the front view, draw horizontal construction lines to the right. Add the outlines of the holes and the center lines in the side view.
4. Draw the right angle 1 1/4 inches above the side view. Draw construction lines from the front view and side view. Complete the top view.
5. Dual-dimension the views omitting the underlined dimensions.
6. Check both drawings and correct errors.
7. Clean the drawing.

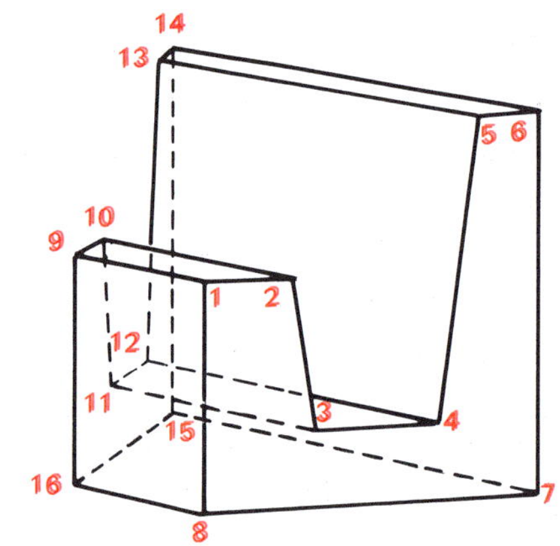

Fig. 6-11 Channel

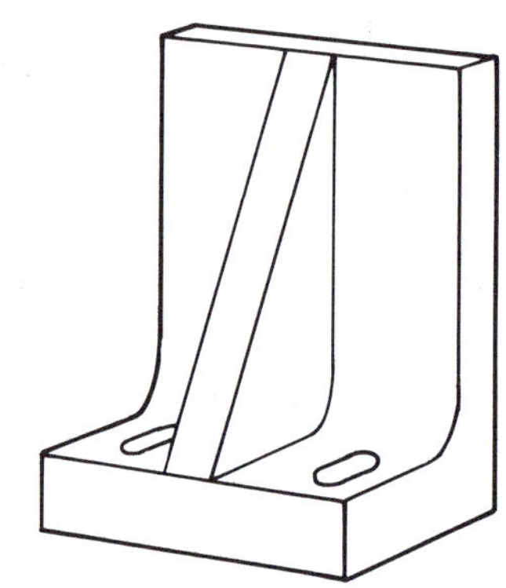

Fig. 6-12 Slotted angle plate

REVIEW QUESTION

Explain how to locate the numbers in the top view of the block.

MISSING FRONT VIEWS

The working drawing, figure 6-13, contains a side and top view of a part called a channel. A pictorial drawing of the channel is shown in figure 6-11. It has sixteen corners. A number is placed in the side view for each of these corners. The same numbers are placed at corresponding corners in the top view. The numbers for the corners of the missing front view are located by the method previously illustrated in figure 6-4.

A side view and top view of a slotted angle plate, figure 6-12, are also given in the working drawing. Locate the missing front view by drawing lines from the two given views using numbers for the corners.

ASSIGNMENT C

1. Draw a three-view working drawing of the channel, figure 6-11.
2. Make a three-view working drawing of the slotted angle plate, figure 6-12.

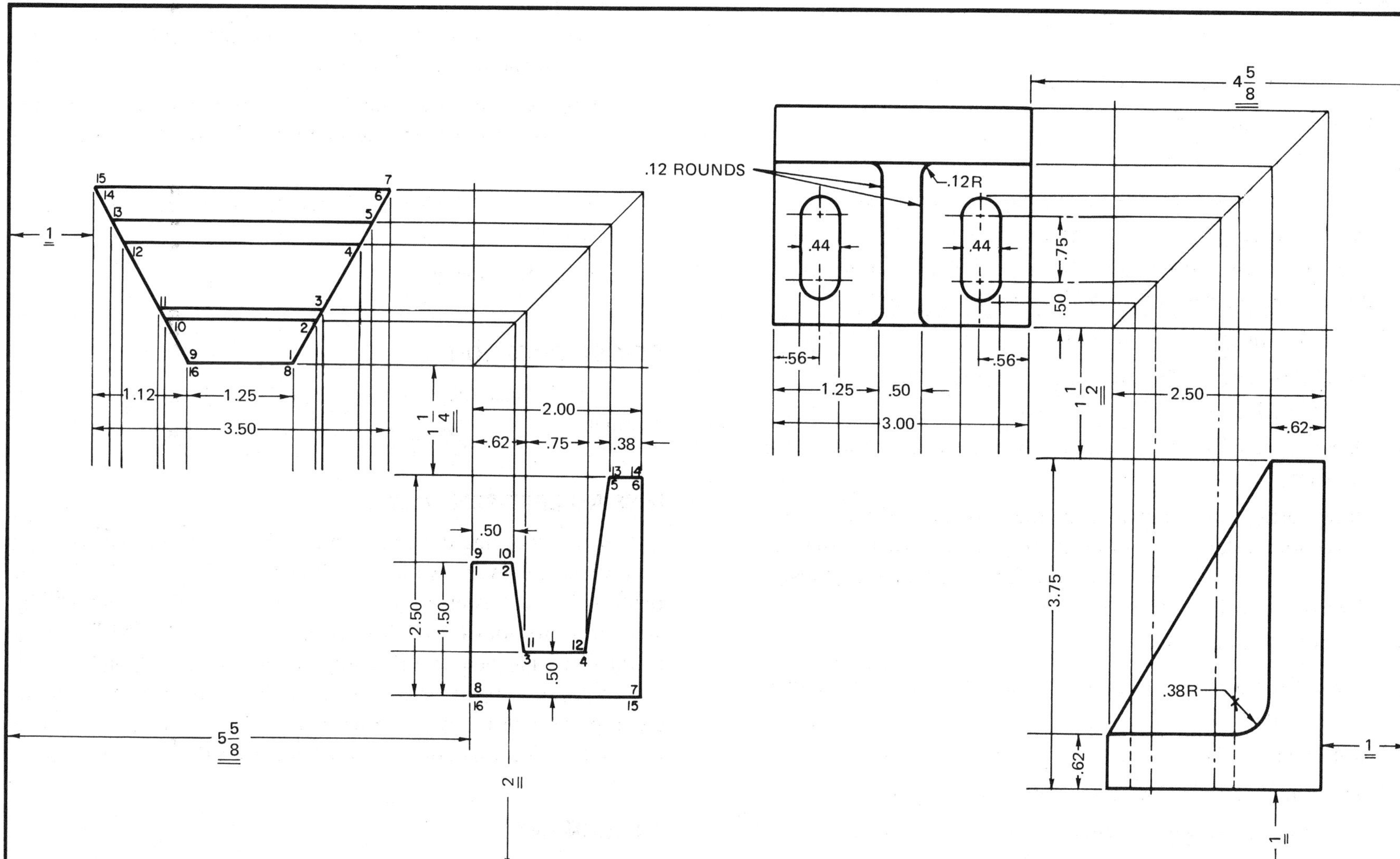

NAME OF SCHOOL CITY OR TOWN	CHANNEL SLOTTING ANGLE PLATE	SHEET No. 6-C NAME	CLASS DATE

Fig. 6-13

Procedure

A. Channel

1. Lay out the margin lines on your drawing sheet.
2. Letter the title strip.
3. Draw the right-side view of the channel.
4. Draw the right angle 1 1/4 inches above the side view.
5. From the corners of the side view, draw vertical construction lines up to the side of the right angle, through an arc of 90 degrees, and continue them horizontally to the left.
6. Draw the top view.
7. Place a number in the side view for each numbered corner of the channel.
8. Place the same number on each corresponding corner in the top view.
9. Draw vertical lines down from the corners of the top view. Draw horizontal lines to the left from each corner in the side view. Locate the corners of the front view at the intersections of these lines. Draw the front view.
10. Dual-dimension the views.
11. Check the accuracy of each view, line, and numbered corner of the channel.

B. Slotted angle plate

1. Draw the right-side view of the slotted angle plate.
2. Draw the right angle 1 1/2 inches above the side view.
3. From each corner of the side view draw a vertical line up to the side of the right angle. Continue these lines through an arc of 90 degrees and extend them horizontally to the left.
4. Draw the top view.
5. From one of the slots in the top view, draw construction lines back to the side view. Draw the outlines and center lines of the slots in the side view.
6. Draw vertical lines down from the top view, and horizontal lines to the left from the side view. Draw the missing front view.
7. Dimension the views. Check the completed drawing for accuracy.
8. Clean the drawing.

REVIEW QUESTION

Explain how the numbers are located in the front view of the channel.

MISSING LEFT-SIDE VIEWS

In the working drawing, figure 6-14, the front and top view of a wedge are given. A pictorial drawing of the wedge is shown in figure 6-15. It has eight corners. A number is placed in the front and top views for each of these corners. Locate the corners of the missing left-side view by using construction lines and numbers.

A top and front view of the universal right angle, figure 6-16, are also given in the working drawing. The missing left-side view is found by drawing construction lines from the two given views.

ASSIGNMENT D

1. Draw the front, top, and left-side views of the wedge, figure 6-15.
2. Draw the top, front, and left-side view of the universal right angle, figure 6-16.

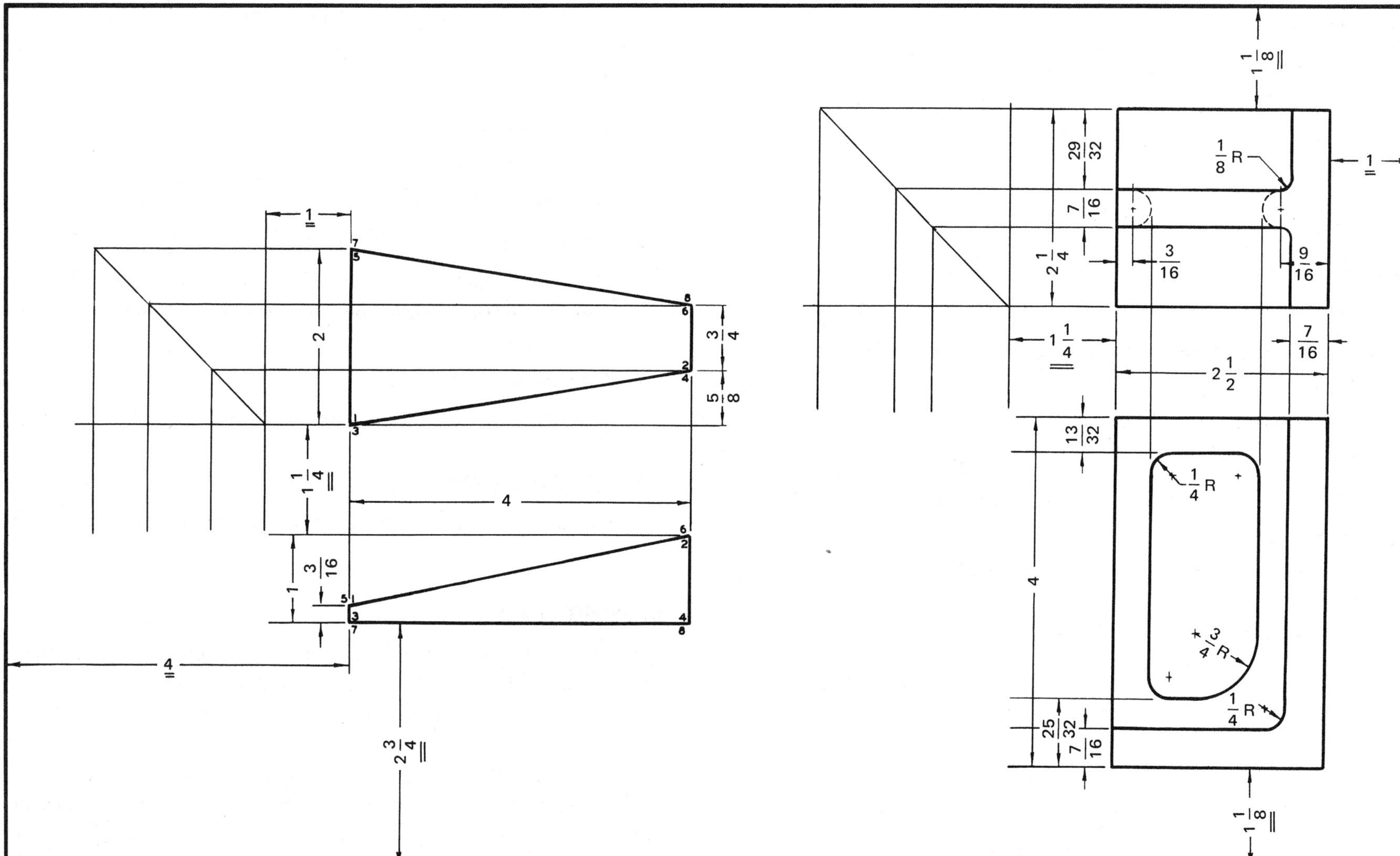

NAME OF SCHOOL CITY OR TOWN	WEDGE UNIVERSAL RIGHT ANGLE	SHEET No. 6-D NAME	CLASS DATE

Fig. 6-14

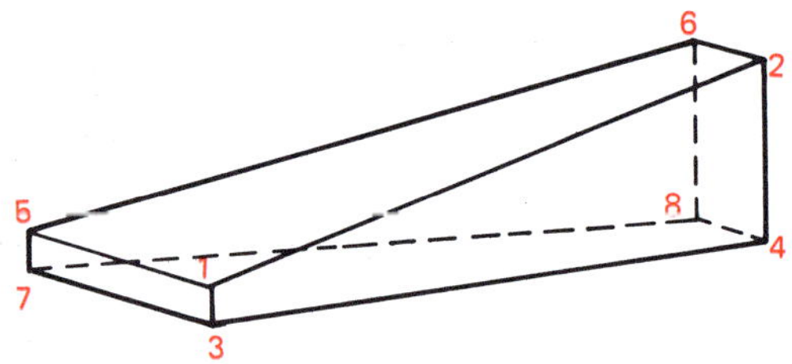

Fig. 6-15 Wedge

Procedure

A. Wedge

1. Lay out the margin lines on the drawing sheet and letter the title strip.
2. Draw the front view of the wedge.
3. Draw the top view.
4. In the front view place a number for each corner of the wedge.
5. Place the same numbers at the corresponding corners in the top view.
6. One inch to the left of the top view, draw a right angle.

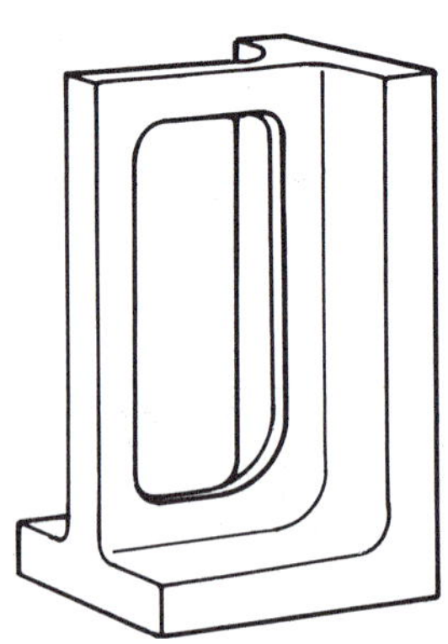

Fig. 6-16 Universal right angle

7. Draw construction lines from the two given views to locate the corners of the left-side view. Draw the left-side view.
8. Dimension the views. Check the accuracy of all lines and views.

B. Universal right angle

1. Draw the top view of the universal right angle.
2. Draw the front view.
3. Draw a right angle 1 1/4 inches to the left of the top view.
4. Draw construction lines from the two given views and complete the left-side view. Be sure to draw the fillets and rounds in this view.
5. Dimension the views.
6. Check the drawings for accuracy of details.

REVIEW QUESTION

Explain how the numbers are located in the left-side view of the wedge.

MISSING LINES

In figure 6-17, several lines have been omitted intentionally from each of the views. Locate the missing lines in each view by drawing construction lines from the corners or the sides of the circle in the other views. For example, in the first drawing, a 1 1/4-inch square hole is shown in the top view. The lines representing the edges of this hole are shown in the side view, but they are missing in the front view. To locate these missing lines, draw construction lines down from the corners of the hole in the top view. Other missing lines are found in the same manner.

In order to complete the views of the second drawing, it may be necessary to place a number for each corner of the object in one view and locate the corresponding corners in the other views. The corners are then connected with lines to complete the views.

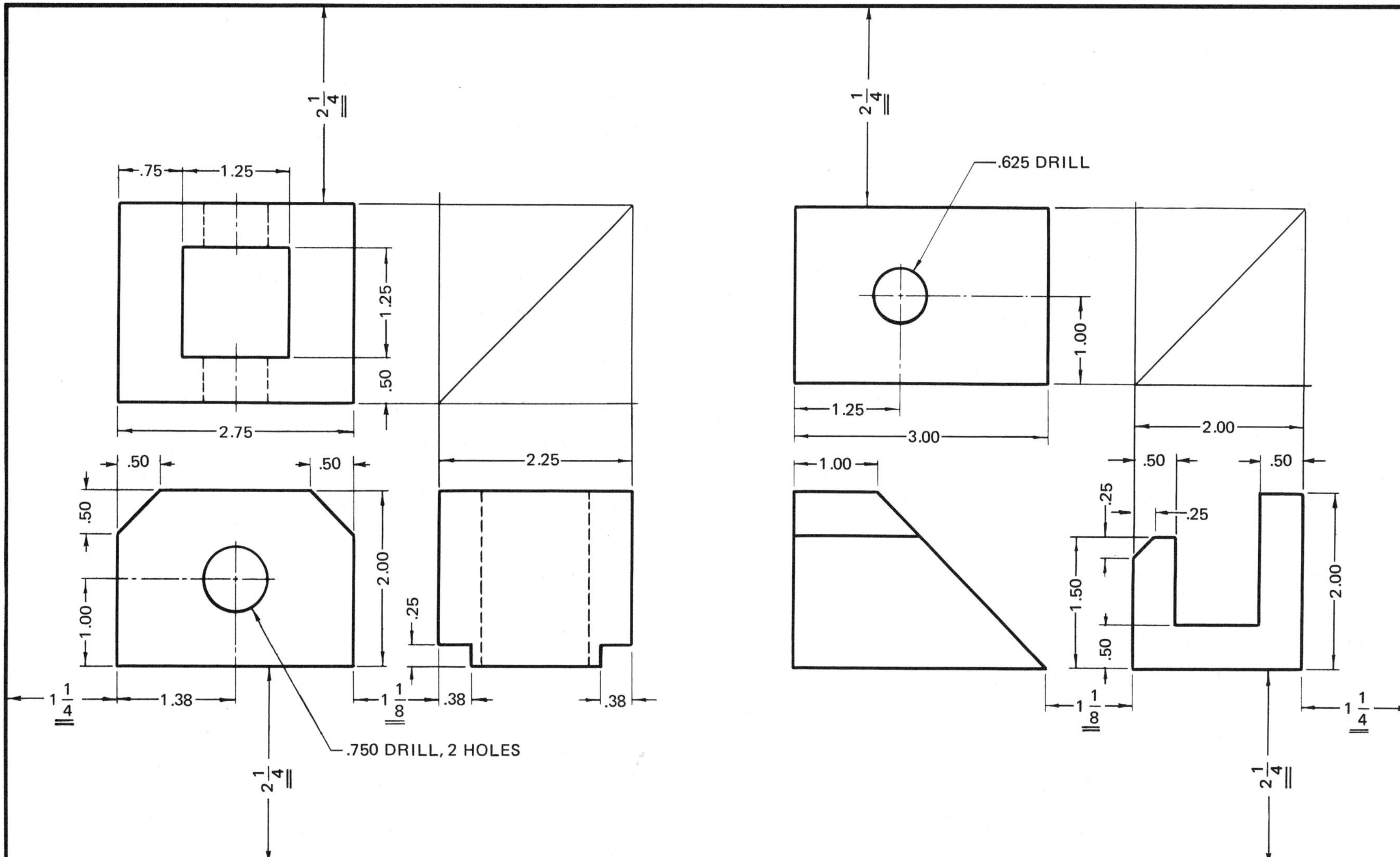

NAME OF SCHOOL CITY OR TOWN	MISSING-LINES PROBLEMS	SHEET No. 6-E NAME	CLASS DATE

Fig. 6-17

ASSIGNMENT E

Draw the two three-view drawings in figure 6-17 and supply the missing lines in each view.

Procedure

1. Lay out the drawing sheet and letter the title strip.
2. Draw the three views of the first object.
3. Draw construction lines as on the preceding drawings. Locate and draw the missing lines in each view.
4. Dimension the views in decimal inches.
5. Draw the three views of the second object.
6. Place a number in the side view for each corner of the block. There are eighteen corners.
7. Draw construction lines to the left from each corner of the side view. Draw the missing lines in the front view.
8. Number the corners in the front view.
9. Draw construction lines from the numbered corners in the front and side views to locate the corners in the top view. Draw the missing lines in the top view.
10. Draw the circle for the .62 inch drilled hole in the top view. Draw the outlines of this hole in the front and side views. The hole goes through to the bottom of the object.
11. Dimension the views in decimal inches.
12. Check the accuracy of the drawing.

INTERSECTING ROUNDS

Figure 6-19 is a dimensioned pictorial drawing of a pedestal bearing. Make a three-view working drawing. Dimension the completed working drawing according to standard dimensioning practice.

The upper edges of the pedestal bearing base are intersecting rounds. Where these rounds meet at the corners, a curved edge is formed. In the top view, these curved edges appear as straight lines which slope at 45 degrees, figure 6-18.

The pedestal bearing also has some fillets which meet to form similar curved edges. These curved edges appear as straight sloping lines in the top view, figure 6-20.

ASSIGNMENT F

Draw a three-view working drawing of the pedestal bearing, figure 6-19. Draw the front, right side, and top views.

Procedure

1. Lay out the drawing sheet and letter the title strip.
2. The front view shows the general shape of the part, so it is drawn first. Draw the front view 1 1/2 inches above the title strip and 3 1/4 inches from the left margin line.
3. Draw the right-side view 3 1/4 inches from the right margin line.

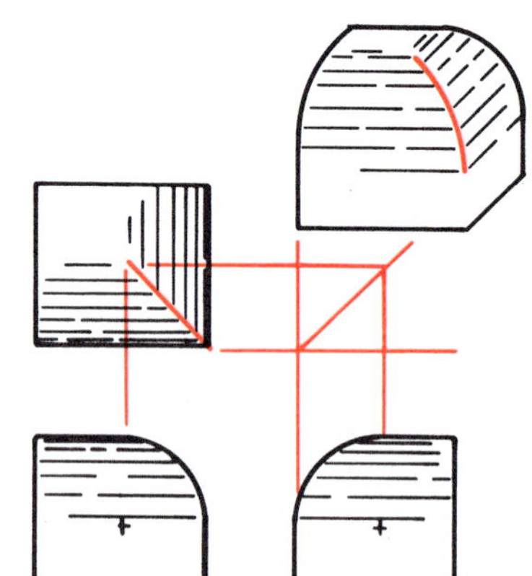

Fig. 6-18 Intersecting rounds

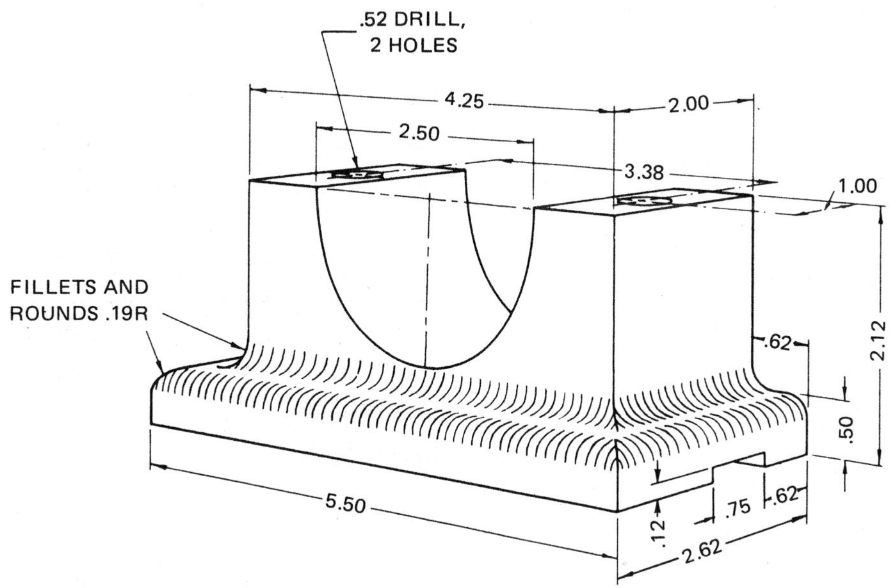

NAME OF SCHOOL CITY OR TOWN	PEDESTAL BEARING	SHEET No. 6-F NAME	CLASS DATE

Fig. 6-19

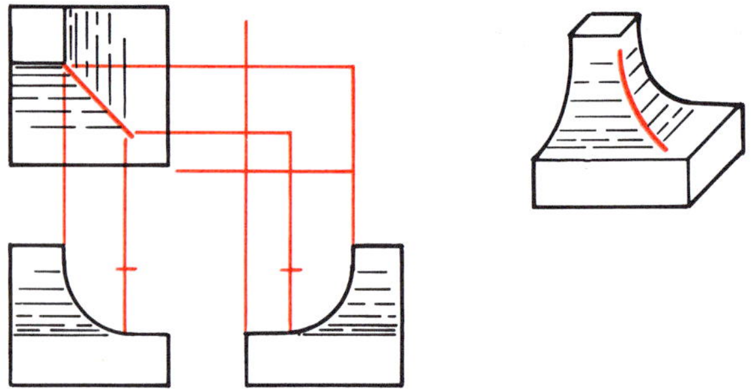

Fig. 6-20 Intersecting fillets

4. Draw the top view 1 1/2 inches below the top margin line. Be sure to draw the 45 degree lines representing the intersections of the rounds and fillets.
5. In the top view, draw the center lines. Draw the circles for the top of the two drilled holes.
6. Draw center lines and outlines of these drilled holes in the front and side views. These holes go through to the slot in the bottom of the bearing.
7. Draw the hidden edges of the .12-inch-by-.75-inch slot in the front and top views. This slot goes through the part.

Dimensioning the Drawing

The following instructions review the rules for dimensioning the drawing. Use dual-dimensioning procedures.

1. Draw dimension lines for the center lines of the circles under and to the right of the top view.
2. Draw dimension line for the diameter of the semicircular opening above the front view.
3. Draw dimension lines for the bearing lengths between the top and the front views.
4. Draw dimension lines for the widths of the bearing at the right of the top view and above the side view.
5. Draw dimension lines for the heights between the front and side views.
6. Draw dimension lines for the slot in the side view.
7. Dimension one of the fillets and one of the rounds in the front view.
8. Dimension one of the drilled holes with a shop note.
9. Check the accuracy of the drawing and correct any errors.

COUNTERBORED HOLES

A dimensioned pictorial drawing of a vise base is illustrated in figure 6-21. A three-view working drawing of the vise base is made from the pictorial drawing. The views are dimensioned according to standard dimensioning practice.

Two holes are drilled in the left side of the vise base. One end of each hole is enlarged with a counterboring tool. A counterbore is a larger, more accurate hole using the same center as the original hole.

CHAMFER

For safety and appearance the sharp upper left corner of the vise base has been cut away. The angular surface resulting from removal of a sharp edge is called a *chamfer*. Chamfers are usually dimensioned with a note stating the size and angle of the chamfer.

ASSIGNMENT G

Draw a three-view working drawing of the vise base, figure 6-21. Draw the front, right side, and top views.

Procedure

1. Lay out the drawing sheet and letter the title strip.
2. Draw the outlines of the front view 2 3/4 inches from the left margin line and 1 3/8 inches above the title strip.
3. Draw the outlines of the side view 2 3/4 inches from the right margin line.

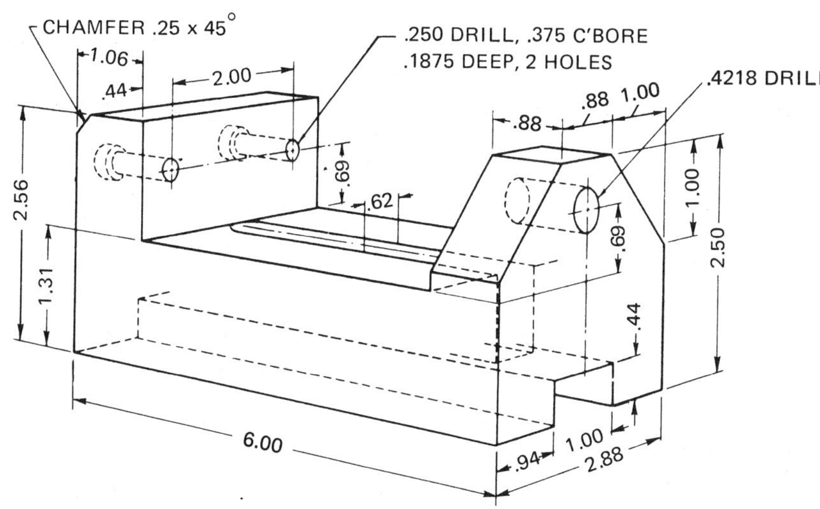

NAME OF SCHOOL CITY OR TOWN	VISE BASE	SHEET No. 6-G NAME	CLASS DATE

Fig. 6-21

4. Draw the outlines of the top view 1 3/8 inches below the top margin line.
5. Draw the hidden edges of the .44 inch by 1.00 inch slot in the front and top views.
6. Draw the .62 inch wide hole in the top view. Draw this hole in the front and side views.
7. In the side view, lay out the center lines for the three holes. Draw the circles representing these holes.
8. Draw these holes in the front and top views.
9. One corner of the front view is chamfered. Draw this chamfer in the top and side views.
10. Two corners are chamfered in the side view. Draw these chamfers in the front and top views.

Dimensioning the Drawing

Use dual-dimensioning procedures.

1. Draw the dimension lines for the lengths of the parts between the top and front views.
2. Draw the dimension lines for the depth at the right of the top view.
3. Draw the dimension lines for the heights at the left and right of the front view.
4. Draw the dimension line for the depth of the .62 inch hole within the hole in the top view.
5. Draw the dimension lines for the .44-inch-by-1.00-inch slot below and at the left of the side view.
6. Place the dimensions for the chamfer in the side view above and to the right of the side view.
7. The dimensions for the center lines of the circles are placed above and to the left of the side view.

   Note: In this case, the rules for dimensioning must be ignored.
8. Dimension the drilled and counterbored holes with a shop note.
9. Dimension the .4218 drilled hole with a shop note.
10. Dimension the chamfer in the front view with a shop note.
11. Check the drawing, giving special attention to dimensioning.
12. Erase extra lines and clean the drawing.

REVIEW QUESTION

Indicate the counterbored holes in the views on the drawing.

FINISHED SURFACES

The die shoe shown in figure 6-22 is a casting.

It has five finished surfaces: the top, the bottom, the tops of the two pads, and the pad on the front. *Finished surfaces* are surfaces that are machined flat to achieve certain dimensions or degrees of smoothness. These finished surfaces are indicated on the drawing with a V-shaped mark.

ASSIGNMENT H

Make a three-view working drawing of the die shoe, figure 6-22. Draw the top, front, and right-side views.

Procedure

1. Lay out your drawing sheet and letter the title strip.
2. Draw the horizontal center line of the top view 3 1/4 inches below the top margin line.
3. Draw a vertical center line 2 1/2 inches from the left margin line.
4. Complete the top view. The two pads are the same size.

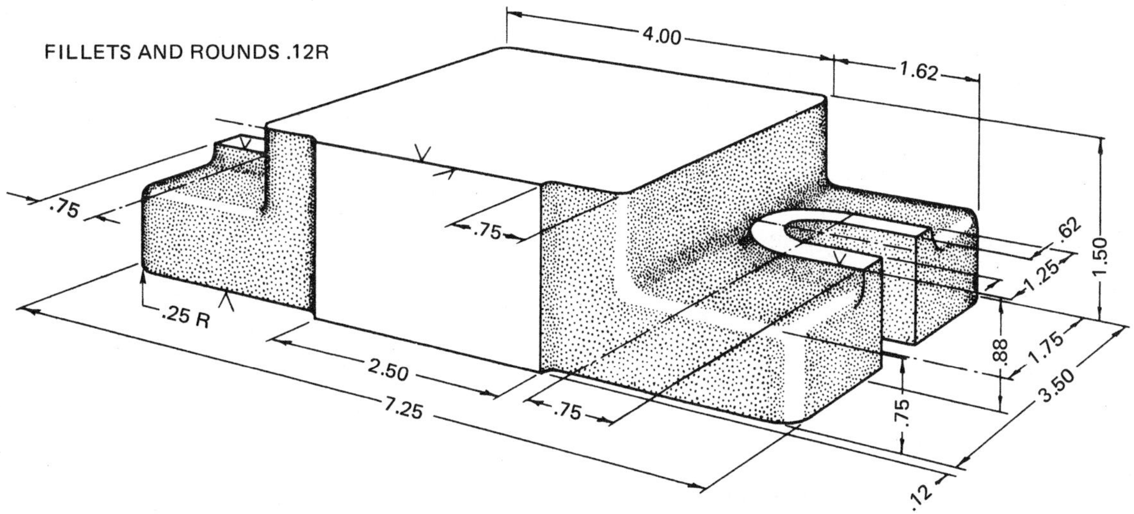

NAME OF SCHOOL CITY OR TOWN	DIE SHOE	SHEET No. 6-H NAME	CLASS DATE

Fig. 6-22

5. Draw the front view 1 3/8 inches above the title strip.
6. Draw the right-side view 1 3/4 inches from the right margin line. Notice that the fillets on each side of the pad run out at the bottom of the .12 inch round.

Dimensioning the Drawing

1. Draw the dimension lines for the vertical center lines, and the dimension lines for the lengths, between the top and front views.
2. Draw the dimension lines for the widths at the right of the top view.
3. Draw the dimension lines for the heights between the front and side views.
4. Dimension the fillets and rounds with a shop note.
5. Place a "V" finish mark on all of the lines representing finished surfaces.
6. Check each view of the drawing and correct any errors.

SHAFT SUPPORT

The shaft support, figure 6-23, is a casting that includes a cored hole and bosses that require special finishing. A *boss* is a circular raised portion above the surface of the casting. Its purpose is to provide a finished bearing surface around a hole. A *cored hole* is a hole deliberately formed in the casting to reduce later machining or drilling.

In order to finish a casting to the dimensions specified on the working drawing, some metal is cut away. The metal that is removed is called the *finish allowance.* Finish marks indicate where the patternmaker must add this allowance for finishing on the pattern. For example, on the shaft support, the bosses are .12 inch high. On the pattern, these bosses are approximately .25 inch high, allowing .12 inch for finishing the top surfaces. Also, the finished casting is to be 2.00 inches high. Allowing .12 inch finish allowance on the top surface of the pattern and .12 inch on the bottom surface makes the pattern 2.25 inches high.

The large hole in the casting, figure 6-24, is the cored hole. The cored hole in the shaft support must be bored to 1.75 inch diameter. To provide for boring, approximately .12 inch finish allowance is left all around in the hole. Cored holes, which are to be finished to accurate dimensions, must be marked with a finish mark on the drawing. Holes drilled from a solid casting do not require a finish allowance. Finish marks are not placed on these holes.

ASSIGNMENT I

Make a three-view working drawing of the shaft support, figure 6-23. Draw the top, front, and right-side views.

Procedure

1. Lay out the drawing sheet and letter the title strip.
2. Draw the horizontal center line for the top view 2 3/4 inches below the top margin line.
3. Draw the vertical center line 5 1/2 inches from the left margin line.
4. Draw vertical center lines for the two drilled holes. Complete the top view.
5. Draw the front view 1 1/4 inches above the title strip. Draw The vertical center line for the right-side view 4 1/4 inches from the right margin line and complete the side view.

Dimensioning the Drawing.

1. Draw the dimension lines locating the center lines of the two drilled holes beneath the top view.
2. Draw the dimension line for the depth (1.50 inch) at the right of the top view.
3. Draw the dimension line for the diameter of the base (3.50 inches) at the right of the top view. When the drawing shows

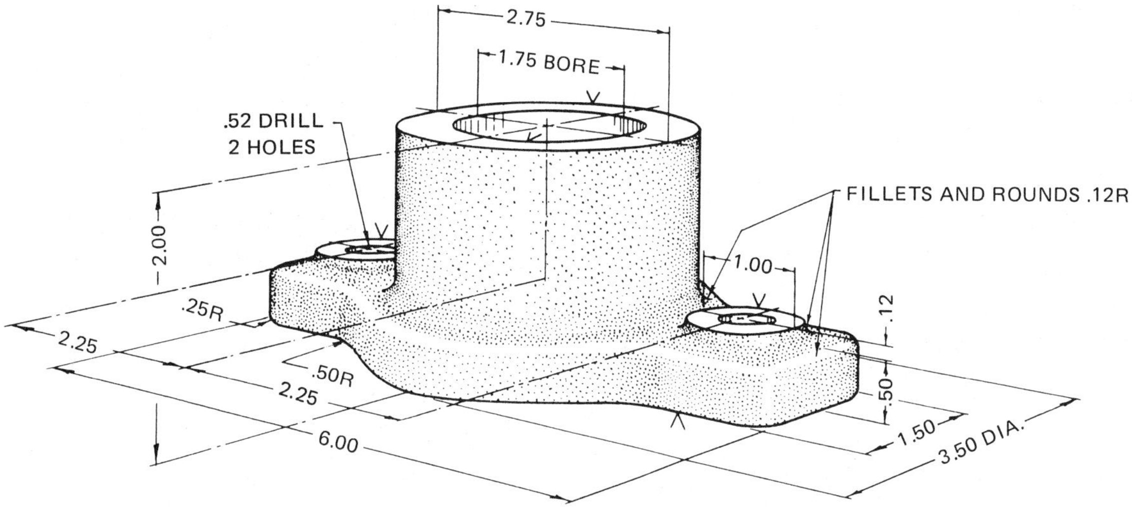

NAME OF SCHOOL CITY OR TOWN	SHAFT SUPPORT	SHEET No. 6-I NAME	CLASS DATE

Fig. 6-23

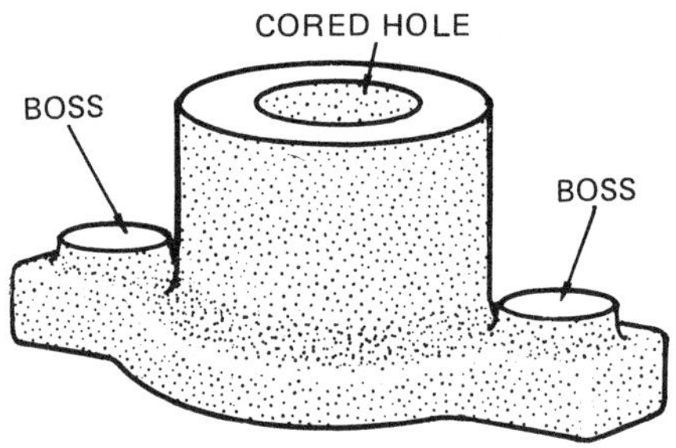

Fig. 6-24 Rough casting for shaft support

clearly that the dimension is for a diameter, the abbreviation D or DIA is omitted from the dimension.

4. Draw the dimension lines for the diameters, 1.75 inches, 2.75 inches, and 1.00 inch above the front view. Dimension the length (6.00 inches) between the top and front views.
5. Show dimensions for heights, .50 inch, .12 inch, and 2.00 inches between the front and side view.
6. Dimension the two drilled holes and the fillets and rounds with a shop note.
7. Add the .25 inch radius dimension at one corner.
8. Place finish marks on all of the lines representing finished surfaces. Only one finish mark is required in the cored hole in the front and side view.
9. Check the accuracy of each view and dimension on the drawing.

REVIEW QUESTIONS

1. What is the finish allowance?
2. Where is the finish allowance added to the pattern?
3. What is a cored hole?
4. How is a cored hole formed in a casting?
5. What is a boss?

DOVETAILS

The cross feed slide, figure 6-25, has a dovetail slide. *Dovetail slides* are used with dovetails in machines where a machine part must move in one direction only. The slide can move back and forth freely on the dovetail member, but its V-shaped sides prevent sidewise or up and down movement.

ASSIGNMENT J

Make a three-view working drawing; top, right side, and front views; of the cross feed slide.

Procedure

1. Lay out the drawing sheet and letter the title strip.
2. Draw the horizontal center line for the top view 3 1/2 inches below the top margin line.
3. Draw the vertical center line 3 7/8 inches from the left margin line.
4. Draw all of the top view except the hidden lines for the dovetail slide.
5. Draw the side view 2 1/2 inches from the right margin and 1 3/4 inches above the title strip.
6. Draw the front view. Complete the top view.
7. Dual-dimension the drawing. Place all the dimensions for the dovetail slide in the side view. Add finish marks.
8. Check each view and all dimensions on the drawing.

REVIEW QUESTIONS

1. Indicate the finished surfaces and the dovetail slide in each view.
2. What is the purpose of a dovetail and a dovetail slide?

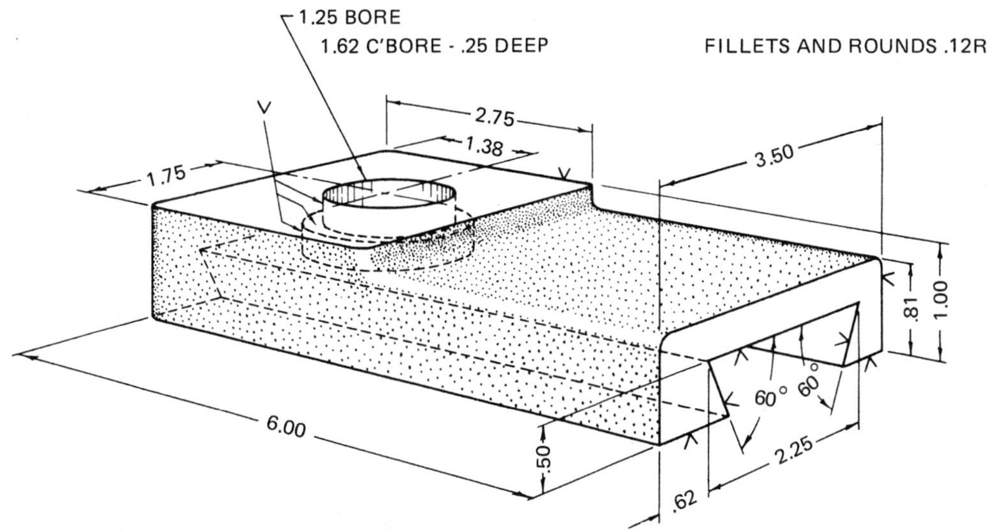

NAME OF SCHOOL CITY OR TOWN	CROSS FEED SLIDE	SHEET No. 6-J NAME	CLASS DATE

Fig. 6-25

DRILLED HOLES AND INTERSECTIONS

The casting of the gear arm, figure 6-27, has two cored holes. One hole is finished to a 1-inch diameter. The other is a long hole which is finished to a width of 3/4 inch. In the different views of the drawing, both of these holes must be marked with finish marks.

The small hole in the center of the boss on the right side is drilled with a letter F drill. An F drill is called a letter size drill. Letter size drills along with wire gauge size drills, (also called number drills) fall between the common fractional size drills. As an example, the F drill is .257-inch diameter. This size is between a 1/4-inch drill and a 17/64-inch drill.

The decimal size of letter drills is given when dimensioning. It is placed in the shop note following the letter or number, figure 6-27, or the decimal alone is given.

The shape of the upright portion of the gear arm is cylindrical. On the right side of the cylinder is a boss. The place where the boss and the cylinder are joined is called an *intersection*. The conventional methods for drawing intersections of cylinders with other parts are shown in figure 6-26. The conventional representations are used to save drafting time and simplify the drawing.

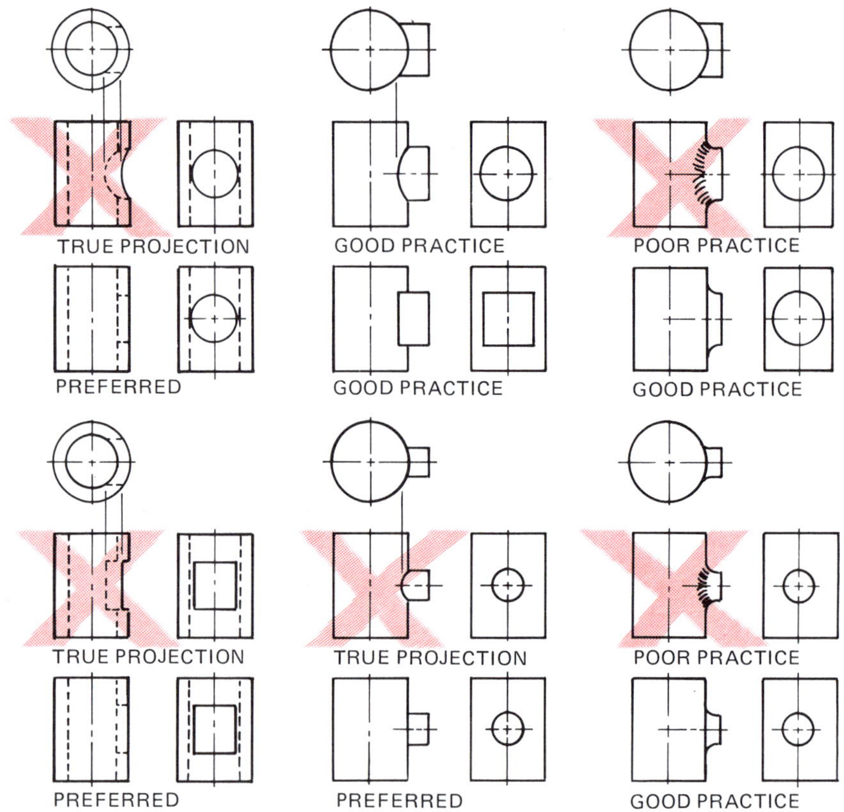

Fig. 6-26 Conventional representation of intersections

ASSIGNMENT K

Draw the top, right side, and front views of the gear arm.

Procedure

1. Lay out the drafting sheet and letter the title strip.
2. Draw the horizontal center line for the top view 3 inches below the top margin line.
3. Draw the first vertical center line 3 1/4 inches from the left margin line.
4. Draw the top view. To show the shape of the top view, a partial view is inserted in the drawing with figure 6-27.
5. Draw the front view 1 3/4 inches above the title strip.
6. Draw the side view 2 inches from the right margin line.
7. Dual-dimension the drawing.
8. Place finish marks on the lines representing finished surfaces.
9. Check all views and dimensions on your drawing for accuracy.

REVIEW QUESTIONS

1. What are letter size drills?
2. What are wire gauge size or number drills?

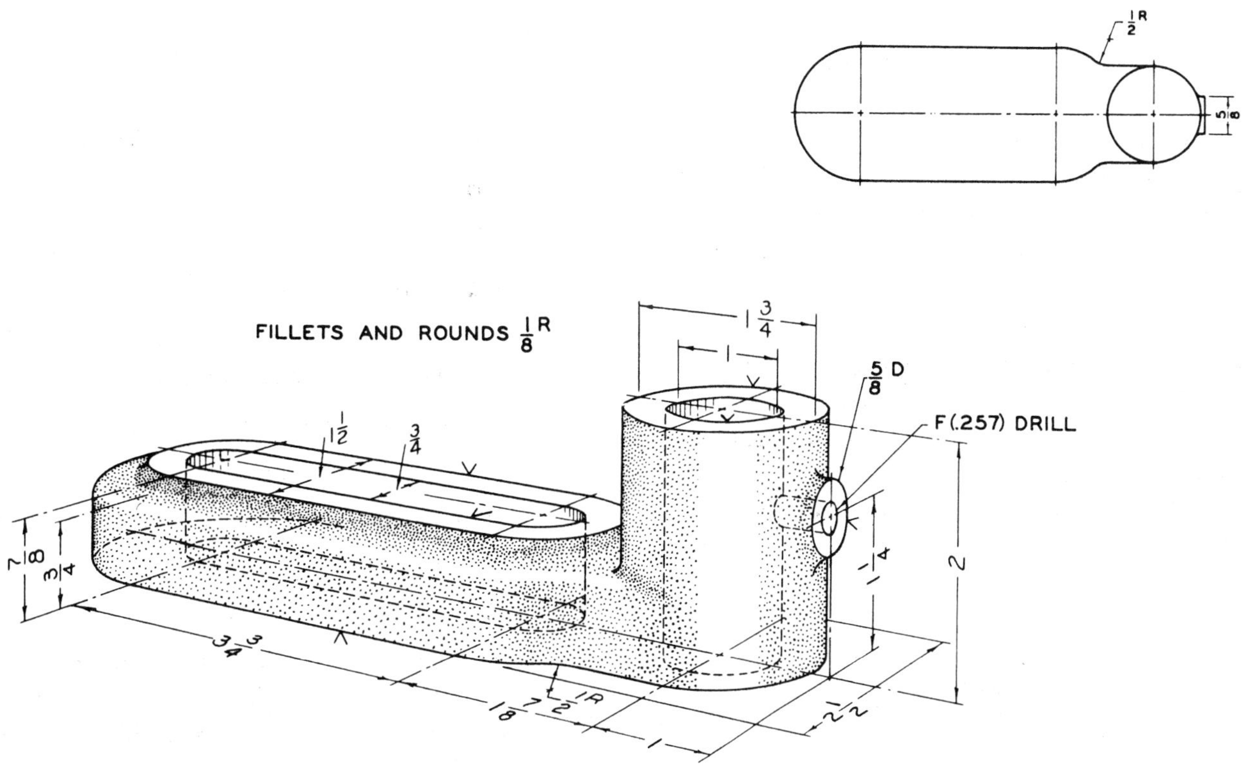

NAME OF SCHOOL CITY OR TOWN	GEAR ARM	SHEET No. 6-K NAME	CLASS DATE

Fig. 6-27

3. Indicate these terms in the views of the drawing: two cored holes, boss, pad, fillet, and round.

REAMED HOLES

The clapper box, figure 6-28, is a casting. All of the surfaces of this casting are finished. When all the surfaces of a casting are to be finished, the notation FINISH ALL OVER or the abbreviation FAO is used in place of the finish marks.

In the pictorial drawing, a piece of the clapper box is broken out to show the counterbored hole more clearly. The missing part is the same as the part directly behind it. It is not necessary to show this section broken out in the views on the drawing.

Three holes in the clapper box are reamed holes. *Reamed holes* are drilled holes machined to a more accurate dimension. One hole is reamed to a .373-inch diameter. This hole is a tight fit for a .375-inch diameter hinge pin that is pressed into the hole. The undersize hole prevents the standard size pin from coming out.

ASSIGNMENT L

Draw the top, front, and right-side views of the clapper box.

Procedure

1. Lay out your drawing sheet and letter the title strip.
2. Draw the horizontal center line for the top views 3 1/4 inches below the top margin line.
3. Draw the vertical center line 6 3/4 inches from the left margin line.
4. Draw the top view.
5. Draw the front view 1 3/8 inch above the title strip.
6. Draw the right-side view 2 1/2 inches from the right margin line.
7. Dual-dimension the drawing. Letter the note "FINISH ALL OVER."
8. Check all views and dimensions on the drawing for accuracy.
9. Erase all unnecessary lines.

REVIEW QUESTIONS

1. What does the note "FINISH ALL OVER" mean to the machinist? To the patternmaker?
2. What is a reamed hole?
3. Why are holes reamed?
4. Why is one of the holes reamed to a .373-inch diameter?

SYMMETRICAL PARTS

The floating punch holder, figure 6-29, is symmetrical. A symmetrical part corresponds in size, shape, and relative position to a part that is on the opposite side of a center line. Symmetrical parts are dimensioned from horizontal and vertical center lines. This method of dimensioning is shown in figure 6-29.

The shape of this T slot is somewhat different from the T slot on the last drawing, yet it serves the same purpose. Another part slides in the T slot, but it cannot be lifted out of the slot. The upper edges of the T slot are chamfered. *Chamfering* cuts the surfaces on a diagonal to remove sharp edges.

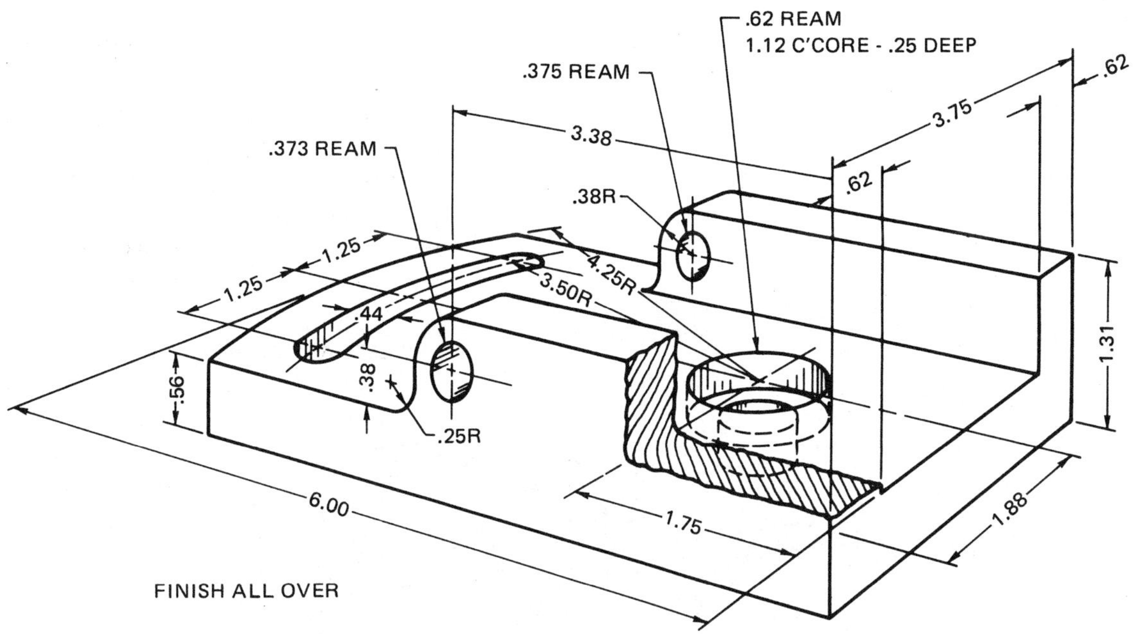

NAME OF SCHOOL CITY OR TOWN	CLAPPER BOX	SHEET No. 6-L NAME	CLASS DATE

Fig. 6-28

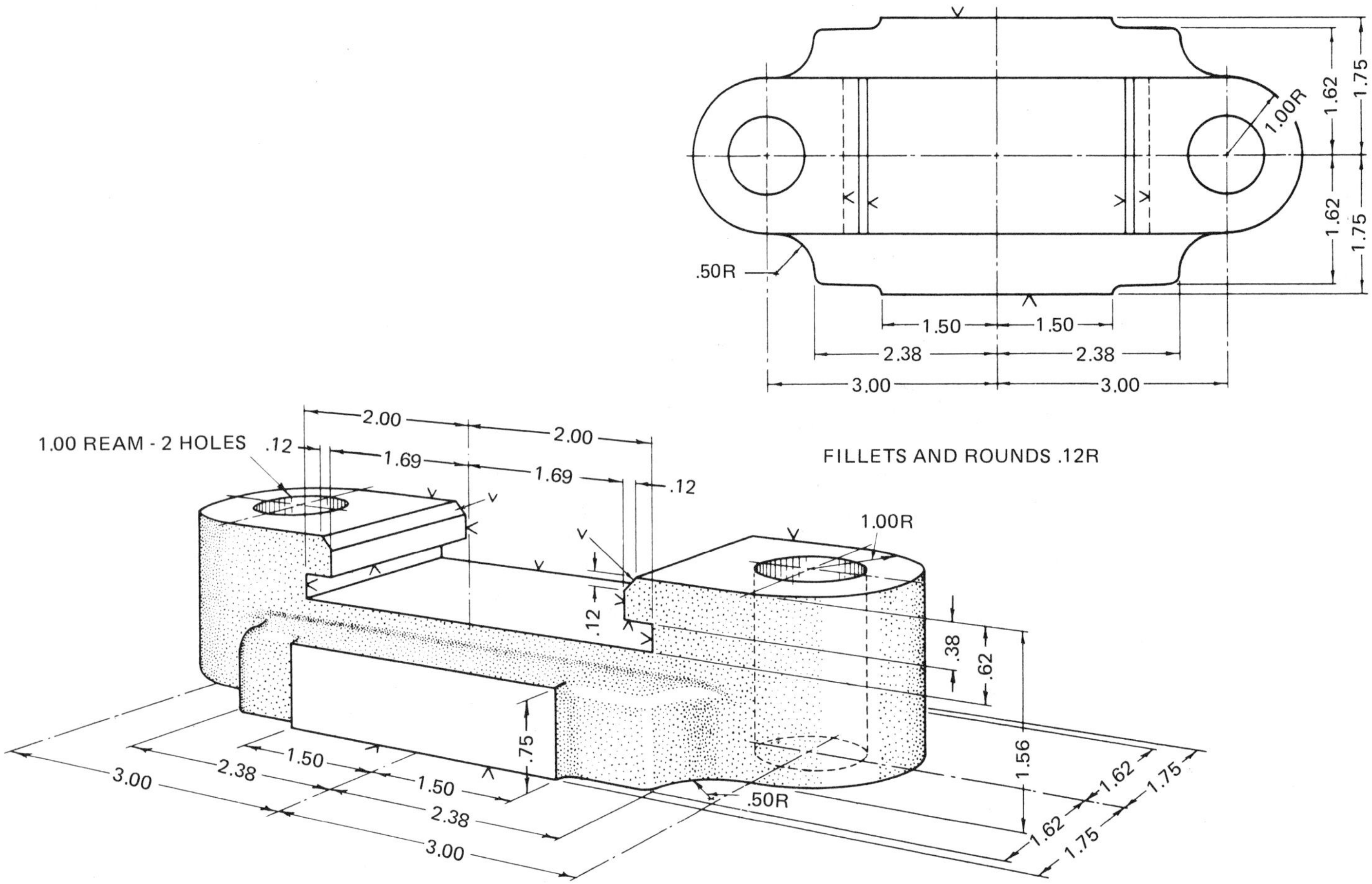

NAME OF SCHOOL CITY OR TOWN	FLOATING PUNCH HOLDER	SHEET No. 6-M NAME	CLASS DATE

Fig. 6-29

ASSIGNMENT M

Draw the top, front, and right-side views of the floating punch holder.

Procedure

1. Lay out the drawing sheet and letter the title strip.
2. Draw the horizontal center line for the top view 2 7/8 inches below the top margin line.
3. Draw the vertical center line 5 1/4 inches from the left margin line.
4. Draw the top view.
5. Draw the front view 1 1/4 inches above the title strip.
6. Draw the vertical center line for the right-side view 3 inches from the right margin line. Draw the side view.
7. The dimensions for the T slot are given in the front view. The dimensions for the lengths and widths are given from the center line in the top view. The dimensions for the heights are placed between the front and side views. Use dual-dimensioning procedures.
8. Place finish marks on all of the lines representing finished surfaces.
9. Letter the notes.
10. Check the accuracy of all lines, views, dimensions, and notes.

REVIEW QUESTIONS

1. How are symmetrical parts dimensioned?
2. What is a T slot?
3. What is the purpose of the T slot?
4. What is a chamfer?
5. What is a reamed hole?

LEFT-SIDE VIEW

In figure 6-30, the top and the front view of a crank are given. The left-side view is missing.

When three-view working drawings are made, it is common practice to draw the front, top, and right-side views. However, if some details of the part can be shown more clearly in the left-side view, then the left-side view should be drawn. The rule that determines the choice of views is to select the view which will have the fewest hidden details. The left-side view of the crank is preferred because it has fewer hidden lines than the right-side view.

SCALE DRAWINGS

A drawing of this crank, if made full size is too large for the 11-by-17-inch sheet. Therefore, the drawing is reduced in size so that it is one-half as large as the actual size of the crank. The ratio of relation between the size of the drawing and the size of the crank is indicated in the title strip by the notation .50 inch = 1.00 inch. This means that .50 inch on the drawing equals 1.00 inch on the crank.

The ratio between the drawing size and the object size is given on all drawings which are reduced or enlarged. The ratio is expressed in a way that indicates the scale used in making the drawing. The first number or fraction always refers to the drawing and the second figure refers to the object.

On scale drawings, the size of the numerals and arrowheads may, if necessary, be made slightly smaller than normal. The thickness of the lines, and the lengths of the dashes in hidden lines and center lines, may be slightly reduced. Regardless of the scale of the drawing, the dimensions must specify the actual size of the object.

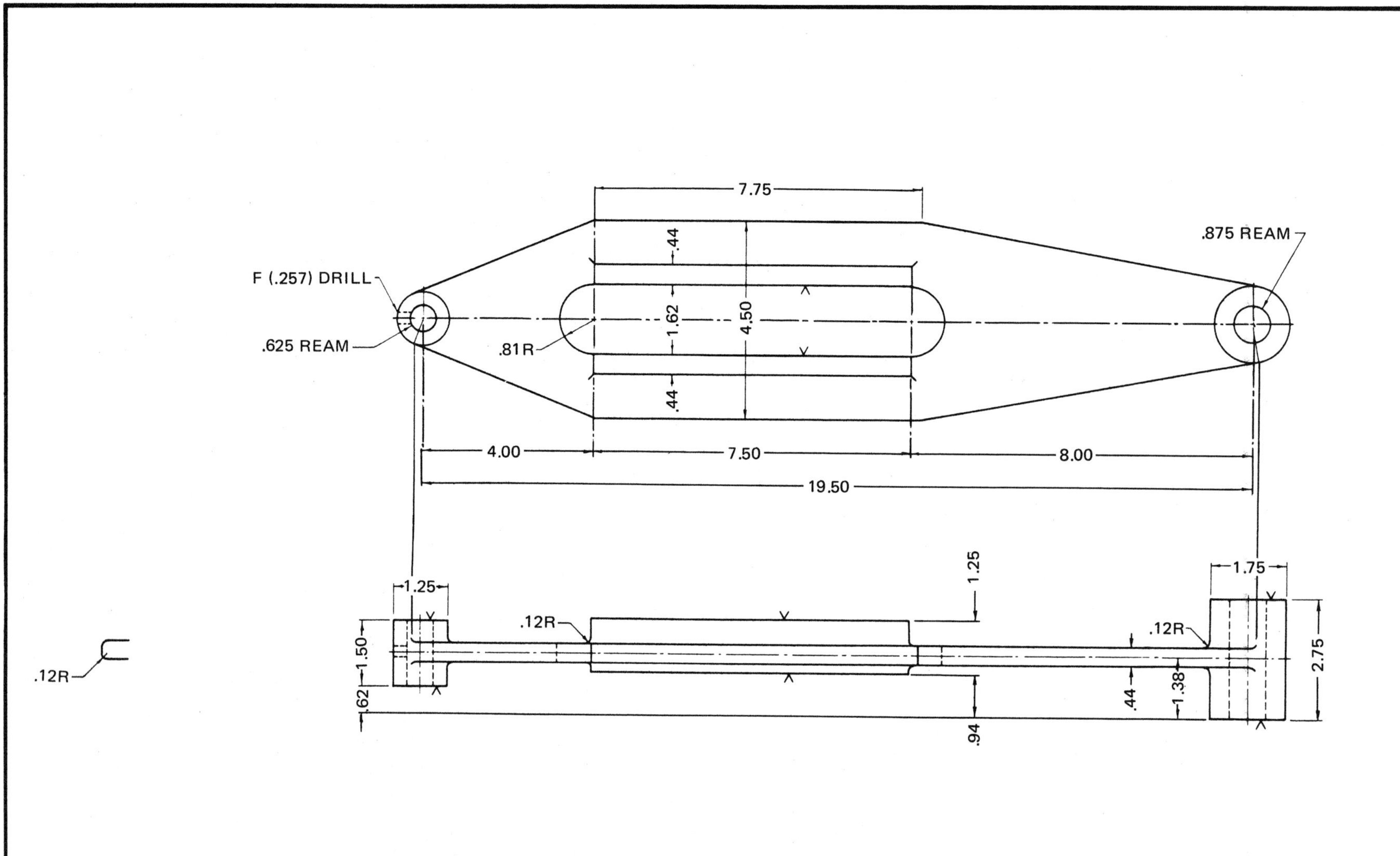

NAME OF SCHOOL CITY OR TOWN	CRANK SCALE: .50 IN. = 1.00 In.	SHEET No. 6-N NAME	CLASS DATE

Fig. 6-30

ASSIGNMENT N

Draw the top, front, and left-side views of the crank to a scale of .50 inch = 1.00 inch.

Procedure

1. Lay out the drawing sheet and letter the title strip.
2. Draw the horizontal center line for the top view 3 1/8 inches below the top margin line.
3. Draw the first vertical center line 4 3/4 inches from the left margin line.
4. Draw the top view.
5. Draw the front view 1 7/8 inches above the title strip.
6. Draw the vertical center line for the side view 2 1/8 inches from the left margin line.
7. Draw the side view beginning with the nearest visible detail and follow with those which lie behind and are partially hidden.
8. Dual-dimension the drawing. Avoid the tendency to over-crowd dimensions on scale drawings. Draw the dimension lines 3/8 inch from the views and leave 1/4 inch between parallel dimension lines.
9. Place finish marks on all lines representing finished surfaces.
10. Letter the notes.
11. Check and correct any errors in the lines, views, or dimensions on the drawing.

REVIEW QUESTIONS

1. Why was the left-side view preferred on this drawing?
2. What does the scale designation mean?
3. To what scale is this drawing made?
4. In which view or views is the crank symmetrical?
5. What is the .257-inch diameter drill called?

TWO-SIDE VIEWS

On this drawing, the top, front, right side, and left-side views of the feed slide, figure 6-31, are drawn. If only one side view is drawn, it would be quite complicated. Therefore, both side views are drawn, each view is a partial view because the details which are clearly described in the left-side view are omitted from the right-side view. Likewise, those details shown in the right-side view are omitted from the left-side view.

In this drawing, the dovetail slide is drawn in the right-side view. The semicircular upright portion is drawn in the left-side view. The circular base is drawn in both views.

Because four views are to be drawn on this sheet, the size of the drawing must be reduced to the scale indicated in the title strip. This scale is .75 inch = 1.00 inch.

DIMENSIONING DOVETAIL SLIDES

The dovetail slide shown in the right-side view, figure 6-32, is accurately measured with the aid of two rods of known diameter. These rods are placed in the dovetail slide. The distance between the rods is measured with gauge blocks. The dimension between the rods is given on the drawing. This dimension (X) is a decimal dimension and is calculated from a formula as follows:

$$\text{Dimension } X = B - D\left(1 + \text{cotangent } \frac{A}{2}\right)$$

Substitute values:

$B = 2.00$, $D = .38$, $A = 60°$;	$X = 2 - .38\left(1 + \text{cotangent } \frac{60°}{2}\right)$
Divide 60° by 2;	$X = 2 - .38\,(1 + \text{cotangent } 30°)$
Cotangent of 30° from table = 1.732	$X = 2 - .38\,(1 + 1.732)$
Add 1 and 1.732;	$X = 2 - .38 \times 2.732$
Multiply .38 by 2.732;	$X = 2 - 1.0382$
Subtract 1.0382 from 2;	$X = .9618$

FILLET RUN OUT

The top and front views of the feed slide are shown in figure 6-33. A .12 inch radius fillet goes partly around the circular base.

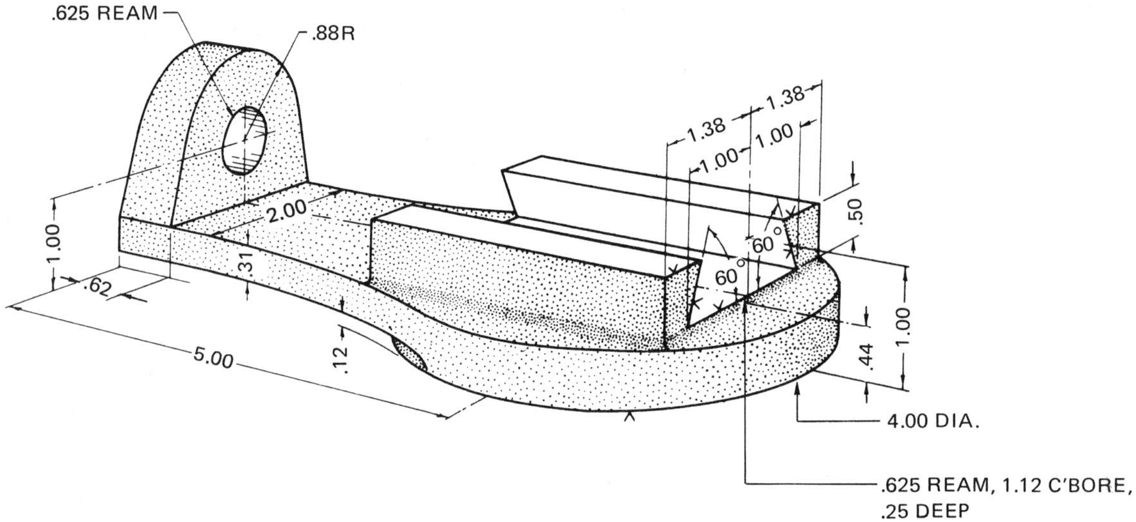

NAME OF SCHOOL CITY OR TOWN	FEED SLIDE SCALE: .75 IN. = 1.00 IN.	SHEET No. 6-O NAME	CLASS DATE

Fig. 6-31

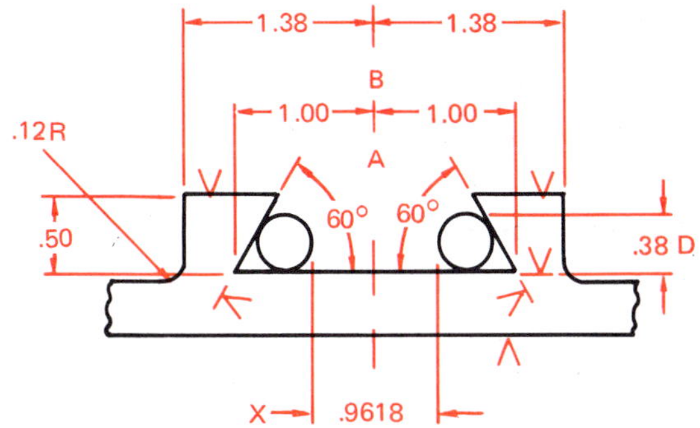

Fig. 6-32

This fillet runs out in the front view and in the left-side view. The construction lines in figure 6-33 suggest how the points are located. These points are used to locate the ends of the runouts in the left-side view.

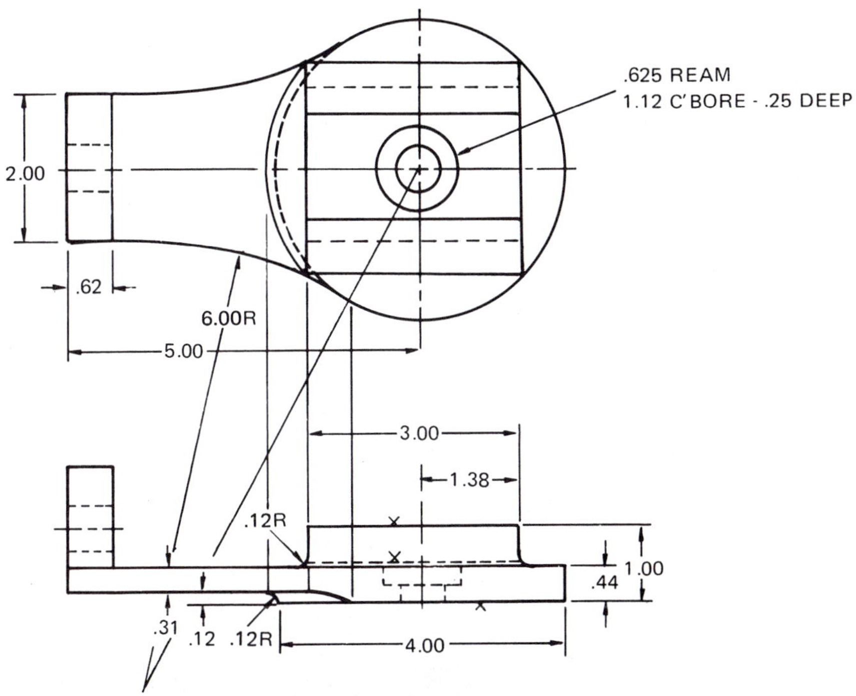

Fig. 6-33

ASSIGNMENT O

Draw the top, front, right side, and left-side views of the feed slide to a scale of .75 inch = 1.00 inch.

Procedure

1. Lay out the margin lines on the drawing sheet and letter the title strip.
2. Draw the horizontal center line for the top view 3 1/8 inch below the top margin line.
3. Draw the vertical center line 7 inches from the right margin line.
4. Draw the outlines of the top view. Each measurement must be 3/4 inch of the actual size of the part.
5. Draw the vertical center line for the right side view 2 5/8 inch from the right margin line. Draw the right-side view 1 5/8 inches above the title strip. In this view, draw only the circular base and the dovetail slide.
6. Draw the center line for the left-side view 2 5/8 inches from the left margin line. Draw the circular base and the upright portion in this view. Draw the runouts of the fillet.
7. Draw the front view and show the runout of the fillet.
8. Complete the top view.
9. Dual-dimension the views. In the left-side view, dimension the .88-inch radius and the .625-inch reamed hole.
10. Place finish marks on all of the lines representing finished surfaces. Some of the finish marks are placed on the extension lines to avoid overcrowding.
11. Check and correct any inaccurately drawn lines, views, or dimensions on the drawing.

REVIEW QUESTIONS

1. For what reason are both side views drawn?

2. Is each side view a complete view or a partial view? Why?
3. Why do you think the scale .75 inch = 1.00 inch is an unusual or odd size for a drawing?
4. Why is the finish mark placed on the extension lines in some places?
5. Using the formula for dovetail slides, find dimension X when distance B is 2.50 inches, angle A is 60 degrees, and D (the diameter of the rods) is .50 inch.

ALTERNATE POSITION OF THE SIDE VIEW

On working drawings, the normal position of the side view is at the right or at the left of the front view. An alternate arrangement of views may be used when the shape and size of a part is such that the three views, in their normal positions, cannot fit into the space on the sheet. When this occurs, the side view is drawn next to, and in line with, the top view.

When the three views of the saddle, figure 6-35, are drawn, it is necessary to use the alternate arrangement of views. The right-side view is drawn at the right of the top view.

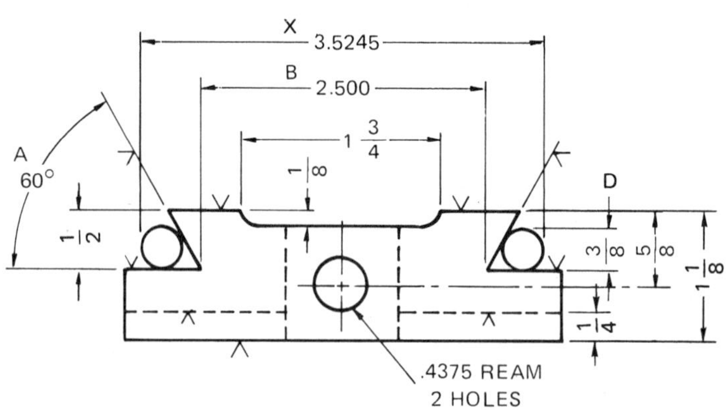

Fig. 6-34

DIMENSIONING DOVETAILS

The upper part of the saddle is a dovetail. Dovetails, like dovetail slides, are accurately measured with the aid of two rods of known diameter. The rods are placed in the dovetail as shown in the side view, figure 6-34. The distance over the rods is measured with a micrometer. This distance is calculated from a formula as follows:

$$\text{Dimension } X = D\left(1 + \text{cotangent } \frac{A}{2}\right) + B$$

Substitute values

B = 2.5″, D = 3/8″, A = 60°	$X = 3/8\left(1 + \text{cotangent } \frac{60°}{2}\right) + 2.5$
Divide 60° by 2;	$X = 3/8\,(1 + \text{cotangent } 30°) + 2.5$
Cotangent of 30° from table = 1.732	$X = 3/8\,(1 + 1.732) + 2.5$
Add 1 and 1.732	$X = 3/8 \times 2.732 + 2.5$
Multiply 3/8 by 2.732	$X = 1.0245 + 2.5$
Add 1.0245 and 2.5	$X = 3.5245$

ASSIGNMENT P

Draw the right side, top, and front views of the saddle. The side view is drawn at the right of the top view.

Procedure

1. Lay out the margin lines on the drawing sheet and letter the title strip.
2. Draw the horizontal center line for the right-side view and the top view 3 3/4 inches below the top margin line.
3. Draw the vertical center line for the side view 2 1/4 inches from the right margin line.
4. Draw the outlines of the side view.

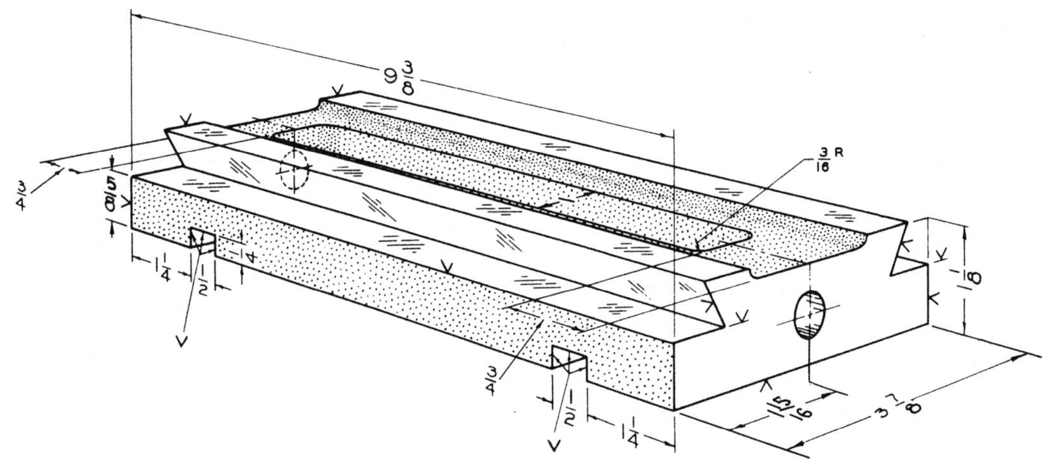

NAME OF SCHOOL CITY OR TOWN	SADDLE	SHEET No. 6-P NAME	CLASS DATE

Fig. 6-35

5. Draw the top view 1 3/4 inch from the left margin line.
6. Draw the front view 1 3/4 inch above the lower margin line.
7. Add the hidden lines in the top view and the side view.
8. Dual-dimension the views.
9. Place finish marks on all of the lines representing finished surfaces. Some of the finish marks are placed on the extension lines to avoid overcrowding.
10. Check the accuracy of all lines, views, and dimensions on the drawing.

REVIEW QUESTIONS

1. Why is the side view drawn at the right of the top view?
2. What is meant by the alternate arrangement of views?
3. What is the scale of the drawing?
4. In which view or views is the saddle symmetrical?
5. Using the formula for dovetails, find dimension X when distance B is 2 3/4 inches, angle A is 60 degrees, and the diameter of the rods is 1/2 inch.

SPOTFACES AND T SLOTS

The tool block shown in the pictorial drawing, figure 6-36, is another casting. On this casting, a circular spot called a spot face is finished around the drilled hole. The depth of the spot face is from .03 to .06 inch below the rough surface.

In the tool block, the slot shaped like an inverted T is call a *T slot.* A T bolt head slides into the T slot and clamps a tool post to the tool block.

ASSIGNMENTQ

Draw the right side, top, and front views of the tool block.

Procedure

1. Lay out the drawing sheet and letter the title strip.
2. Draw the right view 7/8 inch above the title strip and 2 3/4 inches from the right margin line.
3. Draw the top view 5/8 inch below the top margin line and 2 3/4 inches from the left margin line. Draw the front view.
4. Dual-dimension the drawing. Place the dimension for the T slot and the dimension for the .19-by-1.25-inch tongue in the side view.
5. Add the finish marks and letter the notes.

REVIEW QUESTIONS

1. Indicate the spot face and the T slot in the views of the drawing.
2. What is the purpose of the spot face? Of the T slot?

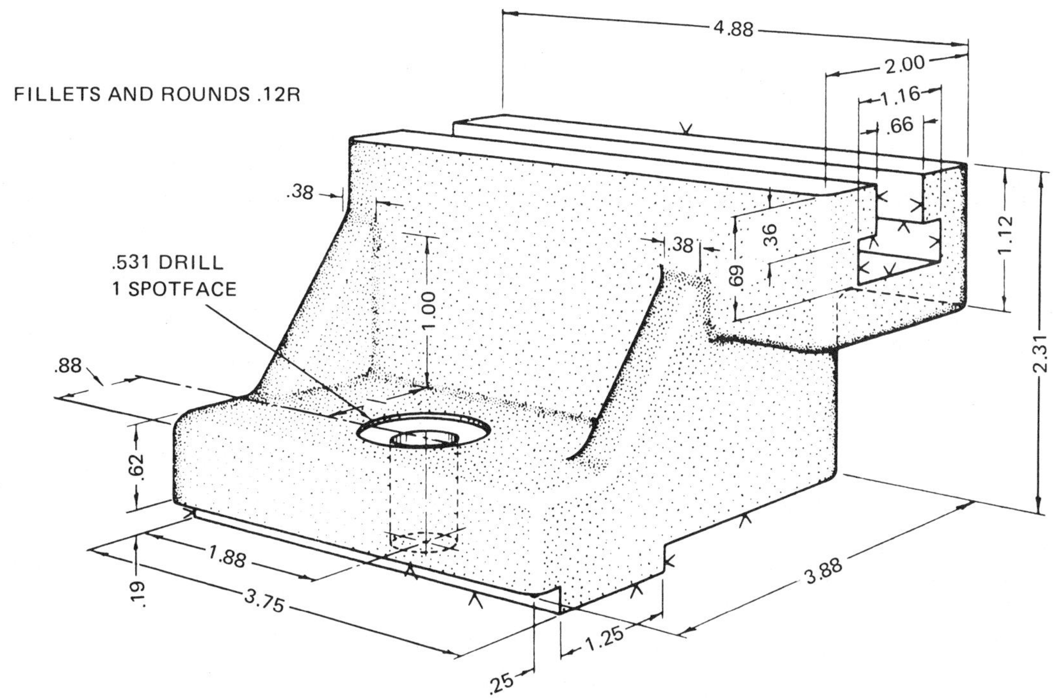

NAME OF SCHOOL CITY OR TOWN	TOOL BLOCK	SHEET No. 6-Q NAME	CLASS DATE

Fig. 6-36

UNIT 7 SECTIONAL DRAWING

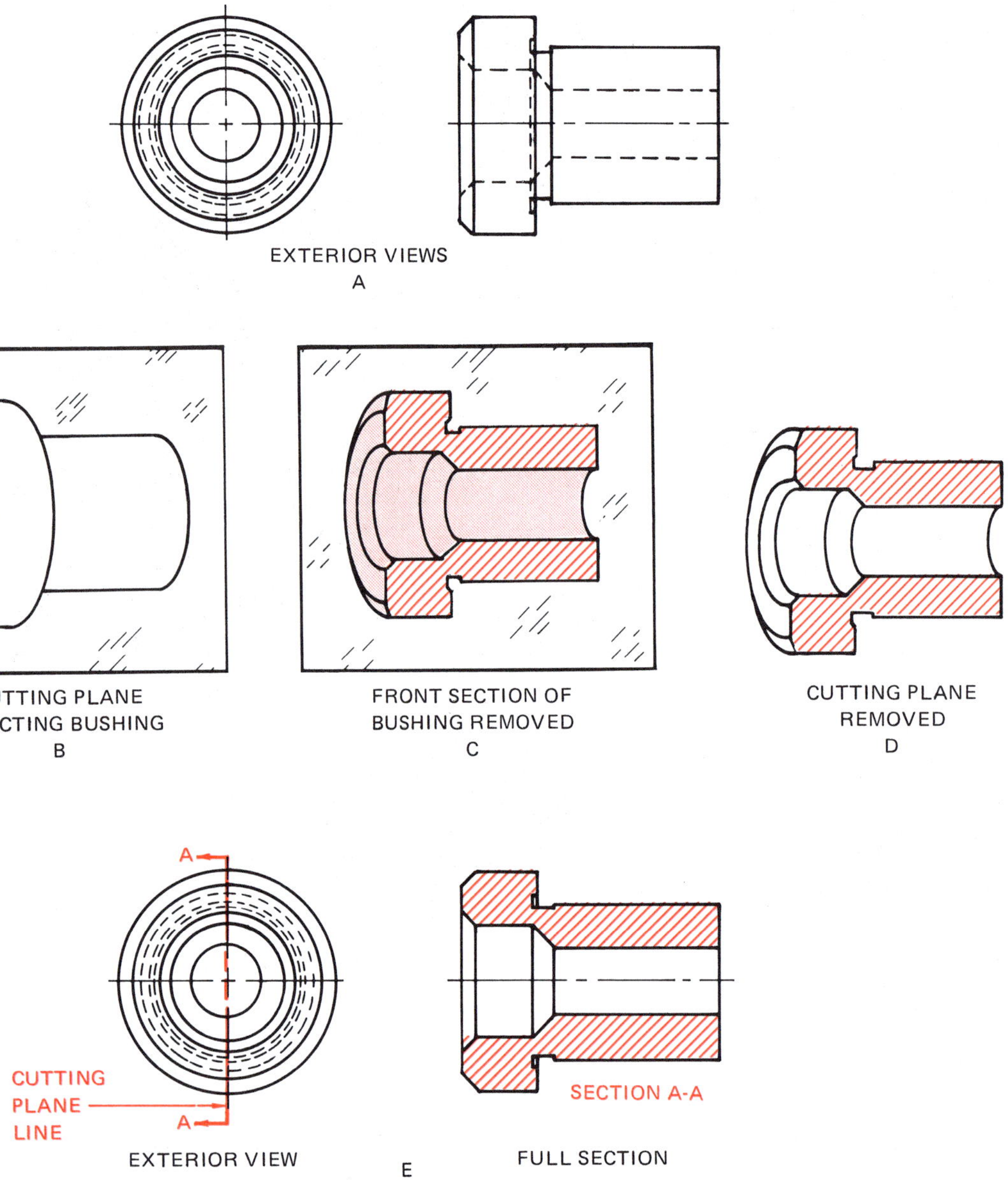

Fig. 7-1 Theoretical construction of a sectional view

All drawings up to this point have been exterior views. Exterior views show the object as it appears to an observer from points located directly above, in front, or at the right or left of the object. The invisible details in the exterior views are shown with hidden lines.

An exterior view of an intricate part may require a confusing number of hidden lines to show its interior details. Such a view is not easy to draw and is more difficult to read. The interior details of a part are seen more clearly in a sectional view. A *sectional view* is one in which a portion of the object is assumed to be removed to reveal the interior details.

The theory of constructing a sectional view is illustrated in figure 7-1. Two exterior views of a special bushing are shown at A. At B, a cutting plane passes through the bushing. At C, the portion between the observer and the cutting plane is removed to reveal the interior details. At D, the cutting plane is removed. The last illustration at E, shows two views of the bushing. The front view is a sectional view, or simply a section.

CUTTING-PLANE LINE

A cutting-plane line is drawn in the left-side view of the bushing as shown at E. The cutting-plane line shows where an imaginary cutting plane passes through an object. The line A-A represents the edge of the cutting plane. Cutting-plane lines are made up of a series of long dashes, 3/4 inch long, alternating with two short dashes, 1/16 inch long. These are drawn as heavy as visible edge lines, figure 7-2.

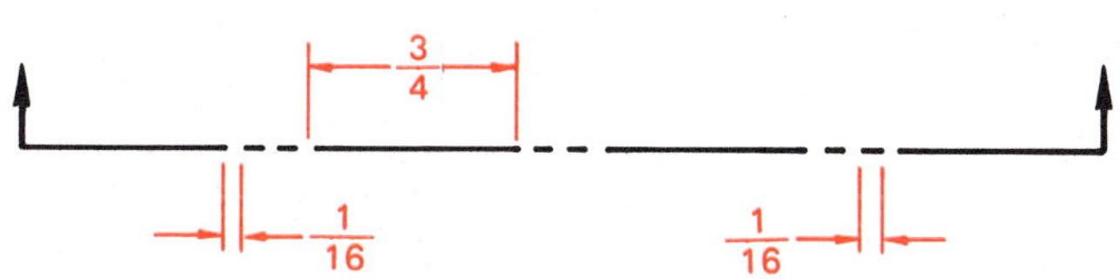

Fig. 7-2 Cutting-plane line

Arrowheads on the ends of the cutting-plane line show the direction the section is viewed. Letters used with cutting-plane lines identify the cutting plane and the corresponding sectional view. The cutting-plane line, the letters, and the arrowheads may be omitted from simple symmetrical views when it is evident that the cutting plane passes through the center of the part.

SECTION LINING OR CROSSHATCHING

In the sectional view, figure 7-1 at E, section lining or crosshatching is drawn across those surfaces which are in contact with the cutting plane. Section lining in sectional views indicates the various construction materials. For the common metals, the section lining is made with thin lines usually drawn at an angle of 45 degrees. Various combinations of lines represent different materials. A few common section-lining symbols are given in figure 7-3.

Section lining is uniformly spaced by eye, without mechanical aid. For small parts, the lines may be 1/32 inch apart; on large parts, they may be 3/16 inch apart. For the sectional views in this series of drawings, the lines should be spaced approximately 3/32 inch apart. *Some manufacturers prefer to use a series of equally spaced, sloping, full lines for section lining all materials. Then, notes specify the materials of which the parts are made.*

VISIBLE DETAILS IN SECTIONAL VIEWS

In sectional views, all visible details behind the cutting plane must be shown.

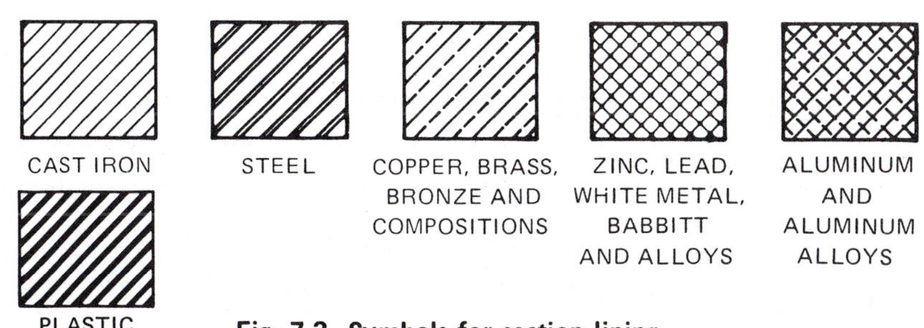

Fig. 7-3 Symbols for section lining

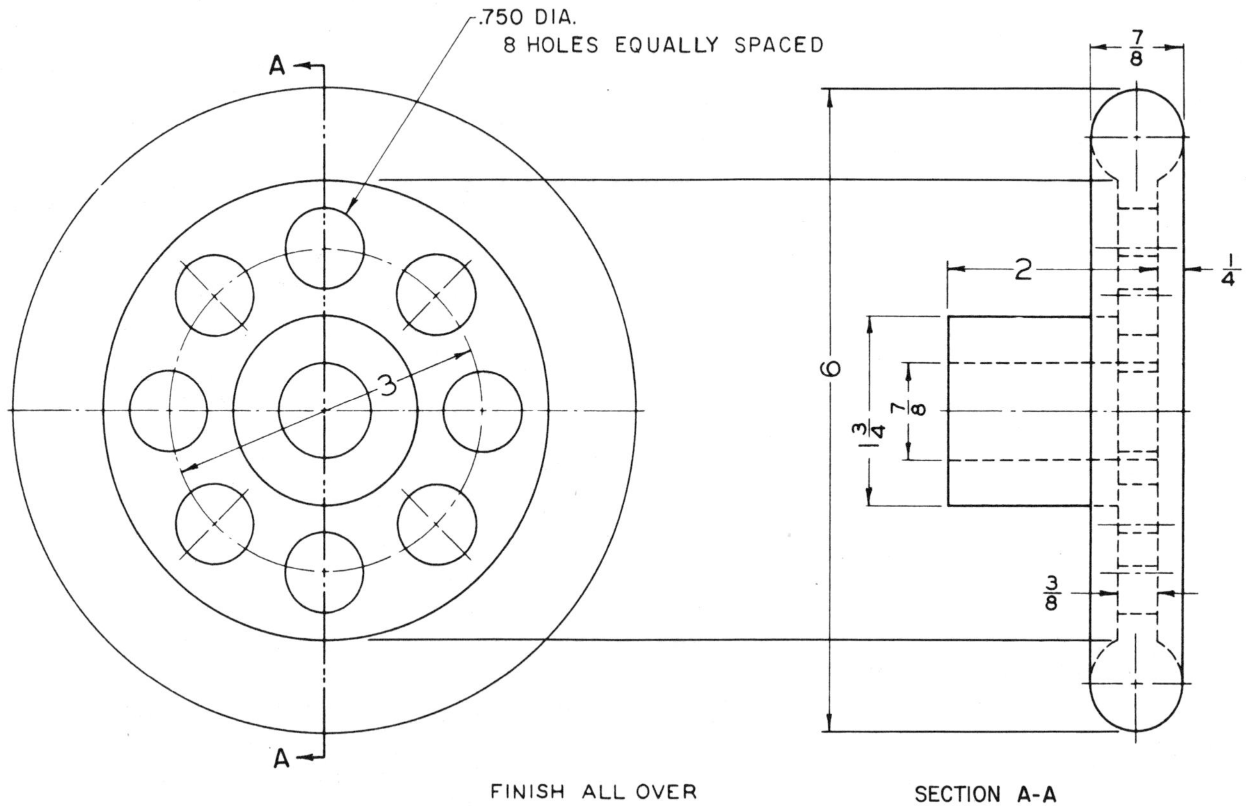

NAME OF SCHOOL CITY OR TOWN	HANDWHEEL	SHEET No. 7-A NAME	CLASS DATE

Fig. 7-4

HIDDEN LINES IN SECTIONAL VIEWS

Hidden lines are usually omitted from sectional views. They are drawn only when it is necessary to show more clearly a hidden part in the section, or to aid in dimensioning the view.

TYPES OF SECTIONAL VIEWS

Sectional views are classified as: full sections, half sections, revolved sections, removed sections, broken-out sections, offset sections, and aligned sections.

REVIEW QUESTIONS

1. Write the definition of a sectional view.
2. Study the symbols for section lining. Describe the various combinations of lines and which material they represent.
3. Section lining is used for two purposes. What are they?
4. For what purpose are cutting-plane lines used? How are they drawn?

FULL SECTIONS

The theory of constructing a sectional view in figure 7-1 shows: the cutting plane passing through the bushing, the front portion of the bushing removed, and a sectional view drawn of that section lying behind the cutting plane. This sectional view is called a full section. A *full section* is a sectional view in which the cutting plane passes entirely through the object.

Figure 7-4 shows two exterior views of a handwheel. In the view at the right, numerous hidden lines are required to show the interior details. These interior details can be seen more clearly in a full section. Therefore, the exterior side view must be changed to a full section.

The cutting-plane line in the view at the left shows where the cutting plane passes through the handwheel. The pictorial drawing, figure 7-5, represents the section of the handwheel that lies behind the cutting plane. This part of the handwheel is drawn in the full section.

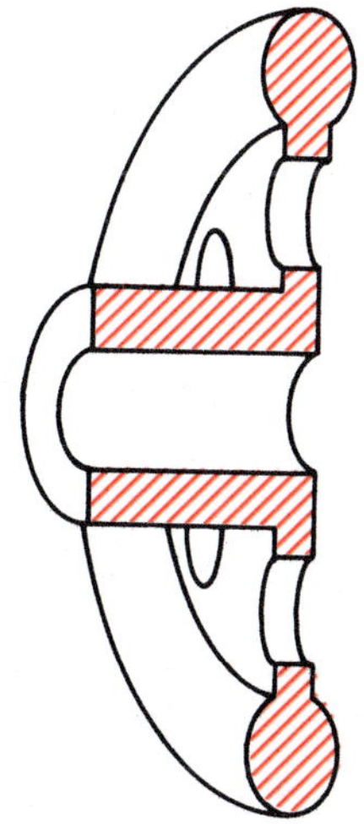

Fig. 7-5 The handwheel (rear half)

ASSIGNMENT A

Draw a two-view working drawing of the handwheel with the side view in full section.

Procedure

1. Lay out the margin lines on the drawing sheet and letter the title strip.
2. Draw the horizontal center line 5 1/8 inches below the top margin line.
3. Draw the vertical center line 5 1/8 inches from the left margin line. This center line replaces the cutting-plane line. The cutting-plane line is not required because it passes through the center of a symmetrical part.
4. Draw all except the second largest of the circles. The diameter of this circle is not given on the drawing. The diameter is found when the sectional view is drawn.
5. Draw the full section 2 3/8 inches from the right margin line. Omit all hidden lines.

6. Draw the section lining for cast iron. Space the lines approximately 3/32 inch apart.
7. Draw the last circle in the other view. The projection lines show how the diameter of this circle is determined.
8. Dimension the drawing using the decimal inch or dual-dimensioning procedure and letter the notes.
9. Check the accuracy of all lines, views, sections, and dimensions on the drawing.

REVIEW QUESTIONS

1. Write the definition of a full section. Give two examples of objects that require a full section.
2. What is the purpose of a cutting-plane line?
3. How are cutting-plane lines drawn?
4. For what two purposes are section lines used?

LONGITUDINAL SECTION

Figure 7-6 shows three exterior views of a die bushing retainer. On the drawing, the front view is to be changed to a full section.

The cutting-plane line in the top view shows that the cutting plane passes entirely through the part lengthwise. This sectional view is called a *longitudinal section*.

GRINDING RECESS

The side and bottom of the 2.25-inch diameter hole are finished by grinding. A .06-inch radius grinding recess is machined in the bottom corner of the hole. This recess is essential to permit the side and bottom to be ground true without leaving a fillet in the corner.

The bushing retainer has a 30-degree sloping surface. The projection lines on the drawing show how to locate this surface in the side view. In actual use, the bushing retainer has another hole in it. This hole is drilled perpendicular to the sloping surface and goes through to the bottom corner of the 2.25-inch diameter hole. Screw threads are cut in this hole and a set screw retains the die bushing in the bushing retainer.

ASSIGNMENT B

Draw a three-view working drawing of the die bushing retainer with the front view in full section.

Procedure

1. Lay out the margin lines on the drawing sheet and letter the title strip.
2. Draw the horizontal center line for the top view 2 5/8 inches below the top margin line. This center line replaces the cutting-plane line. The cutting-plane line is not required because it passes lengthwise through the center of the symmetrical top view.
3. Draw the vertical center line 2 1/2 inches from the left margin line.
4. Draw the top view.
5. Draw the front view 1 5/8 inches above the title strip. Make this view a full section. Draw the section lining for steel.
6. Draw the center line for the side view 2 1/2 inches from the right margin line. Draw the side view.
7. Dimension the drawing using the decimal inch or dual-dimensioning procedure.
8. Place finish marks on the lines representing finished surfaces.
9. Check the accuracy of all lines, views, sections, and dimensions on the drawing.

REVIEW QUESTIONS

1. What is a full section?
2. What is a longitudinal section?

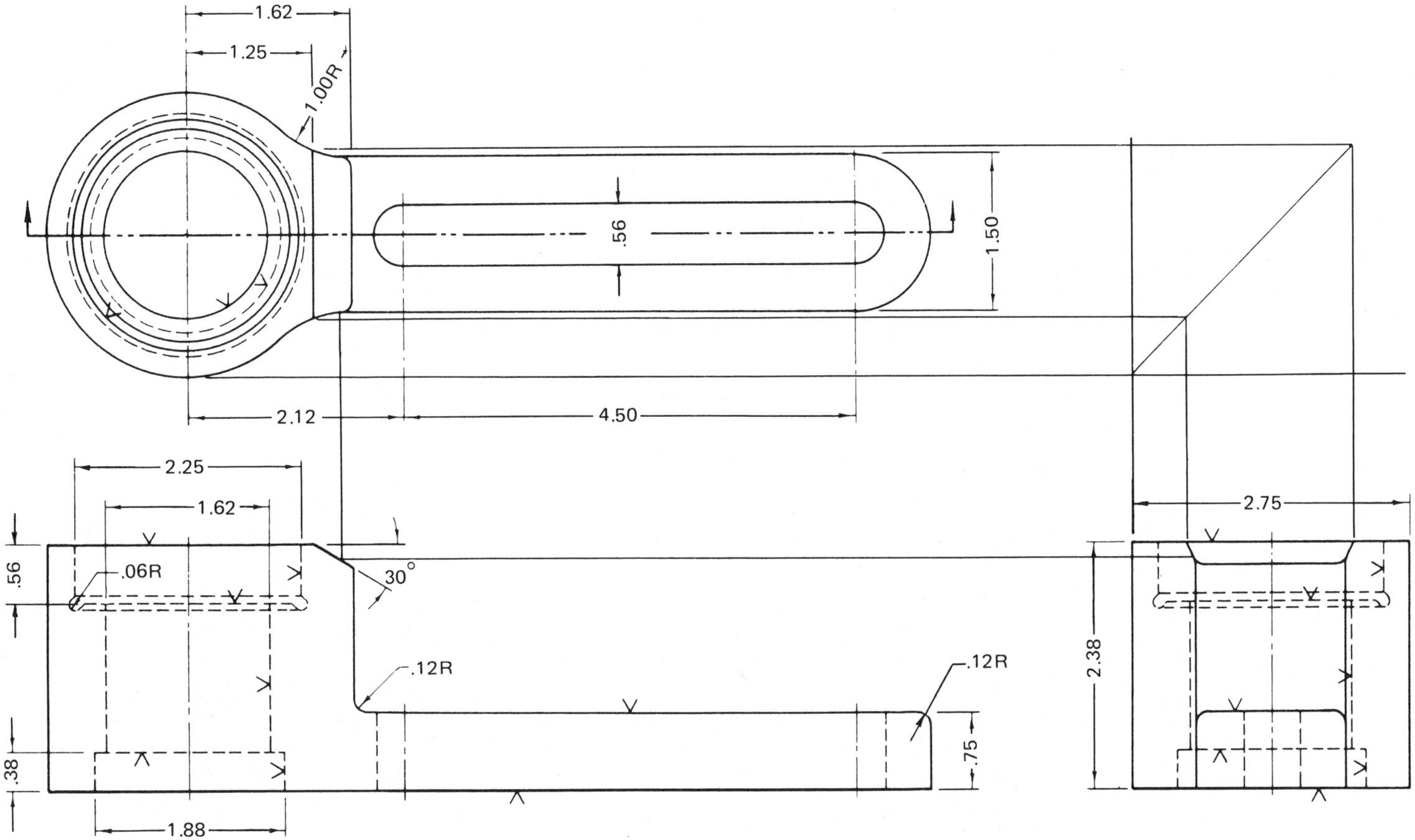

NAME OF SCHOOL CITY OR TOWN	DIE BUSHING RETAINER	SHEET No. 7-B NAME	CLASS DATE

Fig. 7-6

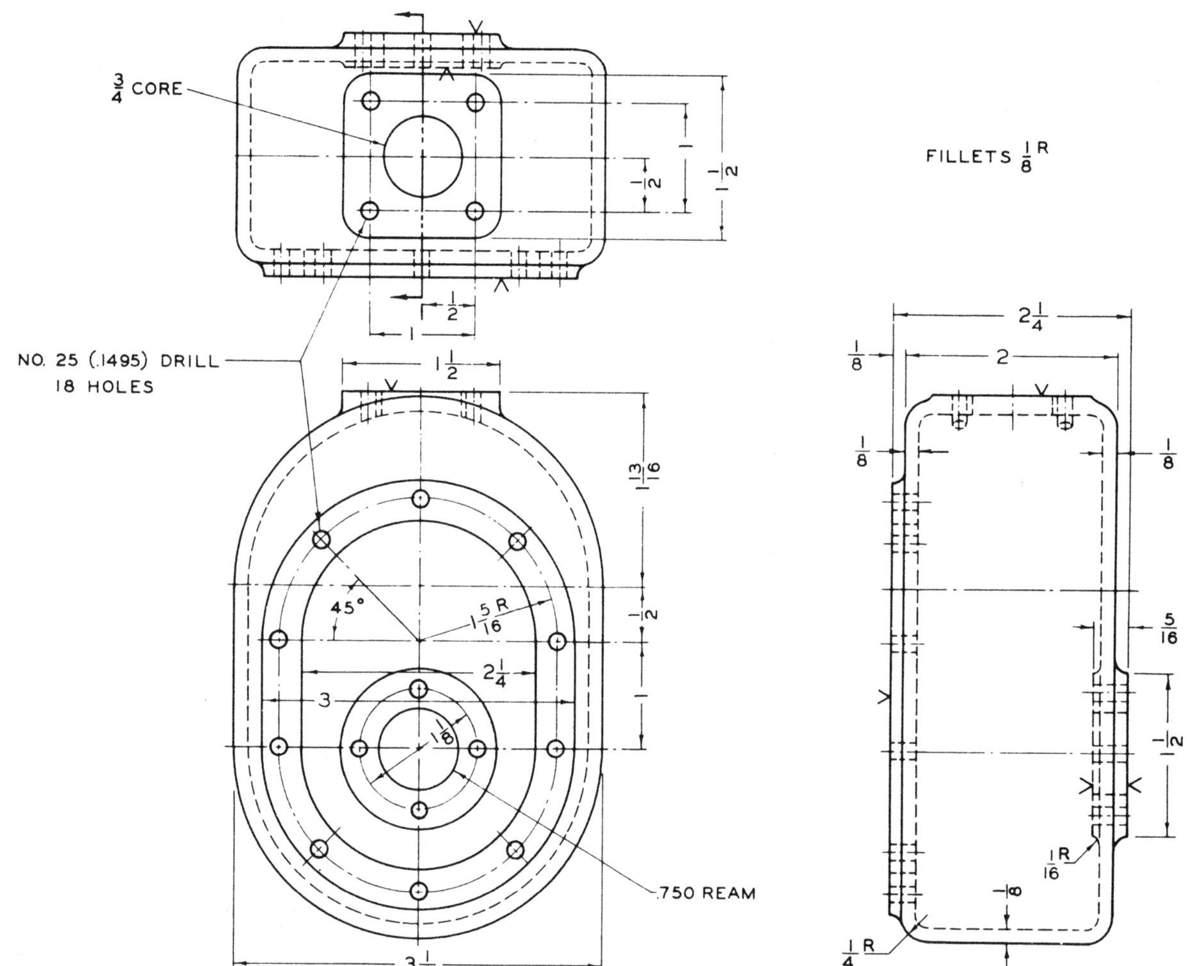

NAME OF SCHOOL CITY OR TOWN	PUMP CASE	SHEET No. 7-C NAME	CLASS DATE

Fig. 7-7

CROSS SECTION

Figure 7-7 shows three views of a pump case. On the drawing, the side view is to be changed to a full section. The cutting-plane line in the top view shows that the cutting plane passes entirely through the part crosswise. This sectional view is called a *cross section.*

ASSIGNMENT C

Make a three-view working drawing of the pump case with the side view in full section.

Procedure

1. Lay out the margin lines on your drawing sheet and letter the title strip.
2. Draw the vertical center line for the front view and the top view. Draw this line 5 1/2 inches from the left margin line. In the top view, this center line replaces the cutting-plane line. The cutting-plane line is not required because it passes through the center of the symmetrical top view.
3. The front view has three horizontal center lines and two circular center lines. Draw the lower horizontal center line 2 1/4 inches above the title strip. Draw the other horizontal and the circular center lines.
4. Complete the front view. Draw the horizontal center line for the top view 1 1/2 inches below the top margin line. Draw the top view.
5. Draw the side view 3 1/2 inches from the right margin line. Make this view a full section. Draw the section lining for brass.
6. Dimension the drawing and letter the notes.
7. Place finish marks on the lines representing the finished surfaces.
8. Check the drawing for accuracy of lines, views, dimensions, and sections.

REVIEW QUESTIONS

1. What is a cross section?
2. What is a longitudinal section?

CROWN

The pictorial drawing of the crown pulley, figure 7-8, is cut entirely across by a vertical cutting plane to reveal its interior construction. At the center of the pulley's face the diameter is larger than at the sides. Consequently, the face has a ridge in the center that is called a *crown.* Pulleys are crowned because a belt has a tendency to climb to the highest point on a pulley.

BORE

Holes through the center of pulleys are bored to make them straight and make sure the pulleys will run true.

KEYWAYS AND KEYSEATS

A keyway, figure 7-9, is cut lengthwise in the hub of the pulley. A keyseat of the same width and depth is cut in a shaft that fits into the hub. When the shaft is placed in the hub, the keyway and keyseat are aligned and a square piece of steel, called a key, is inserted in the space, locking the pulley and shaft together.

The spaces for square and flat keys are dimensioned with a shop note. The width of the keyway or keyseat is given first, followed by the depth, figure 7-10.

WEB

The thin solid section between the hub and the rim is called a *web.* This pulley is made of cast iron and is finished on the face, both sides of the rim, both ends of the hub, and in the hole. The stippled or dotted parts are unfinished.

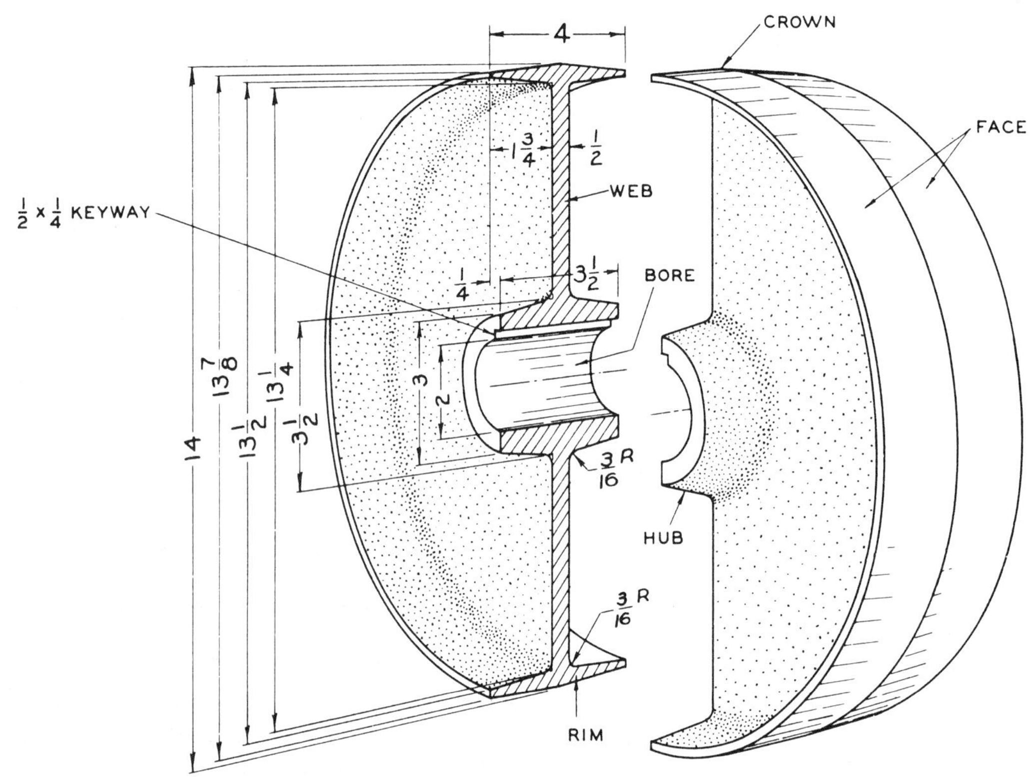

NAME OF SCHOOL CITY OR TOWN	CROWN PULLEY SCALE: $\frac{1}{2}$ IN. = 1 IN.	SHEET No. 7-D NAME	CLASS DATE

Fig. 7-8

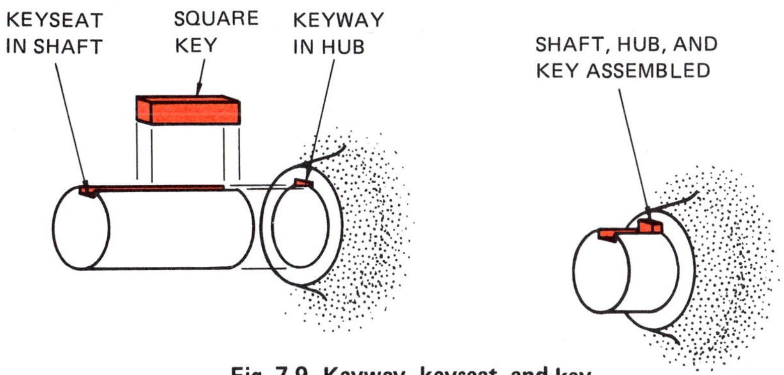

Fig. 7-9 Keyway, keyseat, and key

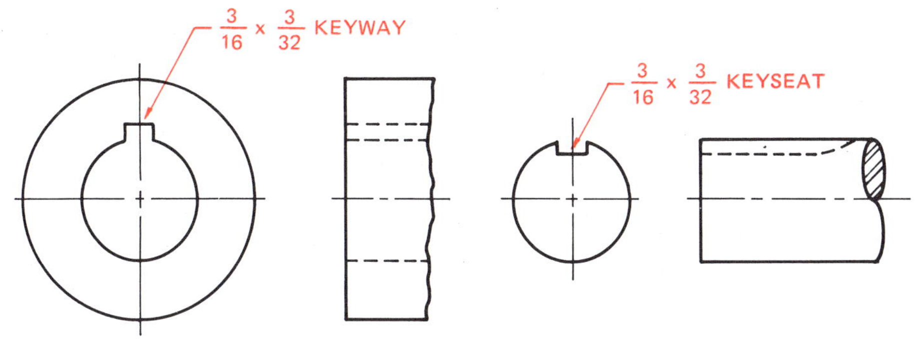

Fig. 7-10 Dimensioning keyways and keyseats

ASSIGNMENT D

Draw a two-view working drawing of the crown pulley figure 7-8, with the front view in full section. Use the scale 1/2 inch = 1 inch.

Procedure

1. Lay out the margin lines on the drawing sheet and letter the title strip.
2. Draw a horizontal center line 4 3/4 inches above the title strip and a vertical center line 5 1/4 inches from the left margin line. Draw the side view of the whole pulley.
3. Draw the full section 1 3/4 inches from the right margin line. Use the section lining for cast iron.
4. Dimension the drawing. A cylindrical-shaped object should have all of its dimensions placed on one view.
5. Place finish marks on the lines representing finished surfaces.
6. Check the accuracy of all lines, views, sections, and dimensions.

REVIEW QUESTIONS

1. Indicate the following parts of the pulley in figure 7-8: face, rim, crown, web, hub, bore, and keyway.
2. What is the purpose of the keyway? Of the crown?
3. Is this sectional view a longitudinal section or a cross section?

HALF SECTIONS

The sectional view for the bearing closure in figure 7-11 is called a half section. Half sections, like full sections, show the interior construction of objects. Half sections have the advantage of showing both the interior and exterior details in one view.

A *half section* is a sectional view in which the cutting plane passes half way through the object. One half of the view is shown in section while the other half shows the exterior.

The theory of constructing a half section is illustrated in figure 7-12. At A, the cutting plane passes half way through a knurled ring gauge. At B, the portion in front of the cutting plane is removed to reveal the interior construction.

A partially dimensioned working drawing is shown in figure 7-13. Here the portion behind the cutting plane is shown in section while the portion below the cutting plane shows exterior details. A center line is usually drawn between the two halves. A visible edge line may be used, but it would denote that a real edge exists between the halves. This is not the case since the part is not actually cut, and the portion in front of the cutting plane is only

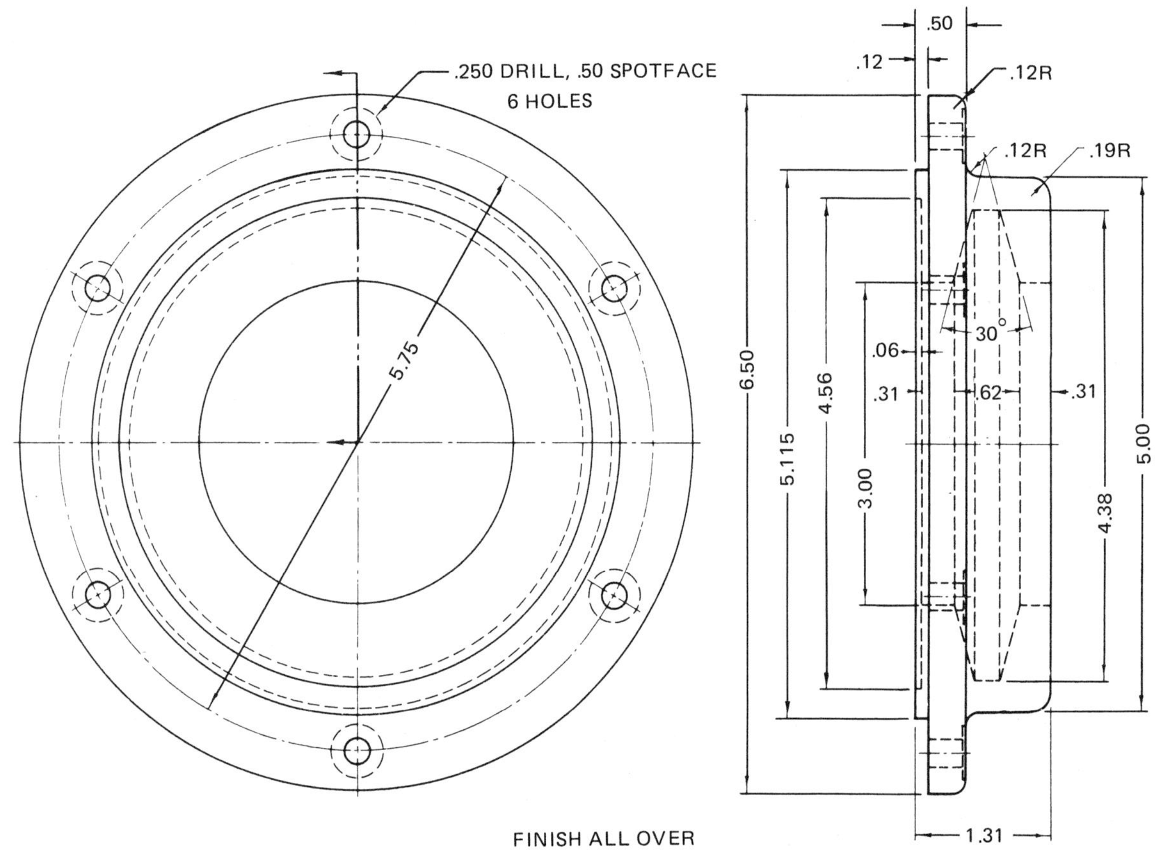

NAME OF SCHOOL CITY OR TOWN	BEARING CLOSURE	SHEET No. 7-E NAME	CLASS DATE

Fig. 7-11

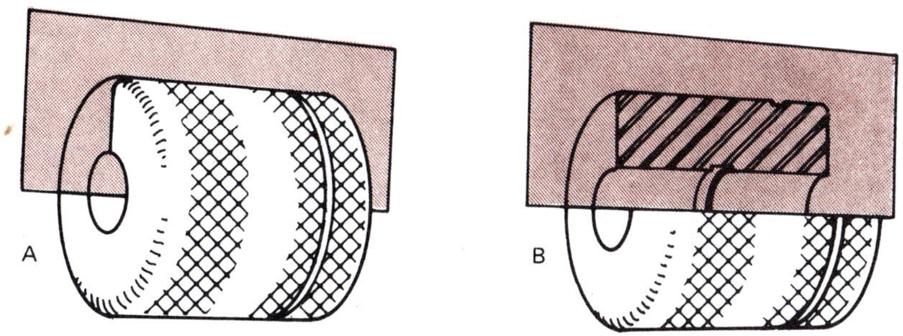

Fig. 7-12 Theoretical construction of a half section

assumed to be removed. Therefore, the center line between the halves is preferred.

HIDDEN LINES IN HALF SECTIONS

Hidden lines are usually omitted from sectional views. When the interior details of half sections are difficult to dimension, hidden lines may be used sparingly to aid in dimensioning the view. For example, in figure 7-13, instead of drawing all of the hidden lines only a few dashes are shown. From these, the extension lines are drawn. Dimension lines for the internal diameters are drawn as usual.

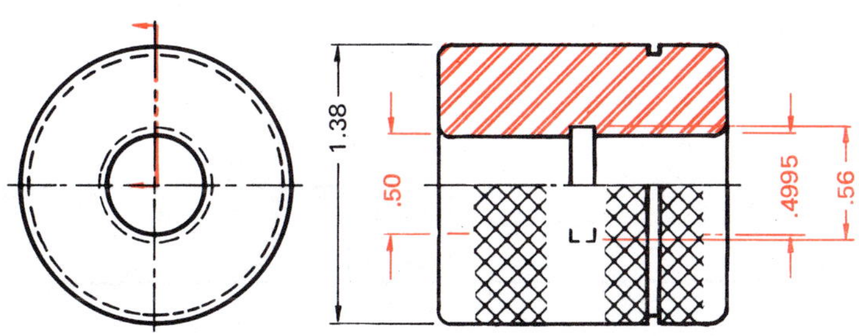

Fig. 7-13

ASSIGNMENT E

Make a two-view working drawing of the bearing closure. Change the front view to a half section.

Procedure

1. Lay out the margin lines on the drawing sheet and letter the title strip.
2. Draw a horizontal center line 4 1/2 inches above the title strip, and a vertical center line 5 5/8 inches from the left margin line. Draw the side view. The cutting-plane line is not required in this view since the cutting plane passes through the center of symmetrical parts.
3. Draw the front view 3 1/8 inches from the right margin line with the upper half in section. Use the section lining for cast iron. Omit hidden lines except for a few dashes from which extension lines are drawn for the 3.00-inch, 4.38-inch, and 4.56-inch diameter dimensions.
4. Dimension the drawing. Check carefully for accuracy.

REVIEW QUESTIONS

1. Write the definition of a half section and give three examples where it may be used.
2. What is the advantage of half sections?
3. Why is a center line used instead of a visible edge line between the two halves of a half section?
4. Why are some hidden lines often necessary?

FLANGE

The pulley in figure 7-14 is called a flanged pulley because it has a raised edge at each side of the face. The flanges keep the belt on the pulley. The thin section of metal between the hub and the rim is called the web. Four large cored holes are made in the web when casting the pulley. The cored holes save metal and reduce the weight of the pulley.

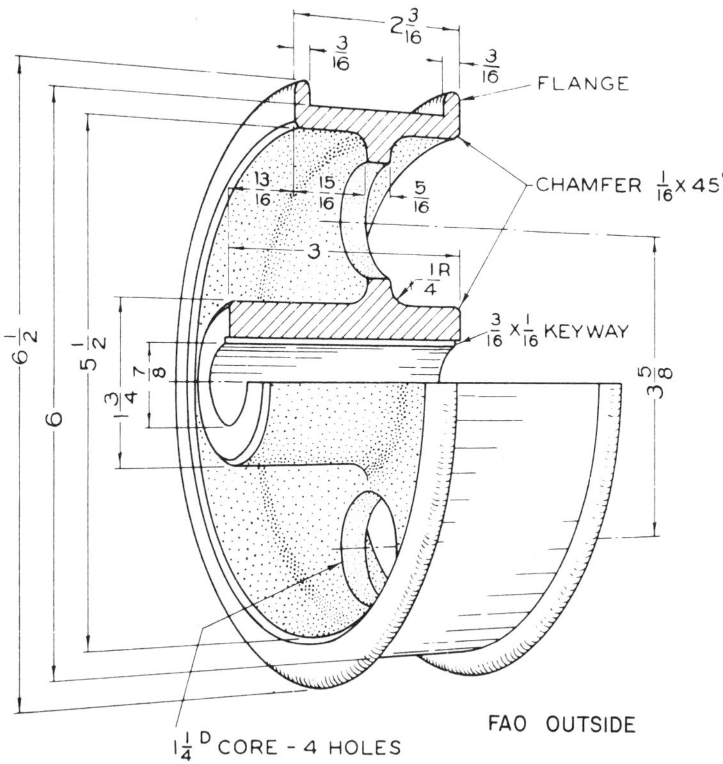

NAME OF SCHOOL CITY OR TOWN	FLANGED PULLEY	SHEET No. 7-F NAME	CLASS DATE

Fig. 7-14

The keyway in the hub of this pulley is made for a flat key instead of a square key. Although the key is called a flat key, its cross section is rectangular. The flat key for this pulley is 3/16 inch wide and 1/8 inch high.

The note *FAO outside* means the pulley is finished all over on the outside. The stippled areas in the pictorial drawing indicate the unfinished surfaces. The sharp edges inside the rim and on the ends of the hub are chamfered.

ASSIGNMENT F

Make a two-view working drawing of the flanged pulley with the front view a half section.

Procedure

1. Lay out the margin lines on the drawing sheet and letter the title strip.
2. Draw a horizontal center line 4 3/4 inches above the title strip, and a vertical center line 5 inches from the left margin line. Draw the side view of the whole pulley.
3. Draw the front view 1 3/4 inches from the right margin line. Make this view a half section. Use the section lining for cast iron. Omit hidden lines except for a few dashes from which the extension lines are drawn for the 5 1/2-inch diameter and 7/8-inch diameter dimensions.
4. Dimension the drawing. A center line circle is drawn through the center of the four holes in the side view. The 3 5/8-inch diameter dimension is placed in this circle.
5. Letter the shop notes and draw the leaders. The notes for the keyway and the cored holes are part of the side view.
6. Check the accuracy of all lines, sections, and dimensions on your drawing.

REVIEW QUESTIONS

1. What is the purpose of the flange?
2. What is a cored hole?
3. What is the difference between a square key and a flat key?
4. What does the abbreviation FAO mean?
5. What does FAO outside mean?
6. What is a half section?

BLIND HOLES

The working drawing, figure 7-15, shows three exterior views of a gear case. The front view and the side view are to be changed to half sections on the drawing.

The pictorial drawing, figure 7-16, illustrates how the gear case appears if a cutting plane passes half way through longitudinally to make the front view a half section and another cutting plane cut half way across to make the side view a half section.

Six holes in the gear case are drilled .75 inch deep. Holes like these, which are drilled into a part but not through it, are called *blind holes*. When blind drilled holes are shown on a drawing, an angle is drawn in the bottom of the holes to correspond approximately to the angle on the point of the drill. This angle is drawn 120 degrees, or 60 degrees on each side of the center line.

DIMENSIONING THE LOCATION OF HOLES

The 1.75-inch and 1.50-inch radius dimensions in the top view indicate how arcs are drawn to locate the centers of the four .25-inch radius tangent arcs. The centers of these arcs are also used as the centers of four blind holes. For the convenience of the machinist, it is better to give horizontal and vertical dimensions to the center of holes. On this drawing, the exact horizontal and vertical dimensions to the four holes are decimals to the third place.

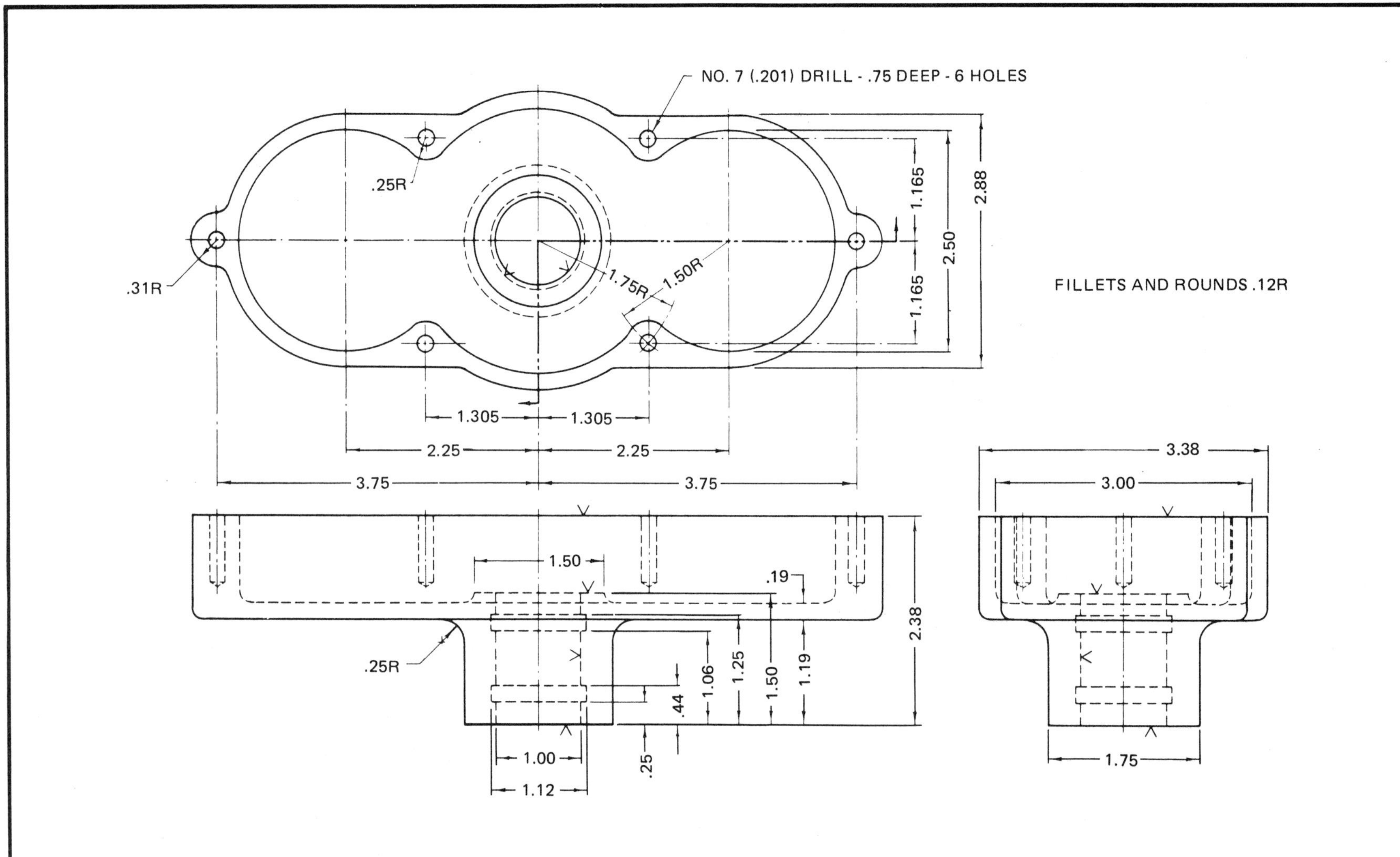

NAME OF SCHOOL CITY OR TOWN	GEAR CASE	SHEET No. 7-G NAME	CLASS DATE

Fig. 7-15

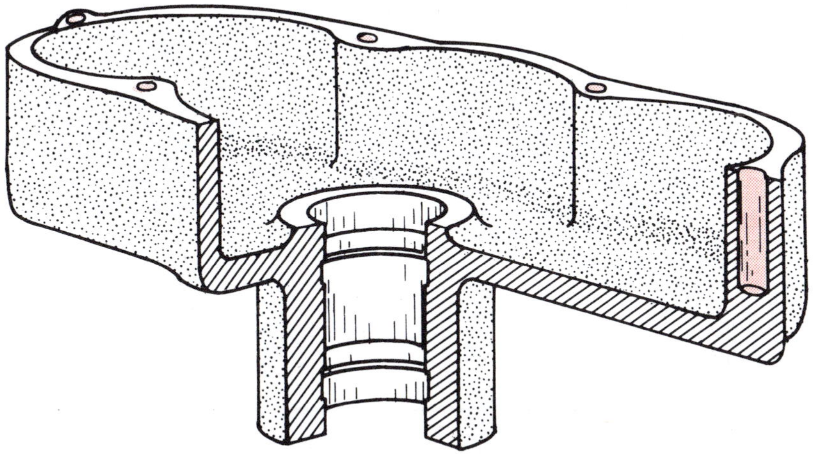

Fig. 7-16 Gear case

ASSIGNMENT G

Make a three-view working drawing of the gear case. The front view and the side view are to be half sections.

Procedure

1. Lay out the margin lines on the drawing sheet and letter the title strip.
2. Draw a horizontal center line 2 1/2 inches below the top margin line, and a vertical center line 5 1/2 inches from the left margin line. These center lines replace the cutting-plane lines. The cutting-plane lines are not required in symmetrical views. Draw the top view.
3. Draw the front view in half section, 1 1/4 inches above the title strip. Omit hidden lines. Draw the section lining for cast iron.
4. Draw a vertical center line 3 1/8 inch from the right margin line. Draw the right-side view in half section. Omit hidden lines.
5. The section lining is drawn slanting in the same direction in both views of the same part.
6. Dual-dimension the drawing. Nominal drill size is dimensioned as shown. A few of the hidden line dashes are drawn to aid in dimensioning the internal diameters.
7. Check the accuracy of all lines, views, sections, and dimensions on drawing.

REVIEW QUESTIONS

1. What is a full section?
2. What is a half section?
3. What is meant by a blind drilled hole?
4. What angle is drawn in the bottom of blind drilled holes?
5. Why is this angle drawn?

OFFSET SECTIONS

In the preceding sectional views, the cutting plane appears to be a single continuous plane. For example, a full section results when a single cutting plane cuts entirely through the part; a half section results when a single cutting plane cuts half way through the part.

For some objects, the cutting plane is bent, or offset in various ways, to make it pass through important parts of the object in order to bring these parts into the sectional view. The use of the offset cutting plane is illustrated in figure 7-17.

At A, the cutting plane passes through a die block. It is offset several times at right angles to make it pass through three holes which could not be shown in a sectional view if the cutting plane were a single continuous plane. The section that lies behind the offset cutting plane is shown at B.

At C, a two-view drawing is shown. The position of the offset cutting plane is indicated by a cutting-plane line in the top view. This line is identified by placing the letter A at each end.

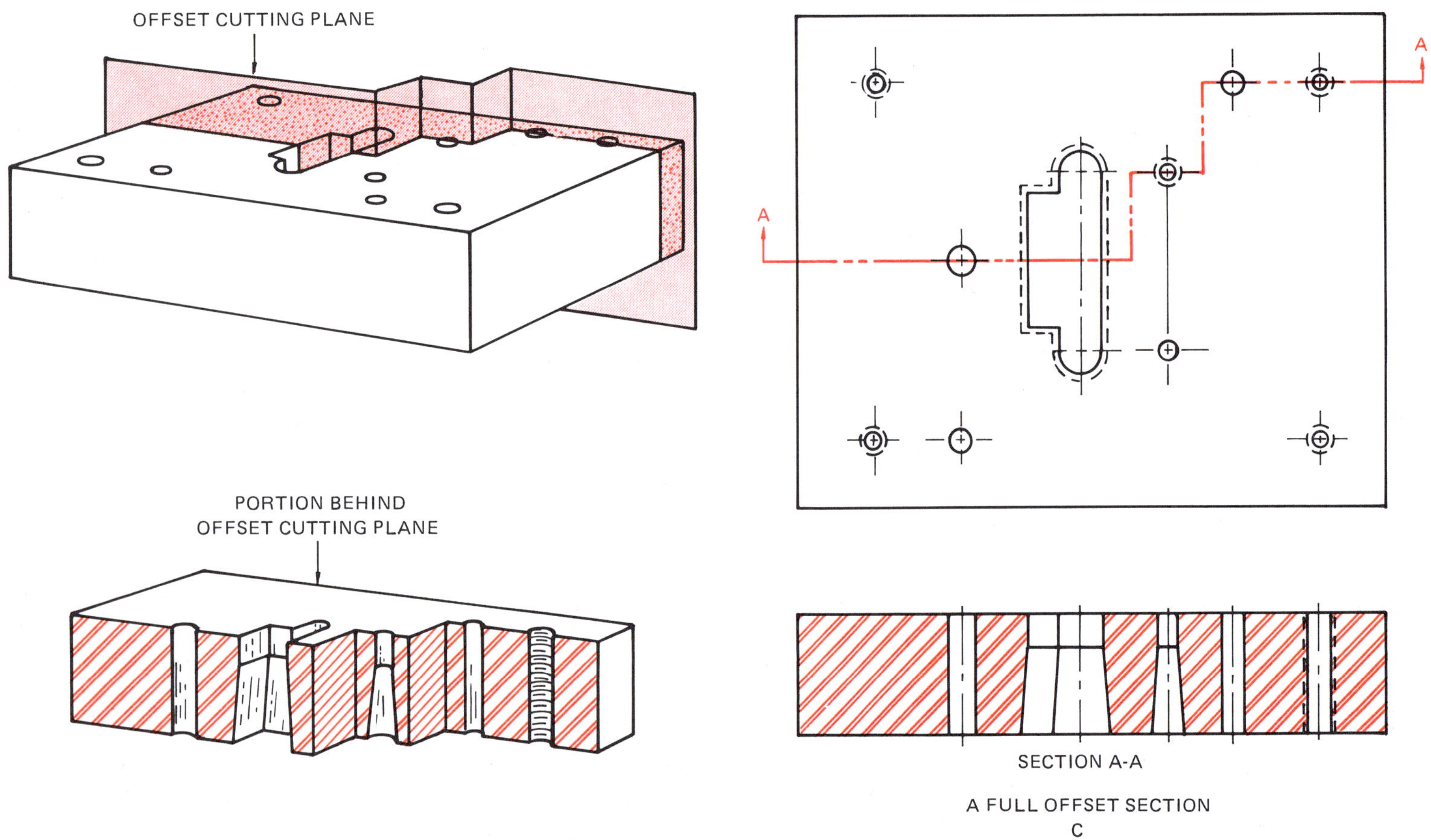

Fig. 7-17 Theoretical construction of an offset section

The arrows on the ends of the line indicate the direction the section is viewed. The front view is a sectional view of that portion lying behind the cutting plane. It is called a full offset section, or simply an offset section.

An *offset section* is a sectional view in which the cutting plane is bent or offset. It is offset to bring important parts into the sectional view which would not otherwise be included if a single continuous cutting plane were used. The offsets are not shown as visible edges in this view.

All visible details beyond the cutting plane are shown in offset sections. Hidden details are omitted unless they make the drawing easier to read. The notation Section A-A or Section B-B is always placed under the offset section to relate the view to the corresponding cutting-plane line. More than one offset cutting plane may be used. More than one offset section may be required to show the interior construction of intricate parts.

ASSIGNMENT H

Make a three-view working drawing of the cover, figure 7-18. The front and side views are to be changed to offset sections on the cutting-plane lines A-A and B-B respectively.

Procedure

1. Lay out the margin lines on the drawing sheet and letter the title strip.
2. Draw the top view. Draw cutting plane lines A-A and B-B.
3. Draw the side view as an offset section on cutting plane line B-B. Use the section lining for cast iron. Show all visible details behind the cutting plane.
4. Draw the front view. Make this an offset section on cutting-plane line A-A. Show all visible details behind the cutting plane.
5. Dual-dimension the drawing. Add notes for the holes as shown.
6. Check lines, views, and dimensions for accuracy.

REVIEW QUESTIONS

1. Define an offset section. Give two examples where it may be used to advantage.
2. What is the purpose of offset sections? How are they identified?
3. Are the bends of the cutting plane shown in the offset section?
4. Why is a letter placed at each end of the cutting-plane line?
5. What is the purpose of the arrowheads on the ends of the cutting-plane line?

REPRESENTING FORESHORTENED PROJECTIONS

The cutting plane in figure 7-19 is an offset cutting plane. It passes half way through the guide ring. It is then offset at 45 degrees to pass through the .375-inch diameter hole. Next, it follows a circular arc back to the center line where it again bends to pass through another hole.

In true projection, a sectional view of the portion behind the cutting plane might be misleading. This is because the distance from the center of the guide ring to the center of the .375 inch hole is foreshortened, while the other two holes are shown at their true distance from the center.

In order to show the .375 inch hole at its true distance from the center of the guide ring, the principles of true projection are violated, and a preferred method of projection is used, figure 7-20.

The true projection at A results in a foreshortened and unsymmetrical view. The preferred projection is shown at B. Here the portion along the offset of the cutting plane is rotated to the center line and then projected to the sectional view. In this sectional view, the holes are shown at their true distance from the

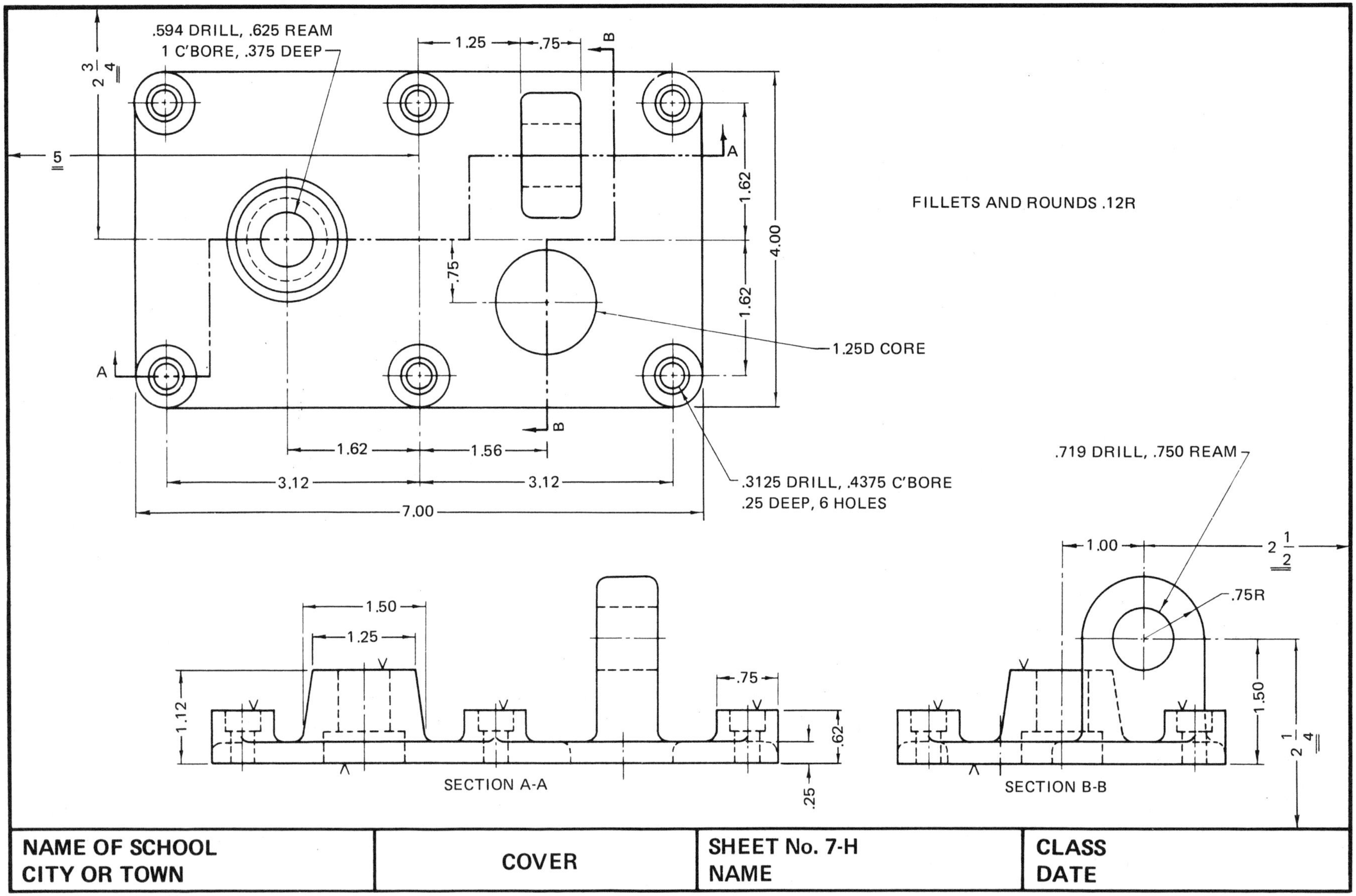
.594 DRILL, .625 REAM
1 C'BORE, .375 DEEP
1.25
.75
B
A
5
4.00
1.62
1.62
.75
FILLETS AND ROUNDS .12R
1.25D CORE
1.62
1.56
3.12
3.12
7.00
.3125 DRILL, .4375 C'BORE
.25 DEEP, 6 HOLES
.719 DRILL, .750 REAM
1.00
.75R
1.50
1.25
1.12
.75
.62
.25
1.50
SECTION A-A
SECTION B-B
NAME OF SCHOOL
CITY OR TOWN
COVER
SHEET No. 7-H
NAME
CLASS
DATE

Fig. 7-18

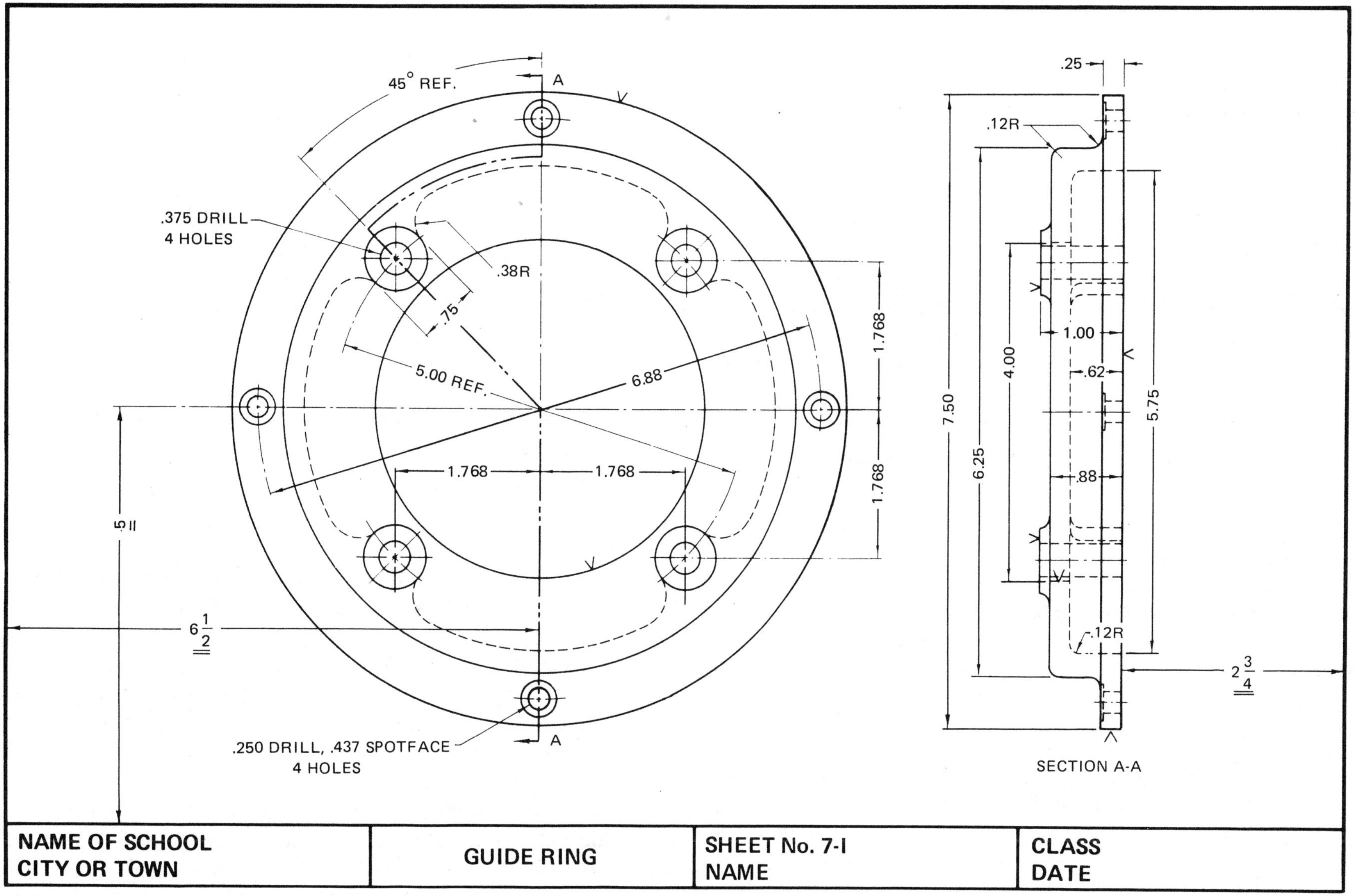
45° REF.
A
.375 DRILL
4 HOLES
.38R
.75
5.00 REF.
6.88
1.768
1.768
1.768
1.768
5 1/2
6 1/2
.250 DRILL, .437 SPOTFACE
4 HOLES
A
.25
.12R
1.00
4.00
.62
7.50
6.25
5.75
.88
.12R
2 3/4
SECTION A-A
NAME OF SCHOOL
CITY OR TOWN
GUIDE RING
SHEET No. 7-I
NAME
CLASS
DATE

Fig. 7-19

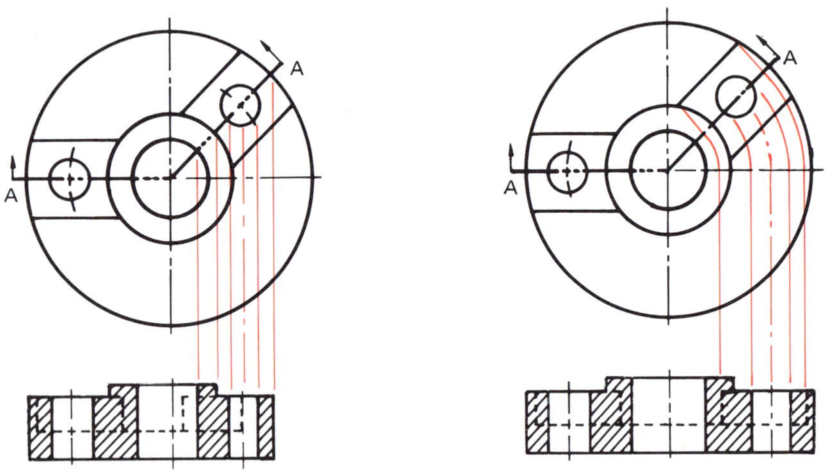

A TRUE PROJECTION VIEW FORESHORTENED AND UNSYMMETRICAL

B PREFERRED PROJECTION VIEW TRUE SIZE AND SYMMETRICAL

Fig. 7-20 Treatment of foreshortened projections

center and the view is symmetrical. Such views are sometimes called *aligned sections.*

Drilled holes in flanges and similar circular parts are shown at their true distances from the center in sectional views and in exterior views. The preferred method of projection is used when necessary.

ASSIGNMENT I

Make a two-view working drawing of the guide ring. The side view is to be changed to an aligned section.

Procedure

1. Lay out the margin lines on the drawing sheet and letter the title strip.
2. Draw the front view. Draw cutting-plane line A-A.
3. Draw the side view making it an offset section on cutting-plane line A-A. Preferred projection is used for the portion along the offset of the cutting plane. Rotate this portion to the center line before projecting to the sectional view. Use true projection for all other details. Show all visible details behind the cutting plane. Use the section lining for cast iron.
4. Dimension the drawing.
5. Check the accuracy of the sectional and regular views on the drawing.

REVIEW QUESTIONS

1. What is an offset section?
2. What is the purpose of offset sections?
3. Why was the part along the offset of the cutting plane rotated to the center line?
4. What is meant by true projection? Preferred projection?

REVOLVED SECTIONS

Revolved sections show the true shape of the cross section of spokes, arms, ribs, bars, and other elongated parts. A *revolved section* is a cross section that has been revolved through 90 degrees.

In theory, a cutting plane cuts across the part as shown in figure 7-21 at A. The cutting plane with the cross section is then revolved through 90 degrees as shown at B. The revolved section is superimposed on the top view at C. The cutting plane is now shown in the drawing.

Revolved sections for various shaped parts are illustrated in figure 7-22. The visible and invisible edge lines of the view are not drawn through the revolved sections. If the original lines of the view conflict with the revolved section, then a portion of the view is broken out to provide a clear space for the revolved section. Freehand break lines are then drawn as shown in the last three revolved sections of figure 7-22.

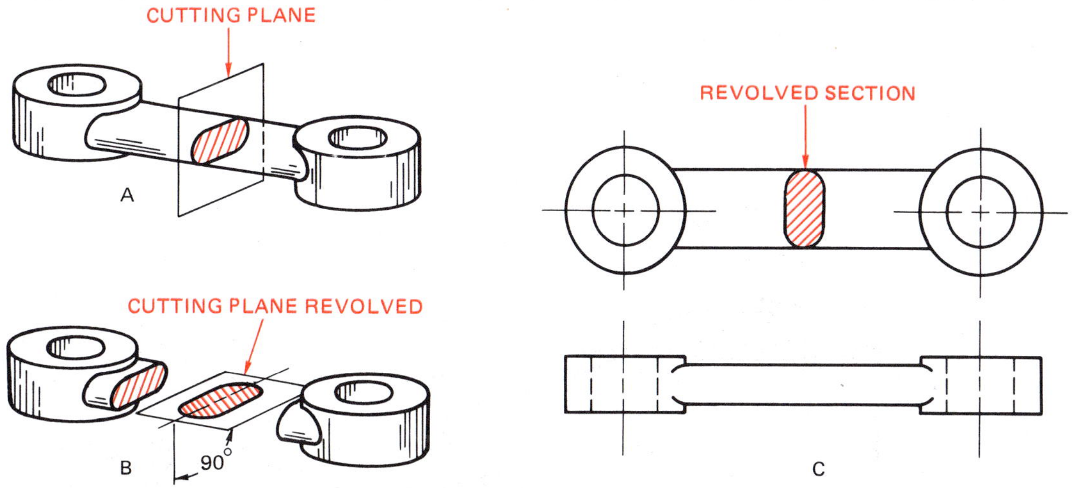

Fig. 7-21 Theoretical construction of a revolved section

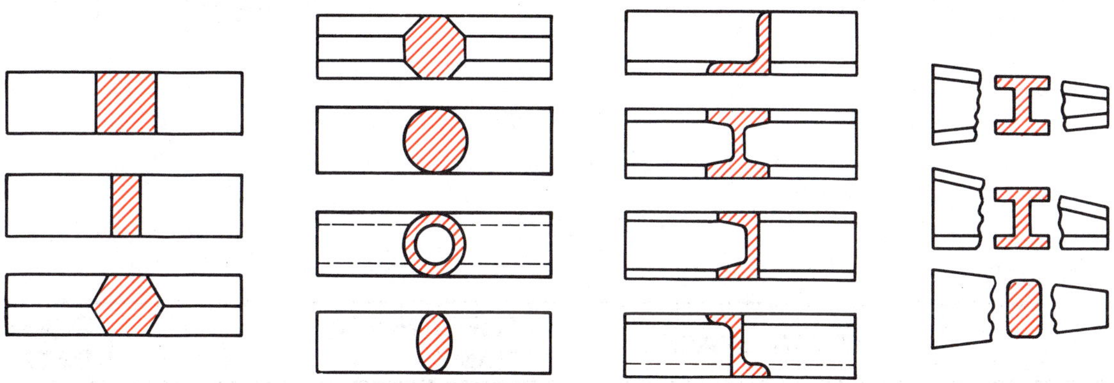

Fig. 7-22 Revolved sections

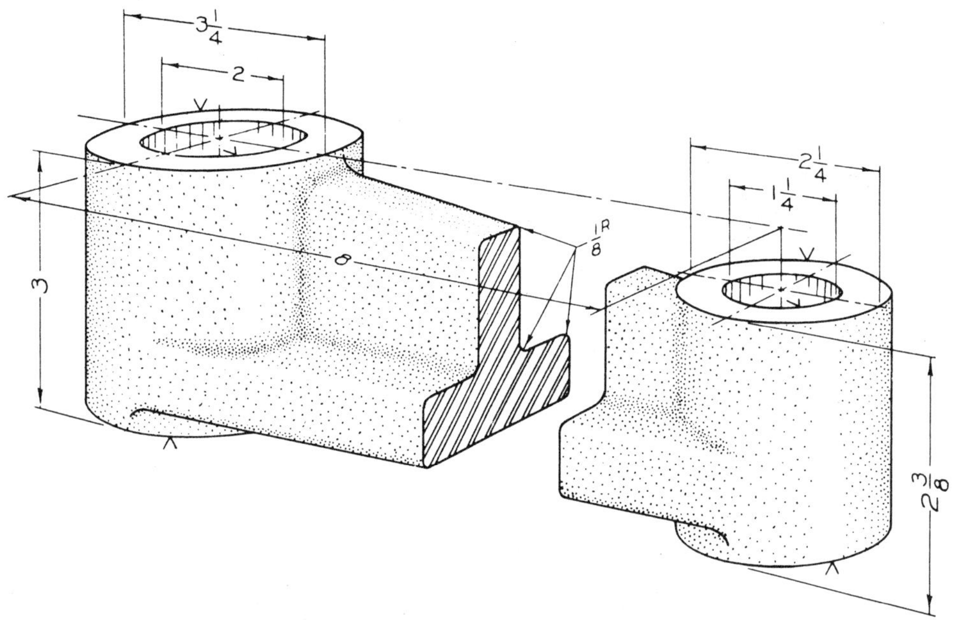

NAME OF SCHOOL CITY OR TOWN	CLAMP BRACKET	SHEET No. 7-J NAME	CLASS DATE

Fig. 7-23

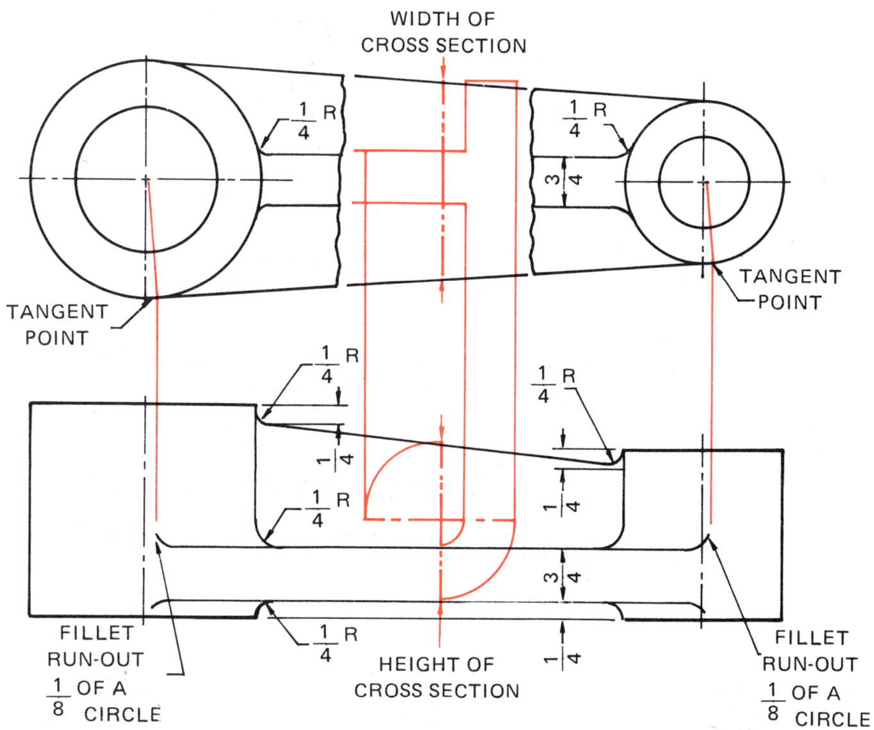

Fig. 7-24 Construction of a revolved section and run-outs of fillets

RUNOUT OF FILLETS

The arm of the clamp bracket, figure 7-23, has filleted intersections. The method of drawing the runouts of these fillets is illustrated in figure 7-24.

ASSIGNMENT J

Make a two-view working drawing of the clamp bracket. On the drawing, a revolved section of the arm of the clamp bracket is to be drawn in the top view according to the procedure in figure 7-21. Here the cross section at the cutting-plane line in the front view is revolved through 90 degrees to the horizontal position. Lines are then projected to the top view to mark the height of the revolved section. The width of the revolved section is the same as the width of the arm at the cutting-plane line in the top view.

Procedure

1. Lay out the margin lines on the drawing sheet and letter the title strip.
2. Draw a horizontal center line 2 3/8 inches below the top margin line; a vertical center line 4 1/8 inches from the left margin line.
3. Draw the top view. Locate the tangent points as shown in figure 7-24.
4. Draw the front view 3/4 inch above the title strip. Draw run-outs of fillets.
5. In the approximate center of the arm in the top and front views, draw a light vertical line to represent the position of the cutting plane. Draw the revolved section in the top view following the method used in figure 7-24. Draw the two break lines freehand. Erase the cutting-plane lines and other lines to leave the revolved section in a clear space.
6. Place dimensions on the drawing in the same position as they appear in the pictorial drawing.
7. Check the accuracy of the drawing.

REVIEW QUESTIONS

1. Define a revolved section.
2. For what purpose are revolved sections used?
3. Are visible or invisible edge lines drawn through revolved sections?
4. Why is a broken out clear space sometimes provided for a revolved section?
5. What metal is represented by the section lining in the revolved section?

REMOVED OR DETAIL SECTIONS

Removed or detail sections are similar to revolved sections since they both show the true shape of the cross section of spokes, arms, ribs, bars, and other elongated parts. The difference between removed and revolved sections is mainly in the place where they are drawn. *Removed sections* are cross sections that are removed from their original positions and placed in any convenient space near the principal view. *Revolved sections* are cross sections that have been revolved through 90 degrees in their original positions.

A removed section is preferred to a revolved section if the revolved section tends to overcrowd the view. Removed sections are often drawn to a larger scale to show details more clearly.

The position of the cutting plane for each removed cross section is indicated by a cutting-plane line in the principal view. These cutting-plane lines are labeled A-A, B-B, and C-C, and the corresponding removed cross sections are labeled SECTION A-A, SECTION B-B, etc., figure 7-25.

Arrowheads on the ends of the cutting-plane lines indicate the direction the section is viewed. However, details visible beyond the cutting plane are not shown in removed sections. Removed sections are sometimes on an extension of a center line, figure 7-25 at B.

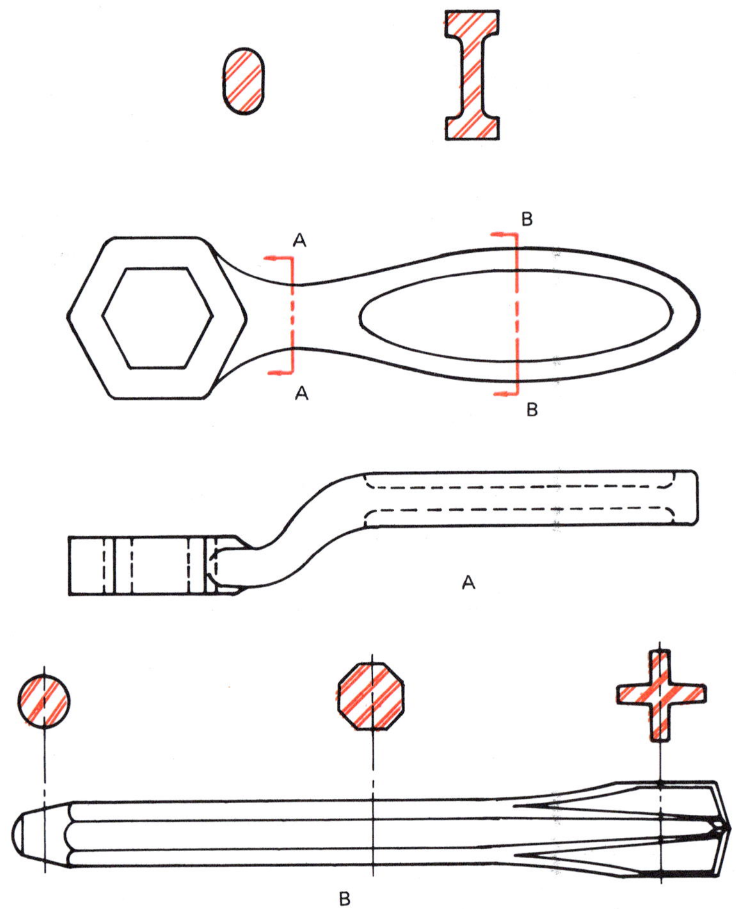

Fig. 7-25 Removed or detail sections

ASSIGNMENT K

Make a two view working drawing of the fixture wrench, figure 7-26. Draw three removed sections as indicated by cutting-plane lines A-A, B-B, and C-C in figure 7-27.

Procedure

1. Lay out the margin lines on the drawing sheet and letter the title strip.
2. Draw a horizontal center line 3 7/8 inches below the top margin line.

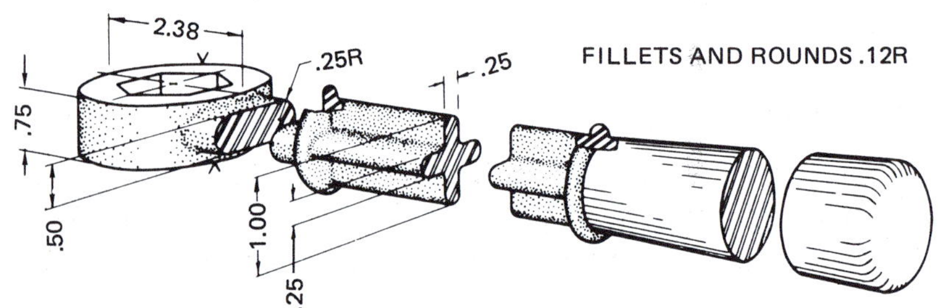

Fig. 7-26

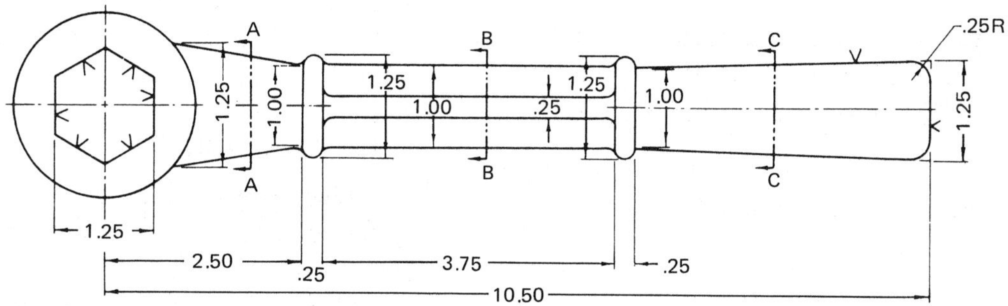

FILLETS & ROUNDS .12 R

NAME OF SCHOOL CITY OR TOWN	FIXTURE WRENCH	SHEET No. 7-K NAME	CLASS DATE

Fig. 7-27

3. Draw a vertical center line 3 1/4 inches from the left margin line.
4. Draw the top view. Draw the hexagon following the same steps as shown in figure 7-28.
5. Draw the horizontal center line for the front view 2 1/8 inches above the title strip. Draw the front view. Draw the intersection of the wrench head and the arm for runouts.
6. In the top view, draw cutting-plane lines A-A, B-B, and C-C as in figure 7-27.
7. In the space above the top view, draw the three removed cross sections. Beneath the removed sections letter the notations SECTION A-A, SECTION B-B, and SECTION C-C.
8. Dimension the drawing.
9. Check your drawing for accuracy of all lines and removed sections.

REVIEW QUESTIONS

1. Write the definition of a removed or detail section.
2. What is a revolved section?
3. Why are removed or detail sections used?
4. What is the main difference between detailed and removed sectional views?
5. What does Section A-A mean when placed under a removed section?

SPOKES IN SECTION

When a pulley with spokes or arms is cut by a cutting plane, the sectional view appears as shown in the first sectional view in figure 7-29.

There is little difference between this full section of a pulley with spokes and the full section of a webbed pulley. In order to distinguish spokes from a web in sectional views, the section lining is omitted from the spokes as shown in the preferred sectional view of figure 7-29. The preferred practice is to omit the section lining from spokes or arms in sectional views of pulleys, wheels, and gears.

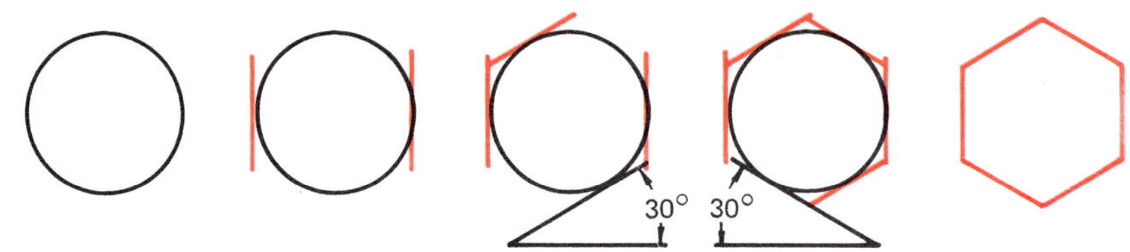

Fig. 7-28 To draw a hexagon about a circle

RIBS IN SECTION

When the cutting plane passes lengthwise through a rib, the preferred practice is to omit the section lining from the rib, figure 7-30.

The V-belt sheave in figure 7-33 has five spokes. A cutting plane passing through this sheave cuts through a spoke in the upper half and through the space between two spokes in the lower half. A sectional view of this sheave is similar to the first sectional

Fig. 7-29 Spokes in section

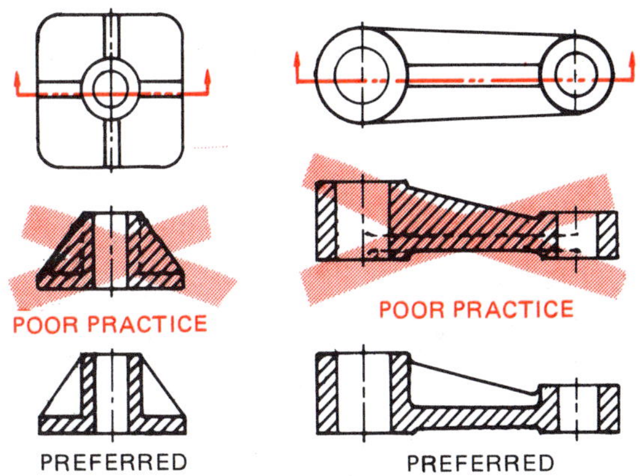

Fig. 7-30 Ribs in section

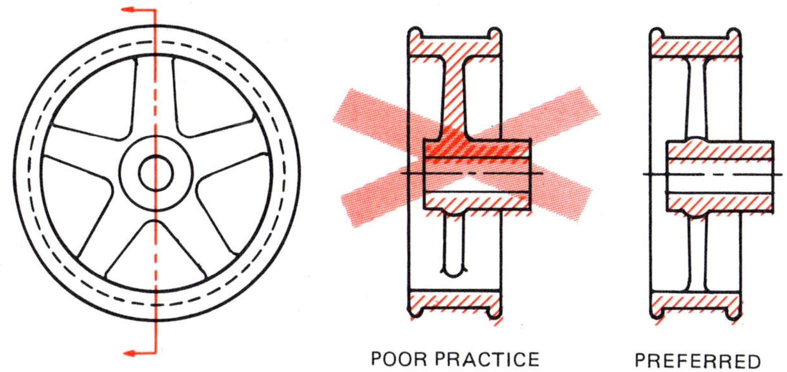

Fig. 7-31 Spoke aligned in section

view in figure 7-31. In this sectional view, the upper spoke is in section, and the lower spoke is foreshortened and not shown in its true shape. Also, the view is not symmetrical. Such a view may be misleading.

ALIGNED SECTIONS

The preferred practice is to assume that the foreshortened spoke is rotated until it is parallel to the cutting plane (plane of the section) and then projected to the sectional view. This is illustrated in the preferred sectional view of figure 7-31. Such a view is called an aligned section. The section lining is omitted from the spokes in the aligned section.

An *aligned section* is a sectional view in which a sloping part is rotated parallel to the cutting plane to show its true shape.

The student is to make an aligned full section of the V-belt sheave. The keyway in this sheave is dimensioned by stating the width of the keyway and the exact distance from one side of the bore to the bottom of the keyway, figure 7-32. This method of dimensioning keyways is preferred when the depth of the keyway is to be checked by gauges.

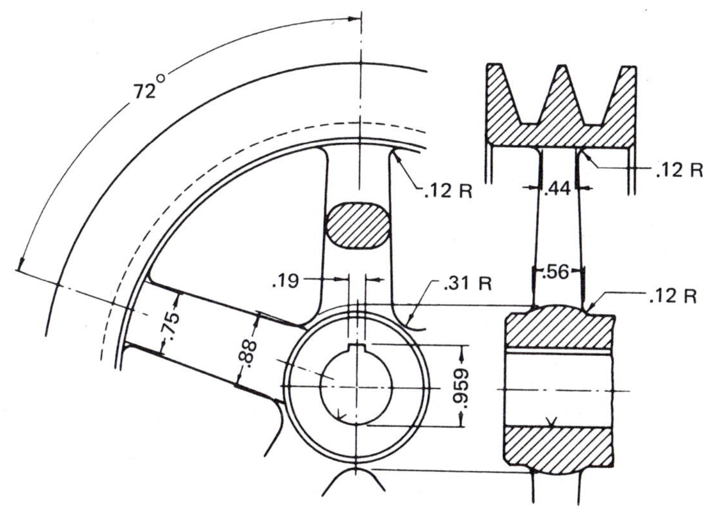

Fig. 7-32

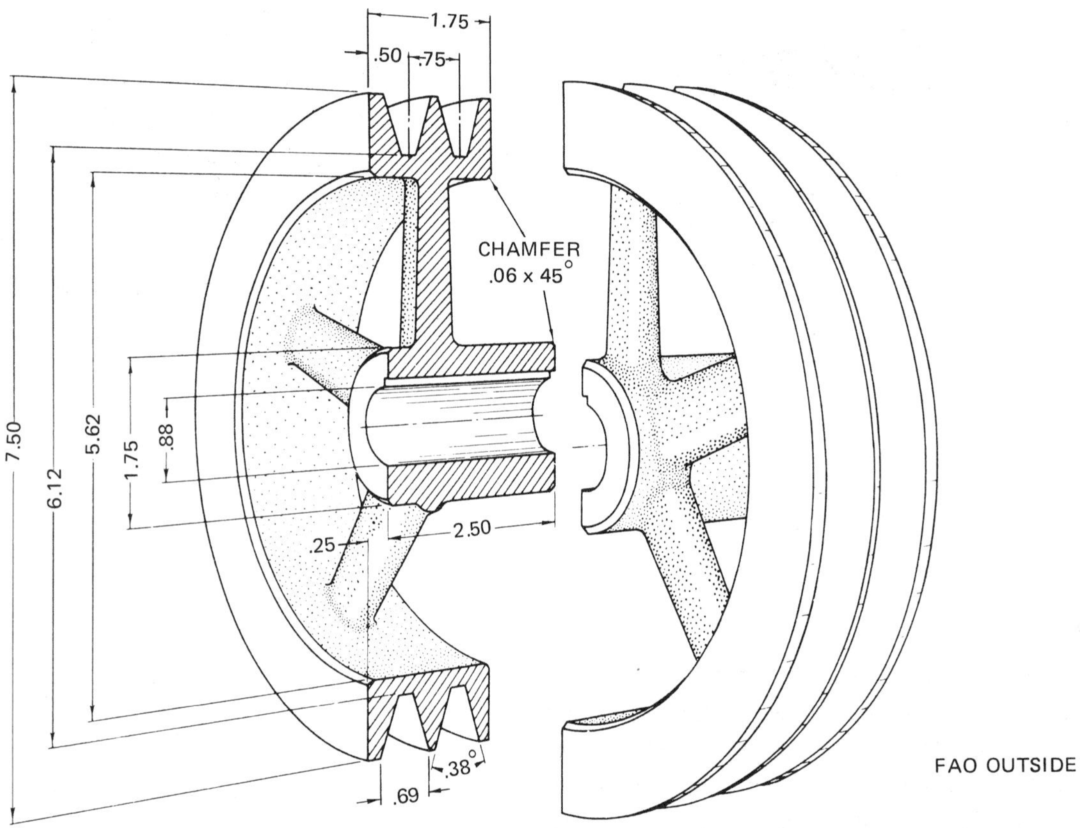

NAME OF SCHOOL CITY OR TOWN	V-BELT SHEAVE	SHEET No. 7-L NAME	CLASS DATE

Fig. 7-33

ASSIGNMENT L

Make a two-view working drawing of the V Belt sheave in figure 7-33. Make the front view a full aligned section and draw a revolved section of one of the spokes.

Procedure

1. Lay out margin lines on the drawing sheet and letter the title strip.
2. Draw a horizontal center line 5 inches below the top margin line and a vertical center line 5 1/4 inches from the left margin line.
3. Draw the side view. The dimensions for the spokes and the keyway are shown in Figure 7-32. Use a protractor to measure the 72 degree angle or divide the circle into five equal parts with dividers.
4. Draw the full aligned section 1 1/2 inches from the right margin line. The 38-degree angles are measured with a protractor. In this view, two humps are drawn on the outside of the hub. These humps represent the cross section of the .31-inch radius fillets which are between the spokes. The projection lines in figure 7-32 show how the height of these fillets is obtained. The radius of the cross section of the fillets is found by trial and error.
5. Draw a revolved section of the upper spoke in the side view.
6. Dimension the views.
7. Check the full aligned section in the front view and the revolved section of the spoke.

BROKEN-OUT SECTIONS

A broken-out section, figure 7-34, is used when less than a half section is sufficient to show some interior detail. Broken-out sections are also used when a full or half section removes some important exterior detail. An irregular break line separates the section from the exterior view.

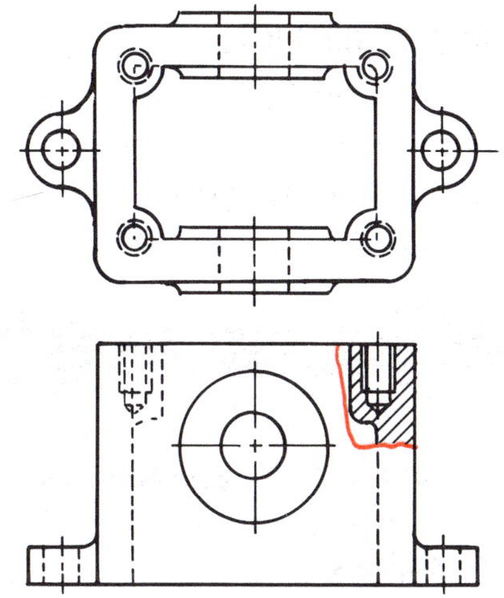

Fig. 7-34 Broken-out section

ASSEMBLY SECTIONS

Assembly drawings are often drawn in section to show how the interior parts fit together. In assembly sections, parts such as shafts, rods, bolts, screws, pins, keys, rivets, balls, and similar parts that lie in the path of the cutting plane are not drawn in section, figure 7-35. The section lining of two adjoining parts is drawn at right angles. The section lining of a third part adjoining the other two is drawn at 30 degrees or 60 degrees.

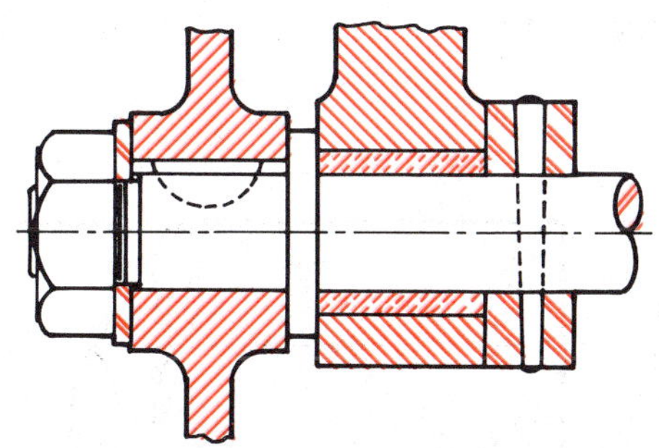

Fig. 7-35 Treatment of shafts, keys, pins and nuts in sectional views

THIN MATERIALS IN SECTION

Materials such as sheet metal, gaskets, and electric insulations that are too thin for conventional section lining, are shown solid in assembly sections. Adjoining parts in sectional views that are shown solid are separated by a small space, figure 7-36.

Fig. 7-36 Thin materials in section

CONVENTIONAL BREAKS

A part that is too long to fit the available drawing space may be shortened to a suitable length by using a conventional break in the view. The solid rod in figure 7-37 is too long to fit all of the views into the available space. Therefore the front view is shortened and a conventional break is drawn in this view. The actual length of the rod must be specified in the dimension when the view is shortened. Figure 7-38 gives the method of constructing conventional breaks for various cross sections.

REVIEW QUESTIONS

1. Write the definition of an aligned section.
2. Why is the section lining omitted from spokes in sectional views?

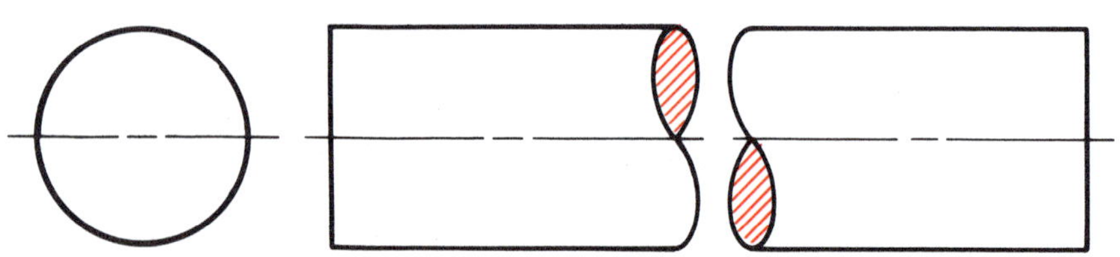

Fig. 7-37

3. Why is the spoke aligned in the sectional view of the V-belt sheave?
4. Why is meant by a broken-out section?
5. In assembly sections some parts are not drawn in section. What are they?
6. How are section lines drawn for two or more adjoining parts in assembly sections?
7. How are thin materials shown in sectional views?
8. What is a revolved section? What is a removed or detailed section?
9. What does FAO outside mean?

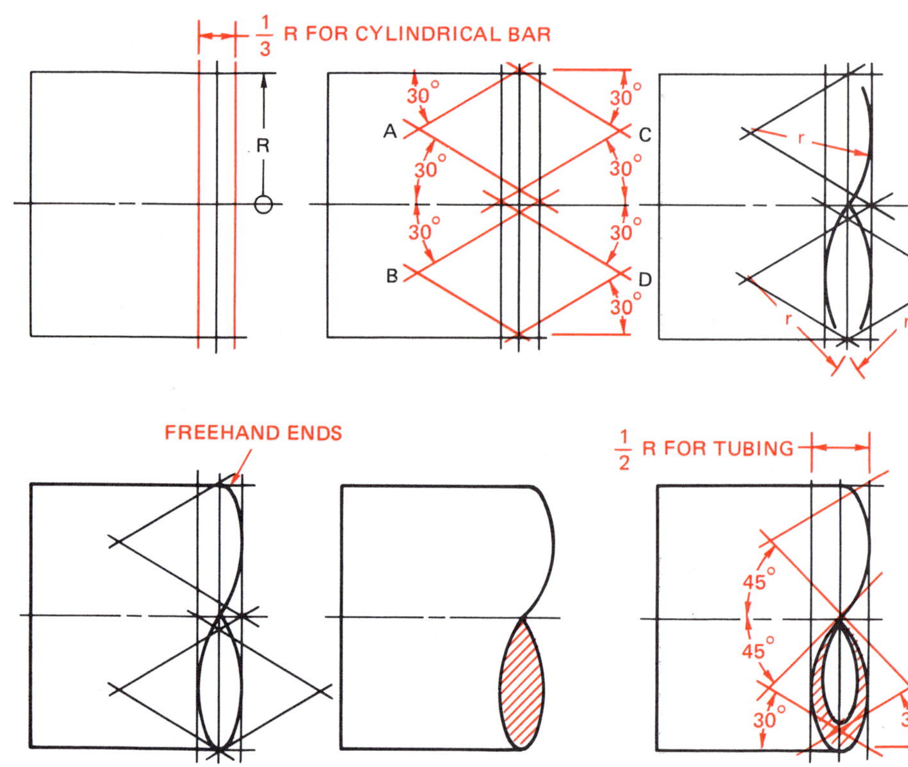

Fig. 7-38 Constructing conventional breaks for various cross sections

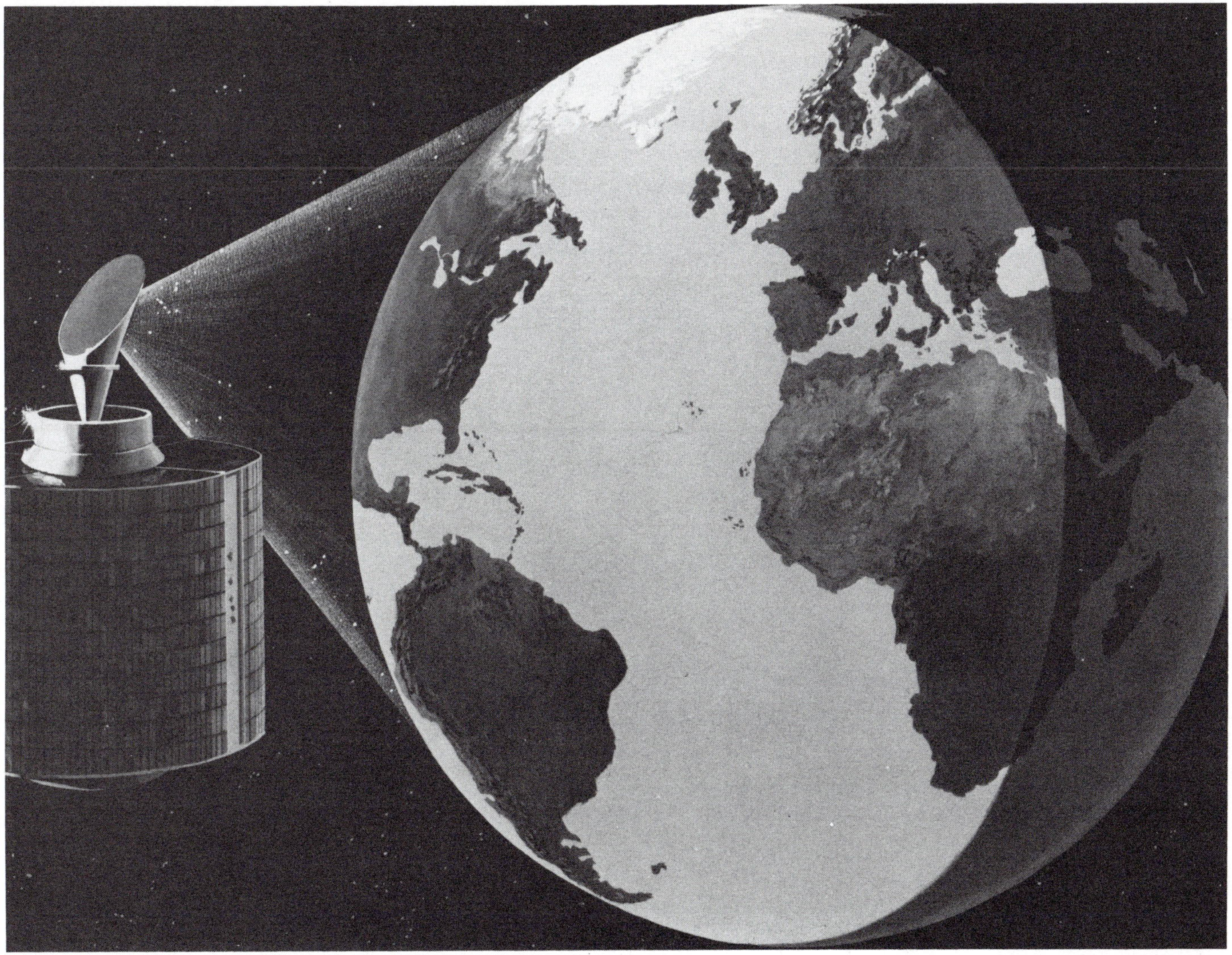

UNIT 8 DESCRIPTIVE GEOMETRY

Descriptive geometry finds points, lines, angles, surfaces, and solids by means of projections. It is based on the same principles orthographic projection uses to make multiview drawings.

The purpose of this unit is to show how the principles of descriptive geometry are used to solve space problems by graphical methods. Points and lines alone, and their relation to each other, are included in the problems.

The drawings in this chapter are done on 11-by-17-inch cross-section paper having four squares to an inch. Each problem has some part given, such as a line or point, and reference lines. The problems must be set up and located on the drawing paper exactly as shown so that the solution is completed in the space allotted.

The given portion of each problem is drawn with dark, sharp lines. Construction lines are light and sharp. The accuracy of the graphical solution to these problems depends upon skill and accuracy in carrying out each step in the procedures.

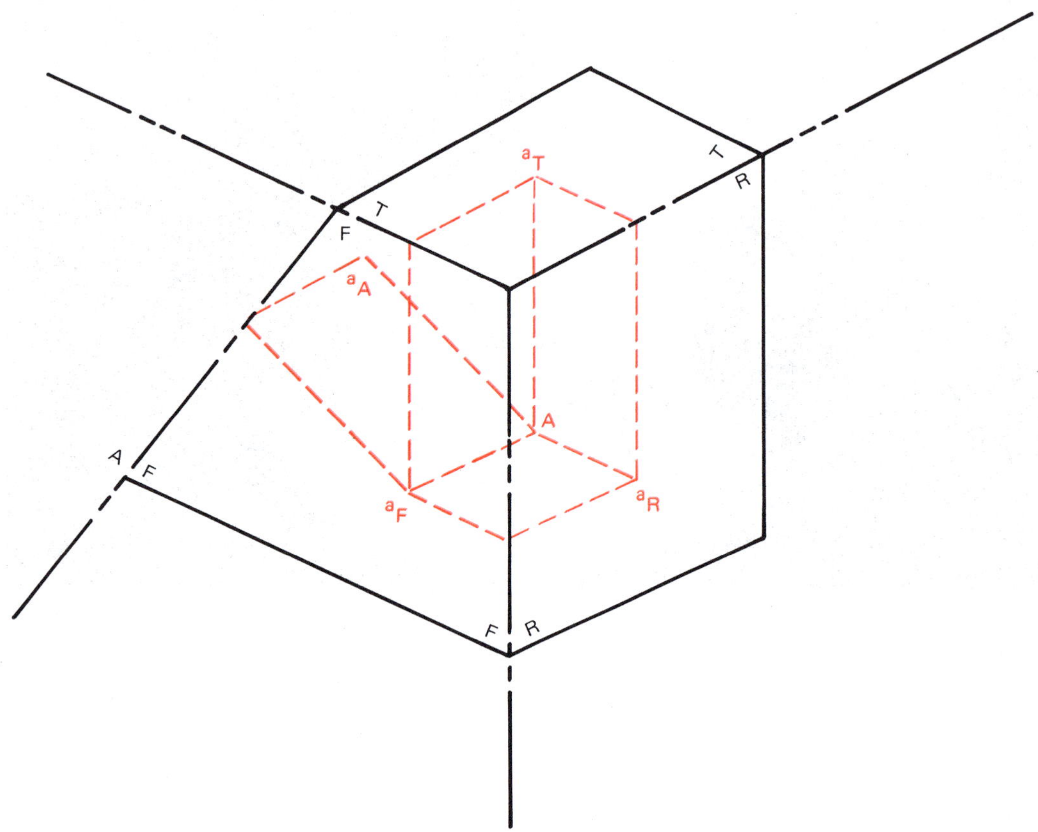

Fig. 8-1

SYSTEM OF NOTATION

Actual points and lines in space are indicated by upper-case letters. Projections of points and lines are indicated by lower-case letters. Projections of points are further labeled with subscripts to indicate the view they are in. The front view of point A is labeled a_F. The top view is labeled a_T. The right side view is labeled a_R.

Reference lines are noted by letters such as T-F, T-R, F-R, and A-F. They are the lines of intersection of the mutually perpendicular projection planes, figure 8-1.

Reference lines, figure 8-2, are made up of a series of dashes 3/4 inches long, alternating with two short dashes 1/16 inch long. These are drawn as heavy as visible edge lines. The distance between a_T and reference line T-F is the distance that point A lies in back of the front vertical plane. The distance between a_F and reference line F-T is the distance that point A lies below the upper horizontal plane.

Figure 8-3 illustrates how the actual projections (top, front, right side and auxiliary view) of point A appear on the drawing. Note that the projectors from one view to another are always perpendicular to the reference lines. Note, too, that the distance X from point A (a_A, a_T, a_R) to the front projection plane is always the same. The distance from a point to any reference plane is always equal in all views in which the point is drawn with reference to the same reference plane.

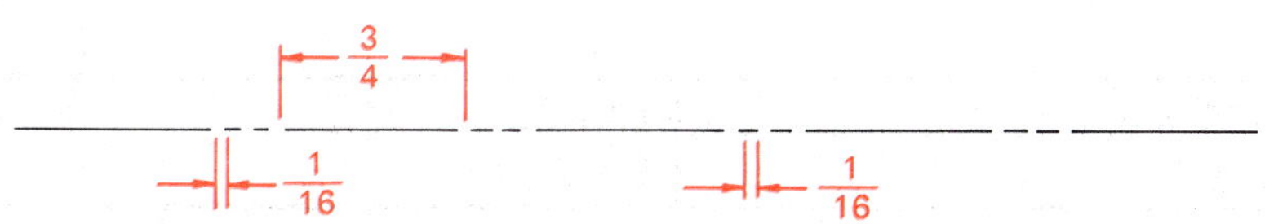

Fig. 8-2 Reference line

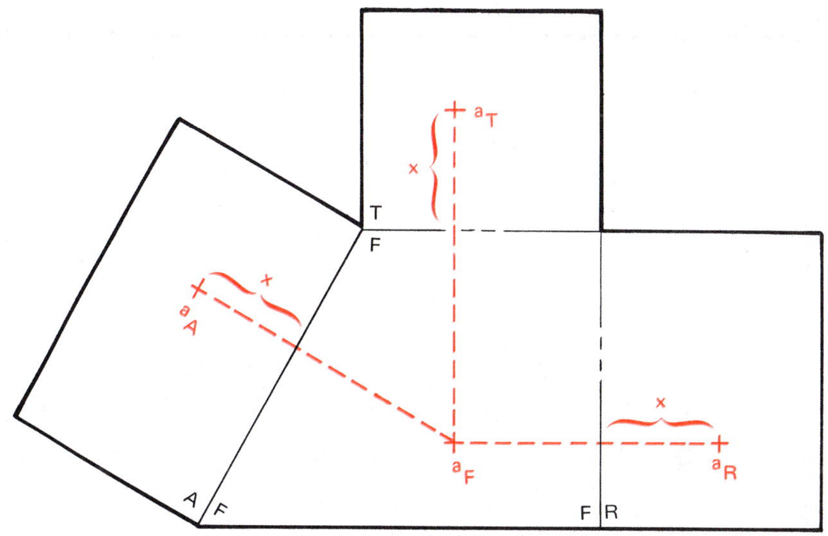

Fig. 8-3

ASSIGNMENT A

Draw and solve the two descriptive geometry problems specified in figure 8-4.

Procedure

1. Lay out the margin lines on the four squares to an inch cross-section drawing paper and letter the title strip.
2. Divide the sheet into two equal spaces.
3. Locate and set up the problems on the drawing exactly as shown. Small errors in this step can cause larger errors in the solution.
4. In problem 1, draw a very light construction line from a_F perpendicular to and through reference line F-T.

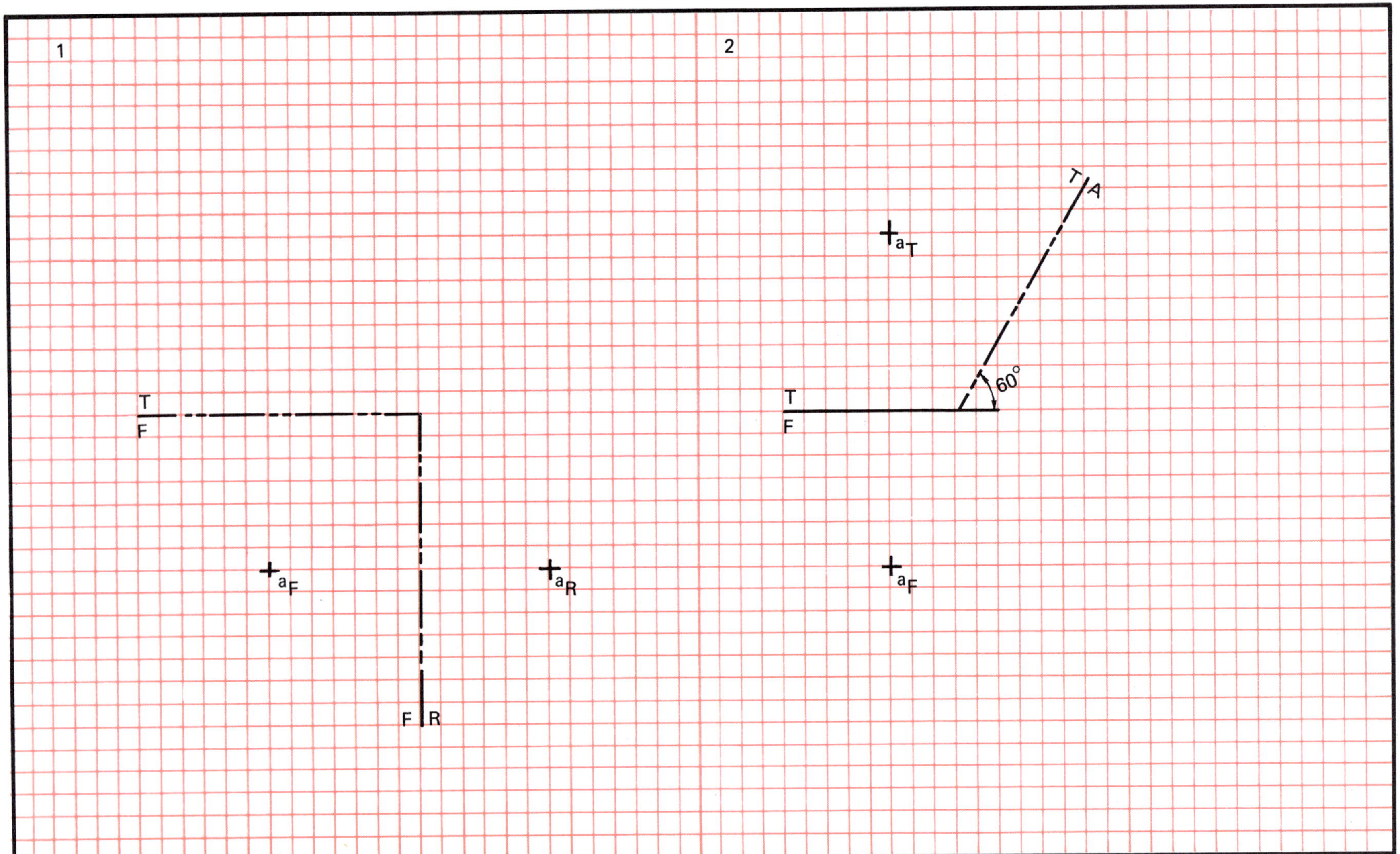

NAME OF SCHOOL CITY OR TOWN	DESCRIPTIVE GEOMETRY	SHEET No. 8-A NAME	CLASS DATE

Fig. 8-4

5. Next, adjust the dividers to the distance from reference line F-R to point a_R. This is the distance that point A lies in back of the front vertical plane (F).
6. Then, with the dividers, point off on the construction line through reference line F-T the distance that point A lies in back of the front vertical plane (F).
7. The point located is a_T.
8. In problem 2, draw a very light construction line from a_T perpendicular to and through reference line T-A.
9. Next, adjust the dividers to the distance from reference line T-F to point a_F. This is the distance that point A lies below the upper horizontal plane (T).
10. Then, with the divider, point off on the construction line through reference line T-A the distance that point A lies below the upper horizontal plane (T).
11. The point located is a_A.
12. Check the accuracy of both problems for location, set up, and solution.

REVIEW QUESTIONS

1. Define descriptive geometry, point, line, and angle.
2. Describe the system of notation used in descriptive geometry.

KEY TO THE SYSTEM OF ORTHOGRAPHIC PROJECTION

The key to the system of projection used in descriptive geometry is pictorially illustrated in figure 8-5. Thoroughly study the details of this figure. Once the principles involved here are understood, the remainder of this subject should be understood without too much difficulty.

Figure 8-5 shows a series of consecutive projection planes where there are two planes perpendicular to a third plane. Most of these projection planes are discussed in the unit on auxiliary views. For example, both the vertical projection plane (F) and the auxiliary elevation plane (A) are perpendicular to the horizontal projection plane (T). The horizontal and inclined auxiliary projection planes (T and B) are perpendicular to the auxiliary elevation projection plane (A).

Note that point A lies X distance below the horizontal projection plane. This distance is indicated on the vertical and the auxiliary elevation projection plane A where the respective projections of point A appear (a_T and a_A).

Also note that point A is Y distance behind the auxiliary

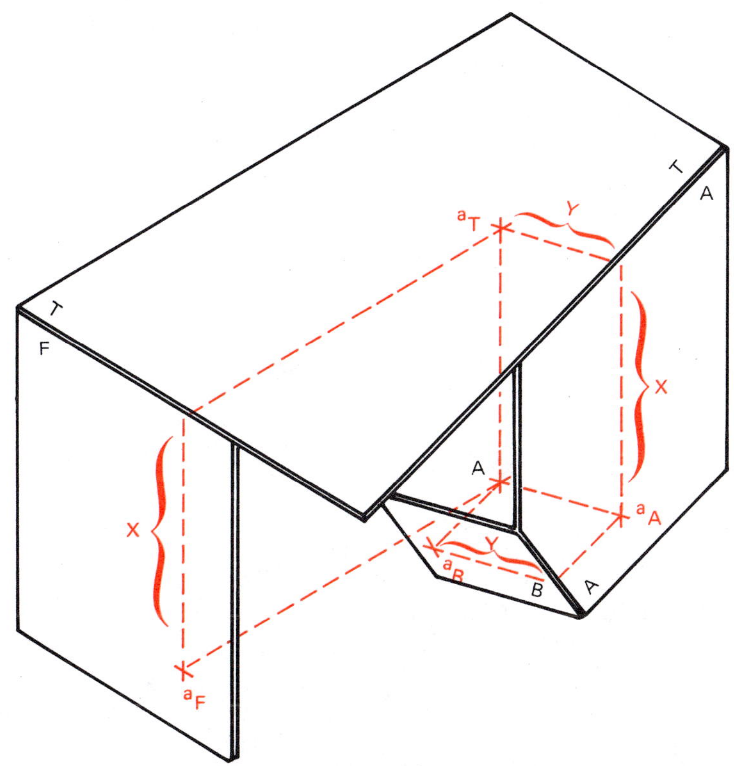

Fig. 8-5

elevation projection plane A. This distance appears on the horizontal and inclined auxiliary projection planes (T and B) where the respective projections of point A appear (a_A and a_B).

When the consecutive projection planes in figure 8-5 are swung into the same plane, they appear as illustrated in figure 8-6.

Figure 8-7 shows the consecutive projection planes as they appear in the plane of a sheet of drawing.

Assume that the horizontal and vertical projections of point A (a_T and a_F) are given. If it is required to find the auxiliary elevation of this point as shown, the distance X is measured with dividers from reference line T-F to the vertical projection a_F. This distance is laid off on the auxiliary elevation plane from reference line T-A on the projector, and thus a_A is located.

In order to find a_A, it is necessary to go back two views (horizontal and vertical) following the projectors to get the measurements which locates a_A.

This procedure applies to all problem solutions.

Apply this procedure to find how a_B is located, keeping in mind the principles involved.

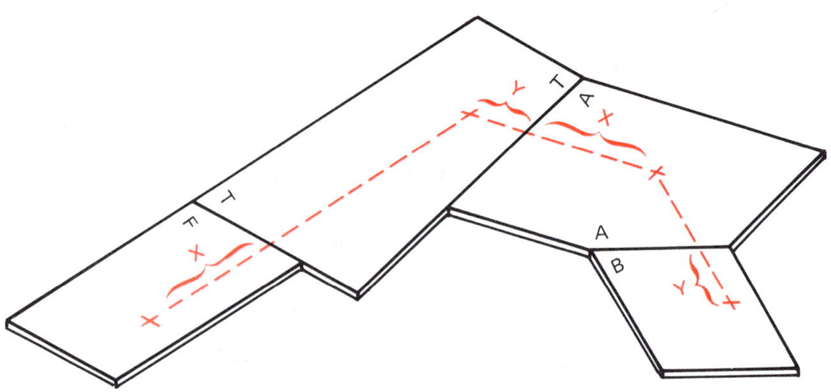

Fig. 8-6

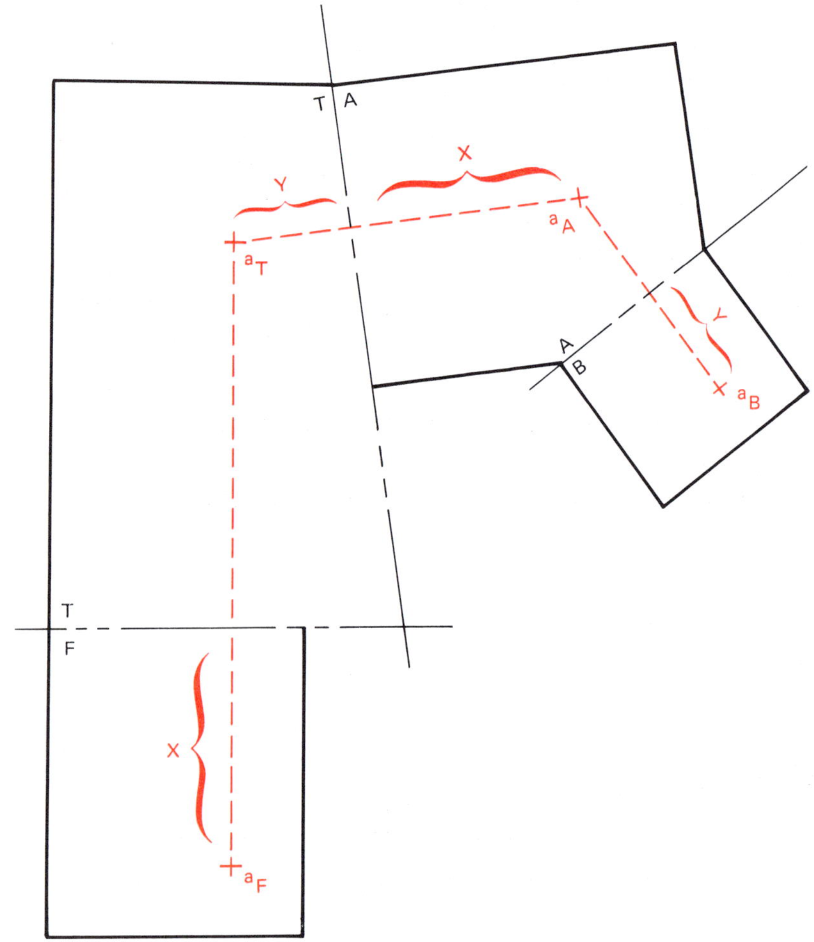

Fig. 8-7

TRUE LENGTH OF A LINE

Line AB in figure 8-8 does not appear true length in any of the three normal views of the object. A line appears true length in a view only when the projection plane is parallel to the line. Drawing a reference line parallel to the line as in figure 8-9, causes the line to be true length in the next view.

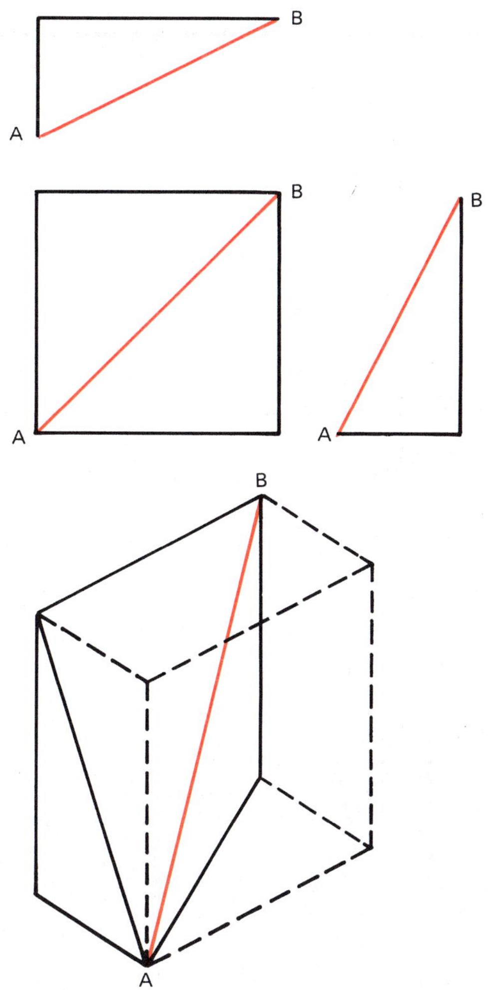

Fig. 8-8

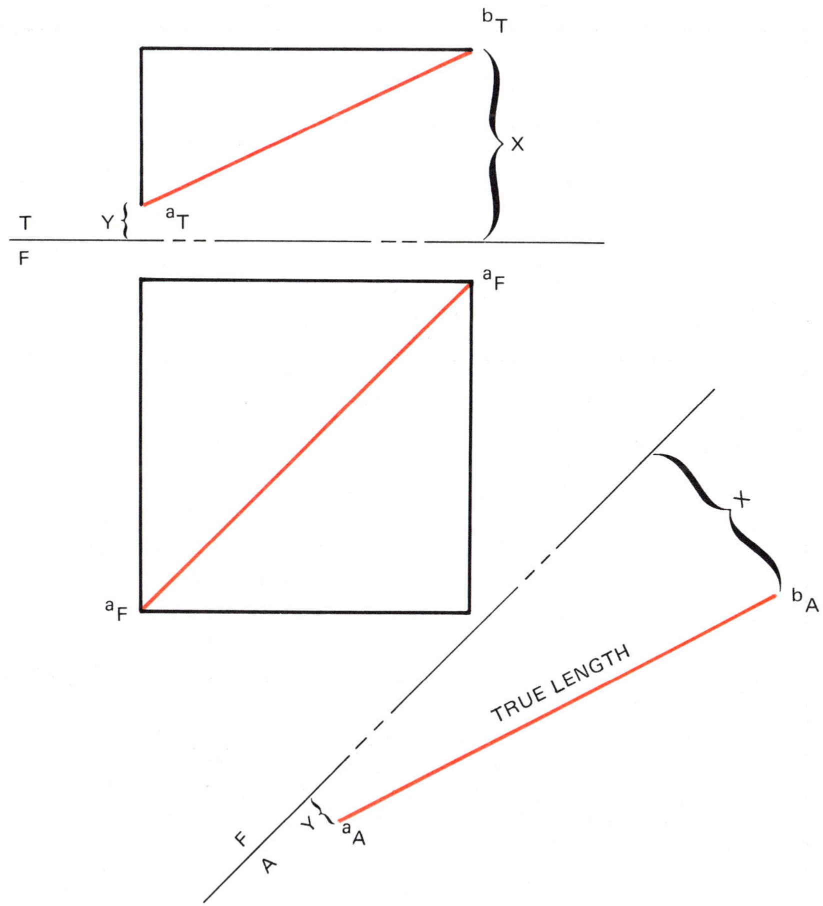

Fig. 8-9

ASSIGNMENT B

Draw and solve the four descriptive geometry problems in figure 8-10.

Procedure

1. Lay out the margin lines on the four squares to an inch cross-section drawing paper and letter the title strip.
2. Divide the sheet into four equal spaces.
3. Locate and set up the problems on the drawing exactly as shown.

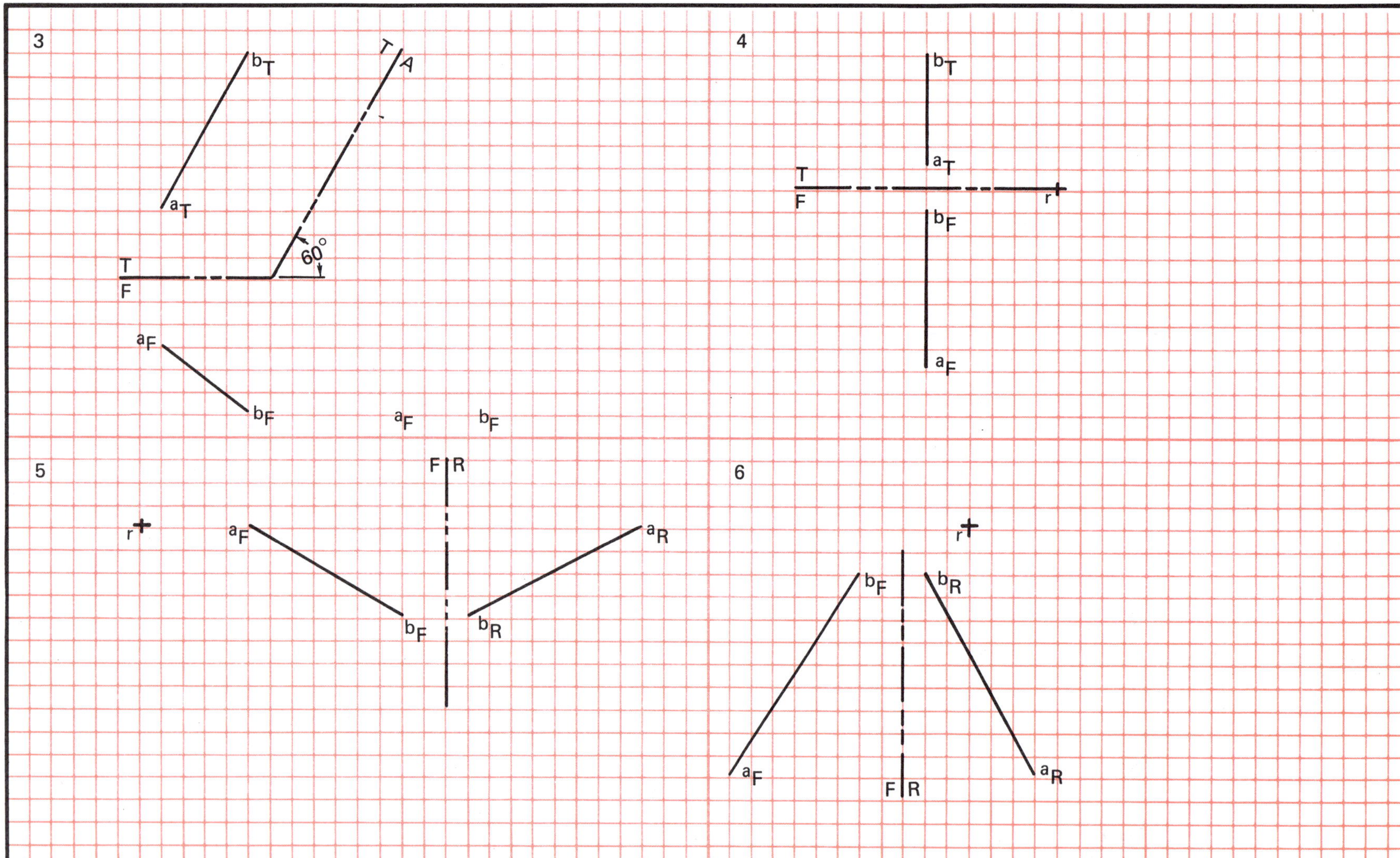

NAME OF SCHOOL CITY OR TOWN	DESCRIPTIVE GEOMETRY	SHEET No. 8-B NAME	CLASS DATE

Fig. 8-10

4. In problem 3, reference line T-A has been given parallel to line a_T b_T so that line AB will appear in true length in View A. Draw very light construction lines from a_T and b_T perpendicular to and through reference line T-A.
5. Adjust dividers to the distance from reference line T-F to a_F. This is the distance that A lies below the upper horizontal plane (T).
6. Transfer this distance to the auxiliary elevation A by pointing it off on the construction line from a_T through reference line T-A.
7. This locates a_A.
8. Find b_A in a similar manner.
9. Connect a_A and b_A with a line. This line a_A b_A is in true length (T.L.).
10. In problem 4, the vertical and horizontal projections of line AB are given. Since AB is a profile line (parallel to profile projection plane R), it will appear in true length in view R.
11. Through reference point r draw a reference line F-R perpendicular to T-F.
12. Following the same procedure used in solving problem 3, find the true length of line AB in problem 4.
13. In problem 5, draw a reference line parallel to one of the given lines and through reference point r. Find the true length of line AB.
14. In problem 6 find the true length of line AB using the same procedure as in problem 5.
15. Check the accuracy of the problems for location, set up and solution.

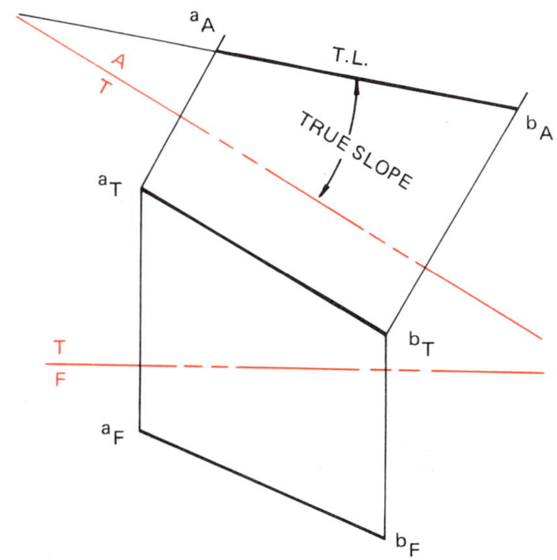

Fig. 8-11

REVIEW QUESTIONS

1. What is the relationship of the projection plane to a line when the line appears in true length on the projection plane?
2. What is a profile line?

TRUE SLOPE OF A LINE

The slope of a line is the angle that the line makes with a horizontal plane. This angle can be seen in true size only in the elevation views in which the line appears in true length, figure 8-11.

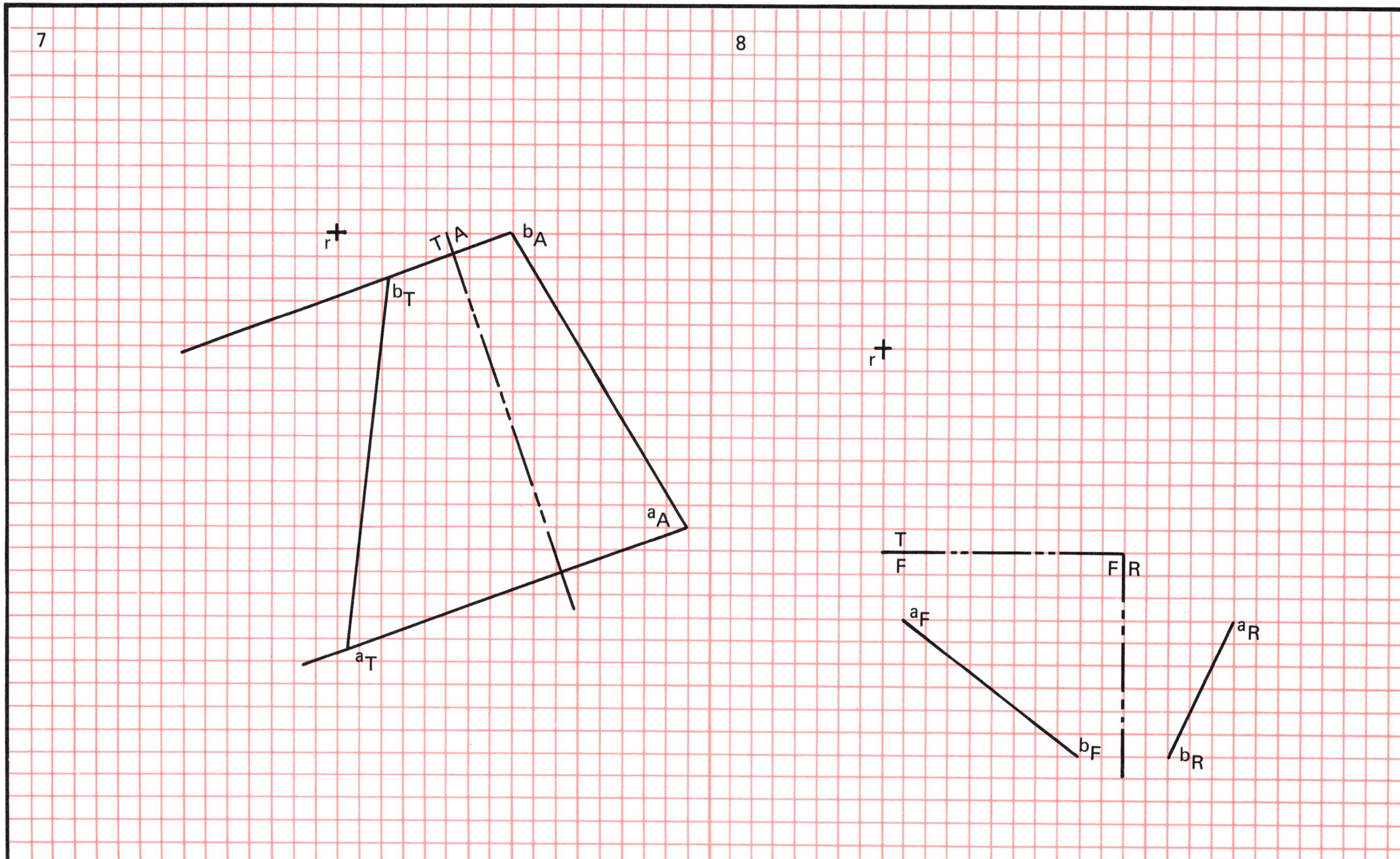

NAME OF SCHOOL CITY OR TOWN	DESCRIPTIVE GEOMETRY	SHEET No. 8-C NAME	CLASS DATE

Fig. 8-12

ASSIGNMENT C

Draw and solve the two descriptive geometry problems specified in figure 8-12.

Procedure

1. Lay out the margin lines on the four squares to an inch cross-section drawing paper and letter the title strip.
2. Divide the sheet into two equal spaces.
3. Locate and set up the problems on the drawing exactly as shown.
4. Since the true slope of a line can be found only in an elevation view in which the line appears in true length, in problem 7 the horizontal projection of line AB ($a_T\ b_T$) must first be found.
5. Through reference point r, draw reference line T-A parallel to $a_T\ b_T$.
6. In view A, $a_A\ b_A$ appears in true length and shows the true slope of line AB.
7. In problem 8 the horizontal projection and an auxiliary elevation projection of line AB are given. Since $a_A\ b_A$ is not in true length, the true slope of AB does not appear in view A.
8. To find the true length of AB and the true slope, draw a reference line T-B through reference point r and complete the projection in view B.
9. Check the accuracy of both problems for location, set up and solution.

REVIEW QUESTIONS

1. What is the relationship of the projection plane to a line when the true slope of the line appears on the projection plane?
2. Can the true slope of a line be found in any view in which the line appears in true length? If not, what are the limitations?

SHORTEST DISTANCE FROM A POINT TO A LINE

The shortest distance from a point to a line is a line that is perpendicular to the given line. This distance is found in the views where the given line appears as a point, figure 8-13.

To project the line of the shortest distance back into the other views, remember that a line appears in true length only in views where the projection plane is parallel to the line. Therefore, the shortest distance line is always parallel to the reference line in any view adjacent to the view in which the shortest line is found in true length.

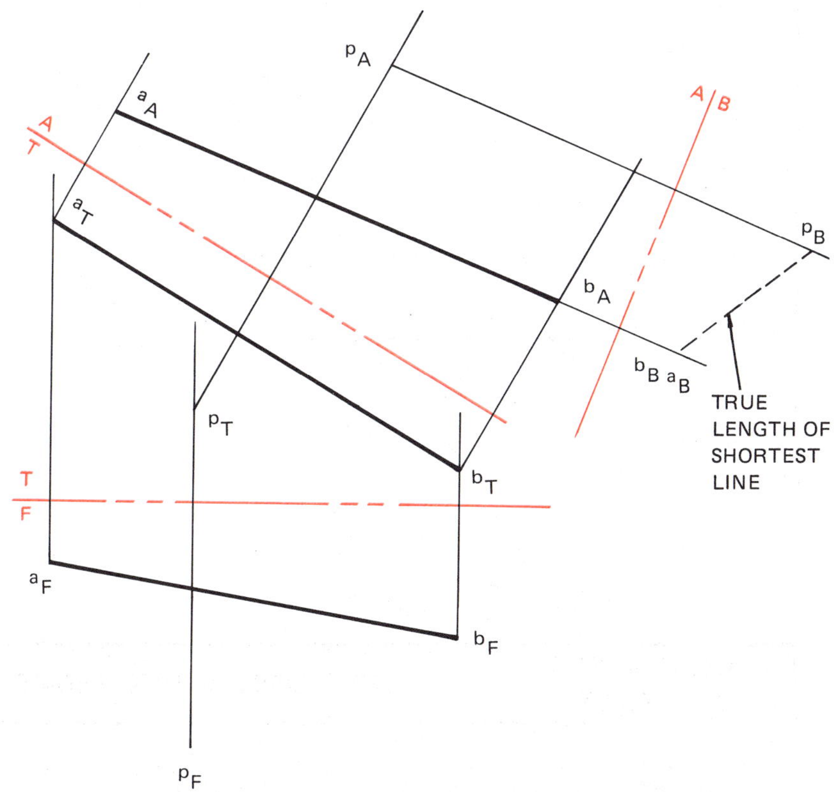

Fig. 8-13

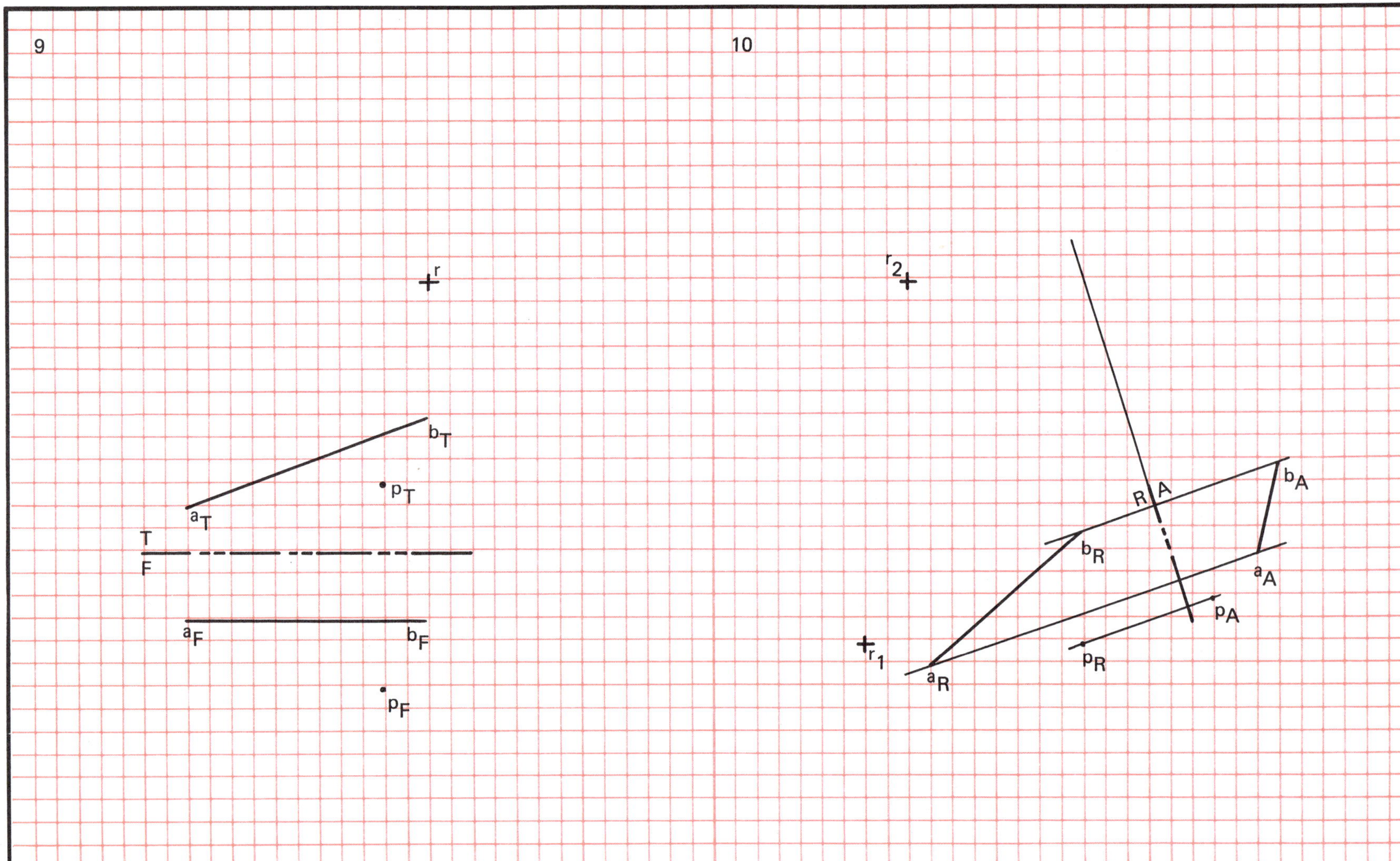

NAME OF SCHOOL CITY OR TOWN	DESCRIPTIVE GEOMETRY	SHEET No. 8-D NAME	CLASS DATE

Fig. 8-14

ASSIGNMENT D

Draw and solve the two descriptive geometry problems specified in figure 8-14.

Procedure

1. Lay out the margin lines on the four squares to an inch cross-section drawing paper and letter the title strip.
2. Divide the sheet into two equal spaces.
3. Locate and set up the problems on the drawing exactly as shown.
4. In problem 9, the front and top views of AB are given. Since AB is a horizontal line (parallel to horizontal projection plane T), it appears in true length in view T.
5. Draw reference line T-A perpendicular to a_T b_T through reference point r.
6. In view A, the line connecting b_A a_A and point p_A is the shortest distance from the point to the line.
7. Project the line representing the shortest distance from the point to the line back into views T and F. Since this line is in true length in view A, it will be parallel to T-A in view T.
8. Find the shortest distance from line AB to point P in problem 10. Draw reference line R-B through reference point r_1, and reference line B-C through reference point r_2.
9. Project the shortest distance from the line to the point back into all views.
10. Check the accuracy of both problems for location, set up, and solution.

REVIEW QUESTIONS

1. What is the relationship of the shortest line from a point to a given line to the given line?
2. What is a horizontal line?
3. What is the relationship of the shortest line from a point to a given line to the projection plane in which the true length of this line is found?

SHORTEST DISTANCE BETWEEN TWO LINES

The shortest distance between two lines is a line that is perpendicular to the given lines. This distance is found in the views where either one of the given lines appears as a point, figure 8-15.

ASSIGNMENT E

Draw and solve the two descriptive geometry problems specified in figure 8-16.

Procedure

1. Lay out the margin lines on the four squares to an inch cross-section drawing paper and letter the title strip.

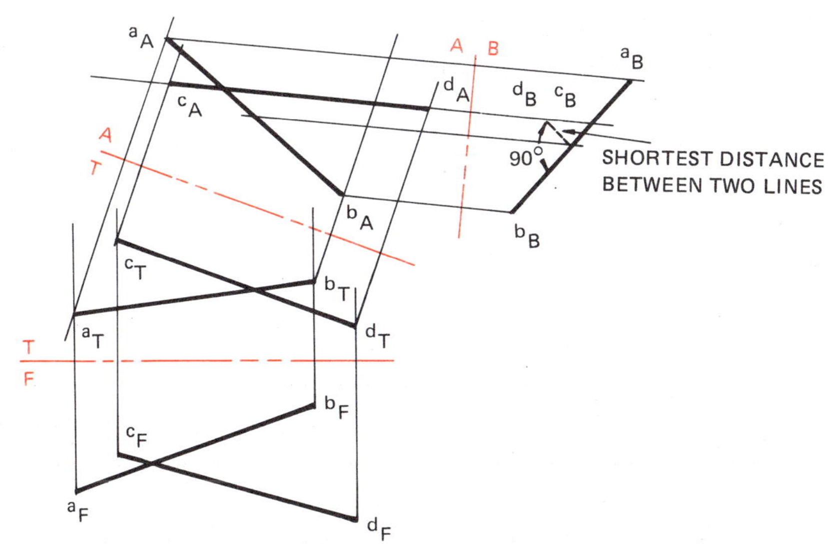

Fig. 8-15

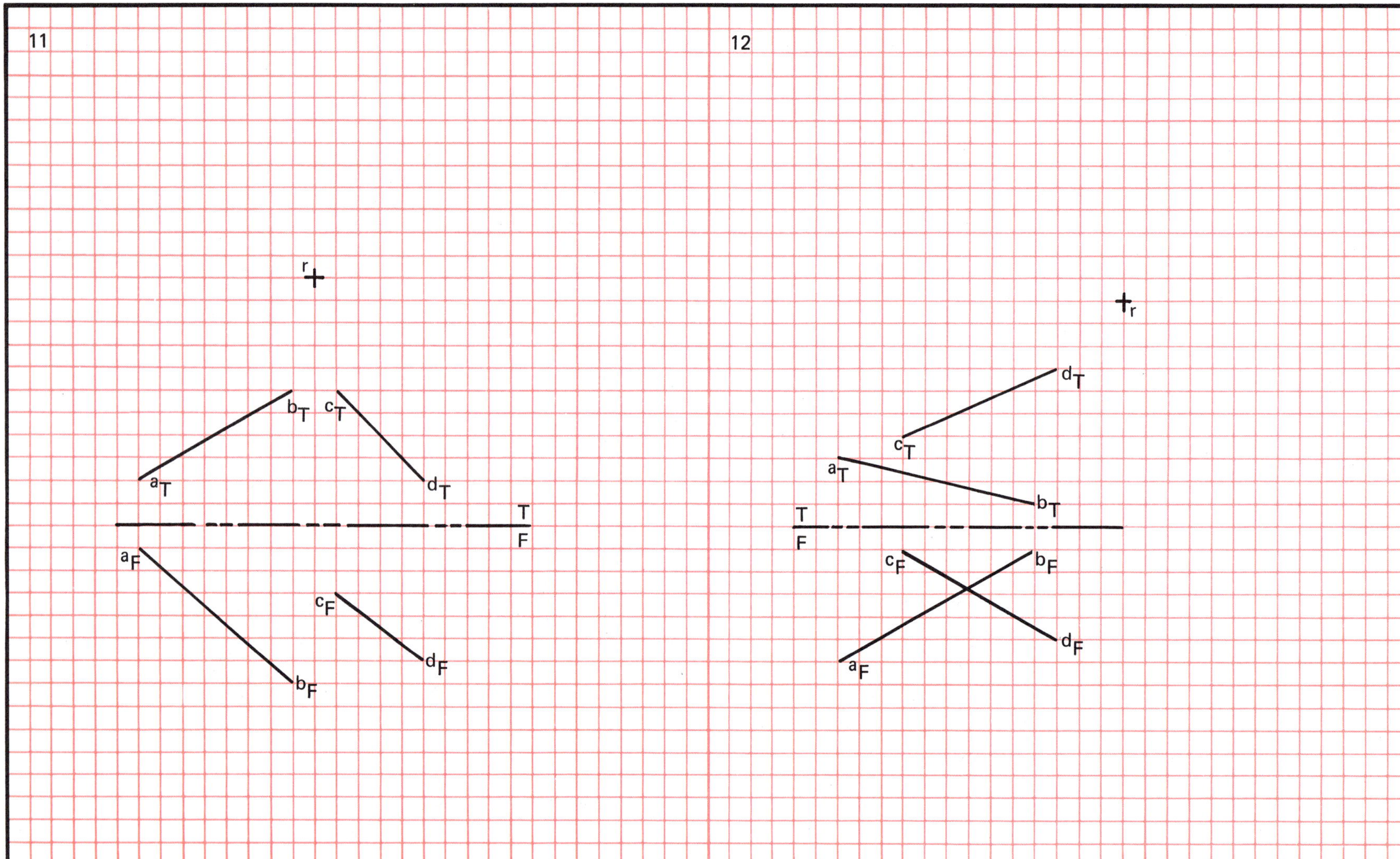

NAME OF SCHOOL CITY OR TOWN	DESCRIPTIVE GEOMETRY	SHEET No. 8-E NAME	CLASS DATE

Fig. 8-16

2. Divide the sheet into two equal spaces.
3. Locate and set up the problems on the drawing exactly as shown.
4. In problem 11, the front and top views of AB and CD are given. To find the shortest distance between these lines, the lines must be projected into a view in which one of the lines will appear as a point.
5. In this problem project line CD into a view where it will appear as a point.
6. Draw a reference line T-A through reference point r and parallel to $c_T\ d_T$.
7. Project $a_T\ b_T$ and $c_T\ d_T$ into view A. Line $c_A\ d_A$ will appear in true length in view A.
8. Draw a reference line A-B through reference point r and perpendicular to true length line $c_A\ d_A$.
9. Project $a_A\ b_A$ and $c_A\ d_A$ into view B. Line $c_B\ d_B$ will appear as a point in view B.
10. Draw a line from point $c_B\ d_B$ perpendicular to $a_B\ b_B$. This line is the shortest distance between the two lines.
11. Project the point where the line joins $a_B\ b_B$ back on to $a_A\ b_A$.
12. Since the shortest distance between the two lines appears true length in view B, it is parallel to reference line A-B in view A. From the projected point on $a_A\ b_A$ draw a line to $c_A\ d_A$ and parallel to A-B.
13. Project the shortest distance between the two lines back into all views.
14. Follow the same procedure for completing problem 12.
15. Check the accuracy of both problems for location, set up, and solution.

REVIEW QUESTIONS

1. Why must a line be perpendicular to both given lines if it is the shortest distance between the lines?
2. Why is a view in which one of the given lines appears as a point used to find the shortest distance between two lines?

ASSIGNMENT F

The two problems in figure 8-17 incorporate the material that has been covered in this unit.

Draw and solve the two descriptive geometry problems specified in figure 8-17.

Procedure for Completing Problems

1. Lay out the margin lines on the four squares to an inch cross-section drawing paper and letter the title strip.
2. Divide the sheet into two equal spaces.
3. Locate and set up the problems on the drawing exactly as shown.
4. In problem 13, the front and top views of AB, CD, and P are given. Determine the true lengths and slopes of AB and CD, and the shortest distance from AB to CD and from P to CD.
5. Construct all reference lines through reference point r.
6. Construct a reference line T-A parallel to $a_T\ b_T$ and project $a_T\ b_T$ and $c_T\ d_T$ into view A. In this view $a_A\ b_A$ will be in true length and the angle that it makes with T-A will be its true slope.
7. Construct a reference line with T-B parallel to $c_T\ d_T$ and project $c_T\ d_T$ and p_T into view B. In this view $c_B\ d_B$ will be in

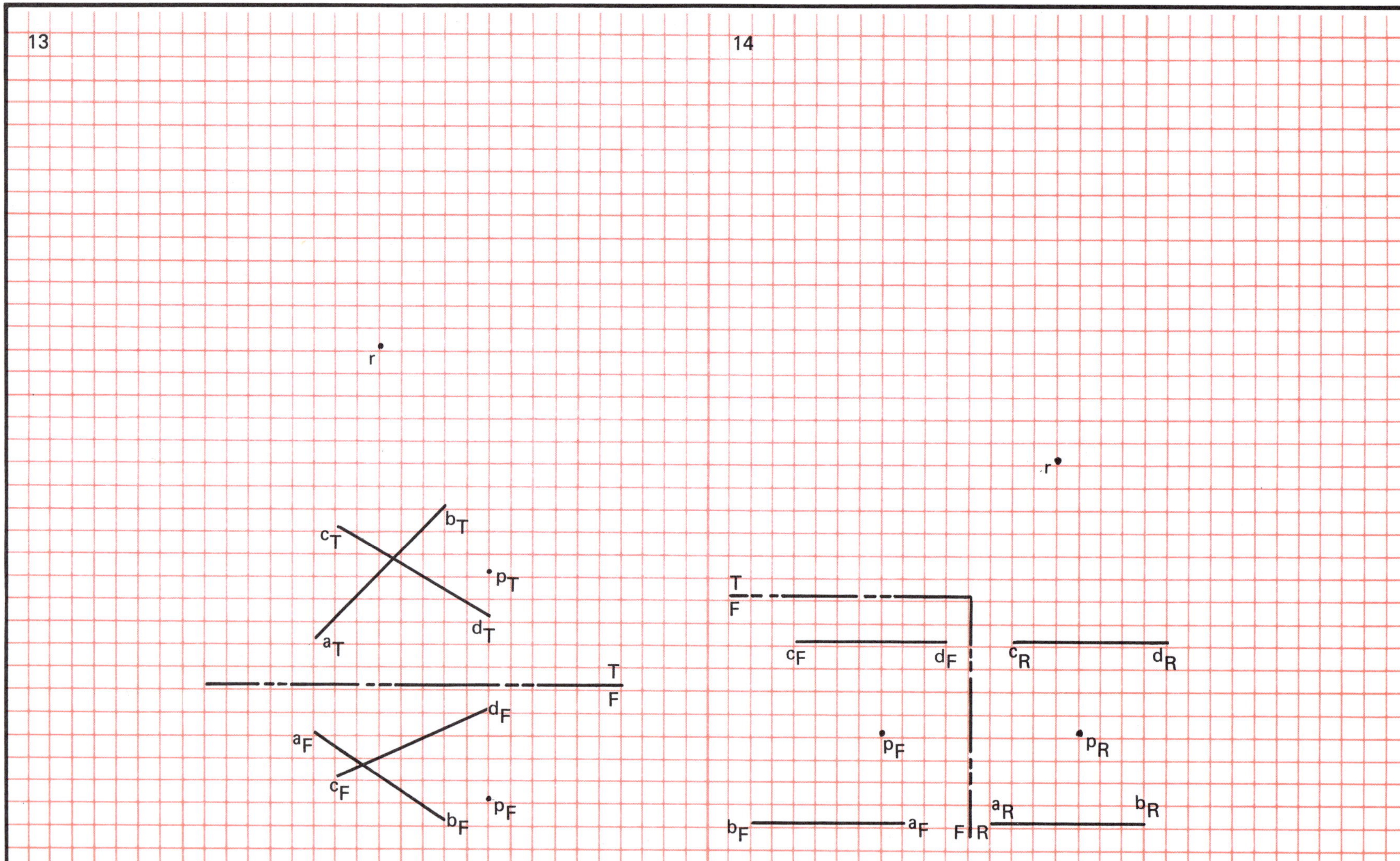

NAME OF SCHOOL CITY OR TOWN	DESCRIPTIVE GEOMETRY	SHEET No. 8-F NAME	CLASS DATE

Fig. 8-17

true length and the angle that it makes with T-B will be its true slope.

8. Construct a reference line A-C perpendicular to a_A b_A and project a_A b_A, c_A d_A and p_A into view C. In this view a_C b_C will appear as a point, and the perpendicular distance from a_C b_C to c_C d_C will be the shortest distance between AB and CD.

9. Project the shortest distance between a_C b_C and c_C d_C back into all views. As the angle between two lines becomes very small, it becomes more difficult to accurately locate their intersection. Accuracy may be increased by crossing the lines and visually bisecting the joint to fix the point of intersection.

10. Construct a reference line B-D perpendicular to c_B d_B and project c_B d_B and p_B into view D. In this view c_D d_D will appear as a point, and the distance from c_D d_D to p_D will be the shortest distance between CD and P.

11. Project the shortest distance between c_D d_D and p_D back into all views. Since the line representing this distance is in true length in view D, it will be drawn parallel to B-D in view B.

12. In problem 14, the front and right-side views of AB, CD, and P are given. Determine the true lengths and slopes of AB and CD, and the shortest distance from AB to CD, from P to AB, and from P to CD.

13. Construct all reference lines through reference point r.

14. Project a_F b_F, c_F d_F, and p_F into view T. In this view both a_T b_T and c_T d_T appear in true length.

15. Construct a reference line T-A perpendicular to c_T d_T and project a_T b_T, c_T d_T, and p_T into view A. In this view c_A d_A will appear as a point, and the perpendicular distance from p_A to c_A d_A will be the shortest distance between P and CD.

16. Project the shortest distance between p_A and c_A d_A back into all views.

17. Construct a reference line A-B perpendicular to a_A b_A and project a_A b_A, c_A d_A, and p_A into view B. In this view a_B b_B will appear as a point, and the perpendicular distance from p_B to a_B b_B will be the shortest distance between P and AB.

18. Project the shortest distance between p_B and a_B b_B back into all views.

19. Check the accuracy of both problems for location, set up, and solution.

REVIEW QUESTIONS

1. In step 15 of the last procedure, why is CD projected into a view as a point before AB?

2. In the same step, what is the difference between "p_A and c_A d_A" and "P and CD"?

UNIT 9 AUXILIARY VIEWS AND REVOLUTIONS

Up until now the objects to be drawn in this book have been suspended within a projection box with their surfaces parallel to the planes of projection. Because each surface of the object is parallel to one of the planes of projection, the view of each surface is projected onto its plane in its true shape and size.

An inclined surface, however, is not parallel to the planes of projection. Therefore, the views of an inclined surface are foreshortened when projected onto the planes to which it is inclined, figure 9-1.

INCLINED SURFACES

In the assortment of blocks with inclined surfaces, the one shown at A has its upper surface inclined to both the horizontal and profile planes. Therefore, it is foreshortened in the top and side views. The foreshortened views are shaded. The block at B has one surface inclined to the vertical and profile planes; this surface is foreshortened in the front and side views. The block shown at A has one surface inclined to the horizontal and vertical planes; this surface is foreshortened in the top and front views.

These examples show that an inclined surface cannot be drawn in its true shape on the ordinary planes of projection. Therefore to show the true shape of an inclined surface, the inclined surface must be projected onto an auxiliary plane which is parallel to it. The view on the auxiliary plane is called an *auxiliary view.*

The purpose of an auxiliary view or a revolution is to show the true shape and size of an inclined surface of an object.

THE FRONTAL AUXILIARY VIEW

In the illustration, figure 9-2, a block with one inclined surface is shown suspended within a transparent box. The transparent box has a vertical, horizontal profile plane, and an auxiliary plane. The auxiliary plane is parallel to the inclined surface of the block.

The inclined surface is foreshortened on the horizontal and profile planes. On the auxiliary plane it is shown in its true shape and size.

The planes of the transparent box are spread out on a flat surface in figure 9-3. The horizontal and profile planes assume their

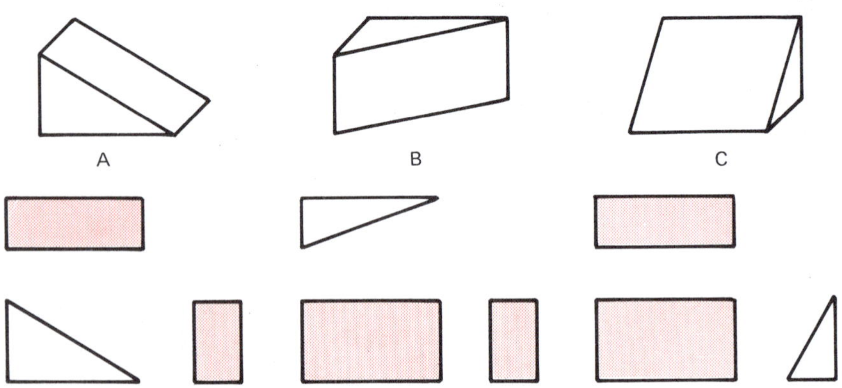

Fig. 9-1 Foreshortened inclined surfaces

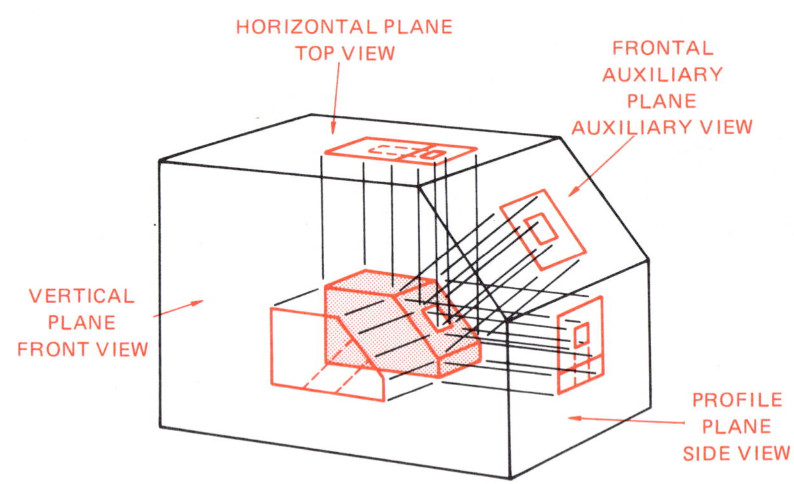

Fig. 9-2 The frontal auxiliary plane

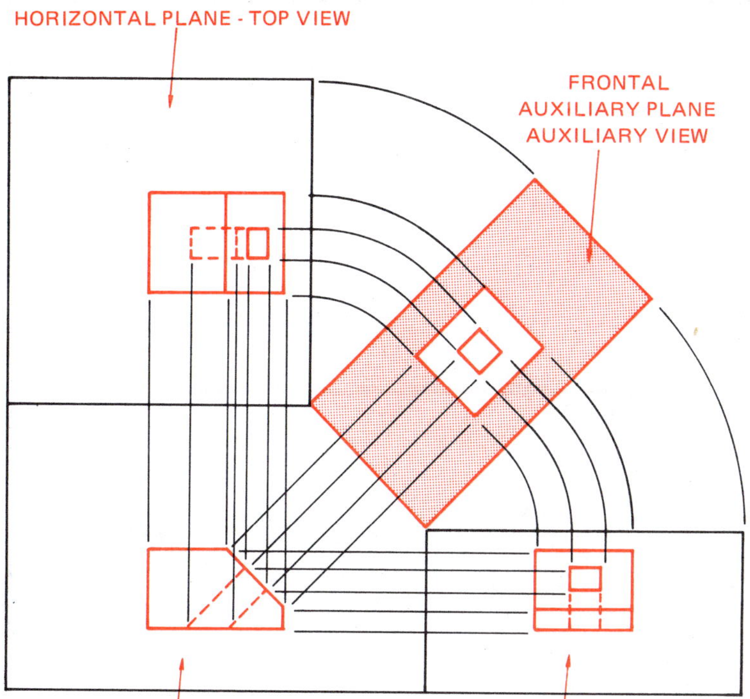

Fig. 9-3 The frontal auxiliary plane

usual positions. The auxiliary plane is attached to the front vertical plane. An auxiliary plane which is perpendicular to the front vertical plane and inclined to the horizontal and profile planes is called a *frontal auxiliary plane.* This view on a frontal auxiliary plane is called a *frontal auxiliary view.*

The slope of a frontal auxiliary plane may be at any angle between the vertical profile planes and the horizontal planes. Frontal auxiliary views may be drawn in any of the four areas between the side views and the top or bottom views, figure 9-4. The frontal auxiliary view, side view, top view, and bottom view all have a common *width.*

Usually only the inclined portion of an object is shown in an auxiliary view. However, drawing the complete auxiliary view is excellent practice.

One purpose of auxiliary views is to show the true shape and size of an inclined surface of an object. Auxiliary views have another important use. Since the true shape and size of an inclined surface is shown only on the auxiliary plane, it is evident that the inclined surface is foreshortened on the other planes to which it is inclined. This means, in the case of a frontal auxiliary view, that the inclined surface is foreshortened in the top view and side view. On many drawings, the foreshortened principal views cannot be completed without first drawing the true shape auxiliary view.

The second purpose of auxiliary views is to aid in the construction of foreshortened principal views.

Figure 9-5 shows the top and front views of a block shaped like the letter H. The top surface of the block is inclined at 45 degrees. The auxiliary view and the side view are missing. Each corner of the block is identified by a number, figure 9-6. These numbers are placed on the correct corners in the top and front views. Lines are projected from the corners of the top view and front view to locate the corners in the auxiliary view and in the side view. For example, a line from point 1 in the top view, and

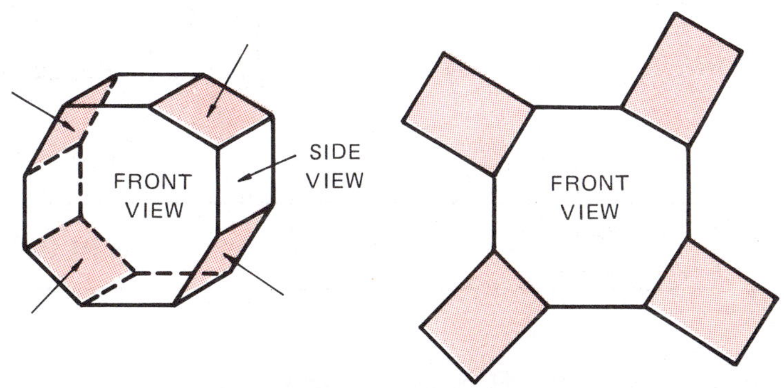

Fig. 9-4 Various locations of frontal auxiliary planes

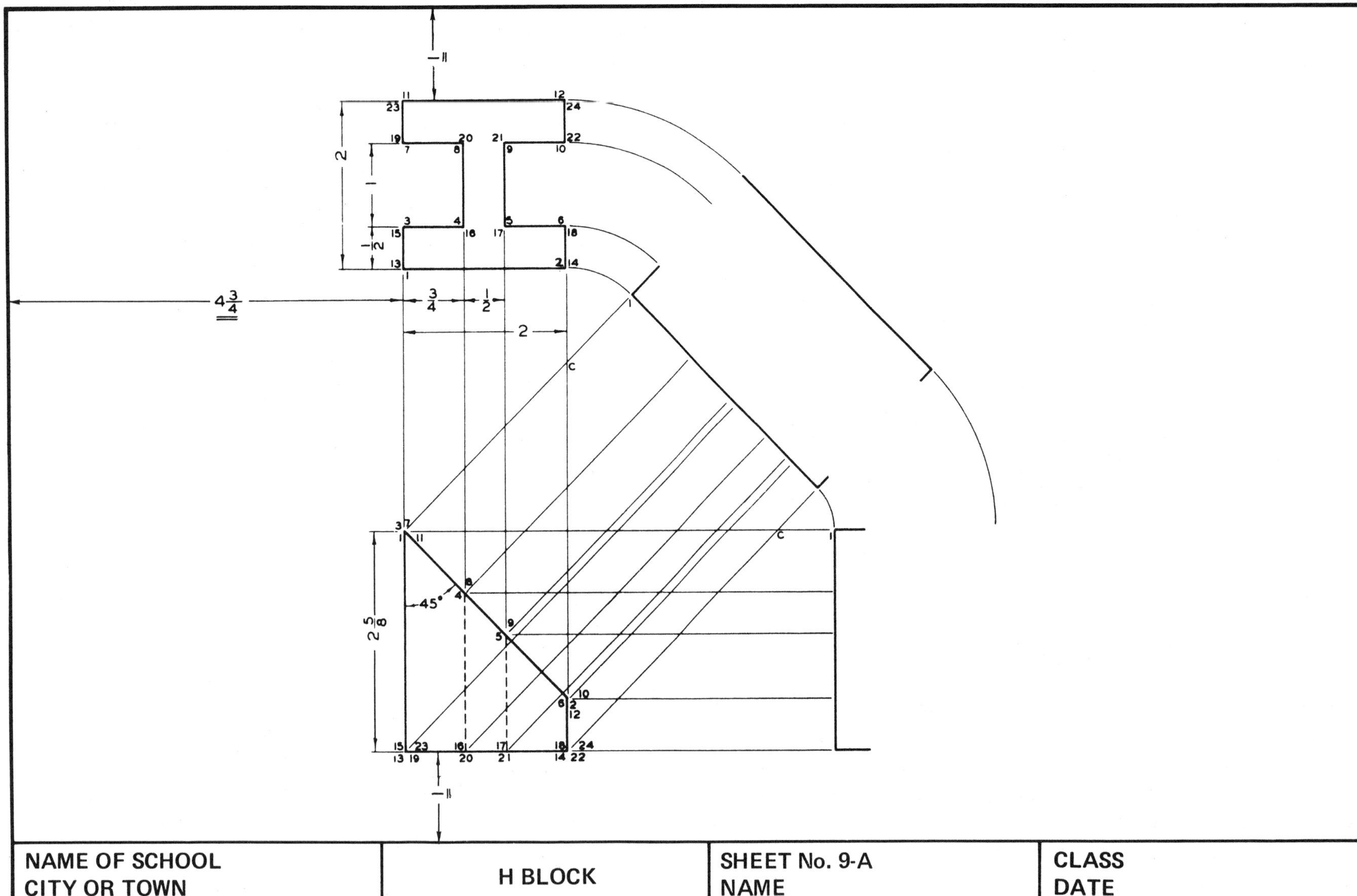
NAME OF SCHOOL
CITY OR TOWN
H BLOCK
SHEET No. 9-A
NAME
CLASS
DATE
4 3/4
3/4
1/2
2
2 5/8
45°

Fig. 9-5

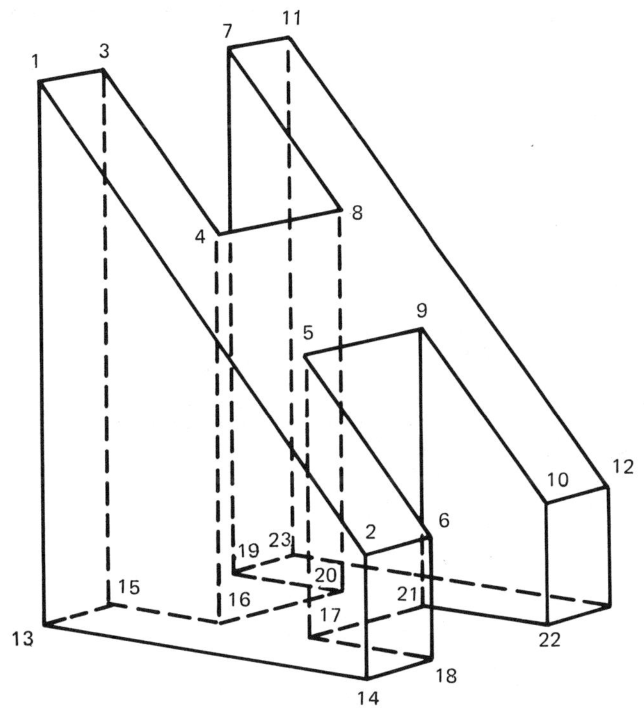

Fig. 9-6 H block

a line from point 1 in the front view, intersect to locate point 1 in the auxiliary view. The other corners are located in the same manner.

ASSIGNMENT A

Draw the top and front views of the H block. Locate and draw the missing frontal auxiliary and side views.

Procedure

1. Lay out the margin lines on the drawing sheet and letter the title strip.
2. Draw the top view and the front view. On these views place a number for each corner as shown.
3. From each corner of the front view, draw lines which are perpendicular to the inclined surface. On this drawing these lines slope 45 degrees.
4. Using point C as a center, draw an arc from each corner of the top view to the upper sloping projection line. Continue these lines at 45 degrees to the lower sloping projection line.
5. Locate and number the corners for the auxiliary view at the intersections of these projection lines. Draw the complete auxiliary view.
6. Project lines to the right from the corners of the front view.
7. Continue the projection lines from the auxiliary view to the side view using lower point C as a center for the arcs.
8. Locate the number for the corners for the side view at the intersections of these projection lines. Draw the side view.
9. Dimension the views. Omit the underlined dimensions. Check for accuracy.

REVIEW QUESTIONS

1. What is the purpose of auxiliary views?
2. What is a frontal auxiliary plane? A frontal auxiliary view?
3. Show how the numbers for the corners in the auxiliary view and the side view are located.

ELLIPTICAL CURVES

The shaft guide, figure 9-7, shows three views: a frontal auxiliary view of an inclined surface, a front view, and a partially drawn top view. The auxiliary view is essential in order to complete the top view.

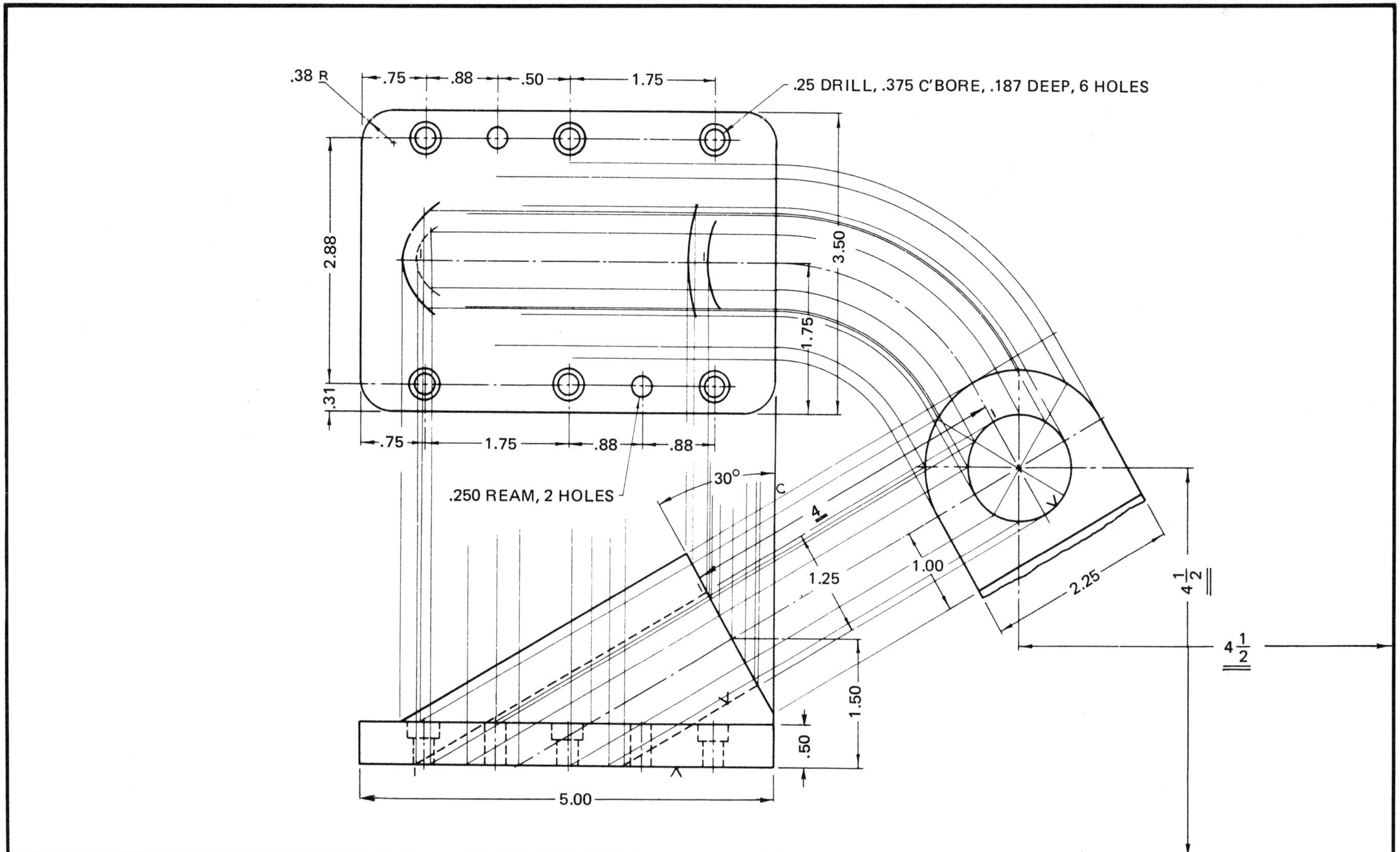

NAME OF SCHOOL CITY OR TOWN	SHAFT GUIDE	SHEET No 9-B NAME	CLASS DATE

Fig. 9-7

The shaft guide, also shown in figure 9-8, has a 1.25-inch diameter hole bored through it at an angle of 30 degrees. In the top view of the working drawing, the ends of this hole are elliptical. The ends of the part through which the hole is bored are semielliptical.

In order to construct these elliptical curves in the top view, a series of points must first be plotted. These plotted points are located at the intersections of lines projected from the front view and from the auxiliary view. For example a line is projected from point 1 in the auxiliary view to locate point 1 in the front view. Lines are then drawn from point 1 in the auxiliary view and from point 1 in the front view. These two lines intersect to locate point 1 in the top view. The other points are located in the same manner. Twelve points are plotted for each ellipse and seven points for each semiellipse.

The points for the elliptical curves are connected by drawing along the edge of an irregular curve. The ellipse in figure 9-9 is drawn as follows: the irregular curve is adjusted until a part of its edge contacts points 1, 2, and 3. To avoid humps in the ellipse, the line is drawn only from A to B. The irregular curve is then placed in the second position to contact points 2, 3, and 4, and the line is drawn from B to 4. This is repeated for each quarter of the ellipse. The irregular curve is then placed in the third position to contact points 12, 1, and 2, and the line drawn from C to A to round off the end of the ellipse. This is repeated at the other end of the ellipse.

Irregular curves, commonly called French curves, are made in a variety of shapes for drawing curved lines that are not true circular arcs. Because irregular curves are made in a variety of shapes, and the curved lines which the draftsman may be required to draw are many and varied, the procedure for drawing curved lines with the irregular curve may not always be the same. The best practice is to contact three or more points along the edge of the irregular curve. Draw the line stopping short of the first and last points. Continue using the next three or more points until the whole curved line is drawn.

ASSIGNMENT B

Make a three-view working drawing of the shaft guide. Draw the frontal auxiliary view, the front view, and the top view.

Procedure

1. Lay out the margin lines on the drawing sheet and letter the title strip.
2. Draw the frontal auxiliary view.
3. Divide the circle into twelve equal parts, and the semicircle into six equal parts. Draw the front view.
4. Project a line from each point in the auxiliary view to locate these points on each end of the hole, and seven points on each end of the semicircular top. Number these points for identification.
5. Use point C as a center and draw the center line from the auxiliary view to the top view. Outline the top view and draw the eight small holes.
6. Draw a projection line from point 1 in the front view and another projection line from point 1 in the auxiliary view. The intersection of these lines is the location of point 1 in the top view. In the same manner, locate twelve points for each ellipse and seven points for each semiellipse.
7. Draw the elliptical curves through these points with an irregular curve.
8. Dimension the view using decimal dimensioning practice and letter the shop notes.
9. Check the accuracy of the drawing.

REVIEW QUESTIONS

1. For what two purposes are auxiliary views used?
2. What is a frontal auxiliary view?
3. For what purpose are irregular curves used?
4. Show how the points are plotted for elliptical curves.

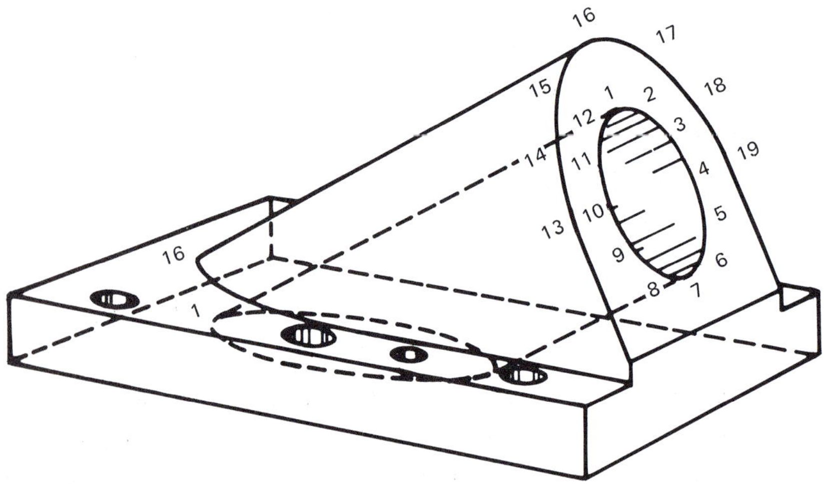

Fig. 9-8 Shaft guide

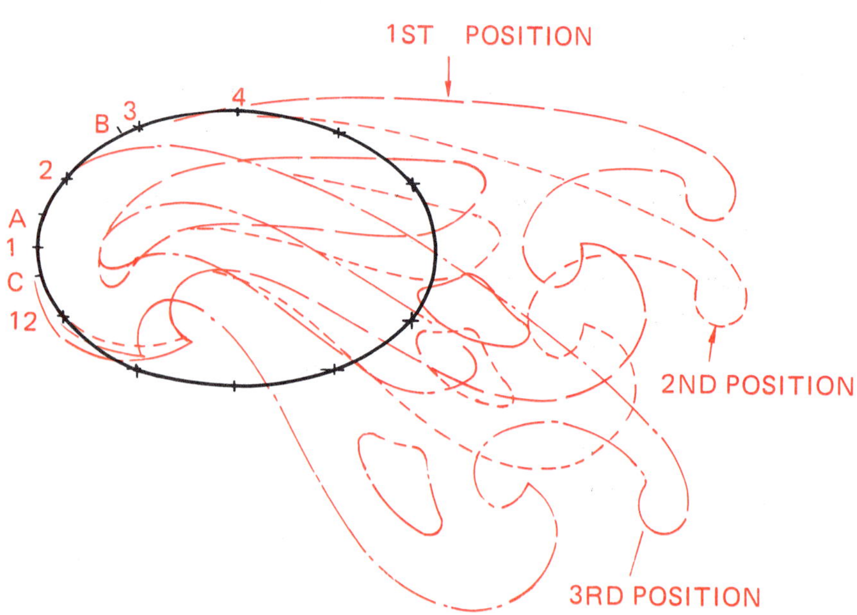

Fig. 9-9 Using the irregular curve

USING REFERENCE PLANES FOR AUXILIARY VIEWS

In previous drawings, the top view, the front auxiliary view, and the side view have a common width from which projection lines are used to project the width from one view to another.

Another method to obtain a common width in the top, auxiliary, and side views is to make use of a reference plane. The reference plane method is used when it is not convenient to draw projection lines between the top, auxiliary, and side views.

In figure 9-10, a reference plane cuts through the center of a symmetrical object. The edge of this reference plane appears as a line when viewed from a point directly above the object. It also appears as a line when viewed from a point directly to the right of the object and from the auxiliary position.

The line representing the edge of the reference plane is shown in the top, auxiliary, and side views. To make these views the same width, the dimensions A and B are measured equally on each side of the reference line in each view, and the views are *symmetrical views.*

In figure 9-11, the reference plane is against the front surface of the object. In the top, auxiliary, and side views, the dimensions A, B, and C are all measured on one side of the reference line, and the views are called *unilateral views.*

In figure 9-12, the reference plane cuts through the object parallel to an inner surface. In the top, auxiliary, and side views, the dimensions A and B are measured on one side of the reference line. Dimension C is measured on the other side. When unequal dimensions are measured on both sides of the reference line, the views are *bilateral views.*

The fixture bracket, figure 9-13, is shown with a reference plane against the finished front surface. In the drawing, figure 9-14, the reference plane is represented by a reference line in the top, auxiliary, and side views. In these views, the common widths are measured from the reference line. Since all of the measurements are made on one side of the reference line, these views are unilateral.

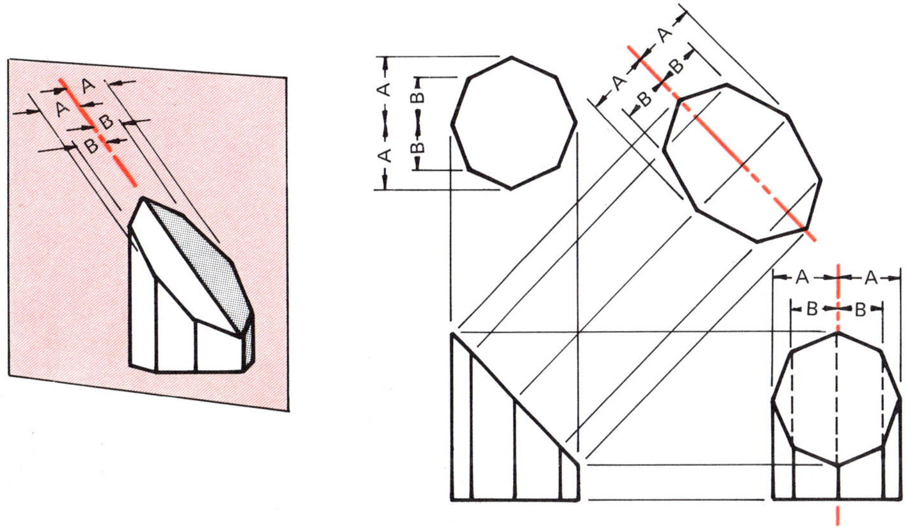

Fig. 9-10 Symmetrical top, auxiliary, and side views

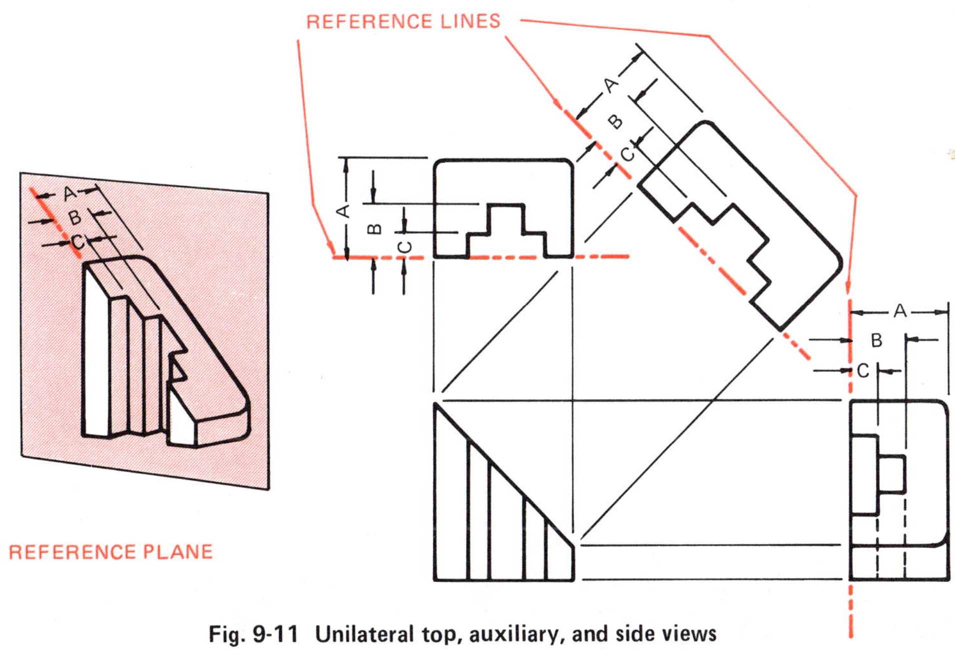

Fig. 9-11 Unilateral top, auxiliary, and side views

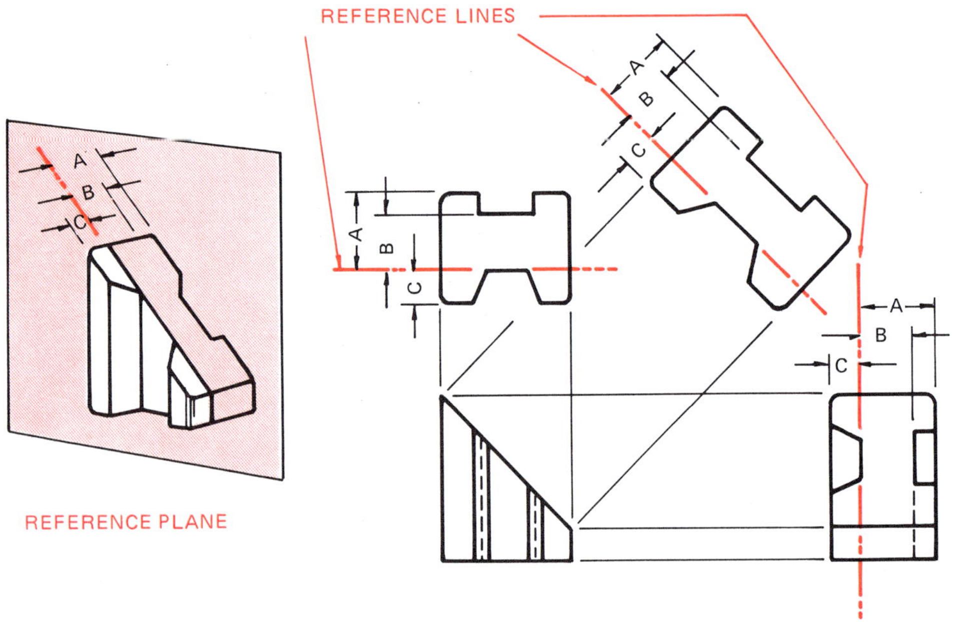

Fig. 9-12 Bilateral top, auxiliary, and side views

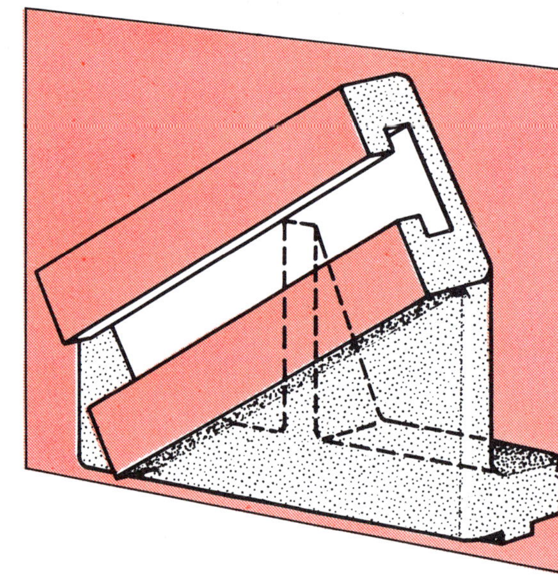

Fig. 9-13 Fixture bracket

In actual drafting practice the reference plane is an imaginary plane and the reference lines are omitted from the finished views.

ASSIGNMENT C

Draw a partial frontal auxiliary view and a front view of the fixture bracket. Complete the side view and top view.

Procedure

1. Lay out the margin lines on the drawing sheet and letter the title strip.
2. Draw the reference lines as shown.
3. Draw the frontal auxiliary view of the inclined portion of the fixture bracket.
4. Draw the front view.
5. Draw the side view. The various heights in this view are projected from the front view, and the widths are measured from the reference line.
6. Draw the top view. The lengths of this view are projected from the front view, and the widths are measured from the reference line.
7. Dimension the views, add the finish marks, and letter the note.
8. Check all views, particularly the top and side views.

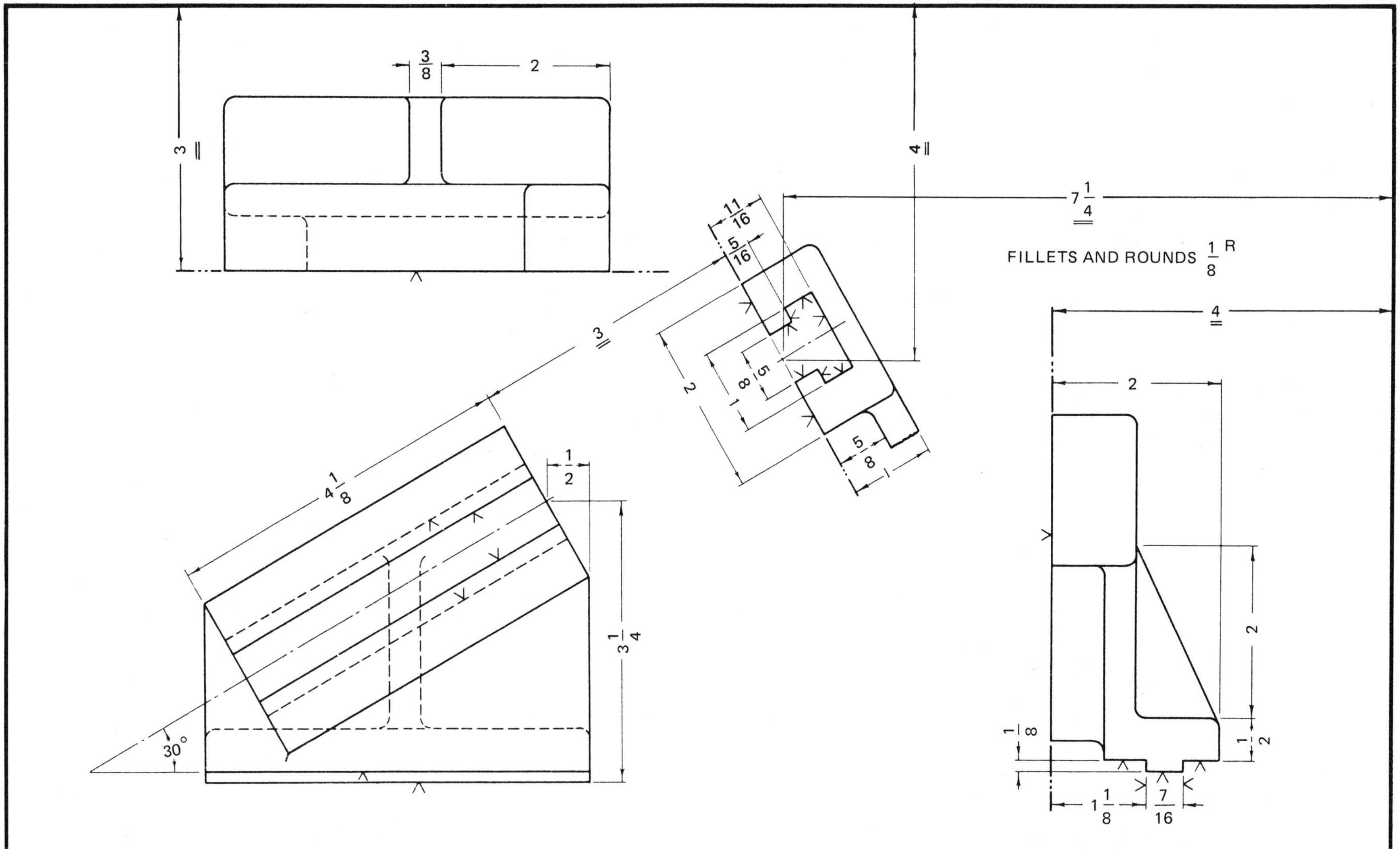

NAME OF SCHOOL CITY OR TOWN	FIXTURE BRACKET	SHEET No. 9-C NAME	CLASS DATE

Fig. 9-14

REVIEW QUESTIONS

1. What is meant by symmetrical top, auxiliary, and side views?
2. What is meant by unilateral top, auxiliary, and side views?
3. What is meant by bilateral top, auxiliary, and side views?

PROFILE AUXILIARY VIEWS

The auxiliary views up to this point are called frontal auxiliary views because the inclined surface of the part is perpendicular to the frontal plane.

In figure 9-15, an object is suspended within a transparent box. The inclined surface of the part is inclined to the horizontal and frontal planes and, consequently, is foreshortened in the top and front views. The auxiliary plane is parallel to the inclined surface. Therefore, the true shape and size of the inclined surface is shown in the auxiliary view.

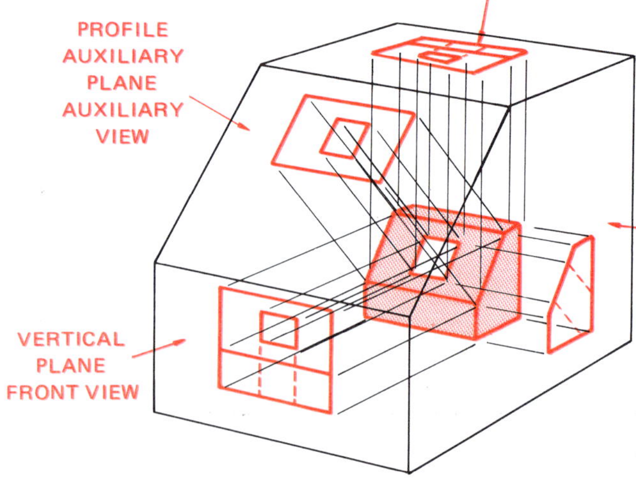

Fig. 9-15 The profile auxiliary plane

In this arrangement, the inclined surface and the auxiliary plane are perpendicular to the profile plane. When the planes of the transparent box are spread out flat, figure 9-16, the principal planes assume their usual positions and the auxiliary plane is attached to the profile plane.

An auxiliary plane which is perpendicualr to the profile plane and inclined to the horizontal and frontal planes is called a *pro-*

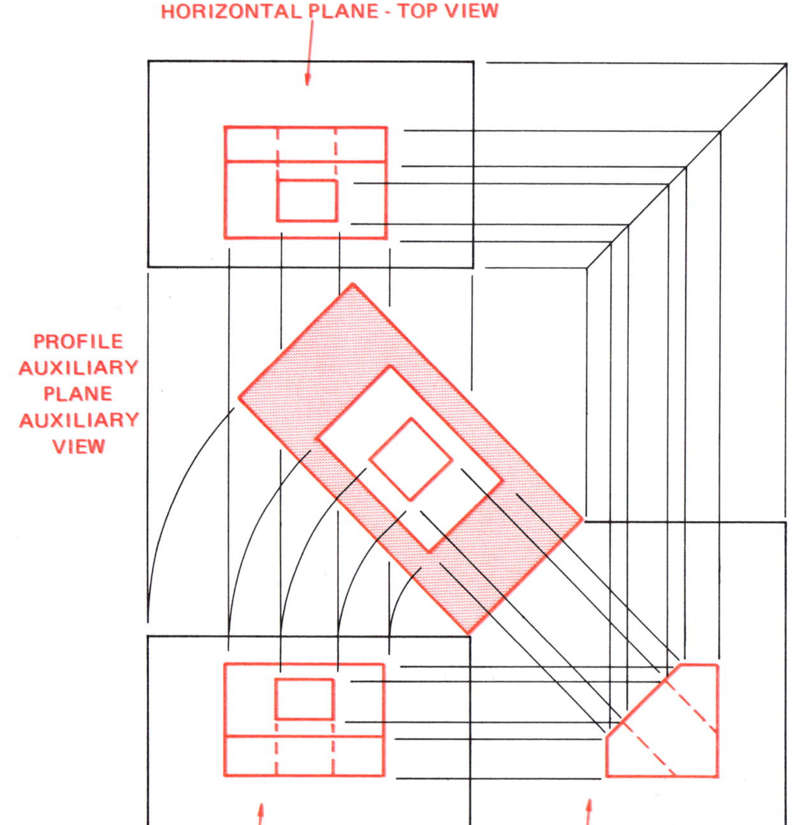

Fig. 9-16 The profile auxiliary plane

file auxiliary plane. The view on the profile auxilary plane is called a *profile auxiliary view.* The profile auxiliary view, the top view, and the front view have a common length, figure 9-16.

The profile auxiliary plane may slope at any angle between the horizontal planes and the front and rear vertical planes as shown in figure 9-17. Also, the profile auxiliary plane may be found in any of the four areas around the profile plane.

The angle pillow block, figure 9-18, shows some surfaces which are inclined to the frontal plane. Consequently, the inclined surfaces are foreshortened in the front view. To draw the foreshortened front view, it is necessary to first draw the side view and the profile auxiliary view. Lines are then projected from the side view and from the auxiliary view to aid in completing the front view.

The circular arcs in the auxiliary view become elliptical curves in the front view. These elliptical curves may be plotted by marking off a number of points on the circles and areas in the auxiliary view. These points are projected to the side view. Next, a line is drawn from each point in the auxiliary view and from each corresponding point in the side view. The intersections of these lines locate the points for the elliptical curves in the front view.

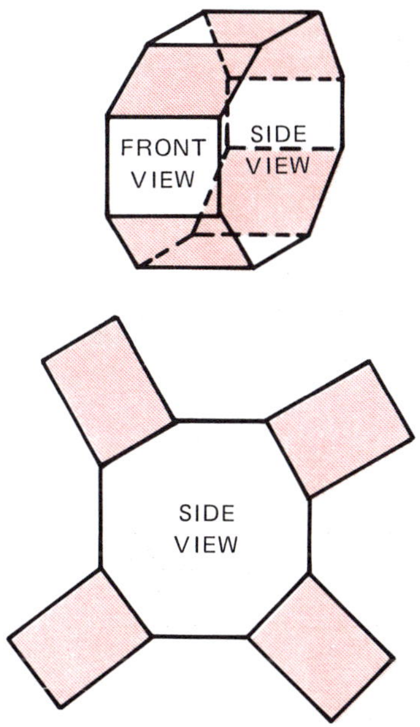

Fig. 9-17 Various positions of profile auxiliary planes

Because many lines must be drawn to locate the points for the elliptical curves, it is necessary to maintain very sharp points on the pencil and compass and to draw very fine light lines.

ASSIGNMENT D

Draw side, profile auxiliary, and front views of the angle pillow block.

Procedure

1. Lay out the margin lines on the drawing sheet and letter the title strip.
2. Draw the side view.
3. Draw the auxiliary view. In the auxiliary view mark off twelve points on each circle and seven points on each semicircle. Project these points to the side view.
4. From these points in the side view, draw lines to the front view.
5. Using point C as a center, project lines from each point in the auxiliary view to the front view. Locate the points for the elliptical curves at the intersections of these lines.
6. Complete the front view.
7. Dual-dimension the views, add the finish marks, and letter the notes. Check all lines, dimensions, and notes on the front, side, and profile auxiliary views for accuracy.

REVIEW QUESTIONS

1. What is a profile auxiliary plane? What is a profile auxiliary view?
2. What is the main difference between a frontal auxiliary view and a profile auxiliary view?

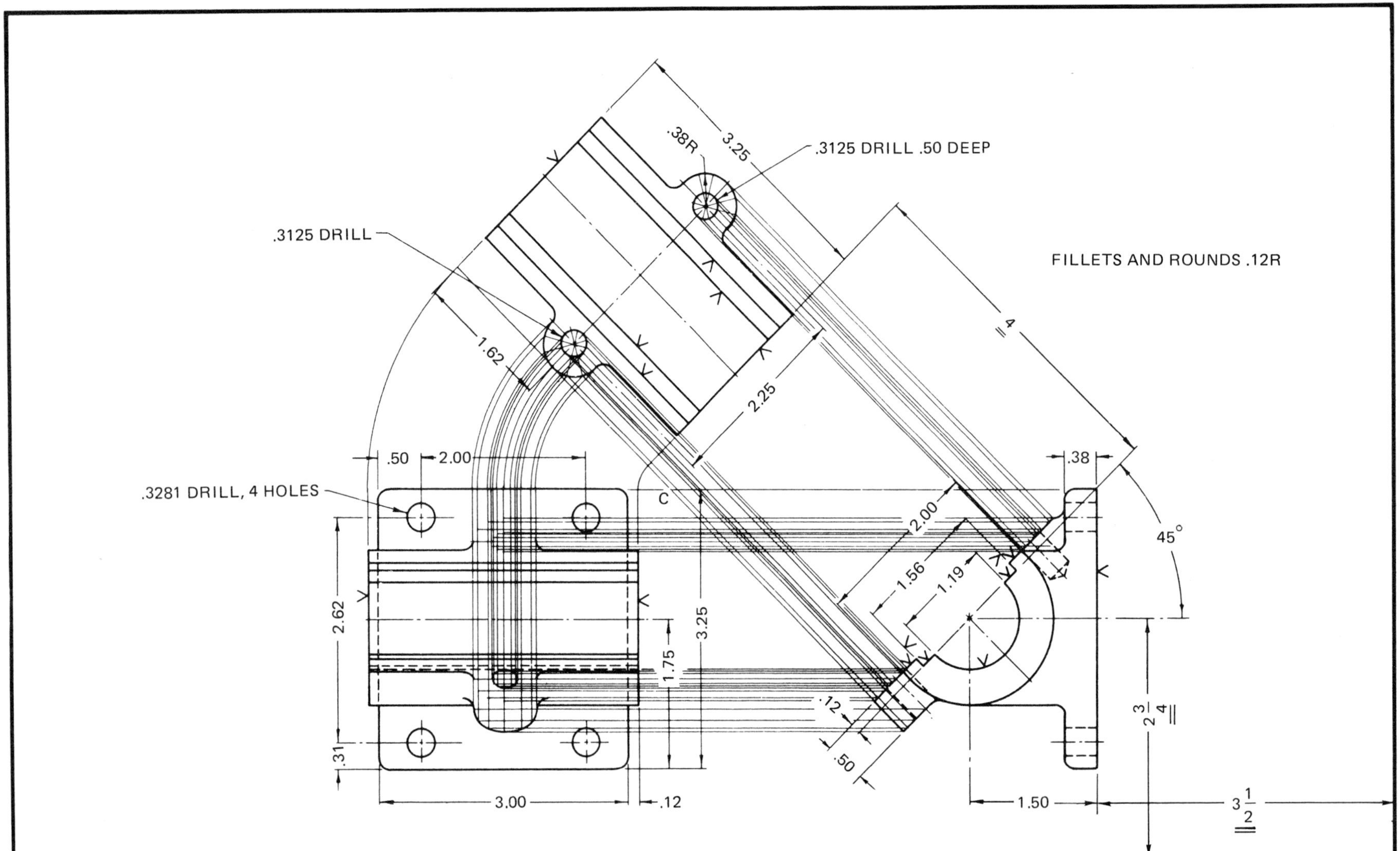

NAME OF SCHOOL CITY OR TOWN	ANGLE PILLOW BLOCK	SHEET No. 9-D NAME	CLASS DATE

Fig. 9-18

TOP AUXILIARY VIEWS

A third type of auxiliary view is pictured in figure 9-19. In this illustration, the inclined surface of the part is inclined to the vertical and profile planes. Consequently, it is foreshortened in the front and side views. The auxiliary plane is parallel to the inclined surface. Therefore, the true shape and size of the inclined surface is shown in this auxiliary view.

In this arrangement, the inclined surface and the auxiliary plane are perpendicualr to the horizontal plane. When the planes are spread out flat, figure 9-20, the principal planes assume their usual positions and the auxiliary plane is attached to the horizontal plane.

An auxiliary plane which is perpendicular to the top horizontal plane and inclined to the vertical and profile planes is called a *top auxiliary plane*. The view on the top auxiliary plane is called a *top auxiliary view*. The top auxiliary view is sometimes called an auxiliary elevation of a plane auxiliary view. The top auxiliary view, the front view, and the side view have a common height.

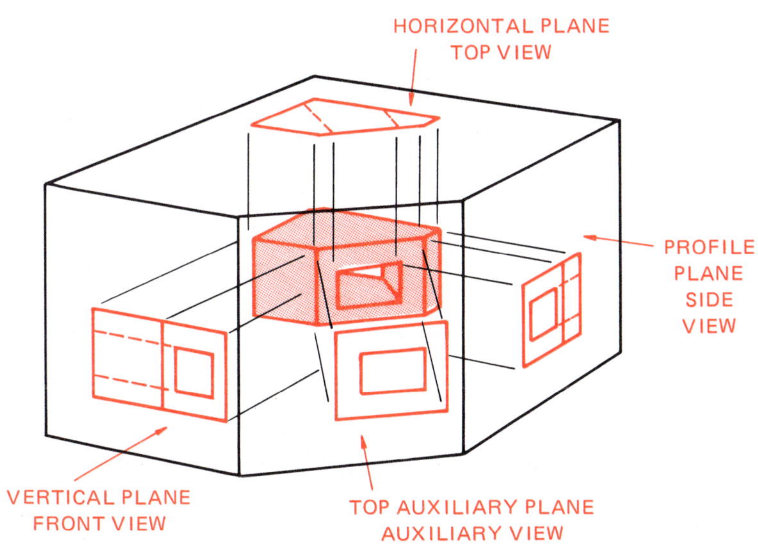

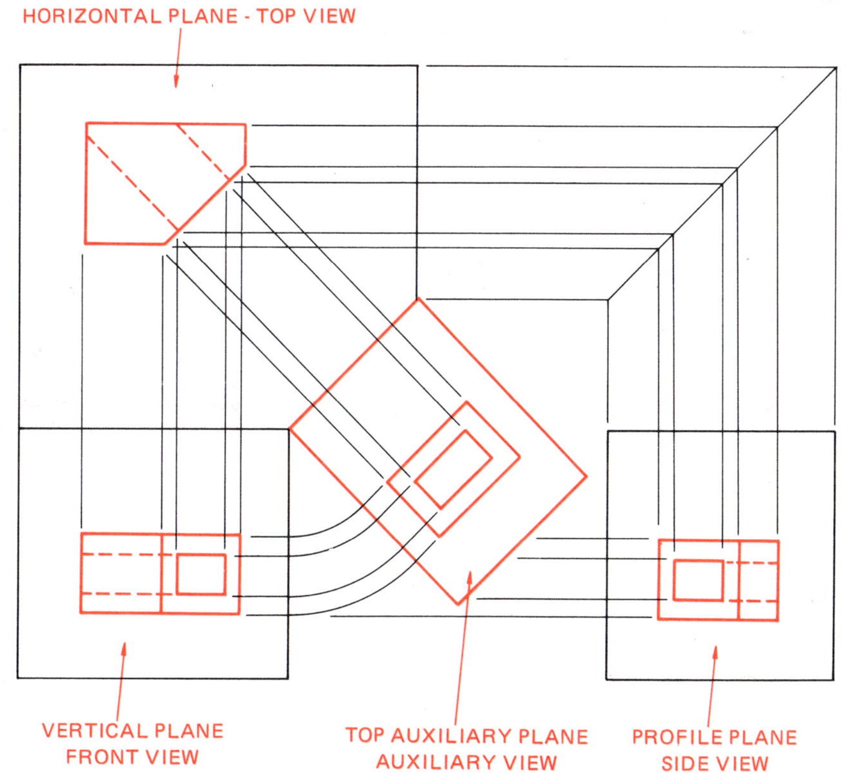

Fig. 9-20 The top auxiliary plane

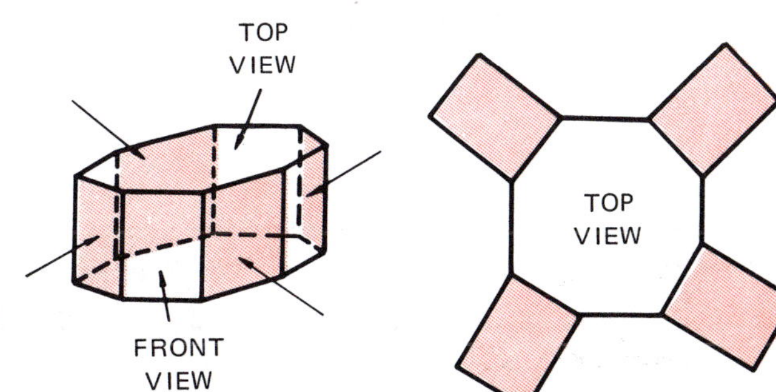

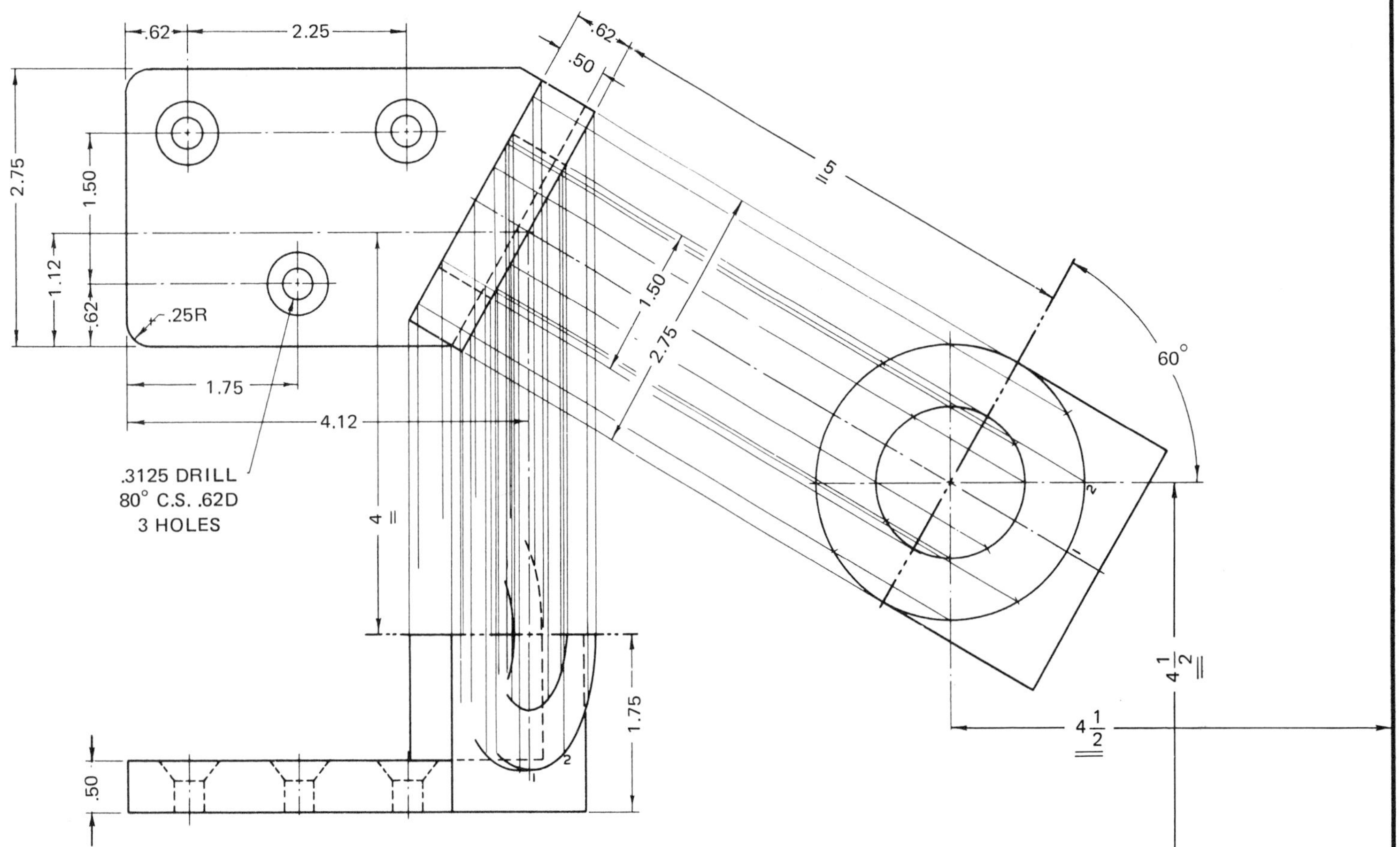

NAME OF SCHOOL CITY OR TOWN	PUMP MOUNTING	SHEET No. 9-E NAME	CLASS DATE

Fig. 9-22

The top auxiliary plane according to figure 9-21 may be at any angle between the profile planes and the front and rear vertical planes. Also, this illustration shows the auxiliary planes attached to the top horizontal plane in any of the four positions.

The pump mounting, figure 9-22, has a surface inclined to the vertical plane, and consequently this surface is foreshortened in the front view. The circles in the auxiliary view become ellipses in the front view. In order to complete the foreshortened front view, it is necessary to first draw the auxiliary and the top view.

Points are marked off on the circles in the auxiliary view. These points are projected to the top view. From the points in the top view, lines are drawn down to the front view.

The auxiliary view and the front view have a common height, so the reference plane method is used to complete the front view.

Figure 9-23 shows a horizontal reference plane cutting through the pump mounting. On the drawing, the position of the reference plane is indicated in the auxiliary view and in the front view by a reference line.

The height of each line in the auxiliary view is measured from the reference line. These heights are then measured from the reference line on each corresponding line in the front view. Then the points are connected with an irregular curve.

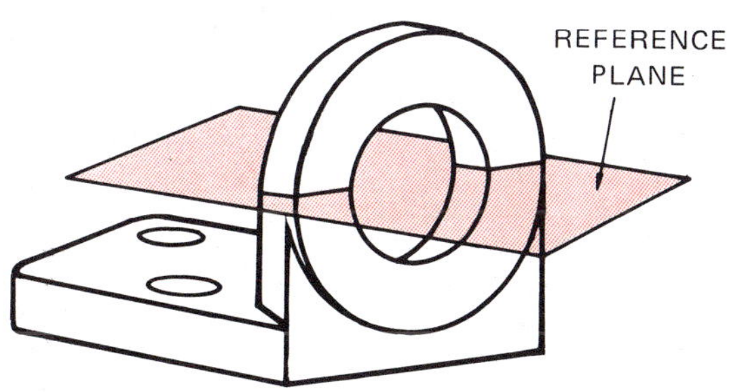

Fig. 9-23 A horizontal reference plane

ASSIGNMENT E

Draw a top auxiliary view, a top view, and a front view of the pump mounting.

Procedure

1. Lay out the margin lines on the drawing sheet and letter the title strip.
2. Draw the auxiliary view.
3. Draw the top view.
4. Draw the reference lines in the auxiliary view and front view.
5. Divide the circles in the auxiliary view into twelve equal parts.
6. From each point on the circles draw a projection line to locate the same points in the top view. The points may be numbered for identification.
7. From each point in the top view, draw a line down to the front view. Because the auxiliary view and front view have a common height, the height from the reference line to each point in the auxiliary view is transferred to the proper line in the front view. The dividers are used for this purpose.
8. Begin with the large circle and measure the distance from the reference line to point 1. Trace the line from point 1 to the top view and then downward to the front view. On the line in the front view, measure from the reference line the distance to point 1. Locate the other points in the same manner.
9. When the twelve points of the circle are plotted, connect them with an irregular curve.
10. Continue this procedure for the other elliptical curves. Some of the elliptical curves are invisible and are not necessary in this view. Draw only the invisible curve indicated on the drawing.
11. Complete the front view.

12. Dual-dimension the views and letter the note.
13. Check the accuracy of all lines, dimensions, and notes in the top, front, and top auxiliary views on the drawing.

REVIEW PROBLEMS

1. What is a top auxiliary plane?
2. What is a top auxiliary view?
3. What is the main difference between a frontal auxiliary view, a profile auxiliary view, and a top auxiliary view?
4. What is the purpose of the countersunk holes?

OBLIQUE AUXILIARY VIEWS

The three types of auxiliary views drawn show the true shape of surfaces which are inclined to two of the principal planes and perpendicular to one. In each case, the auxiliary plane is hinged to the principal plane to which it is perpendicular.

In figure 9-24, an object is shown suspended within a transparent box. One surface of this object is inclined to all of the principal planes and perpendicular to none. A surface that is inclined to all of the principal planes is called an *oblique* surface. In theory, the true shape of an oblique surface may be projected onto an oblique plane which is parallel to the oblique surface as shown in figure 9-25.

In actual practice, an auxiliary view must first be drawn on an auxiliary plane which is perpendicular to one of the principal planes. Also, this auxiliary plane must be perpendicular to one

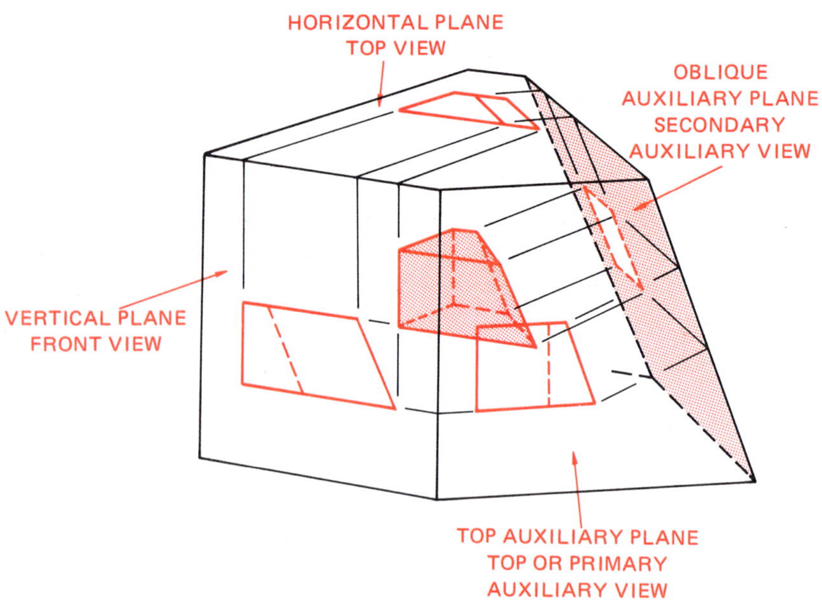

Fig. 9-24 The oblique auxiliary plane

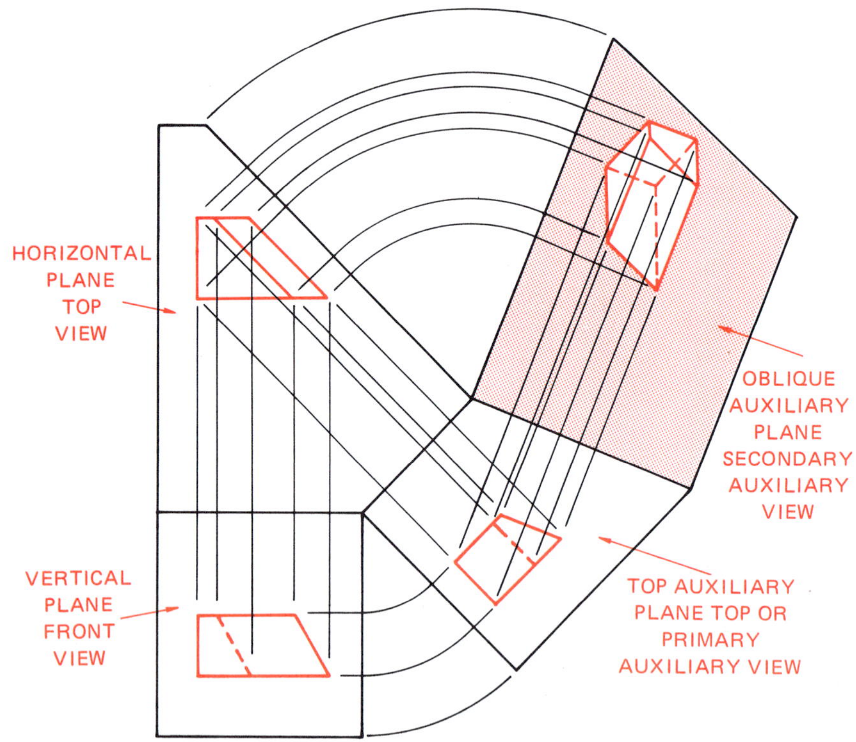

Fig. 9-25 Secondary (double) auxiliary plane

edge of the oblique surface. The oblique surface then appears as a line on this auxiliary plane. See the top auxiliary plane in figure 9-24. Then, a second auxiliary view is drawn on an oblique plane which is perpendicular to the auxiliary plane.

For example, in figure 9-24, a top auxiliary view is drawn on an auxiliary plane which is perpendicular to the top horizontal plane. In this view, the oblique surface is represented by a line. Then another auxiliary view is drawn on an oblique plane perpendicular to the auxiliary plane in order to show the true shape of the oblique surface. The planes are spread out flat in figure 9-25. The auxiliary plane is attached to the horizontal plane and the oblique plane is attached to the auxiliary plane. They are attached to the planes to which they are perpendicular.

Because a second auxiliary view is necessary to show the true shape of the oblique surface, it is called a *secondary auxiliary view*, and the first auxiliary view is called a *primary auxiliary view*. They are also referred to as *double auxiliary views*.

The height of the secondary auxiliary view, figure 9-25, is the same as the oblique height of the primary auxiliary view. Therefore, lines perpendicular to the oblique surface are projected from the primary auxiliary view to obtain the height of the secondary auxiliary view.

The length of the secondary view, figure 9-25, is the same as the oblique length of the top view. Lines are projected from the top view to obtain the length of the secondary auxiliary view.

In figure 9-26, the same views are drawn without the projection planes. A reference plane is used in the top view and the secondary auxiliary view to obtain the common lengths. The reference plane appears as a line when viewed from the top and from the auxiliary positions, so a reference line is drawn in the top view and in the second auxiliary view. The lengths are measured from the reference line in the top view and transferred to the secondary auxiliary view.

The tool grinding fixture, illustrated in figure 9-27 and partly drawn in figure 9-28, has some of its upper surfaces and edges in-

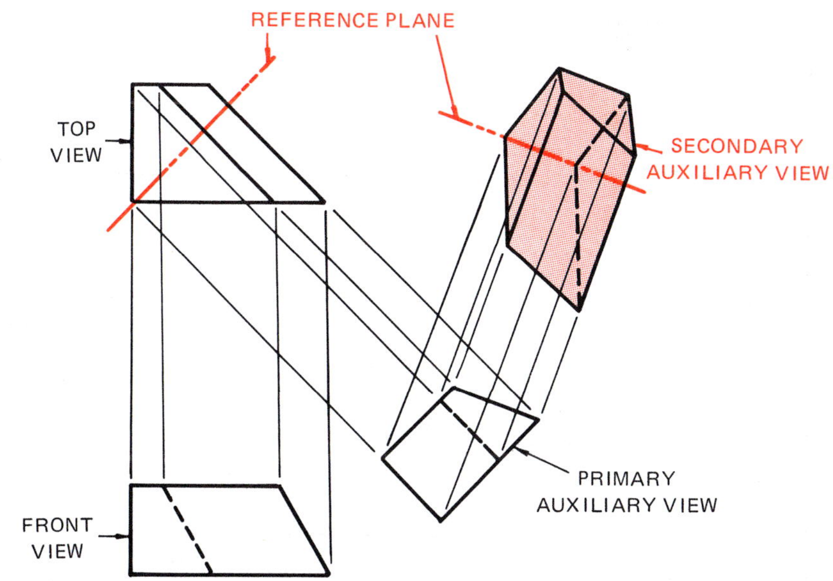

Fig. 9-26 Secondary auxiliary view with reference plane

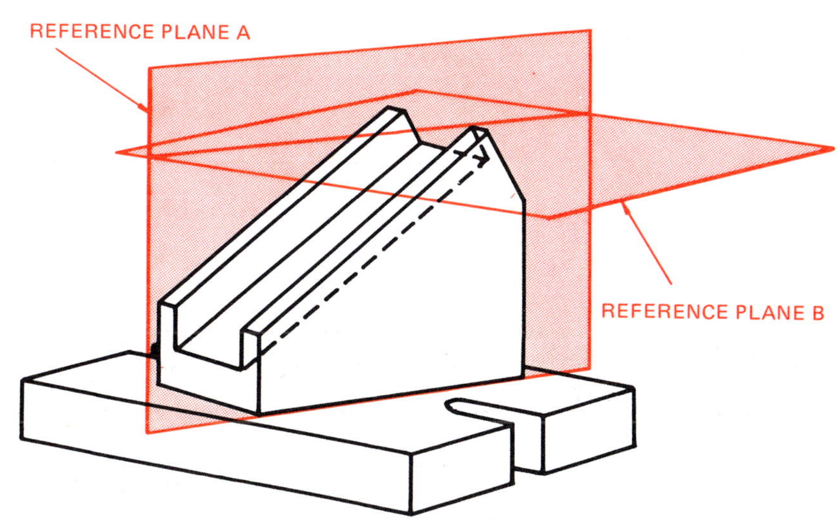

Fig. 9-27 Tool grinding fixture

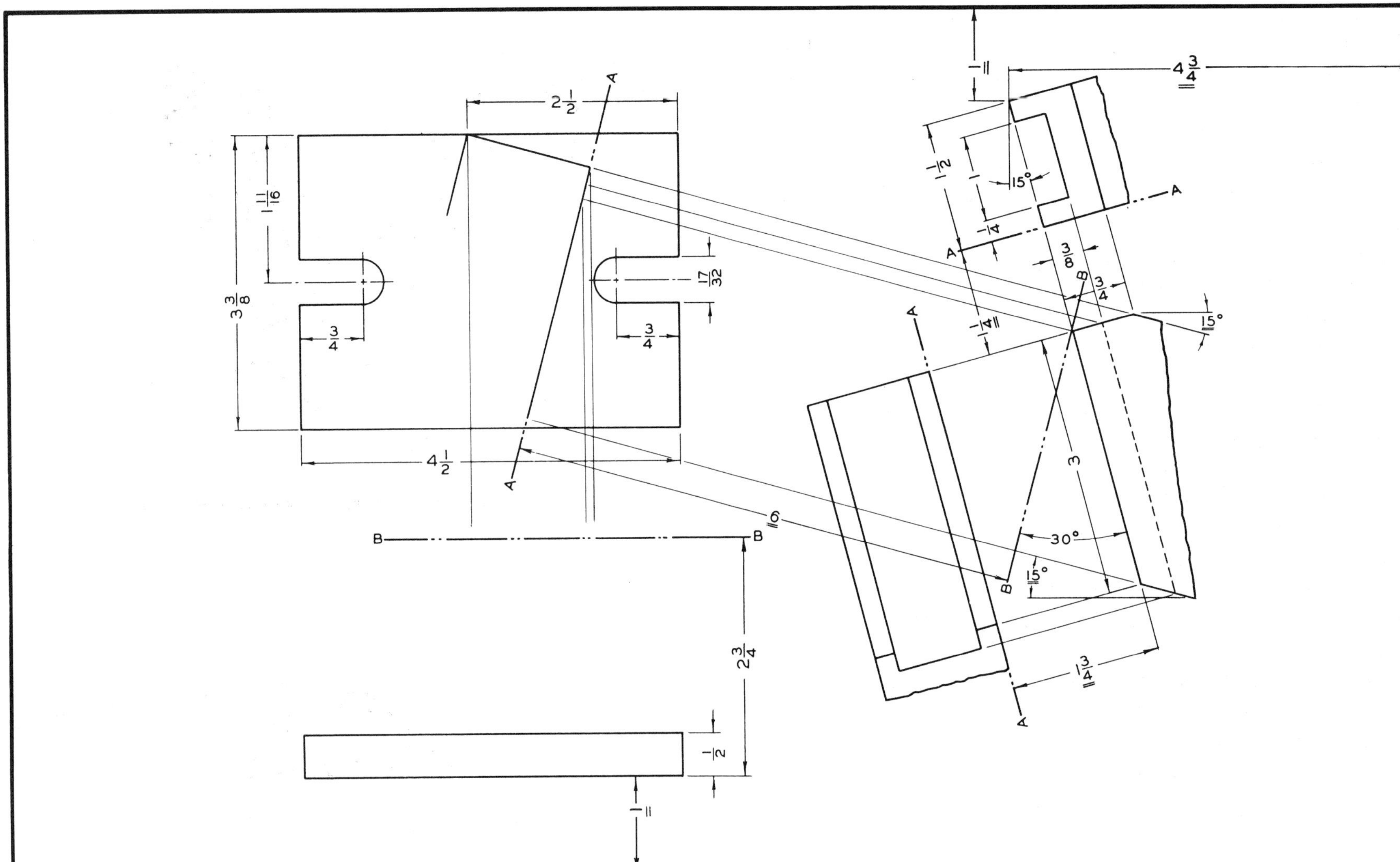

NAME OF SCHOOL CITY OR TOWN	TOOL GRINDING FIXTURE	SHEET No. 9-F NAME	CLASS DATE

Fig. 9-28

clined 30 degrees from the horizontal plane. The upper end is inclined 60 degrees from the horizontal plane. The fixture is turned at an angle of 15 degrees with the ends of the base. The upper surfaces are oblique surfaces.

A top auxiliary view is necessary to show the true shape of the surface which is inclined 15 degrees from the profile plane and perpendicular to the horizontal plane. Two secondary auxiliary views are needed to show true shapes of the oblique surfaces.

The two secondary auxiliary views and the top view have a common width, so measurements are made from the vertical reference plane A as in figure 9-27. The front view and the primary auxiliary view have a common height, so measurements are made from the horizontal reference plane B.

ASSIGNMENT F

Draw two secondary auxiliary views, one primary auxiliary view, a top view, and a front view of the tool grinding fixture.

Procedure

1. Lay out the margin lines on the drawing sheet and letter the title strip.
2. Draw the secondary auxiliary view of the upper end of the fixture and reference line A-A.
3. Draw the primary auxiliary view and reference line B-B.
4. Draw the secondary auxiliary view of the upper surfaces. The length is projected from the primary auxiliary view and the widths are measured from reference line A-A.
5. Project lines from the primary auxiliary view to obtain the lengths of the oblique surfaces in the top view.
6. Measure the widths of the oblique surfaces in the top view from reference line A-A.
7. Complete the top view. From each corner of the top view project a line downward to the front view.
8. The height (or depth) of each point is measured from reference line B-B in the primary auxiliary view and transferred to the proper line in the front view.
9. Complete the front view. Dual-dimension the views.
10. Check all lines and dimensions in the top, front, primary, and secondary auxiliary views.

REVIEW QUESTIONS

1. What is an oblique surface? What is an oblique plane?
2. What is the purpose of the secondary auxiliary view?
3. For what purpose are auxiliary views used?

REVOLUTION

The same results as auxiliary views may be obtained by a method of drawing called revolution. Instead of projecting the inclined surface onto an auxiliary plane parallel to it, the object is revolved until the inclined surface is parallel with one of the principal planes. Then, the inclined surface is shown in its true shape and correct size on a principal plane.

The auxiliary view and revolution methods are compared in figure 9-29. Here, the object is first drawn in its normal position. The inclined surface is perpendicular to the frontal plane and inclined to the horizontal and profile planes, and its true shape and size are shown in the frontal auxiliary view.

Next, the object is shown revolved about a point called the axis of revolution. The object is revolved until the inclined surface is parallel to the profile plane. The true shape and size of the inclined surface is now shown in the side view. This side view and the auxiliary view are identical.

When the object is revolved, the front surface remains parallel to the frontal plane. Therefore, the revolved front view does not change in shape or size. Only its position is changed. The revolved

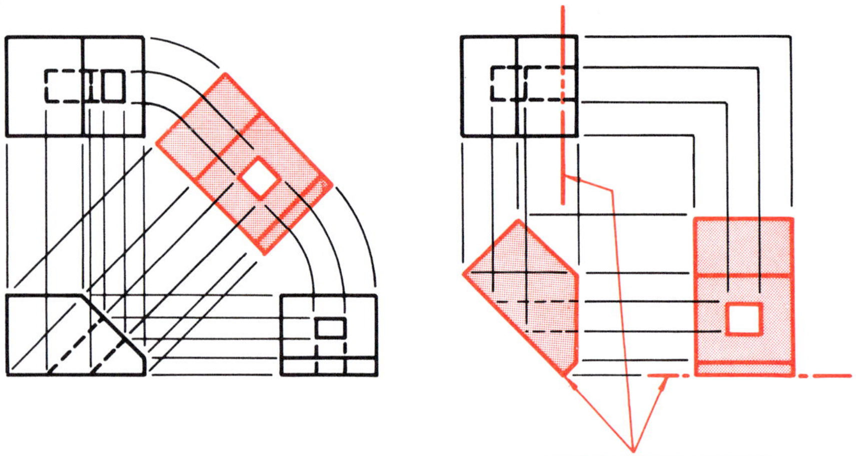

Fig. 9-29 Comparison of auxiliary and revolution methods

front view is copied from the first front view. The revolved top view and the side view are found in the usual manner.

The axis of revolution is an imaginary axis on which the object is revolved. In the illustration, figure 9-29, the axis of revolution is perpendicular to the frontal plane and parallel to the horizontal and profile planes. In the top and side views, the axis of revolution is represented by a line. In the front view it appears as a point.

On the drawings which follow, the object, or at least a part of it, is first drawn in the simplest position and then revolved to the required position.

ASSIGNMENT G

Make a three-view working drawing of the H block in its natural or simple position, figure 9-30. The H block is revolved clockwise 45 degrees to bring the inclined surface parallel with the profile plane. Then, another three-view drawing is made.

Procedure

1. Lay out the margin lines on the drawing sheet and letter the title strip.
2. Draw the top and front views as shown and then complete the side view. A number placed on each corner identifies these corners in each view.
3. Draw the second front view revolved clockwise 45 degrees. Remember that the front surface remains parallel to the frontal plane. Therefore, the front view retains the same size and shape as the first front view, only its position is changed.
4. The revolved top view is the same width as the first top view. Project a line to the right from each corner of the first top view and a line up from each corresponding corner of the revolved front view. The corners of the new top view are located at the intersections of these lines. For example: A line from point 1 in the top view and a line from point 1 in the revolved front view intersect to locate point 1 in the revolved top view. The other corners are located in the same manner. Locate the corners. Draw the revolved top view.
5. From each corner of the revolved front view, project a line to the right. From each corner of the revolved top view, project a line to the right and continue these lines around one-fourth of a circle and downward. The corners of the revolved side view are located at the intersections of these lines. Locate the corners and then draw the revolved side view.
6. Dual-dimension the views.
7. Check each regular and revolved view, and all dimensions on the drawing.

REVIEW QUESTIONS

1. What is the difference between the auxiliary view method of drawing and the revolution method?

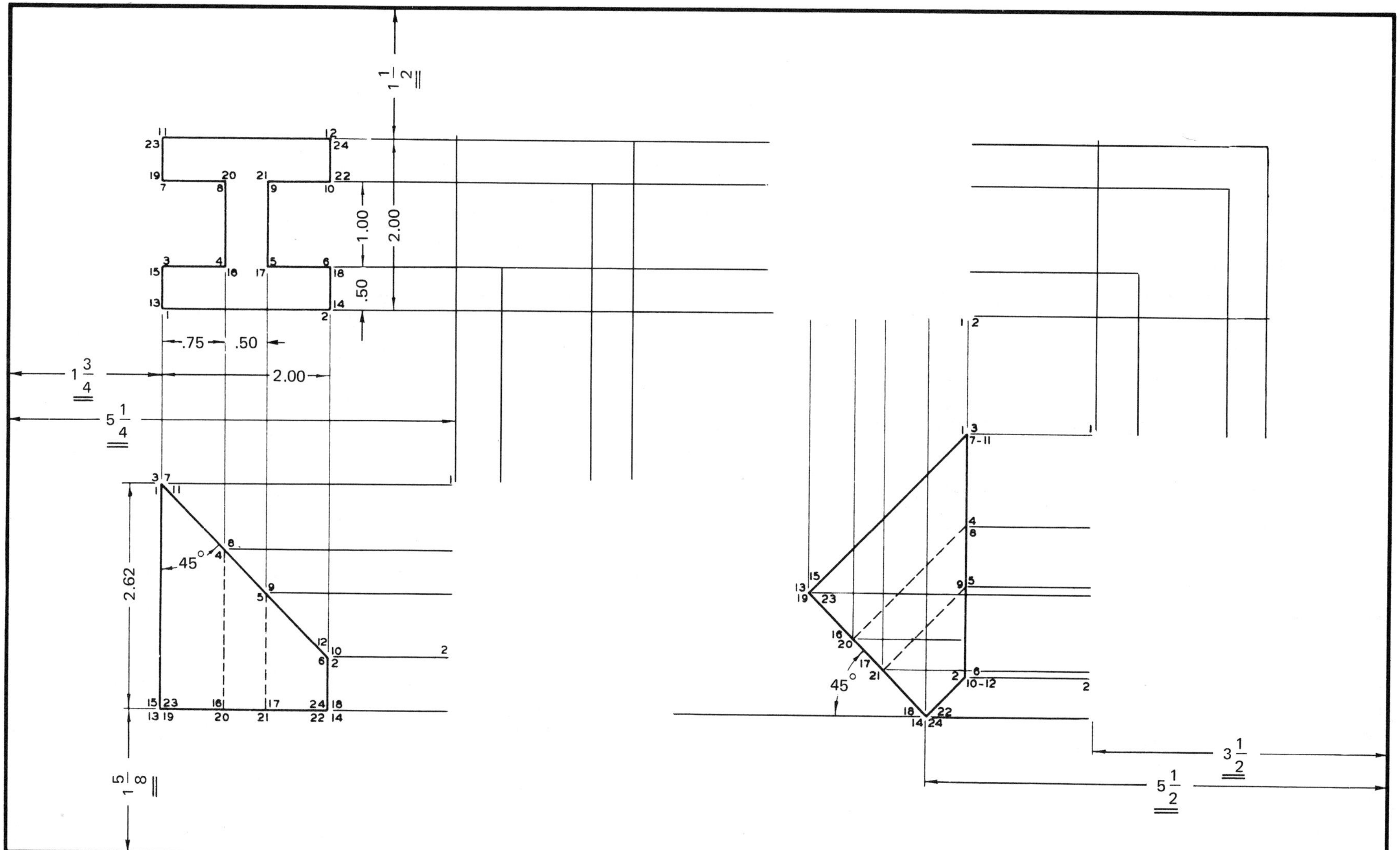

NAME OF SCHOOL CITY OR TOWN	H BLOCK	SHEET No. 9-G NAME	CLASS DATE

Fig. 9-30

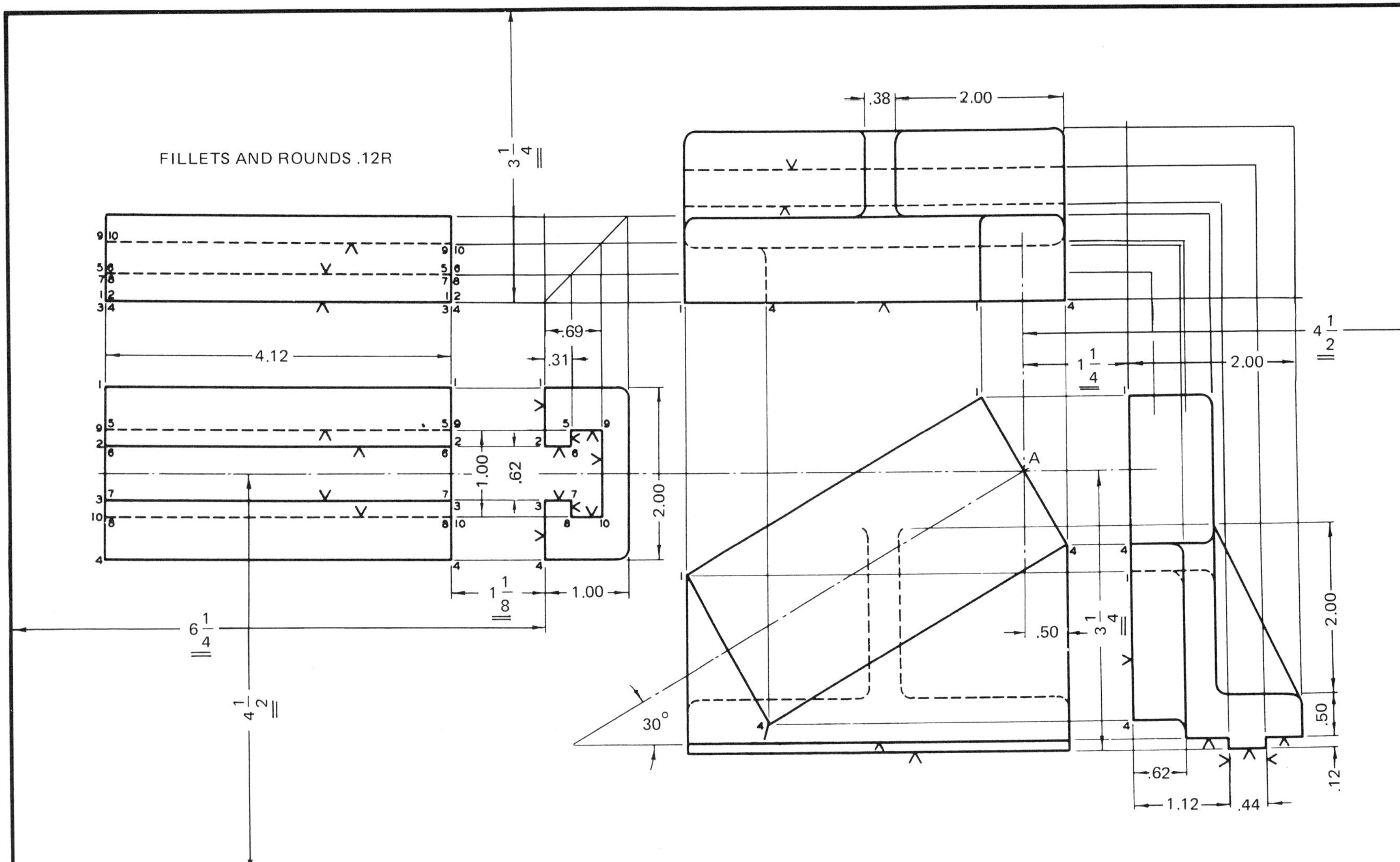

NAME OF SCHOOL CITY OR TOWN	FIXTURE BRACKET	SHEET No. 9-H NAME	CLASS DATE

Fig. 9-31

2. What is the axis of revolution?
3. In figure 9-29, which views in the two drawings are identical in size and shape?
4. Which corresponding views in the two drawings are different?
5. Which views are the same width?

REVOLUTIONS USED TO AID IN CONSTRUCTING PRINCIPAL VIEWS

Revolution, like auxiliary views, may be used to help construct principal views. The inclined portion of the object is first drawn in a simple position; that is, with its surfaces parallel to the ordinary planes of projection.

The front view is then revolved to the required position on an axis of revolution which is perpendicular to the frontal plane. The side and top views can then be projected from the front view.

ASSIGNMENT H

Draw three views of the inclined portion of the fixture bracket, figure 9-31, in the simple position. Revolve the inclined portion 30 degrees counterclockwise and draw a complete three-view drawing.

Procedure

1. Lay out the margin lines on the drawing sheet and letter the title strip.
2. Draw the three views in the simple position as shown. Draw the side view first. Draw the front view and the top view.
3. Number each corner in the three views as shown.
4. Draw the front view of the inclined portion revolved 30 degrees counterclockwise using point A as the axis of revolution. This view is the same shape and size as the simple view. Only its position is changed. Copy the numbers at each corner.
5. From each corner of the first top view project a line to the right. From each corresponding corner of the revolved front view project a line up. Locate the corners for the revolved top view at the intersections of these lines. Draw the revolved top view.
6. From each corner of the revolved front view, project a line to the right. From each corner of the revolved top view, project a line to the right. Continue these lines around one-fourth of a circle and then downward. Locate the corners for the revolved side view at the intersections of these lines. Draw the revolved side view and then complete this view.
7. Complete the front and top views.
8. Dual-dimension the views, add the finish marks, and letter the note.
9. Check each regular and revolved view and all lines and dimensions for accuracy.

REVIEW QUESTIONS

1. What view shows the true shape of the T slot?
2. In which views of the completed three-view drawing are the inclined surfaces foreshortened?
3. What is meant by counterclockwise?

REVOLUTION ABOUT AN AXIS PERPENDICULAR TO THE PROFILE PLANE

The objects in the last two drawings are revolved about a horizontal axis which is perpendicular to the frontal plane. When the axis of revolution is perpendicular to the frontal plane, the revolved front view is the same shape and size as front view in simple position.

An object having an inclined surface perpendicular to the profile plane is shown in figure 9-32. This object is revolved counterclockwise about a horizontal axis of revolution which is perpendicular to the profile plane. The true shape and size of the inclined surface is then shown in the revolved front view.

When the axis of revolution is perpendicular to the profile plane, the revolved side view is the same shape and size as the side view in simple position.

REVOLUTION ABOUT A VERTICAL AXIS

Figure 9-33 shows an object having an inclined surface which is perpendicular to the top horizontal plane. This object is revolved clockwise about a vertical axis of revolution. The true shape and size of the inclined surface is shown in the revolved front view.

When the axis of revolution is perpendicular to the horizontal plane, the revolved top view is the same shape and size as the top view in simple position.

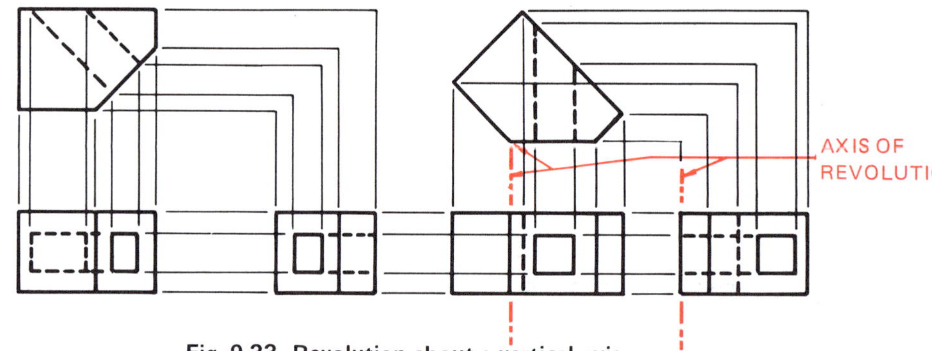

Fig. 9-33 Revolution about a vertical axis

CIRCLES IN REVOLUTION

An object whose views contain circles or circular arcs is first drawn in the simple position to show the true curves. When the object is revolved to the required position, the circles and arcs are foreshortened and become elliptical curves.

The pump mounting shown in figure 9-34 contains some circles and arcs which are foreshortened in the revolved front view. In the simple front view, points are marked off on the circles. These points are projected to the simple top view.

The top view is then revolved to the required position and the points are carefully transferred to this view. Lines are then projected to the right from the points in the simple front view and down from the corresponding points in the revolved top view in order to locate the points for the elliptical curves in the revolved front view.

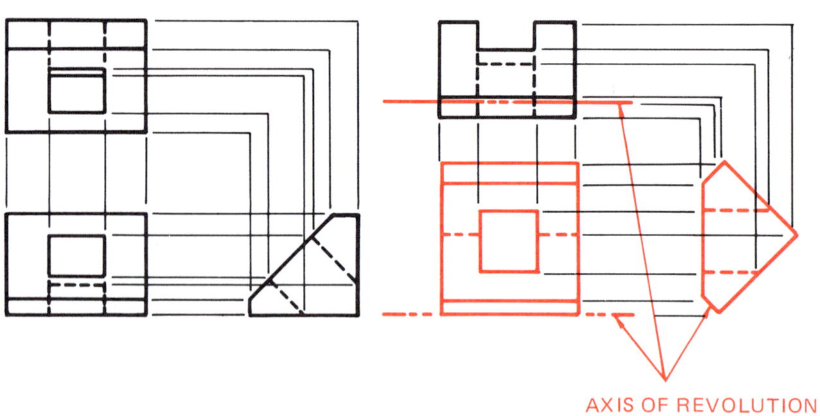

Fig. 9-32 Revolution about an axis perpendicular to the profile plane

ASSIGNMENT I

Draw in the simple position a front view and a top view of the circular part. Draw a top view and a front view of this part revolved counterclockwise about a vertical axis, and complete the top and front views.

Procedure

1. Lay out the margin lines on the drawing sheet and letter the title strip.

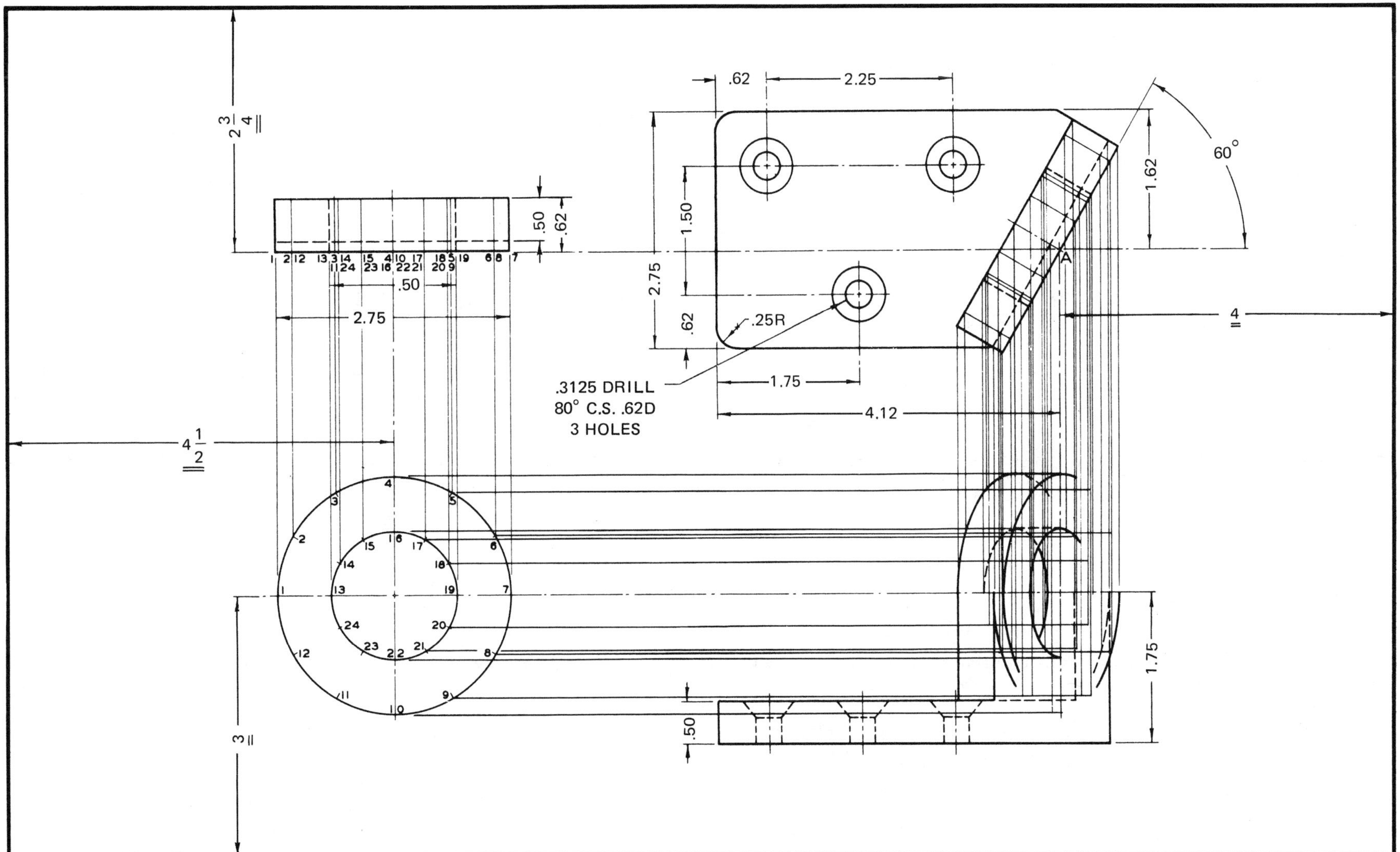

NAME OF SCHOOL CITY OR TOWN	PUMP MOUNTING	SHEET No. 9-I NAME	CLASS DATE

Fig. 9-34

2. Draw the front view as shown. Divide the circles into twelve equal parts. Number each point. Draw the top view.
3. From each point in the front view, draw a line up to locate the points in the top view. The points should be marked on the front edge, on the invisible edge, and on the rear edge. Because of the many points, the numbers are marked only on the front edge.
4. Draw the top view revolved counterclockwise 60 degrees about the vertical axis of revolution point A.
5. Using the dividers or compass, transfer the points to the revolved top view. These many points cannot be numbered without cluttering up the view.
6. Draw light lines to the right from the first twelve points in the front view.
7. From the same twelve points on the front edge of the revolved top view draw very fine light lines down.
8. Locate the points at the intersections of these lines; and then connect them with an irregular curve. The resulting curve is an ellipse.
9. Draw lines down from points 1 to 7 on the rear edge to locate the points for the semielliptical rear edge. Connect these points with an irregular curve.
10. From the invisible edge in the top view draw lines down from points 7 to 1 to locate points for semielliptical front edge. Connect these points.
11. Draw lines to the right from each point of the round hole.
12. Project lines down from the same points on the front edge of the revolved top view to locate points for front edge of hole. Connect these points.
13. Draw lines down from the points on the rear edge of the hole and draw this invisible edge in the front view.
14. Complete the top view and the front view.
15. Dimension the views and letter the note. Check the accuracy of the regular and revolved views, lines, and dimensions.

REVIEW QUESTIONS

1. Which two views are the same when the axis of revolution is perpendicular to the frontal plane? To the profile plane?
2. Which two views are the same when the axis of revolution is perpendicular to the horizontal plane?

SUCCESSIVE REVOLUTION

The objects in the three preceding drawings are revolved once about a single axis. This is called *simple revolution.* If the object is revolved about an axis and then revolved again about another axis, it is called *successive revolution.*

Successive revolution may replace primary and secondary auxiliary views to show the shape of a surface which is inclined to all of the principal planes, such as oblique surface. The illustrations in figure 9-35 show how the true shape of an oblique surface is drawn by the double auxiliary and successive revolution methods.

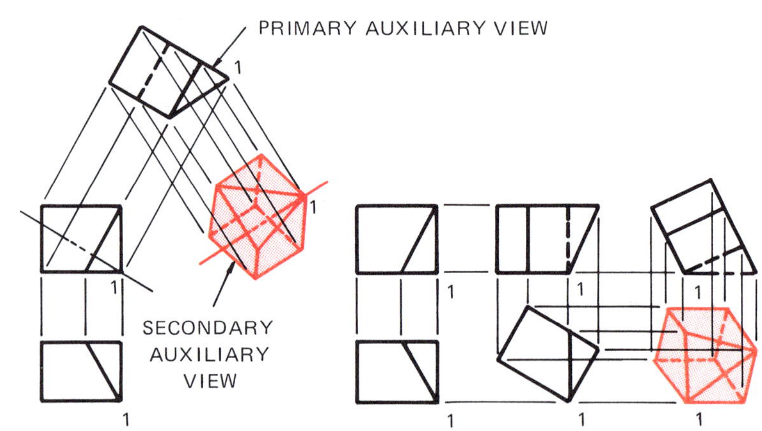

Fig. 9-35 Comparison of double auxiliary and successive revolution methods

The first drawing shows a top and front view of a rectangular prism with one corner sliced off to provide an oblique surface. A primary top auxiliary view is projected from the top view. A secondary auxiliary view, which shows the true shape of the oblique surface, is projected from the primary auxiliary view.

The second illustration shows a top view and a front view of the same object in simple position. The object is revolved about a horizontal axis until the oblique surface is perpendicular to the top horizontal plane. The revolved top view is again revolved about a vertical axis until the oblique surface is parallel to the frontal plane, and another revolved front view is drawn. Since the oblique surface is now parallel to the frontal plane, this surface is shown in its true shape in the second revolved front view. This view is the same as the secondary auxiliary view.

Another practical use for successive revolution is to help construct the principal view when a portion of the object is an oblique position. The tool grinding fixture is drawn using the successive revolution method. Figure 9-36 shows a front and top view of the oblique portion drawn in the simple position. These views show the true shape and size of the oblique portion.

The front view is then revolved on a horizontal axis until the upper surface is at the required angle (30 degrees) from the horizontal plane, and the revolved top view is drawn. The revolved top view is now revolved again to the required position which is 75 degrees from the frontal plane. The required front view is then drawn. After the oblique portion is drawn in the required position, the remainder of the tool grinding fixture is added to the views.

ASSIGNMENT J

Make a two-view working drawing of the tool grinding fixture using successive revolution for the oblique portion.

Procedure

1. Lay out the margin lines on the drawing sheet and letter the title strip.
2. Draw the front view of the oblique portion in the simple position as shown. Draw the top view. Number the corners in both views.
3. Draw the front view revolved clockwise 60 degrees about horizontal axis A. Copy the numbers from the first front view.
4. From each corner of the revolved front view project a light line up. From each corner of the top view project a light line to the right. Locate the numbers for the revolved top view at the intersections of these lines and then draw the revolved top view.
5. The revolved top view is again revolved counterclockwise 75 degrees about vertical axis B to the required position. Copy the numbers.
6. From each corner of this top view, project a light line down. From each corner of the revolved front view project a light line to the right. Locate the numbers for the required front view at the intersections of these lines. Then, draw the required front view.
7. Complete the top view. Complete the front view.
8. Dual-dimension the views. Check the regular and successive revolution views.

REVIEW QUESTIONS

1. What is meant by a simple revolution?
2. What is meant by successive revolution?

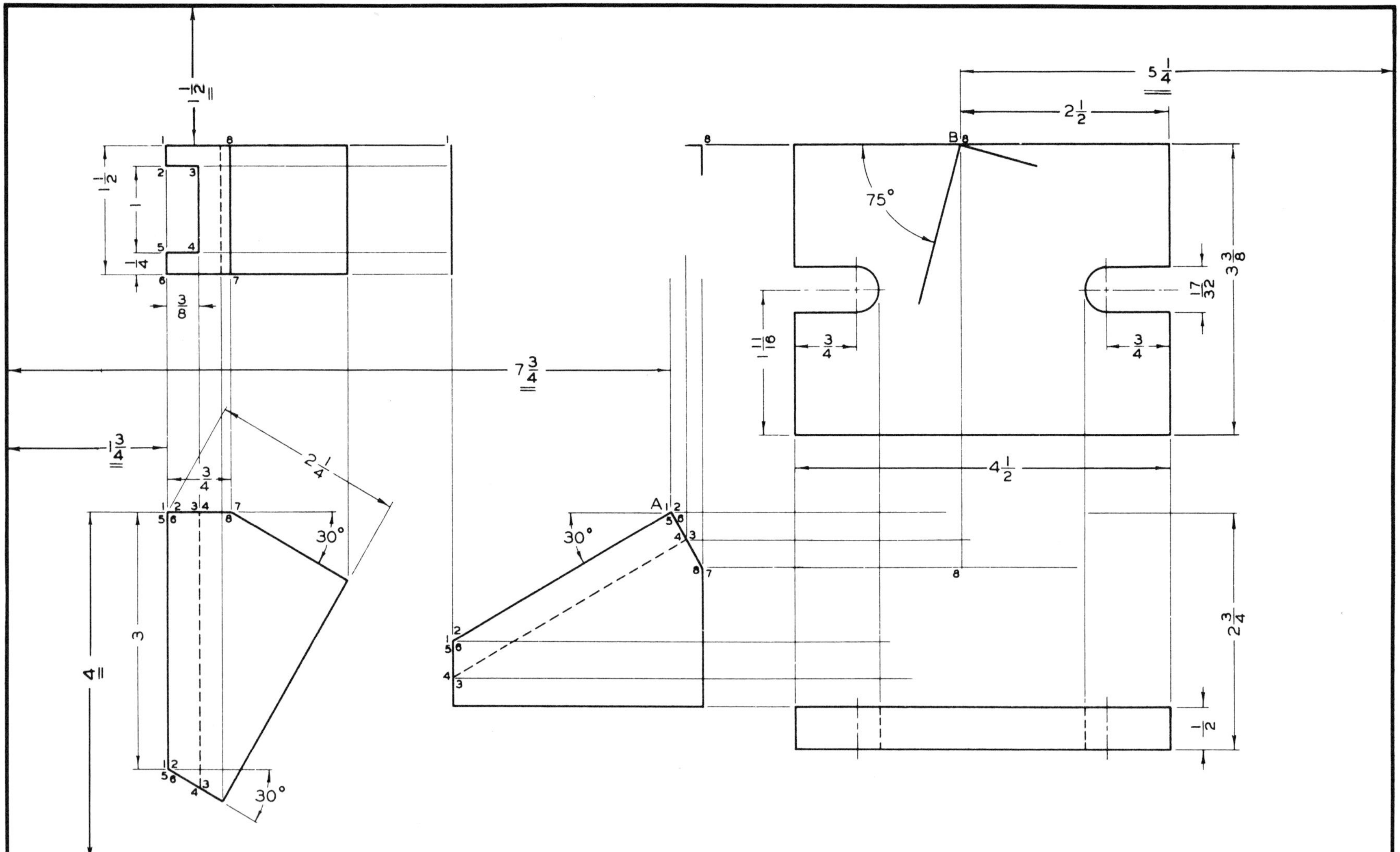

NAME OF SCHOOL CITY OR TOWN	TOOL GRINDING FIXTURE	SHEET No. 9-J NAME	CLASS DATE

Fig. 9-36

UNIT 10 PICTORIAL DRAWINGS

Pictorial drawings are drawings that picture the objects in one view only. Pictorial drawings are often used in place of multiview drawings to describe objects and convey information to those who are not trained to read and interpret orthographic drawings. Pictorial drawings generally show how parts are assembled, for piping diagrams, electrical wiring diagrams, catalogue illustrations, and similar applications. They illustrate patent office drawings and furniture design. The methods used in making pictorial drawings are especially helpful to the drafter when making freehand sketches.

ISOMETRIC DRAWING

The use of pictorial drawings is limited because they are more difficult to draw, their intricate parts cannot be pictured clearly, and they are difficult to dimension. One method of making pictorial drawings is called *isometric drawing.*

The theory of isometric drawing is illustrated in figure 10-1. In this illustration, the principal views of a cube are projected onto the principal planes of projection, and an oblique auxiliary view is projected onto an oblique auxiliary plane. The angles between

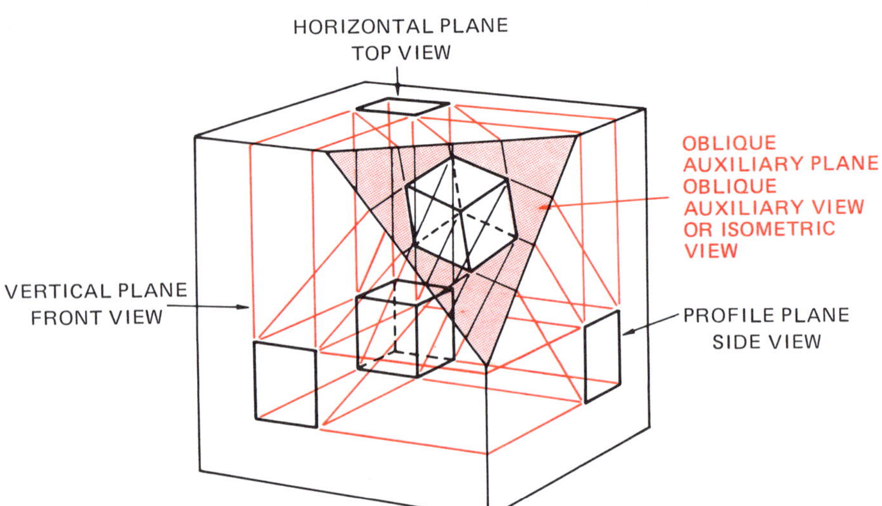

Fig. 10-1 Isometric view projected onto an isometric plane of projection

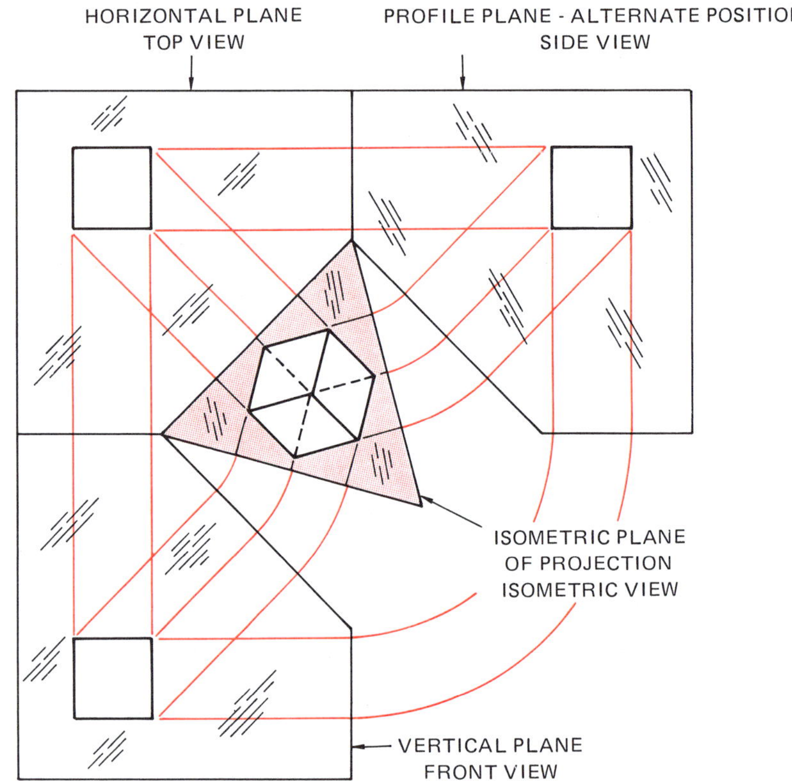

Fig. 10-2 The position of the isometric plane of projection

the oblique plane and each of the principal planes are equal. Because these angles are equal, the angles between the edges of the cube and the oblique plane also are equal. Likewise, all of the edges are foreshortened equally in the auxiliary view.

When the edges of a cube in three directions are foreshortened equally in a view, the view is called an *isometric projection.* The word *isometric* means equal measure. The plane on which the isometric projection is drawn is called an *isometric plane of projection.*

The planes of projection are spread out on a plane in figure 10-2 to show the relative position of the views. The isometric plane of projection is at an angle of 45 degrees with each principal plane. Note that the edges of the cube in the isometric view are foreshortened equally.

Since the isometric view in figure 10-2 is an oblique auxiliary view, it can be drawn by the double auxiliary view method previously explained.

When double auxiliary views are drawn, a primary auxiliary view is first projected from one of the principal views, then a secondary auxiliary view is projected from the primary auxiliary view. This method is used to draw the oblique auxiliary view in figure 10-3.

The primary auxiliary view is projected perpendicular to diagonal A-B passing through the body of the cube. Then, a secondary auxiliary view is projected parallel with the same body diagonal A-B. The secondary auxiliary view in figure 10-3 is the same as the oblique auxiliary view in figures 10-1 and 10-2. The edges are foreshortened equally so this view is an isometric view.

The oblique auxiliary view may be drawn by still another method called *successive revolution.* A drawing of a cube using successive revolution to obtain the isometric view is shown in figure 10-4.

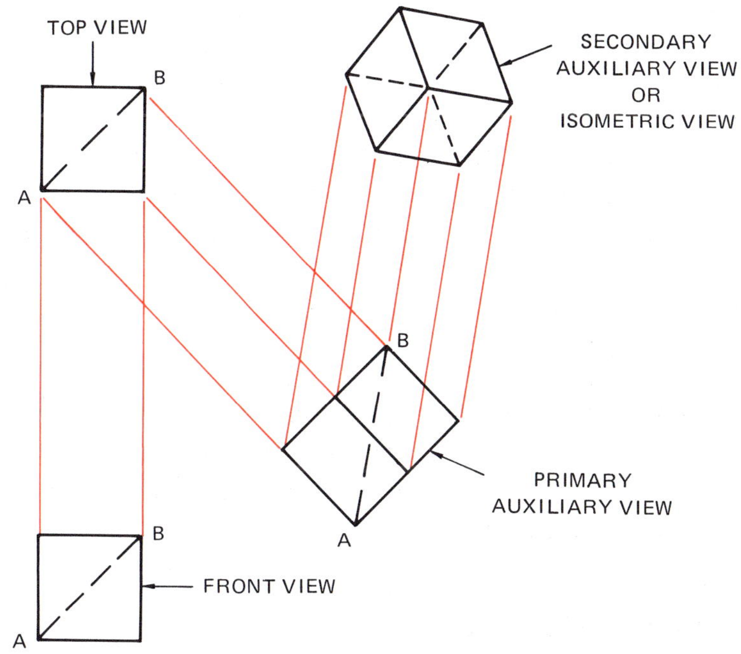

Fig. 10-3 **Isometric view of a cube obtained by the double auxiliary method**

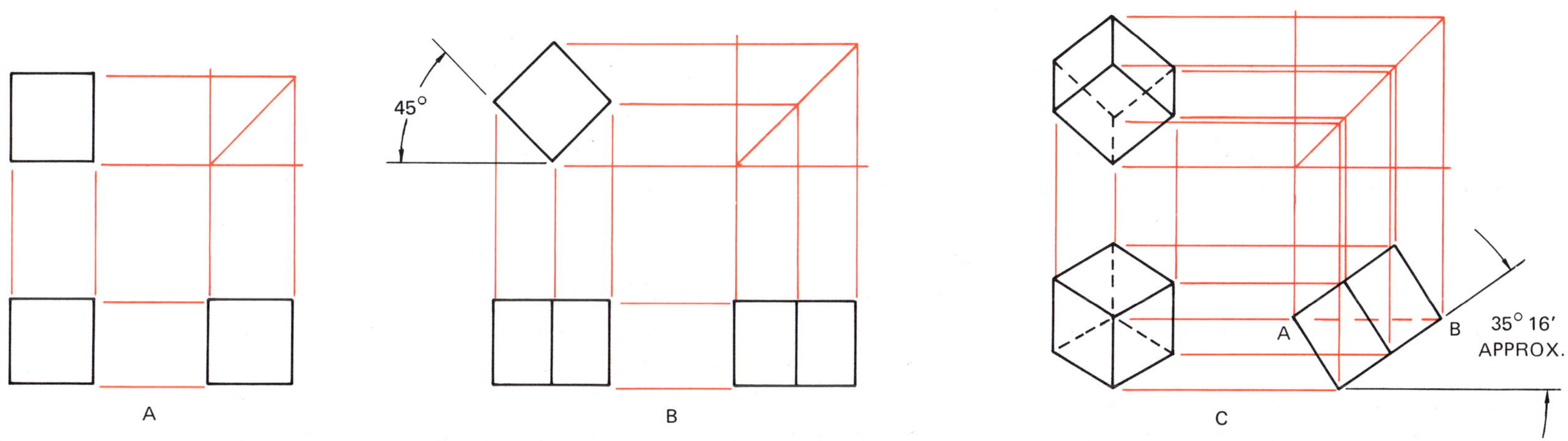

Fig. 10-4 **Isometric view of a cube obtained by revolution**

In this illustration at A, the faces of the cube are parallel to the principal planes of projection. The top view at B has been revolved 45 degrees and the front and side views are then drawn. At C, the side view has been tilted so that the body diagonal A-B is horizontal. Then the top and front views are drawn. The body diagonal is perpendicular to the front vertical plane.

The front view at C is identical (except in size) to the oblique auxiliary view in figure 10-2 and to the secondary auxiliary view in figure 10-3. The edges are foreshortened equally, so this view is an isometric view.

In the front view at C, the leading edges of the cube (the edges nearest the observer) make equal angles with the vertical plane. These angles are approximately 35 degrees 16 minutes. The leading edges are 120 degrees apart in the isometric view and are foreshortened equally.

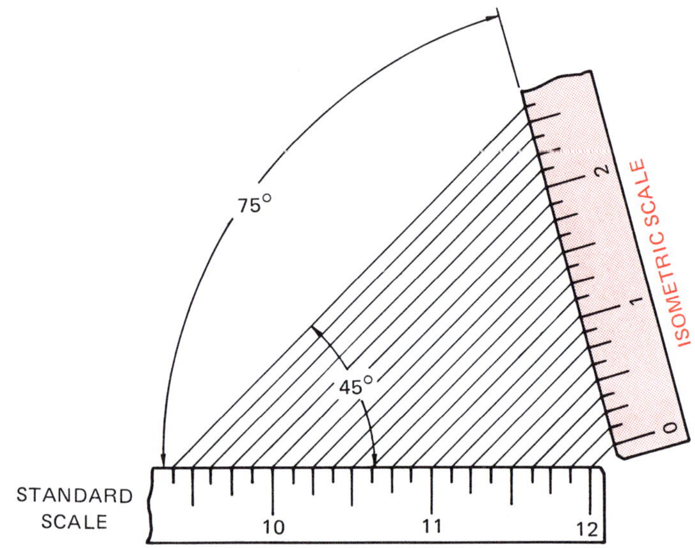

Fig. 10-5 An isometric scale

ISOMETRIC PROJECTION

A view projected onto a vertical plane in which all of the edges are foreshortened equally is called an *isometric projection.* The edges in the view which are 120 degrees apart are called the *isometric axes.* The three surfaces shown in the view are called *isometric planes.* Lines parallel to isometric axes are called *isometric lines.*

Because the isometric axes are always 120 degrees apart and the edges are foreshortened equally, isometric views are drawn without resorting to secondary auxiliary views or to successive revolution. The isometric axes and lines parallel to them can be drawn with the 30-60 degree triangle, and the edges can be measured with an isometric scale. An isometric scale for measuring the foreshortened lines may be constructed as shown in figure 10-5.

Drawings in isometric are usually measured with a scale having standard graduations rather than with an isometric scale. This results in a drawing which is considerably larger than an isometric drawing made by projection. An isometric projection and an isometric drawing are compared in figure 10-6.

DRAWING IN ISOMETRIC

When an isometric drawing is made, the part to be drawn is first enclosed in a box. This box is then drawn in isometric. The axes are made 120 degrees apart and all measurements are made on isometric lines. Figure 10-7 shows step-by-step how to make an isometric drawing. The measurements for the length, width and height of the box must be made on the isometric axes. Other features of the object are then drawn and all measurements made on isometric lines. In isometric drawings, invisible edges are usually omitted.

In isometric drawings, all measurements must be made on the isometric axes or on lines parallel to them.

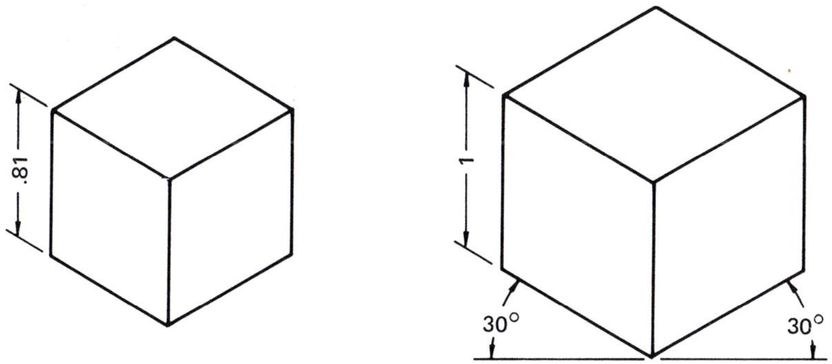

Fig. 10-6 An isometric projection and an isometric drawing compared

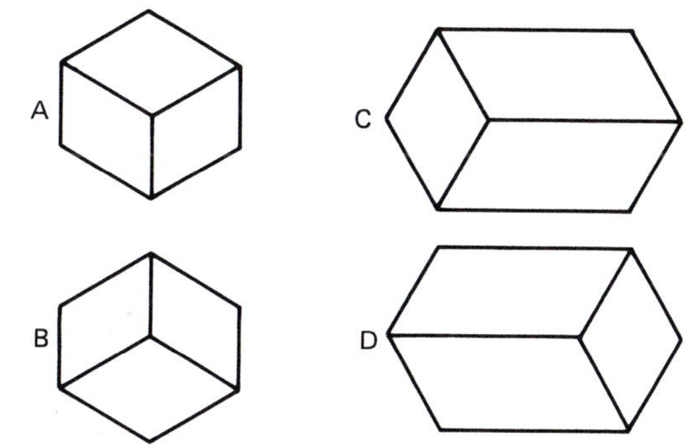

Fig. 10-8 Four positions of the isometric axes

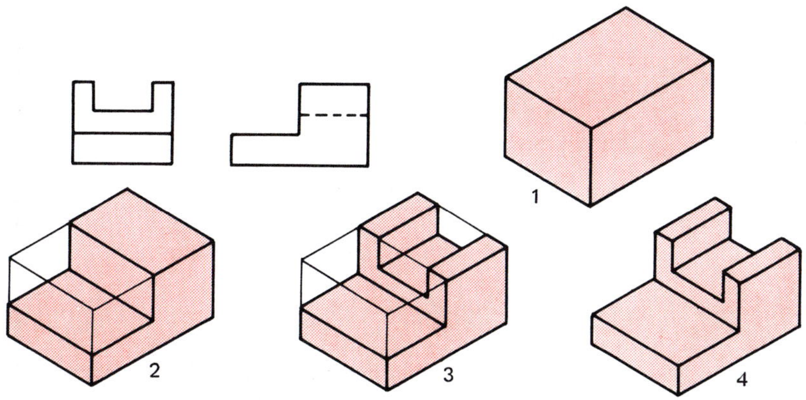

Fig. 10-7 Steps in making an isometric drawing

POSITIONS OF THE ISOMETRIC AXES

The isometric axes may be drawn in four different positions, figure 10-8. At A, one axis is in a vertical position and the other two are slanting upward to show the top. At B, these axes are reversed to show the bottom. At C, one axis is drawn horizontal and the other two slant toward the left to show the left end. These axes are reversed at D to show the right end.

ASSIGNMENT A

Draw the top, front, and left-side views of the jig body, figure 10-9. Make an isometric drawing from the orthographic views.

Procedure

1. Lay out the margin lines on the drawing sheet and letter the title strip.
2. Draw the three views as shown.
3. Enclose each view in a rectangle.
4. Draw the rectangular box in isometric.
5. Draw the isometric drawing of the jig body within the isometric box.
6. Complete the drawing. Check the accuracy of the top, front, and left-side views.

REVIEW QUESTIONS

1. What is a pictorial drawing?
2. For what purposes are they used?
3. What is meant by a primary auxiliary view?

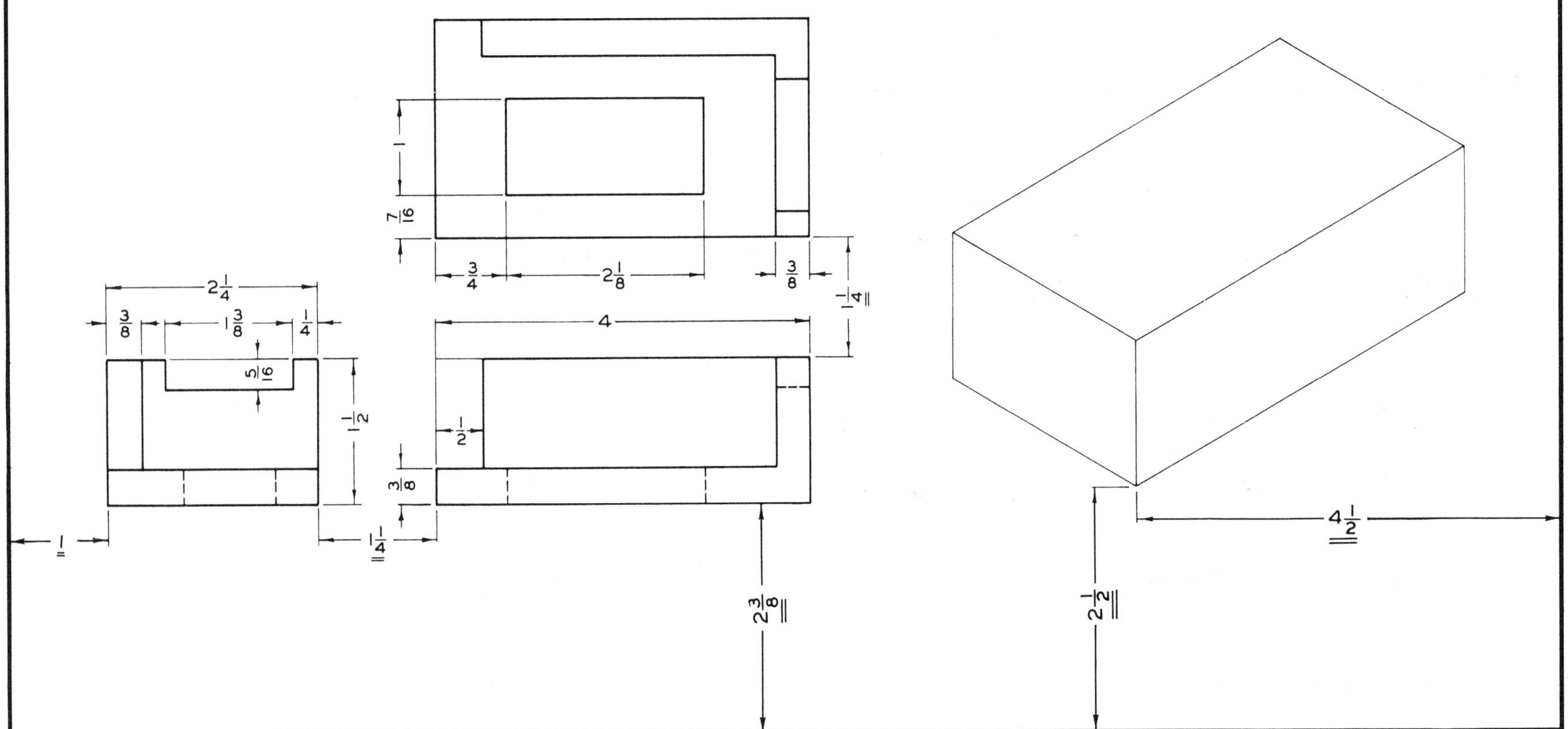

NAME OF SCHOOL CITY OR TOWN	JIG BODY	SHEET No. 10-A NAME	CLASS DATE

Fig. 10-9

4. What is meant by a secondary auxiliary view?
5. What is meant by successive revolution?
6. What is meant by isometric projection?
7. How does an isometric projection differ from an isometric drawing?

NONISOMETRIC LINES

Straight lines not parallel to the isometric axes are called *nonisometric lines.* Nonisometric lines are not shown in their true lengths in isometric drawings. They are drawn by first locating their end points on isometric lines and then connecting the end points with lines, figure 10-10.

The miter box in figure 10-11 has some nonisometric lines. The orthographic views of the miter box are first enclosed in a box. The box is drawn in isometric. The end points of the nonisometric lines are located on the isometric lines, and the end points are then connected with lines.

ASSIGNMENT B

Draw the top, front, and side views of the miter box. Make an isometric drawing of the miter box.

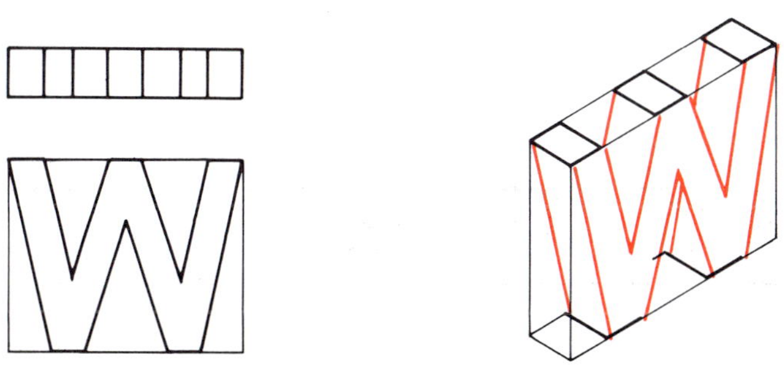

Fig. 10-10 Isometric drawing with nonisometric lines

Procedure

1. Lay out the margin lines on the drawing sheet and letter the title strip.
2. Draw the three views of the miter box.
3. Enclose the views in a box, then draw the enclosing box in isometric.
4. Locate the end points of the nonisometric lines.
5. Draw the nonisometric lines. Complete the drawing.
6. Check the regular views and the isometric drawing for accuracy.

REVIEW QUESTIONS

1. What are isometric lines?
2. What are nonisometric lines?

COORDINATE CONSTRUCTION METHOD

Nonisometric lines which do not lie on isometric planes are drawn using the *coordinate construction method.*

In this method, the views are enclosed in a box as usual. The ends of the sloping lines on the part are projected to one side of the box. Horizontal and vertical lines, called coordinates, are drawn from the ends of the projectors to the edges of the box as shown at points 1, 2, and 3 in figure 10-12.

The box is drawn in isometric and the coordinates are drawn to locate points 1, 2, and 3. Isometric lines are drawn from points 1, 2, and 3. The end points of the nonisometric lines can then be measured on the isometric lines. The end points are connected with lines to complete the drawing.

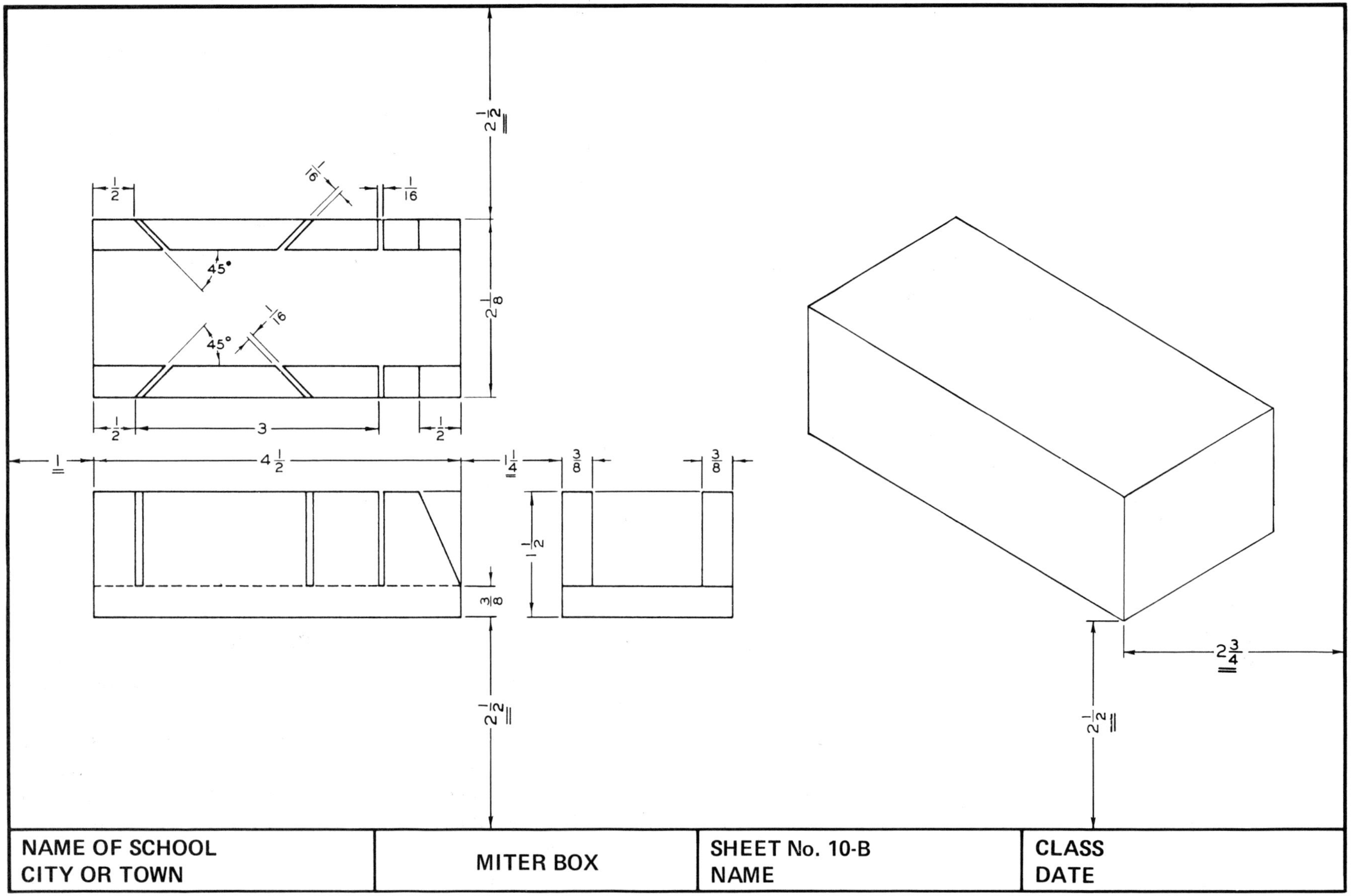

1/2
1/16
1/16
45°
45°
2 1/8
3
1/2
1/2
1
4 1/2
1 1/4
3/8
3/8
1 1/2
3/8
2 1/2
2 1/2
2 1/2
2 3/4
NAME OF SCHOOL
CITY OR TOWN
MITER BOX
SHEET No. 10-B
NAME
CLASS
DATE

Fig. 10-11

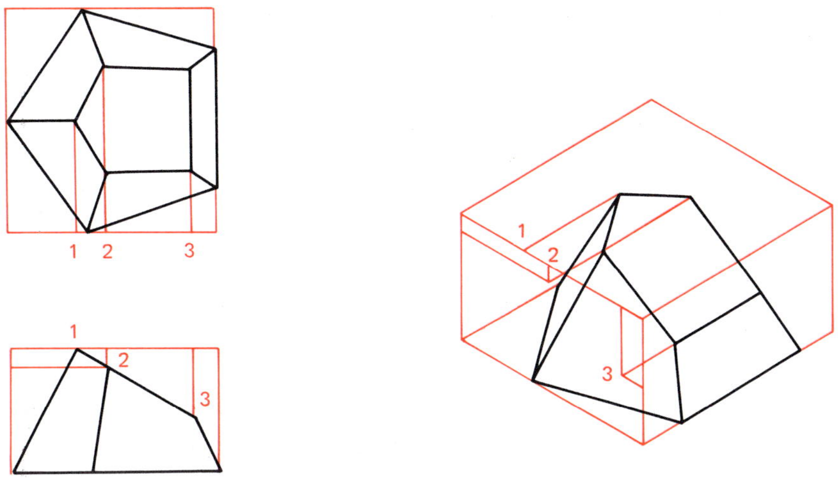

Fig. 10-12 Co-ordinate construction method

ASSIGNMENT C

Draw the front, side, and top views of the inclined guide, figure 10-13. Make an isometric drawing of the inclined guide.

Procedure

1. Lay out the margin lines on the drawing sheet and letter the title strip.
2. Draw the front view. Draw the side view. Draw the top view.
3. Enclose the views in a rectangular box and draw the coordinates.
4. Draw the enclosing box in isometric. Locate and draw the coordinates.
5. Locate the end points of the nonisometric lines.
6. Draw the nonisometric lines and complete the isometric view.
7. Complete the drawing.
8. Check the three regular views and the isometric drawing of the inclined guide.

REVIEW QUESTIONS

1. What are coordinates in isometric drawings?
2. For what purpose are coordinates used?

CIRCLES IN ISOMETRIC

A circle in an orthographic drawing becomes an ellipse in the isometric drawing. The ellipses in isometric drawings usually are drawn with the compass, using four different centers. The method of locating these four centers is shown in figure 10-14.

The circle in the view is first enclosed in a square. The square is drawn in isometric. Two lines are drawn from the vertex of each obtuse angle in the isometric square. These lines are 30 degrees from the sides of the isometric square. Each line bisects and is perpendicular to one side of the isometric square. The intersections of these perpendiculars are the centers of the small arcs forming the ends of the ellipse. The vertex of each obtuse angle is the center of the large arcs forming the sides of the ellipse.

Circles on isometric planes may be drawn in six different positions as illustrated in figure 10-15.

CYLINDERS IN ISOMETRIC

Cylindrical shaped parts, like all other parts drawn in isometric, are first enclosed in a box. The enclosing box is drawn in isometric. Then, the cylinder is drawn within the box.

When cylinders are drawn in isometric, they are drawn with the axis of the cylinder in either a vertical or horizontal position, figure 10-16. The flange, the hub, and the holes in figure 10-17 are cylindrical shapes and all of the circles are ellipses in the isometric drawing.

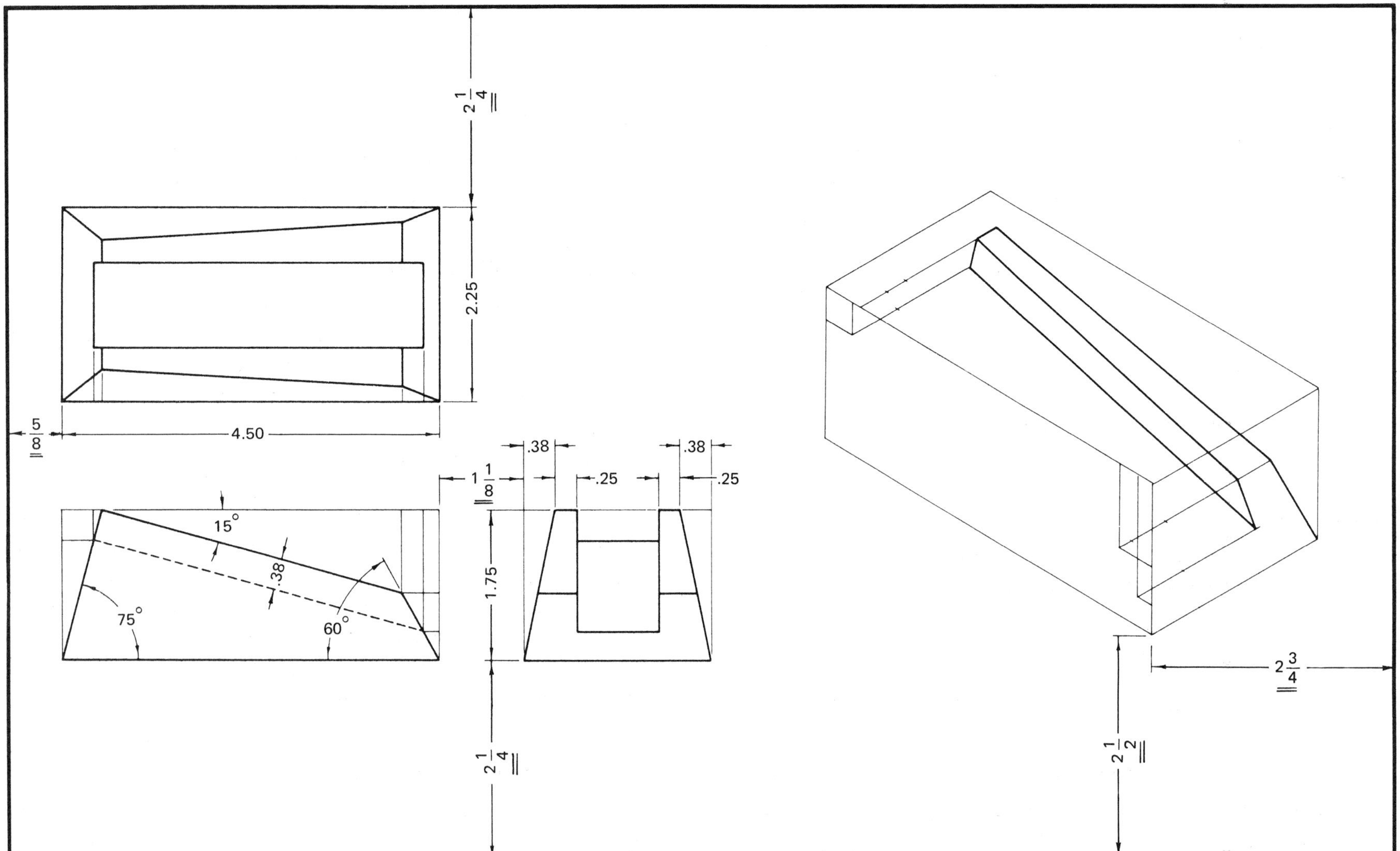

NAME OF SCHOOL CITY OR TOWN	INCLINED GUIDE	SHEET No. 10-C NAME	CLASS DATE

Fig. 10-13

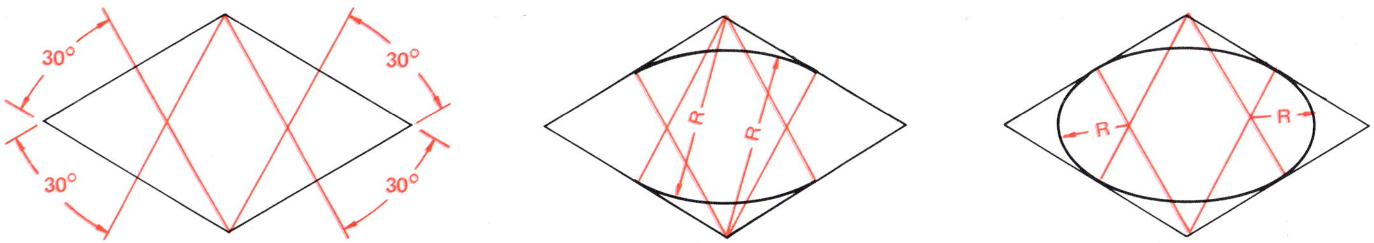

Fig. 10-14 Drawing a circle in isometric

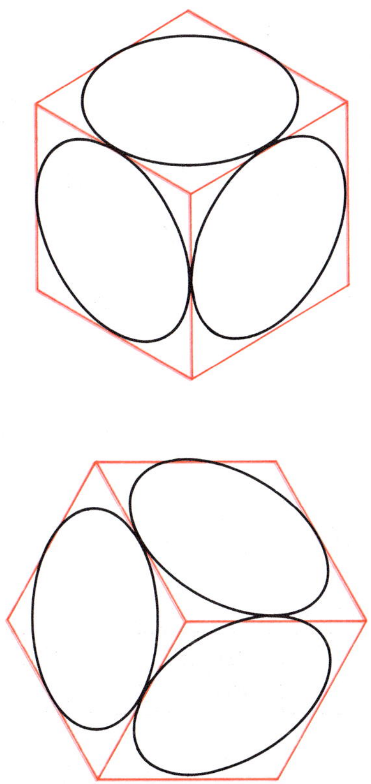

Fig. 10-15 Circles on isometric planes

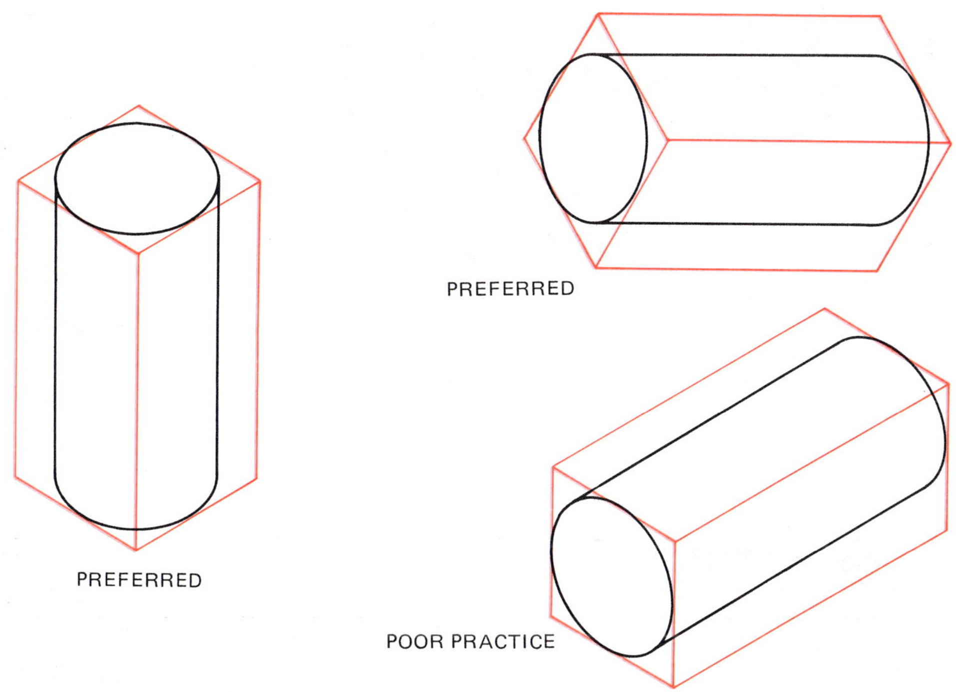

Fig. 10-16 Cylindrical shapes in isometric

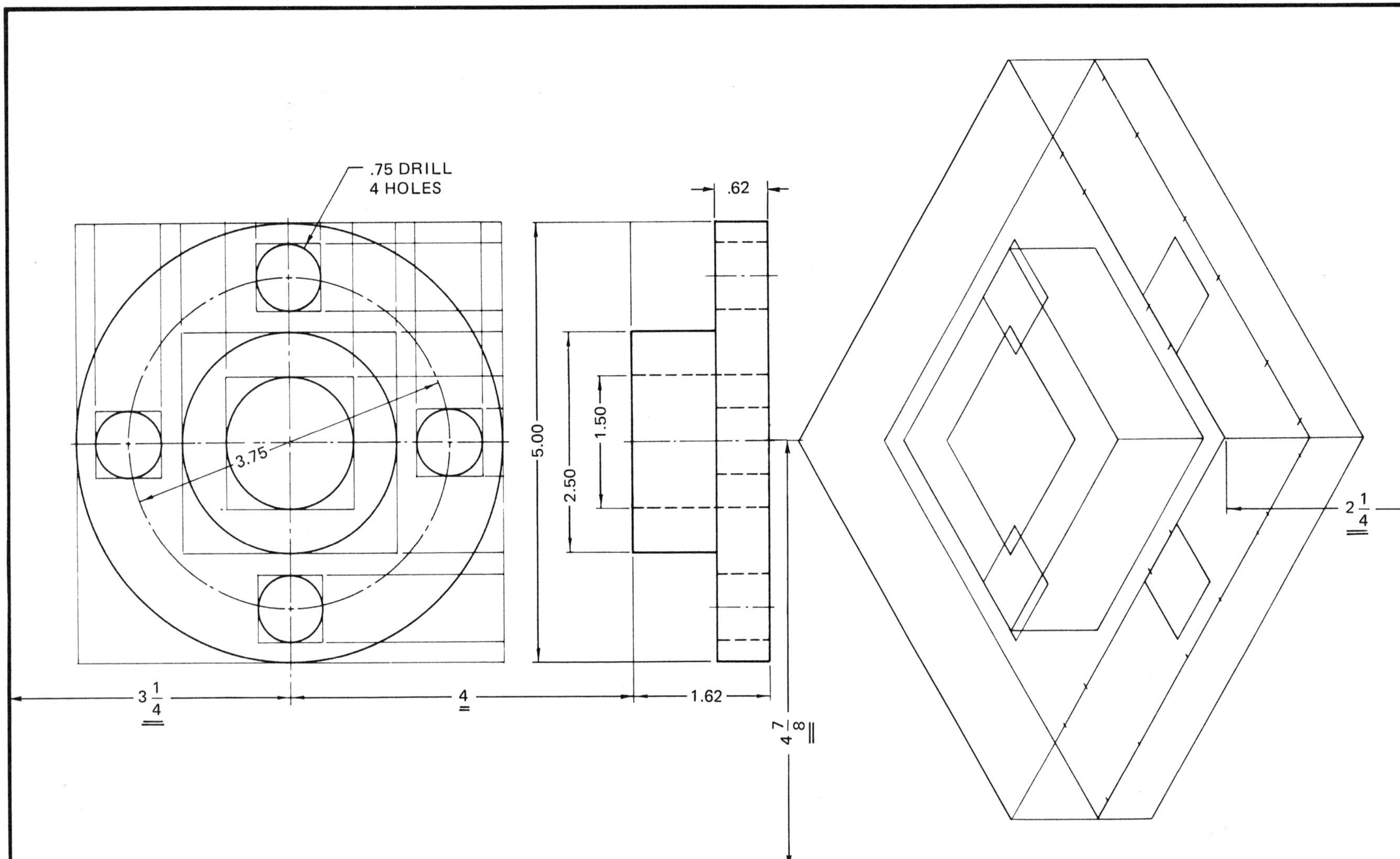

NAME OF SCHOOL CITY OR TOWN	FLANGE	SHEET No. 10-D NAME	CLASS DATE

Fig. 10-17

ASSIGNMENT D

Draw the front and side views of the flange. Make an isometric drawing of the flange.

Procedure

1. Lay out the margin lines on the drawing sheet and letter the title strip.
2. Draw the front and side views.
3. Enclose each circle in a square and enclose the side view in a box.
4. Draw the enclosing box in isometric.
5. Locate and draw each square on the isometric planes.
6. Proceed as shown in figure 10-14 and draw the circles in isometric.
7. Complete the drawing. Check the regular views and the isometric drawing of the flange.

REVIEW QUESTION

1. Explain in detail the step-by-step method used to draw circles in isometric.

ARCS IN ISOMETRIC

Circular arcs are drawn in isometric with a compass. Since an arc is part of a circle, the four-center method used in drawing circles in isometric is used for drawing the arcs, figure 10-18.

The arcs in the orthographic view are continued around to make full circles, and a square is drawn around each circle as shown in figure 10-14.

The support in figure 10-19 has several arcs and circles.

ASSIGNMENT E

Draw the top, front, and right-side views of the support. Make an isometric drawing of the support.

Procedure

1. Lay out the margin lines on the drawing sheet and letter the title strip.
2. Draw the top, front, and right-side views. Enclose the views in a box.
3. Extend the arcs to make full circles. Draw a square around all of the circles.
4. Draw the enclosing box in isometric.

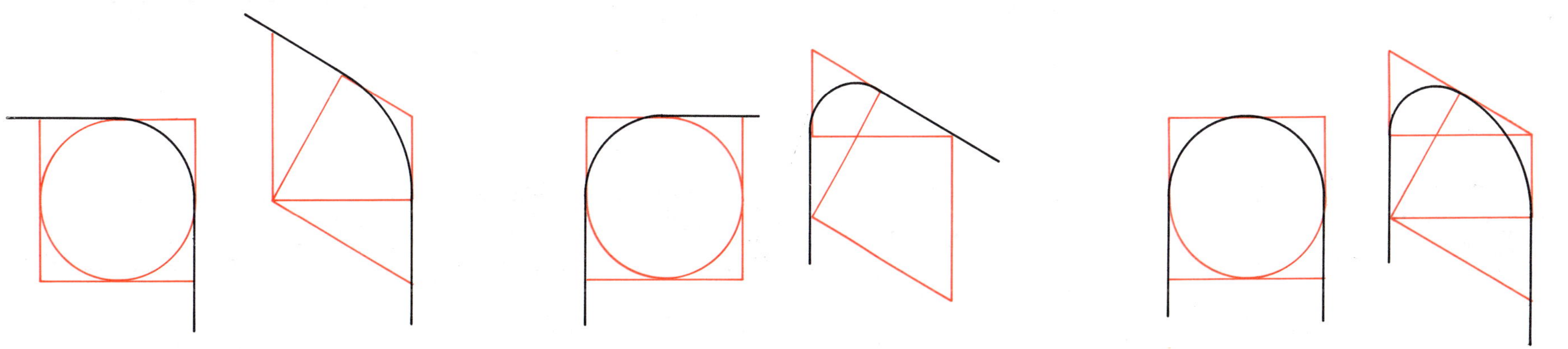

Fig. 10-18 Arcs in isometric

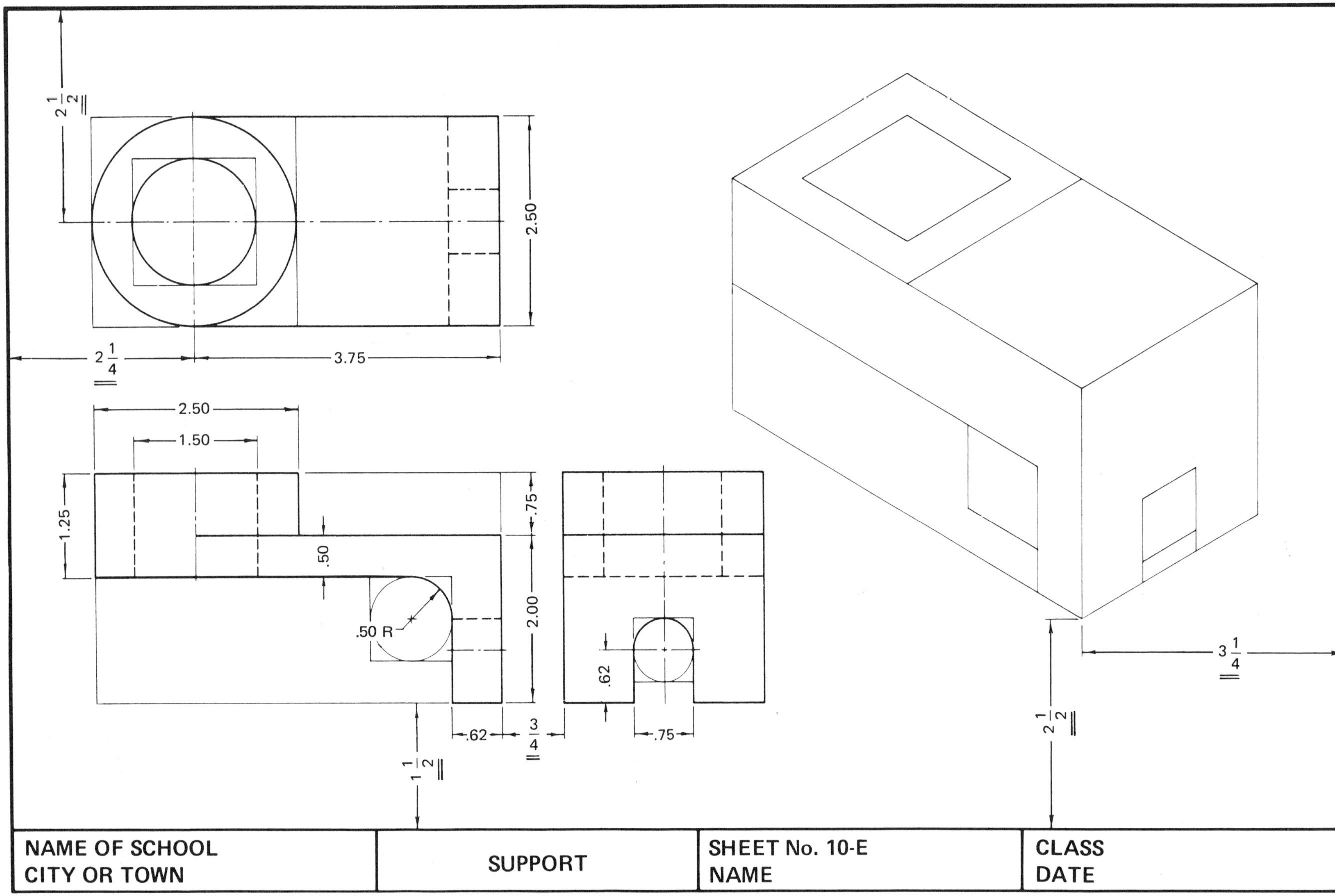
NAME OF SCHOOL
CITY OR TOWN
SUPPORT
SHEET No. 10-E
NAME
CLASS
DATE

Fig. 10-19

5. Locate and draw each isometric square on its isometric plane.
6. Locate the centers for the isometric circles and arcs following the method illustrated in figure 10-14.
7. Complete the isometric view.
8. Check the regular views and the isometric drawing of the support.

CURVED LINES IN ISOMETRIC

Curved lines, arcs, and circles may be drawn in isometric by another method using coordinate lines.

In figure 10-20, coordinate lines are drawn from selected points on the curved line to a horizontal reference line. This reference line is drawn in the isometric view. The spaces between the coordinate lines are pointed off on the isometric reference line and the coordinate lines are drawn. The length of each coordinate line is transferred to the corresponding line in the isometric view. The points are then connected with an irregular curve.

The molding in figure 10-21 has a curved line on each end which is drawn in the isometric view using coordinate lines.

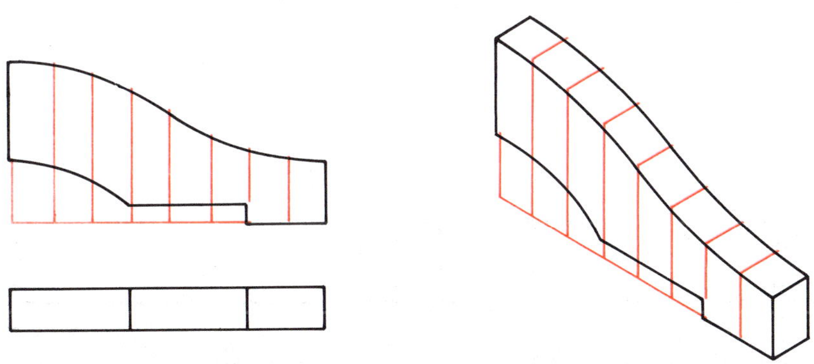

Fig. 10-20 Curved lines in isometric

LONG OBJECTS IN ISOMETRIC

Long objects, when drawn in isometric, look better if the long axis is made horizontal as shown in figure 10-22.

Conventional Breaks

A part that is too long to fit the available space on the drawing paper may be shortened to a suitable length if a conventional break is drawn in the view. The molding is too long to fit all of the views into the available space. Therefore, the front view is shortened and a conventional break is drawn in this view. The actual length of the part must be specified in the dimension when the view is shortened. See figure 10-23 for conventional breaks for various cross sections.

ASSIGNMENT F

Draw the front and side views of the molding. Make an isometric drawing of the molding.

Procedure

1. Lay out the margin lines on the drawing sheet and letter the title strip.
2. Draw the side view.
3. Draw the front view. Shorten this view to 2 inches. Draw the conventional break for wood freehand to resemble the edges of a broken piece of wood.
4. Enclose the side view in a rectangle and draw the coordinate lines.
5. Draw the enclosing box its full length in isometric (long axis horizontal).
6. Draw coordinate lines. Locate the points on these lines. Draw curved line.

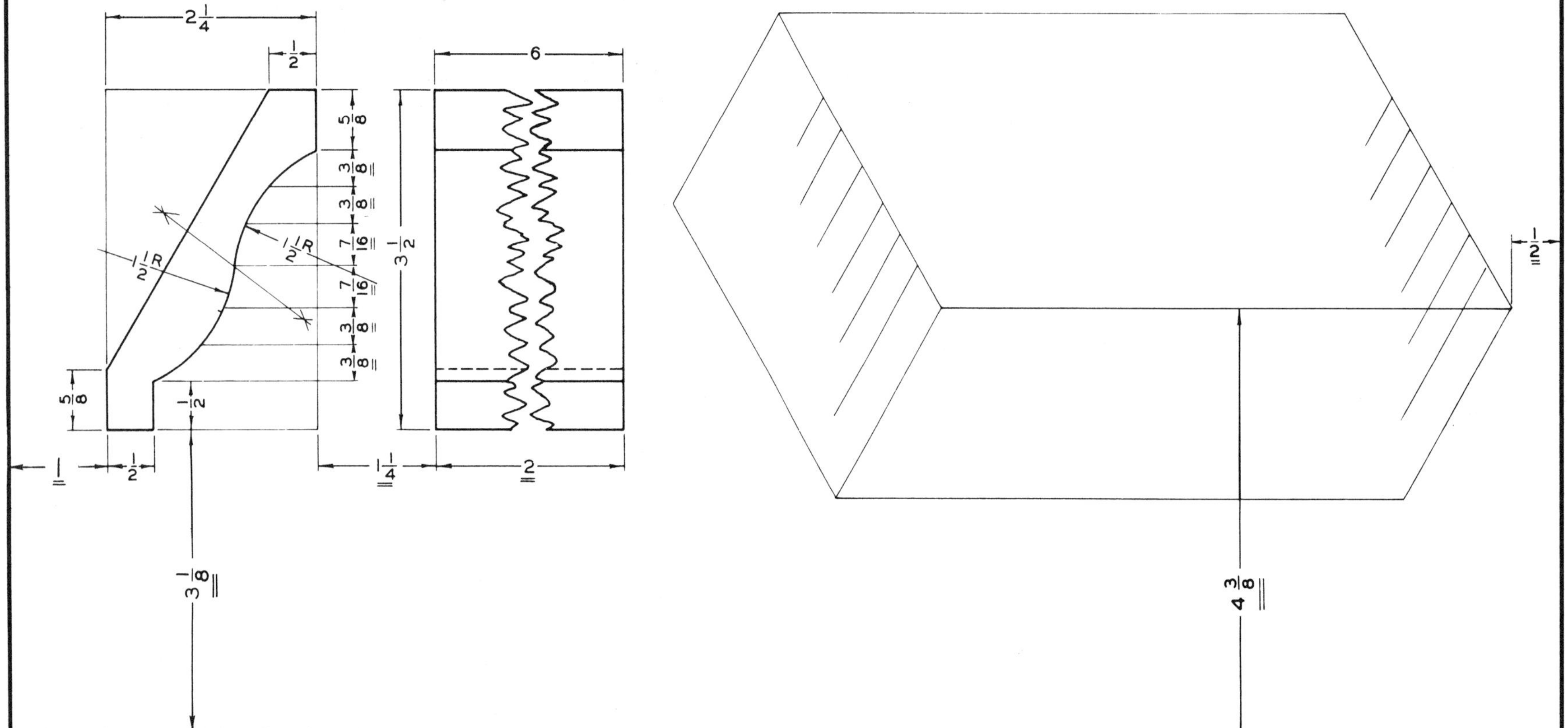

NAME OF SCHOOL CITY OR TOWN	MOLDING	SHEET No. 10-F NAME	CLASS DATE

Fig. 10-21

7. Complete the isometric view.
8. Check the accuracy of the front and side views and the isometric drawing of the molding.

ISOMETRIC SECTIONAL VIEWS

Isometric drawings are sometimes made in section to show the interior details more clearly. The cutting plane for isometric sections should be an isometric plane. When an isometric full section is drawn, the section in contact with the cutting plane is drawn first. The details behind the cutting plane are then filled in. Section lining in isometric sections is usually drawn at a 60 degree angle, figure 10-24.

LINES TANGENT TO ARCS IN ISOMETRIC

The rod guide, figure 10-25, has some lines which are tangent to arcs. In the orthographic view, the arcs are continued around to make full circles. These circles are drawn in the isometric view. The lines are then drawn tangent to the isometric circles.

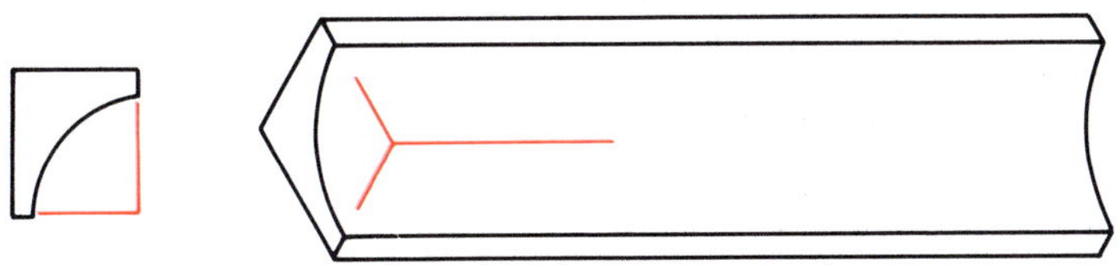

Fig. 10-22 Long axis horizontal

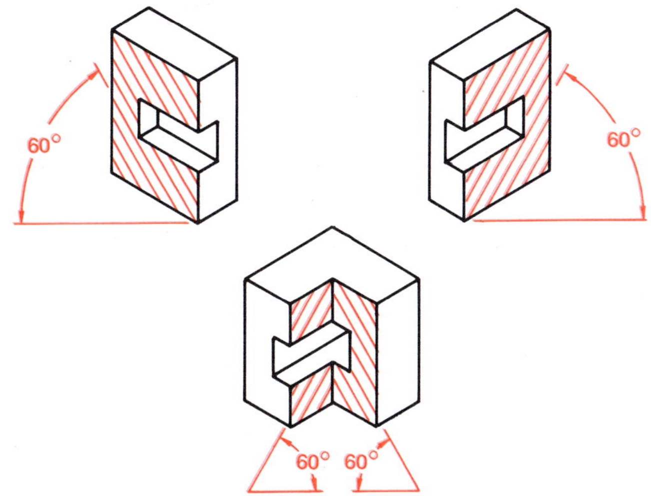

Fig. 10-24 Section lining in isometric sections

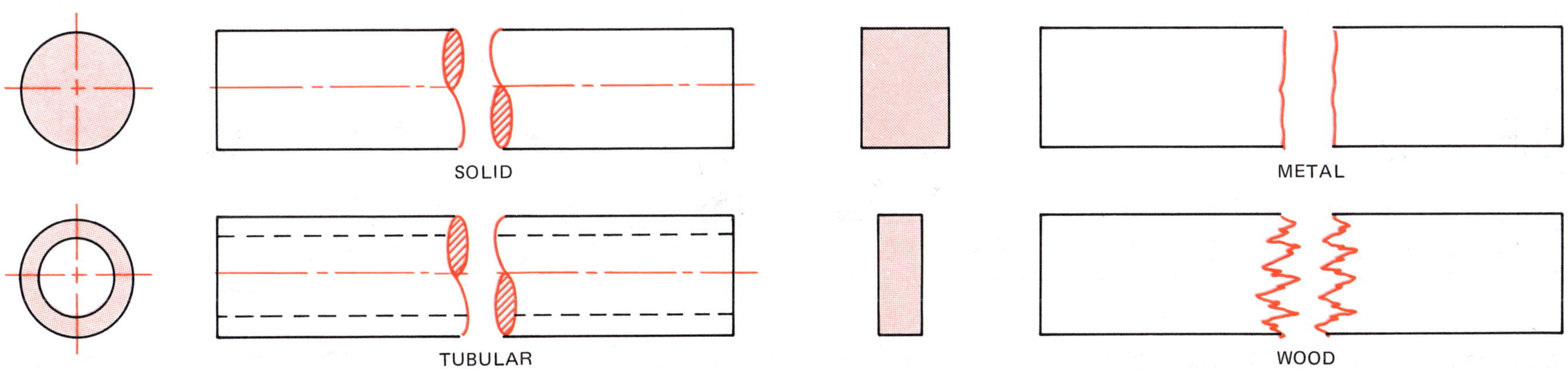

Fig. 10-23 Conventional breaks

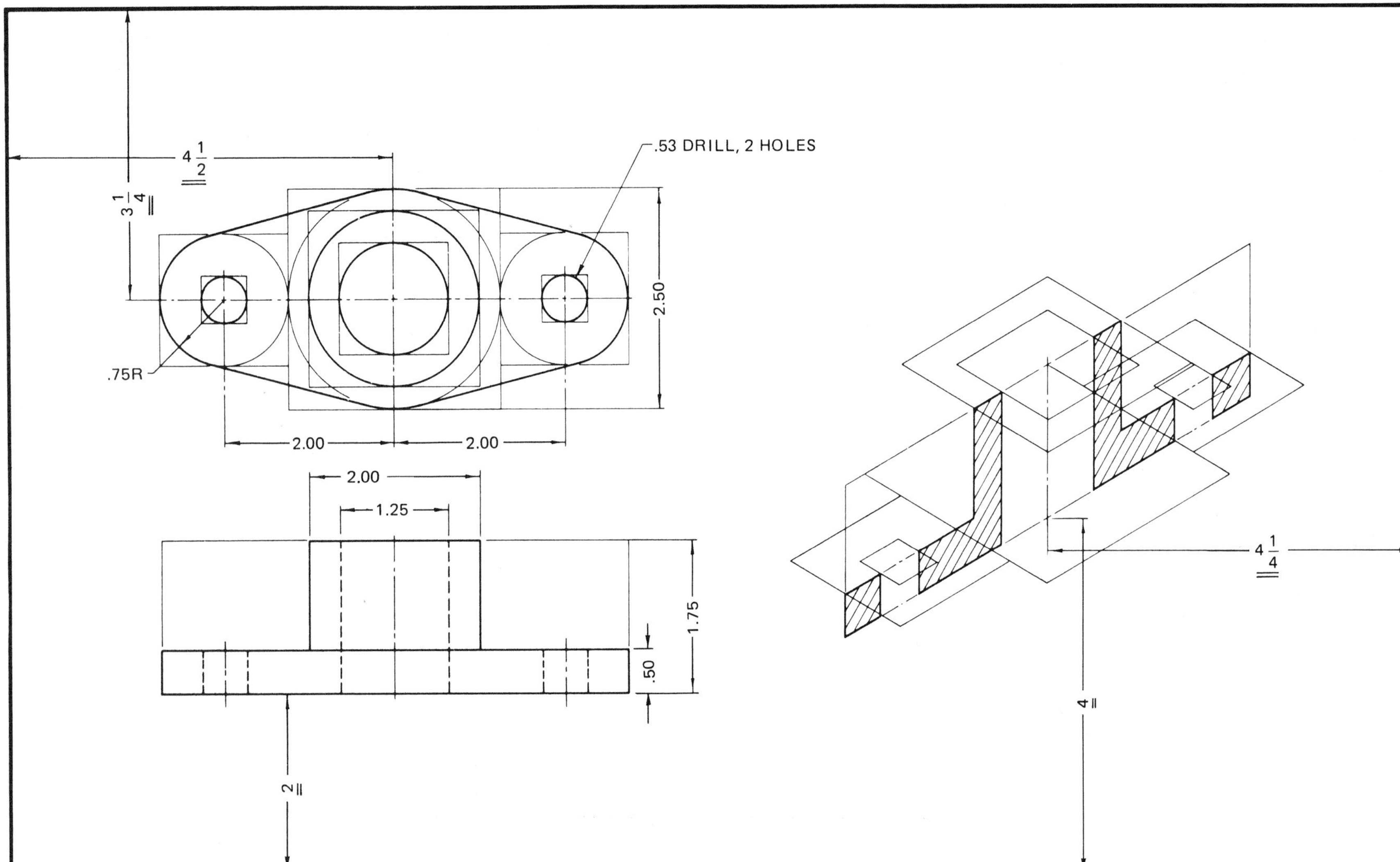

NAME OF SCHOOL CITY OR TOWN	ROD GUIDE	SHEET No. 10-G NAME	CLASS DATE

Fig. 10-25

ASSIGNMENT G

Draw the top and front views of the rod guide. Make an isometric full section of the rod guide.

Procedure

1. Lay out the margin lines on the drawing sheet and letter the title strip.
2. Draw the top view and enclose each circle in a square.
3. Continue the tangent arcs around to make full circles. Enclose these circles in squares.
4. Draw the front view and enclose it in a rectangle.
5. Draw this rectangle in the isometric view. This rectangle represents the cutting plane.
6. Draw the part of the rod guide which is in contact with the cutting plane.
7. Draw the isometric square for each circle on its isometric plane.
8. Use the four-center method and draw all of the semicircles which lie behind the cutting plane.
9. Draw the lines which are tangent to the arcs.
10. Complete the isometric view. Use section lining for cast iron.
11. Check the accuracy of the top and front views and the isometric full section.

REVIEW QUESTIONS

1. What is a full section view?
2. For what purpose is section lining used?
3. What is the usual angle for section lining in isometric?

ISOMETRIC HALF SECTIONS

Isometric half sections are drawn with the front quarter removed. Both cutting planes should be isometric planes.

When isometric half sections are required, the practice is to draw the entire part in isometric and then remove the front quarter from the view.

The section lining in isometric half sections should slant 60 degrees, but in opposite directions for each cutting plane, as shown in figure 10-24.

ASSIGNMENT H

Draw the top and front views of the roller end plate, figure 10-26. Draw an isometric half section.

Procedure

1. Lay out the margin lines on the drawing sheet and letter the title strip.
2. Draw the top view. Draw the front view.
3. Draw the isometric half section. Use section lining for cast iron.
4. Check the accuracy of the two regular views and the isometric half section of the roller end plate.

REVIEW QUESTIONS

1. What is a half-section view?
2. What is an isometric half section?
3. Why are half sections sometimes preferred to full sections?

DIMENSIONING ISOMETRIC DRAWINGS

Isometric drawings are dimensioned according to the following rules:

- Visible surfaces are dimensioned in preference to hidden surfaces.
- Both extension lines for a dimension are extended from isometric visible edges.

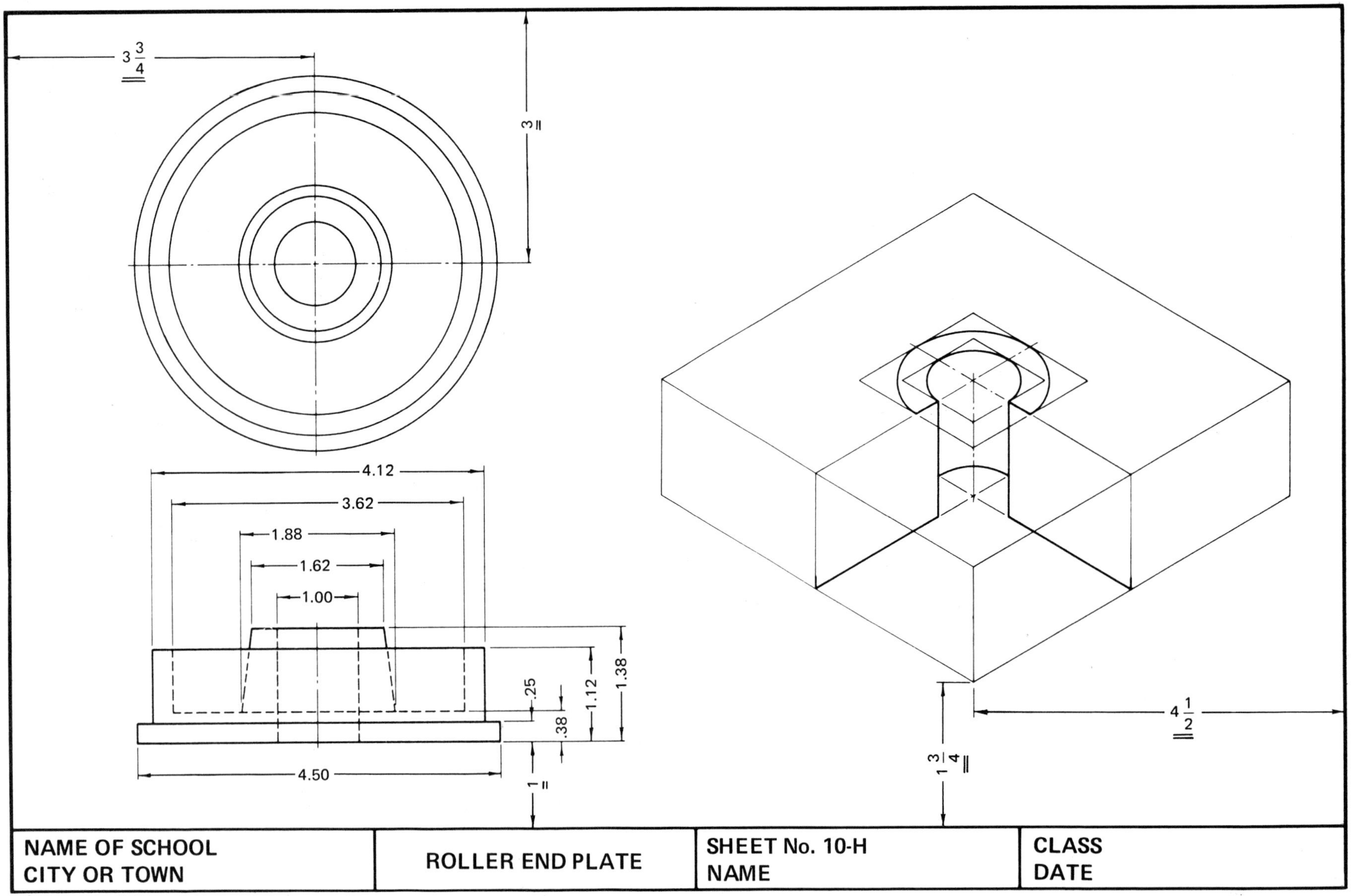
3 3/4
3
4.12
3.62
1.88
1.62
1.00
.25
1.12
1.38
.38
4.50
1
1 3/4
4 1/2
NAME OF SCHOOL
CITY OR TOWN
ROLLER END PLATE
SHEET No. 10-H
NAME
CLASS
DATE

Fig. 10-26

- Horizontal and vertical dimension lines on the orthographic views are parallel to isometric axes in the isometric view.
- Vertical numerals are used in dimensioning. The numerals and the arrowheads are made in isometric.
- The dimensions locating the center lines of round holes are drawn in the same plane with the circles.
- The radius dimension of an arc is drawn from the center of the enclosing isometric square to the arc.
- Dimension lines are kept off the view wherever possible.
- Holes are best dimensioned with a note stating the size of the hole and the machining operation involved.
- Notes need not be lettered in isometric.

The correct application of these rules is shown in the isometric drawing, figure 10-27. In contrast to this, figure 10-28 gives an isometric drawing incorrectly dimensioned.

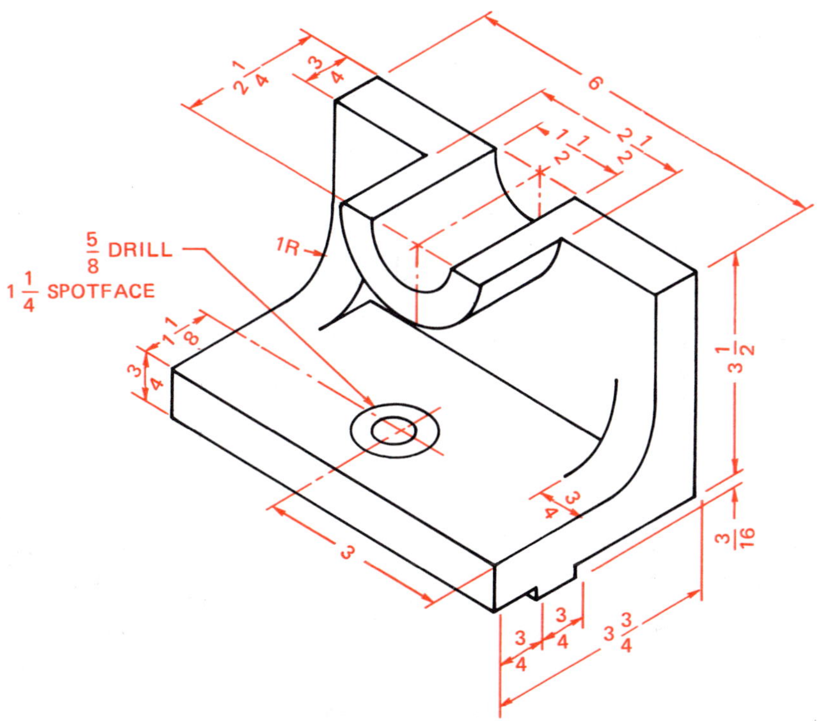

Fig. 10-27 An isometric drawing dimensioned correctly

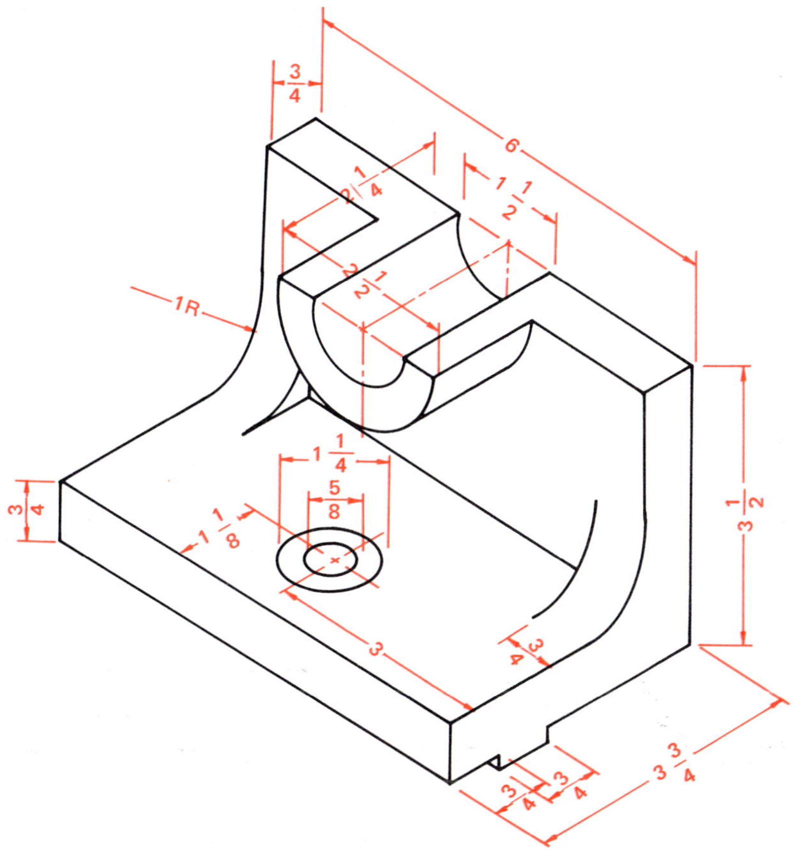

Fig. 10-28 An isometric drawing dimensioned incorrectly

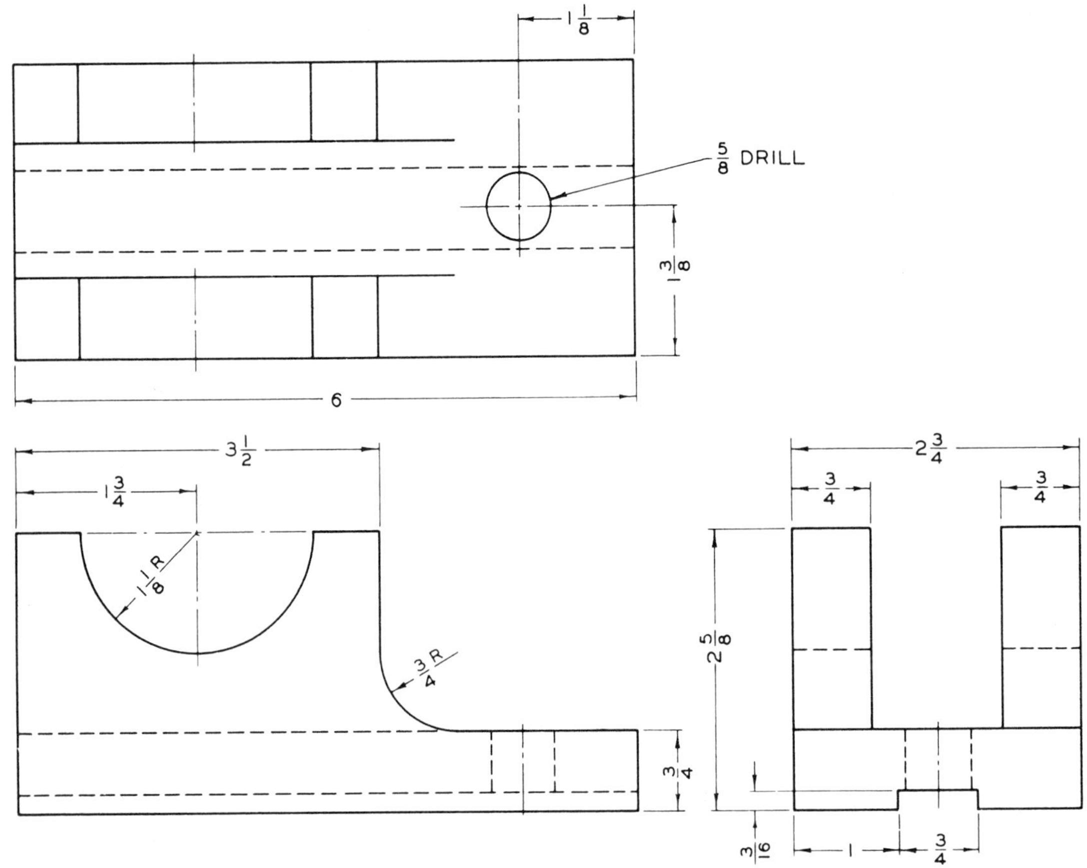

NAME OF SCHOOL CITY OR TOWN	CRADLE	SHEET No. 10-I NAME	CLASS DATE

Fig. 10-29

ASSIGNMENT I

Make an isometric drawing of the cradle, figure 10-29. Dimension the drawing.

Procedure

1. Lay out the margin lines on the drawing sheet and letter the title strip.
2. Draw the isometric view. Locate the lower corner 1 1/2-inches above the lower margin line and 7 inches from the right margin line.
3. Dimension the drawing according to the rules given for dimensioning isometric drawings.
4. Complete the drawing.
5. Check the isometric drawing and the dimensions.

AXONOMETRIC PROJECTION

An isometric drawing is a projection of an object that has been rotated and then tilted until the three leading edges make equal angles with the plane of projection. These one-view projections in which the projectors are perpendicular to the plane of projection are called *axonometric projections*.

OBLIQUE PROJECTION

In orthographic projection and axonometric projection, the projectors are perpendicular to the plane of projection. *Oblique projection* is a one-view drawing where the projectors may be at any angle (except 90 degrees) with the plane of projection.

A comparison of an oblique projection and an orthographic projection of a cube is shown pictorially in figure 10-30. The orthographic view is seen from a point directly in front. The oblique view is seen when looking at the cube with the line of sight parallel with the hypotenuse of a right triangle.

The line of sight, or the oblique projectors, may be at any

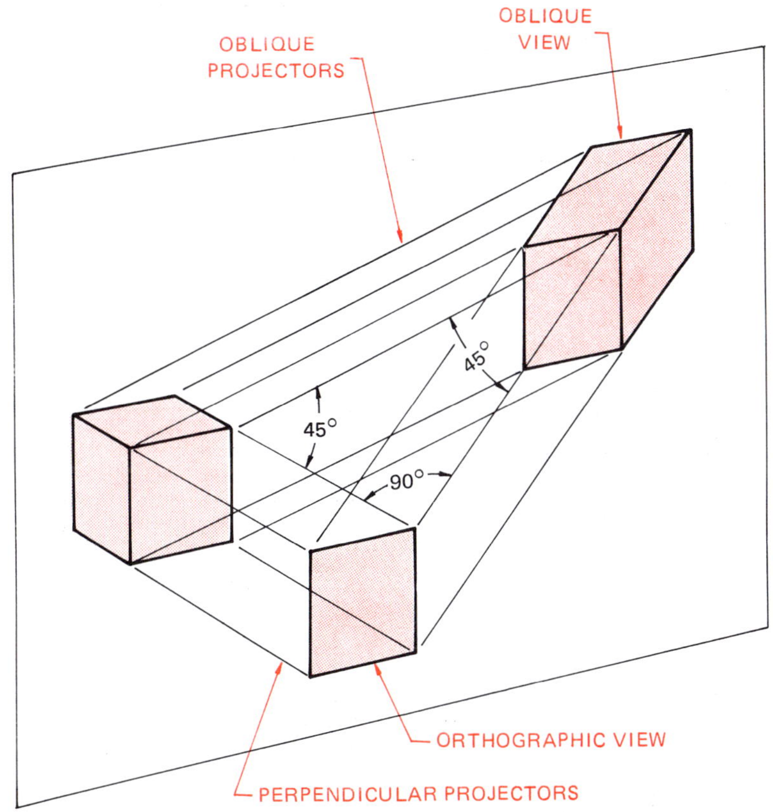

Fig. 10-30 Comparison of an oblique projection and an orthographic projection.

angle other than 90 degrees. However, when the oblique projectors are at 45 degrees with the plane of projection all of the edges of the cube are their true lengths in the oblique projection, figure 10-31. For this reason, the 45-degree angle projectors generally are used.

When the angle of the oblique projectors is 45 degrees, the projection is called a *cavalier projection*.

The receding axis of the cube may be drawn at various angles as pictured in figure 10-32. The angles 30, 45, and 60 degrees usually are used.

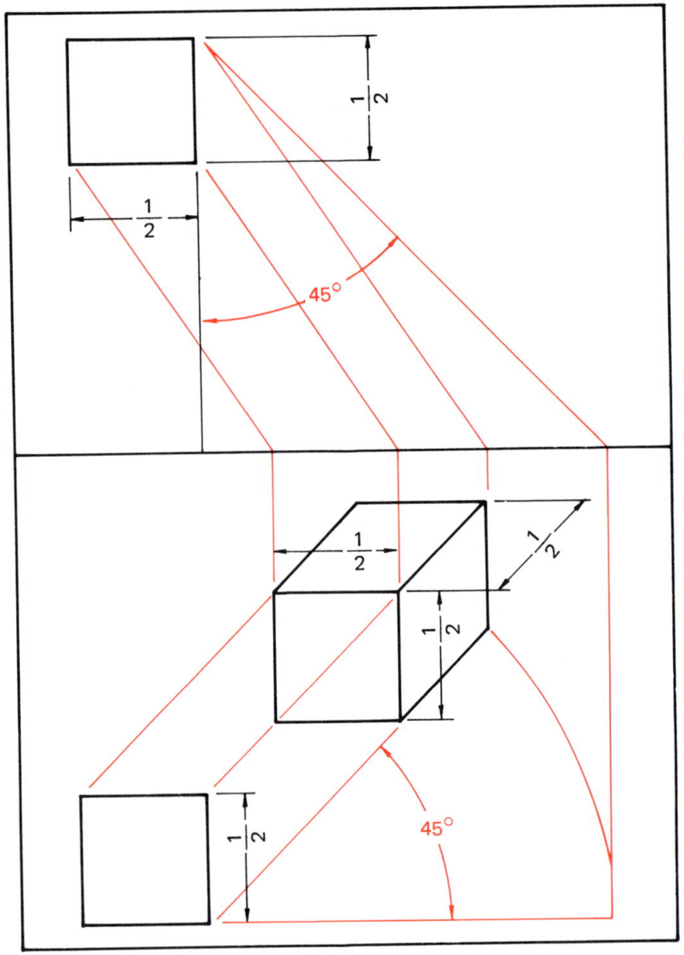

Fig. 10-31 An oblique projection of a cube

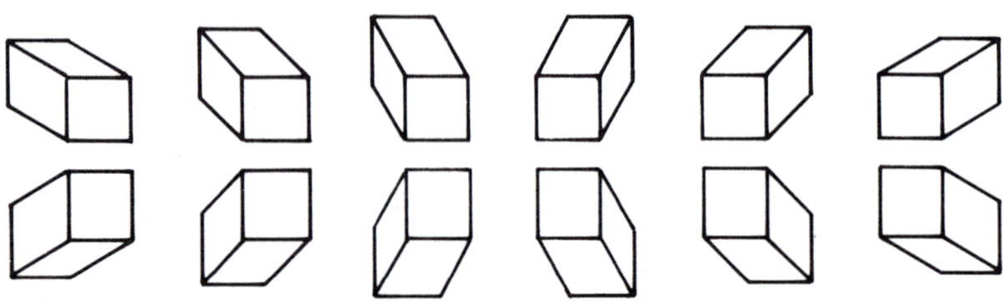

Fig. 10-32 Various angles of the receding axis

ADVANTAGES OF OBLIQUE DRAWING

In isometric projection, the faces of the object and the three axes are at an angle to the plane of projection. In oblique projection, one face of the object and two of the axes are drawn parallel to the plane of projection, and the parallel face is always projected in its true shape and size in the oblique view. Hence, the first point to consider in making an oblique drawing is: the face of the object that shows the contour or irregular shape, or the circular outline, should be parallel to the plane of projection. This is illustrated in figure 10-33.

DISADVANTAGES OF OBLIQUE DRAWING

Oblique drawings have a distorted and unnatural appearance because the receding lines are parallel and seem to diverge, while to the eye receding lines seem to converge toward a point in the distance. This phenomenon is illustrated in figure 10-34.

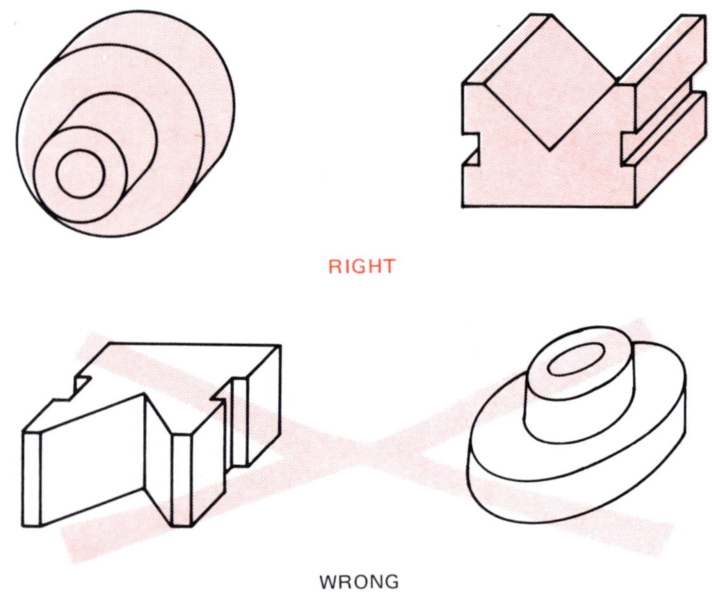

Fig. 10-33 Contour parallel to picture plane

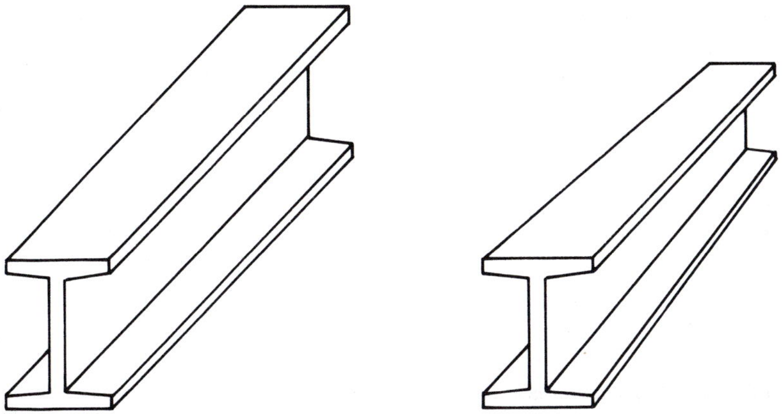

Fig. 10-34 Oblique and perspective views

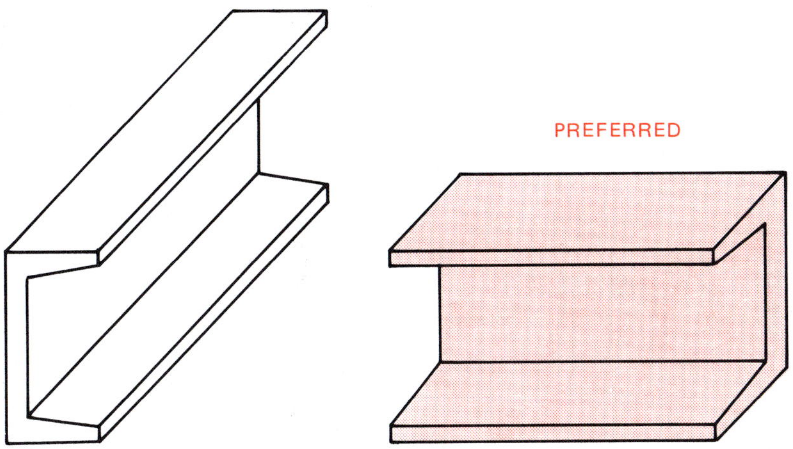

Fig. 10-35 Longest axis parallel to picture plane

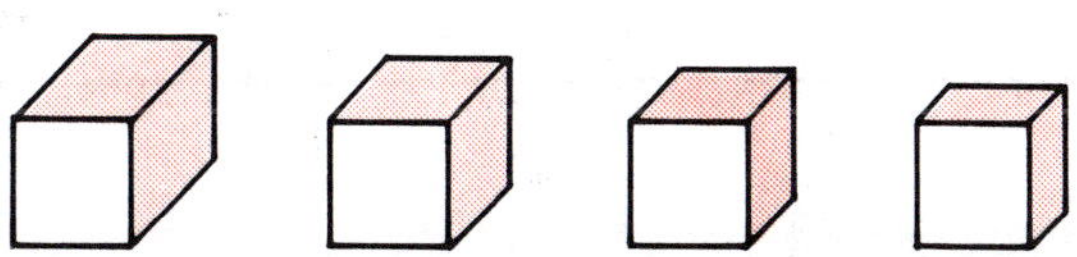

Fig. 10-36 Reduction of receding axis

The appearance of distortion may be lessened, especially for long objects, if the long axis is parallel to the plane of projection as illustrated in figure 10-35. Hence, the second point to consider is: the longest axis of the object should be parallel to the plane of projection. If the shape of the part is such that this conflicts with the first rule, then preference should be given to rule one.

The appearance of distortion also may be lessened if the length of the receding axis is reduced. Any suitable scale of reduction may be chosen, as illustrated by the four oblique drawings of a cube in figure 10-36. When the receding edges of an oblique drawing are reduced one-half, it is called a *cabinet drawing*.

DIMENSIONING OBLIQUE DRAWINGS

The same rules that were given for dimensioning isometric drawings are followed when dimensioning oblique drawings.

REVIEW QUESTIONS

1. What is an isometric projection? A dimetric projection? A trimetric projection?
2. What is the principal difference between orthographic projection and oblique projection?
3. Under what conditions is an oblique projection called a cavalier projection?
4. What is the first point or rule to consider when making an oblique drawing?
5. What is the second point or rule?
6. What is a cabinet drawing?

HOW TO MAKE AN OBLIQUE DRAWING

The first drawing to be made in this group is an oblique drawing of a box jig. This box has surfaces and edges which are parallel to the oblique planes, and some surfaces and edges which are inclined to the oblique planes, figure 10-37.

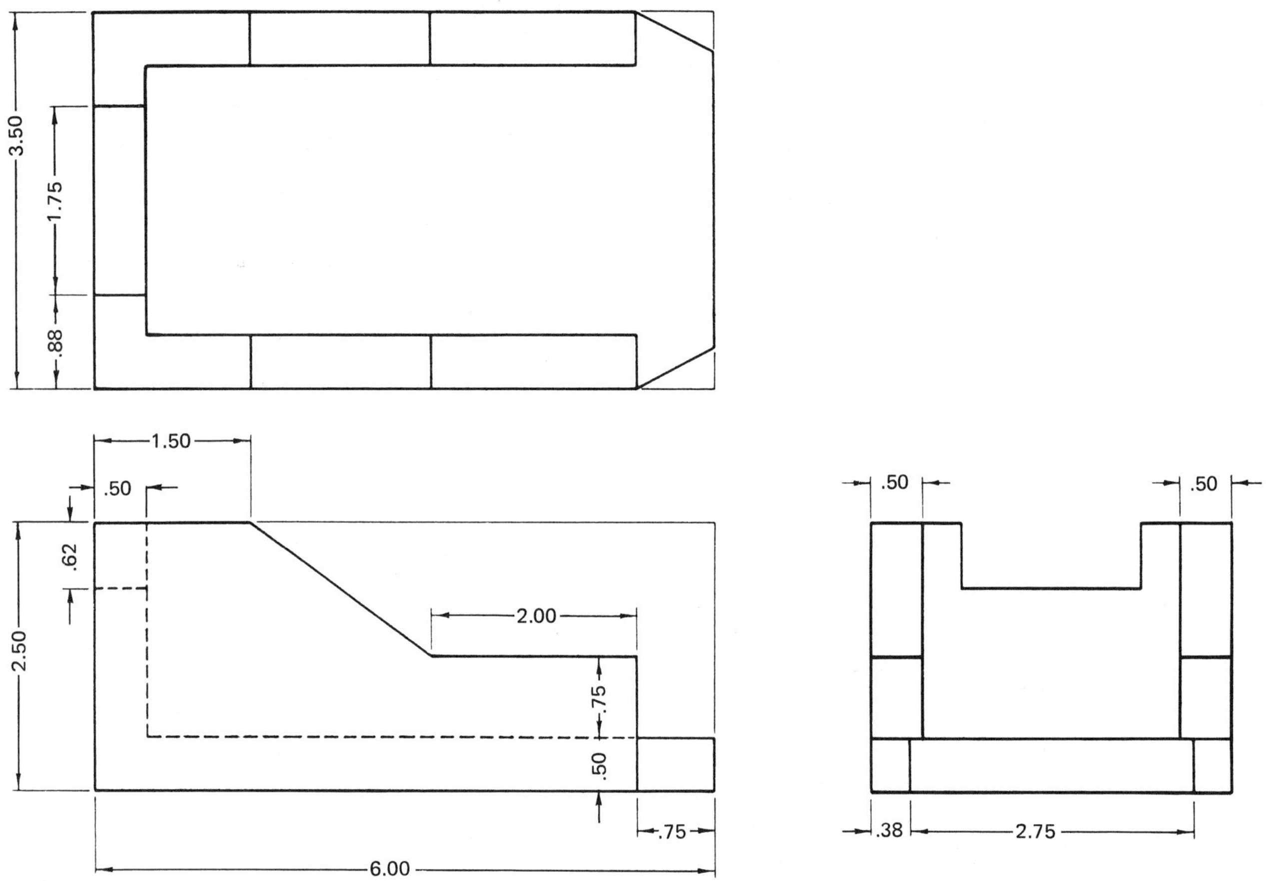

NAME OF SCHOOL CITY OR TOWN	BOX JIG	SHEET No. 10-J NAME	CLASS DATE

Fig. 10-37

Oblique drawings, like isometric drawings, are made by first enclosing the object in a box and then drawing the box in oblique. All measurements are made full size on the oblique axes.

The box jig on this drawing is enclosed in a box. Because the front view of the box jig has an irregular outline, follow "rule one" and place this face parallel to the picture plane. The enclosing box is then drawn in oblique. All measurements are made on the oblique axes. The end points of the inclined edges are located on the oblique axes. The end points are then joined with inclined lines.

Invisible edges generally are omitted from oblique drawings unless their use makes the drawing clearer.

Dimensioning Oblique Drawings

The rules for dimensioning isometric drawings are followed when dimensioning oblique drawings. The dimensions on this orthographic drawing are placed in the same positions and on the same planes as they will appear in the oblique view. For example, the front view is a vertical plane. This plane is also a vertical plane in the oblique drawing. Therefore, the dimensions shown in the front view will be in the same position on the vertical planes in the oblique drawing.

Likewise, the top view is on a horizontal plane. The dimensions shown on the top view will be placed on a horizontal plane in the oblique drawing. Again, the side view is on a vertical profile plane. The dimensions shown on this side view will be placed on the receding vertical planes of the oblique drawing.

ASSIGNMENT J

Make an oblique drawing of the box jig.

Procedure

1. Lay out the margin lines on the drawing sheet and letter the title strip.
2. Draw a rectangle for the front plane of the enclosing box. Draw this rectangle 4 1/2 inches from the left margin line and 2 1/2 inches above the title strip.
3. Complete the enclosing box. The angle of the receding axes are drawn at 45 degrees.
4. Draw the oblique drawing within the box.
5. Dimension the drawing.
6. Complete the drawing.
7. Check the accuracy of the oblique drawing of the box jig and the positioning of dimensions.

NONOBLIQUE EDGES AND SURFACES

The post footing in figure 10-38 has some edges and surfaces which are not parallel to any of the oblique planes. Such edges and surfaces are called nonoblique edges and surfaces. The end points of nonoblique lines are located by measuring on the oblique axes and then drawing coordinate lines as in isometric drawing. The coordinate lines are shown in the top view on this drawing.

ASSIGNMENT K

Make an oblique drawing of the post footing.

Procedure

1. Lay out the margin lines on the drawing sheet and letter the title strip.
2. Draw the enclosing box. The front plane of this box should be 4 1/4 inches from the left margin line and 2 1/2 inches above the title strip. The angle of the receding axes is 45 degrees.
3. Draw the oblique drawing within the box. Measure and draw the coordinate lines to locate the end points of the nonoblique lines.

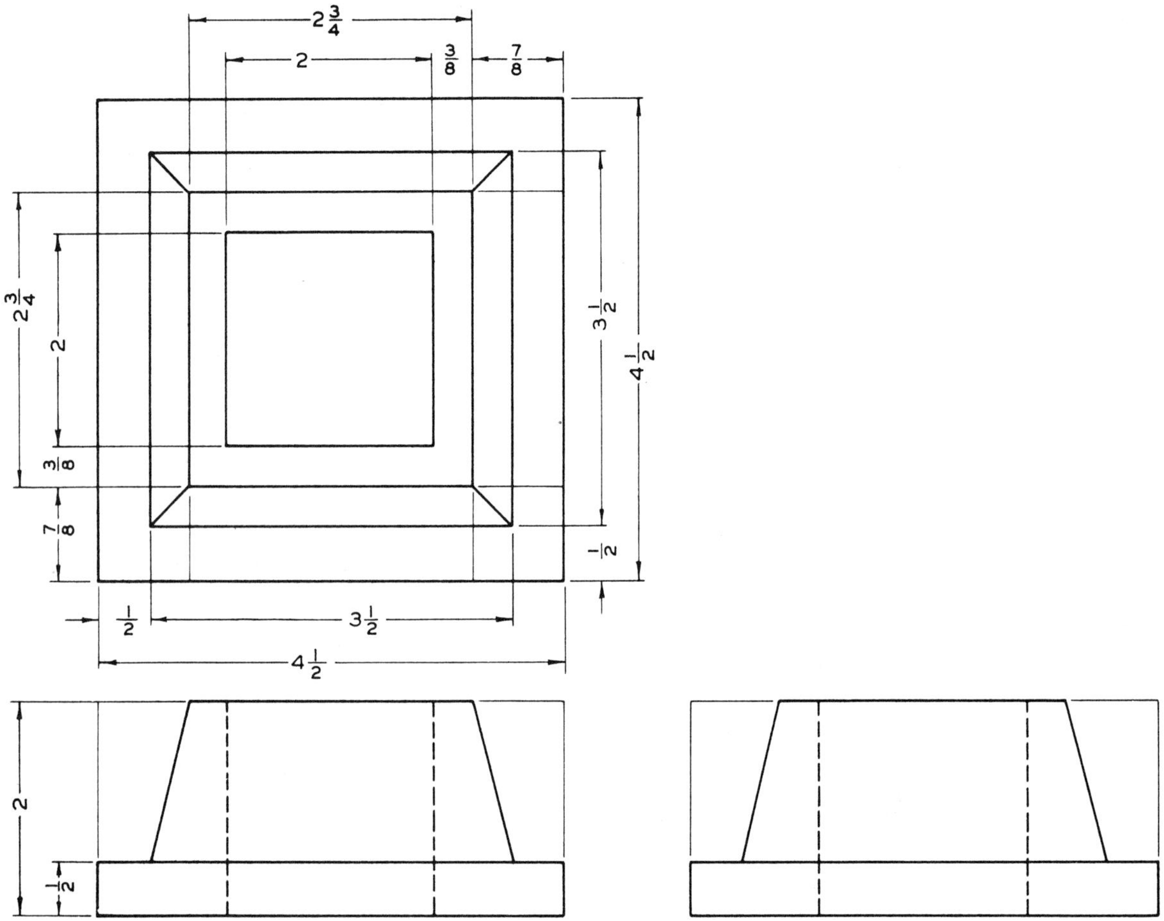

NAME OF SCHOOL CITY OR TOWN	POST FOOTING	SHEET No. 10-K NAME	CLASS DATE

Fig. 10-38

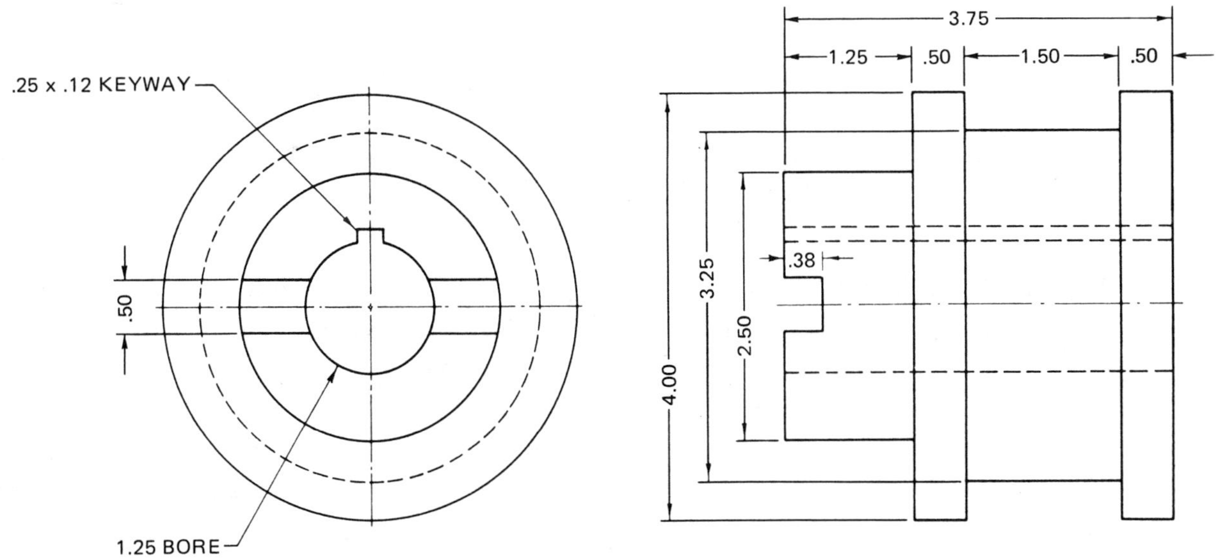

NAME OF SCHOOL CITY OR TOWN	CLUTCH	SHEET No. 10-L NAME	CLASS DATE

Fig. 10-39

4. Dimension the drawing. Dimensions on the orthographic views are placed in the same positions and on the same planes as they appear in the oblique drawing. The dimensions in the front view are on the front vertical plane; in the top view, on the horizontal planes.
5. Check the accuracy of the oblique drawing of the post footing and the location of all dimensions.

REVIEW QUESTIONS

1. Explain how an oblique drawing differs from an orthographic drawing.
2. Describe how to locate the end points of nonoblique lines.

DIMENSIONING OF OBLIQUE DRAWINGS OF CYLINDERS

When an oblique drawing is made of a cylindrical shaped part, the circular outline should be parallel to the picture plane, figure 10-33. The centers for the other circular edges on the cylinder lie one behind the other on the axis of the cylinder. Therefore, in oblique drawings of cylinders, the axis is drawn first and the centers of the circles are located on the axis.

The clutch in figure 10-39 is cylindrical and the circular face is drawn parallel to the picture plane.

Oblique drawings of cylinders are dimensioned as shown in figure 10-40. The dimension for the length is placed on the drawing parallel with the axis of the cylinder. The diameter is given in a note.

ASSIGNMENT L

1. Make an oblique drawing of the clutch.

Procedure

1. Lay out the margin lines on the drawing sheet and letter the title strip.
2. Draw a vertical center line 6 1/2 inches from the left margin line.
3. Draw a horizontal center line 4 inches above the title strip.
4. Draw the two circles on the front face of the clutch.
5. From the center of these two circles draw the receding axis of the clutch. The angle of the receding axis on this drawing is 30 degrees.
6. On the receding axis, measure from the center 1.25 inches, .50 inch, 1.50 inches, and .50 inch. These points are the centers for the other circles. Draw these circles omitting the invisible portions.
7. Draw the keyway. Draw the slot which cuts across the hub.
8. Dimension the drawing. Complete the drawing.
9. Check the accuracy of all lines and the position of dimensions on the clutch.

REVIEW QUESTIONS

1. What is the first or "rule one" when making oblique drawings?
2. What is the second or "rule two" to follow for oblique drawings?

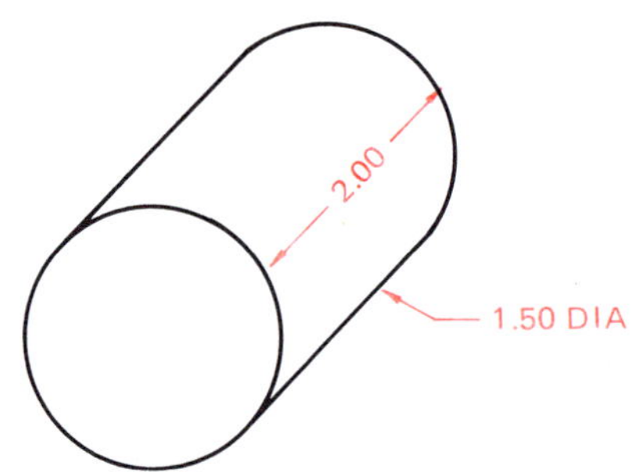

Fig. 10-40 Dimensioning an oblique drawing of a cylinder

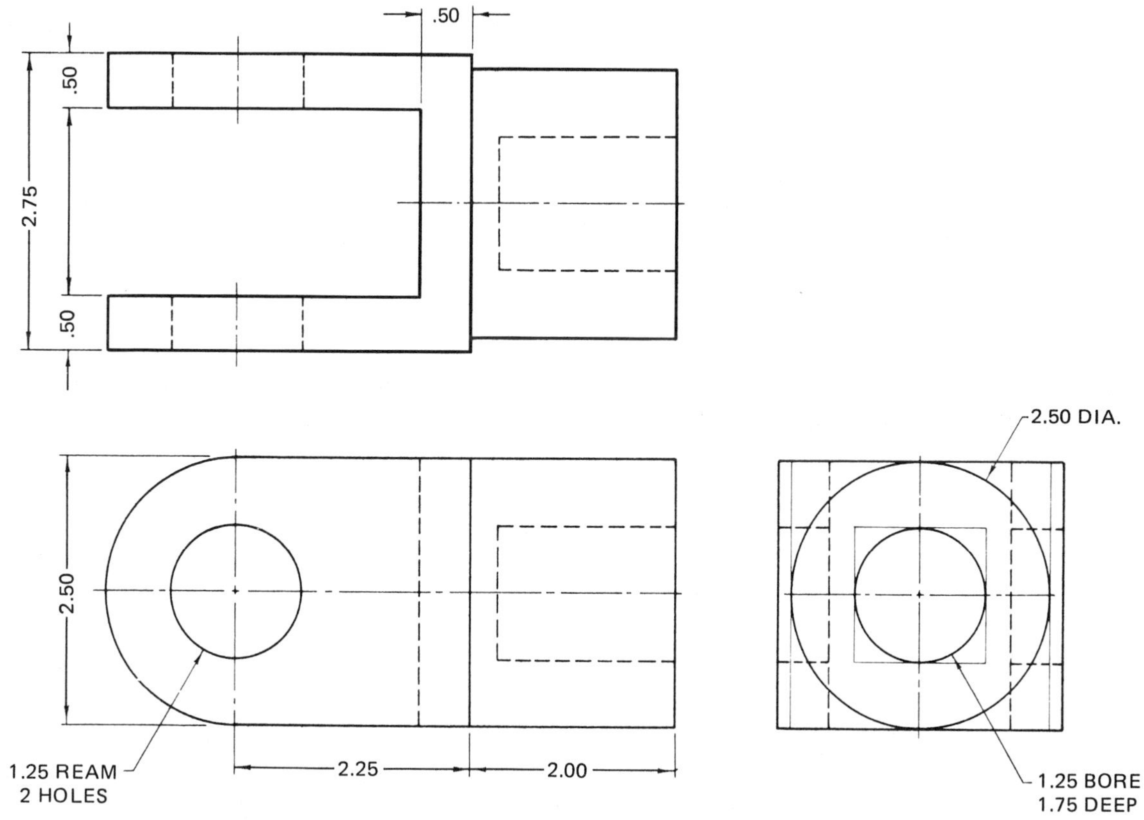

NAME OF SCHOOL CITY OR TOWN	YOKE	SHEET No. 10-M NAME	CLASS DATE

Fig. 10-41

OBLIQUE ELLIPSES

The yoke in figure 10-41 has a circle and a semicircle on the front face. It also has two circles on the right end. Since both circular faces cannot be parallel to the picture plane, "rule two" is used for making this oblique drawing. Rule two states that the longest axis should be parallel to the picture plane. Therefore, the front face of the yoke, having the longest axis and some circular arcs, is drawn parallel to the picture plane.

The two circles on the right end are on the receding vertical plane. Circles on the receding plane in oblique drawings are ellipses. These ellipses can be drawn approximately by the four-center method pictured in figure 10-42. The circles are enclosed in a square and the square is drawn in oblique. A perpendicular is drawn from the midpoint of each side of the oblique square. The intersections of these perpendiculars are the centers for the arcs.

ASSIGNMENT M

Make an oblique drawing of the yoke.

Procedure

1. Lay out the margin lines on the drawing sheet and letter the title strip.
2. Draw a vertical center line 5 1/2 inches from the left margin line.
3. Draw a horizontal center line 4 1/4 inches above the title strip.
4. Draw the front face of the yoke.
5. From the center of the circle draw the receding axis. The angle of this axis is 45 degrees.
6. On this axis locate the centers for the three circles and semicircles which lie behind the front face.

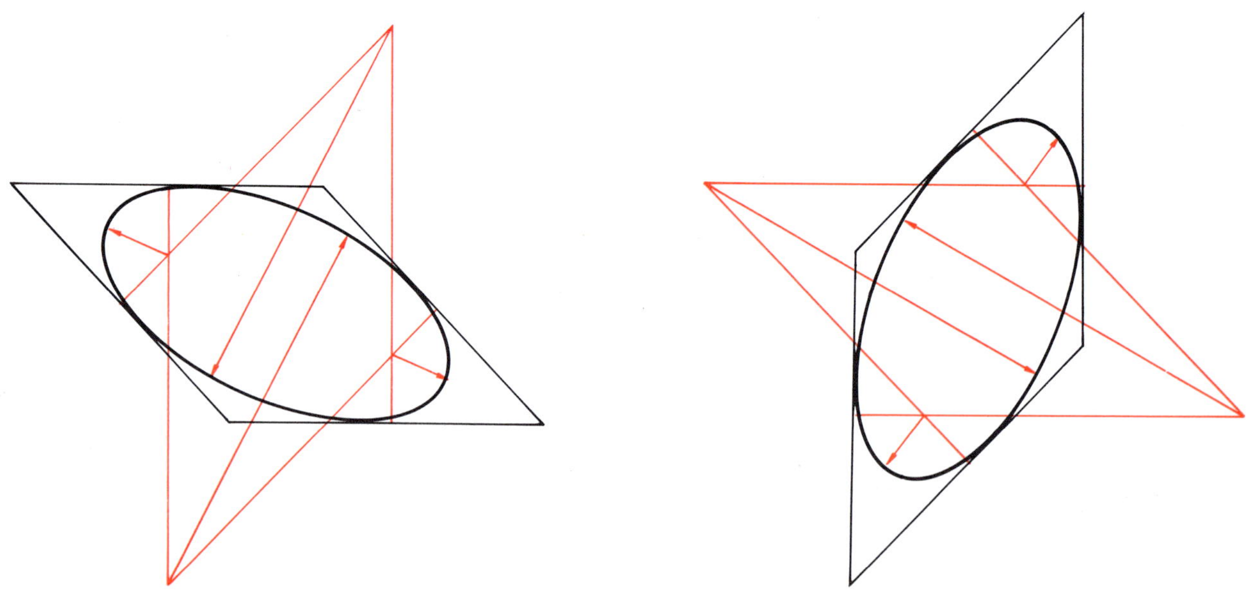

Fig. 10-42 Oblique circles on receding planes

7. Draw these circles and semicircles omitting the invisible portions. Complete this part of the view.
8. The hub of the yoke is a cylinder. On the drawing, this cylinder is enclosed in a box as in isometric drawing. Locate and draw the enclosing box in the oblique view.
9. Draw perpendiculars from the midpoints of each side of the oblique squares to locate the centers for the ellipses.
10. Draw the ellipses on the end and the partial ellipse where the cylinder joins the other part of the yoke.
11. Dimension the drawing. The dimensions in the orthographic views are in the same positions and on the same planes as they appear in the oblique view.
12. Complete the drawing.
13. Check the accuracy of all lines, circles, and parts of circles and the position of dimensions in the yoke.

OBLIQUE CABINET DRAWING

The drill jig in figure 10-44 has a semicircle in the front end, and a round hole and a semicircle in the rear end. It also has a hole and a semicircle in the right side. The planes having the most circles are parallel to the picture plane. So, in this case, the front and rear ends will be parallel to the picture plane.

This arrangement makes the longest axis the receding axis which is contrary to rule two. To reduce the effect of distortion, the receding axis is reduced by making a *cabinet drawing.* In a cabinet drawing the receding axis is reduced one-half.

The circle and the semicircle in the side view is an ellipse, and a semiellipse on the receding plane of the cabinet drawing. However, because the receding plane is reduced one-half, the four-center method for drawing an ellipse cannot be used. Instead, points are plotted for the curved lines of the ellipse and the points are joined with an irregular curve. Carefully review this method as illustrated in figure 10-43.

ASSIGNMENT N

Make a cabinet drawing of the drill jig.

Procedure

1. Lay out the margin lines on the drawing sheet and letter the title strip.
2. Draw a vertical center line 6 1/2 inches from the left margin line.
3. Draw a horizontal center line 4 inches above the lower margin line.
4. Draw the semicircle on the front plane.
5. From the center of this semicircle draw the receding axis at an angle of 45 degrees.
6. On this axis locate the centers for the other circles and semicircles. Remember to measure one-half of the full-size dimension.

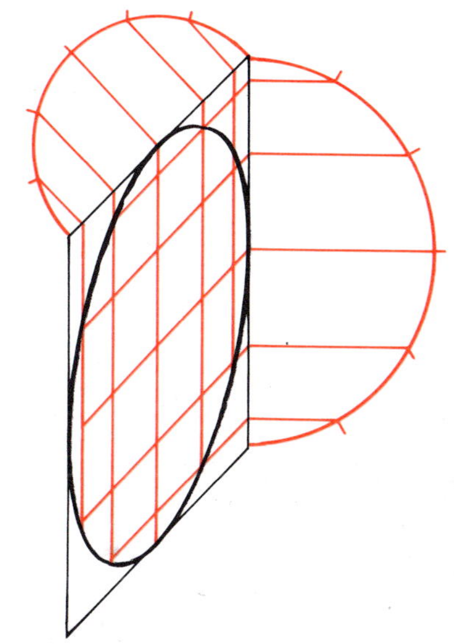

Fig. 10-43 Circles on the receding plane when axis is reduced

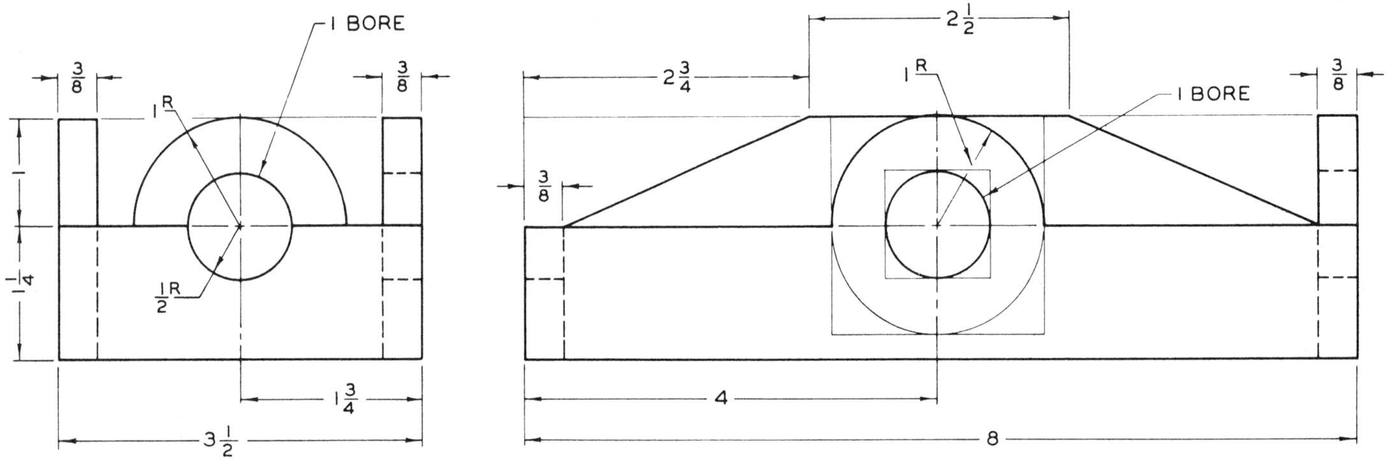

NAME OF SCHOOL CITY OR TOWN	DRILL JIG	SHEET No. 10-N NAME	CLASS DATE

Fig. 10-44

7. Draw the other circles and semicircles omitting the invisible portions.
8. Make a cabinet drawing of the enclosing box for the drill jig. The receding axis is one-half the full-size dimension.
9. The circles in the side view are enclosed in squares. Draw these shapes in the cabinet drawing. Draw the ellipse and the semiellipse.
10. Locate and draw the remaining parts of the drill jig.
11. Dimension the drawing.
12. Check the accuracy of all circles, parts of circles, and dimensions in the cabinet drawing of the drill jig.

REVIEW QUESTIONS

1. Explain how oblique circles are drawn on cabinet drawings.
2. Explain the difference between an oblique drawing and a cabinet drawing.

PERSPECTIVE DRAWINGS

Perspective drawings, like isometric and oblique drawings, are one-view pictorial drawings.

In isometric drawing, the projectors are perpendicular to the picture plane. The projectors in oblique drawing are oblique to the picture plane. The projectors in both systems are parallel.

In perspective drawing, the projectors are not parallel. They converge at a point corresponding to the eye of the observer. A perspective drawing is more like a photograph; it shows the object as it appears to the eye of the observer.

Figure 10-45 illustrates the principle of perspective drawing. It shows an eye of the observer looking through a transparent picture plane at a cube. Projectors, extending from each corner of the cube to the eye, pierce the picture plane. The piercing points are joined with lines, and the result is a perspective drawing on the picture plane. Notice that the projectors, called *visual rays* in perspective drawing, are not parallel.

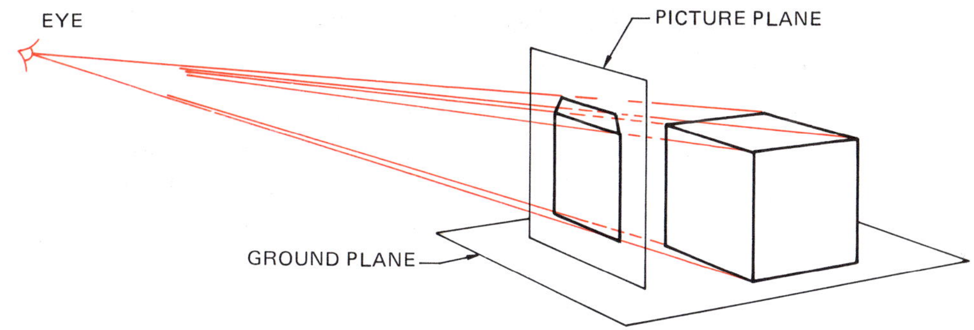

Fig 10-45 A perspective drawing on a picture plane

Figure 10-46 shows a perspective drawing obtained by orthographic projection. A top view or plan is shown. Visual rays drawn from the corners of the plan to the station point pierce the picture plane.

A side view or elevation also is shown with visual rays drawn from the corners of the elevation to the station point, piercing the picture plane. The perspective drawing on the picture plane is shown next. Projectors drawn from the piercing points on both picture planes locate the corners of the cube in the perspective view.

TERMS

The technical terms used in perspective drawing are illustrated pictorially in figure 10-46.

- The top view is called the *plan;* the front and side view, the *elevations*.
- The *picture plane* (PP) is a vertical plane containing the perspective drawing.
- The *station point* (SP) represents the point of view of the observer.

- The *ground line* (GL) represents the edge of the ground plane on which the object rests.
- The *eye level* (EL) or horizon is a line at the level of the eye.
- The *center of vision* (CV) is a point in the distance at which receding parallel lines converge. It is always at the level of the eye and in line vertically with the station point.

PARALLEL PERSPECTIVE

A perspective drawing of an object that has its front face parallel to the picture plane is called a *parallel perspective* drawing. It is also called a *one-point perspective* because the receding lines converge at one point.

Parallel perspective is best suited for objects having an irregular outline or a circular face. The face having the irregular or circular outline is placed parallel with the picture plane.

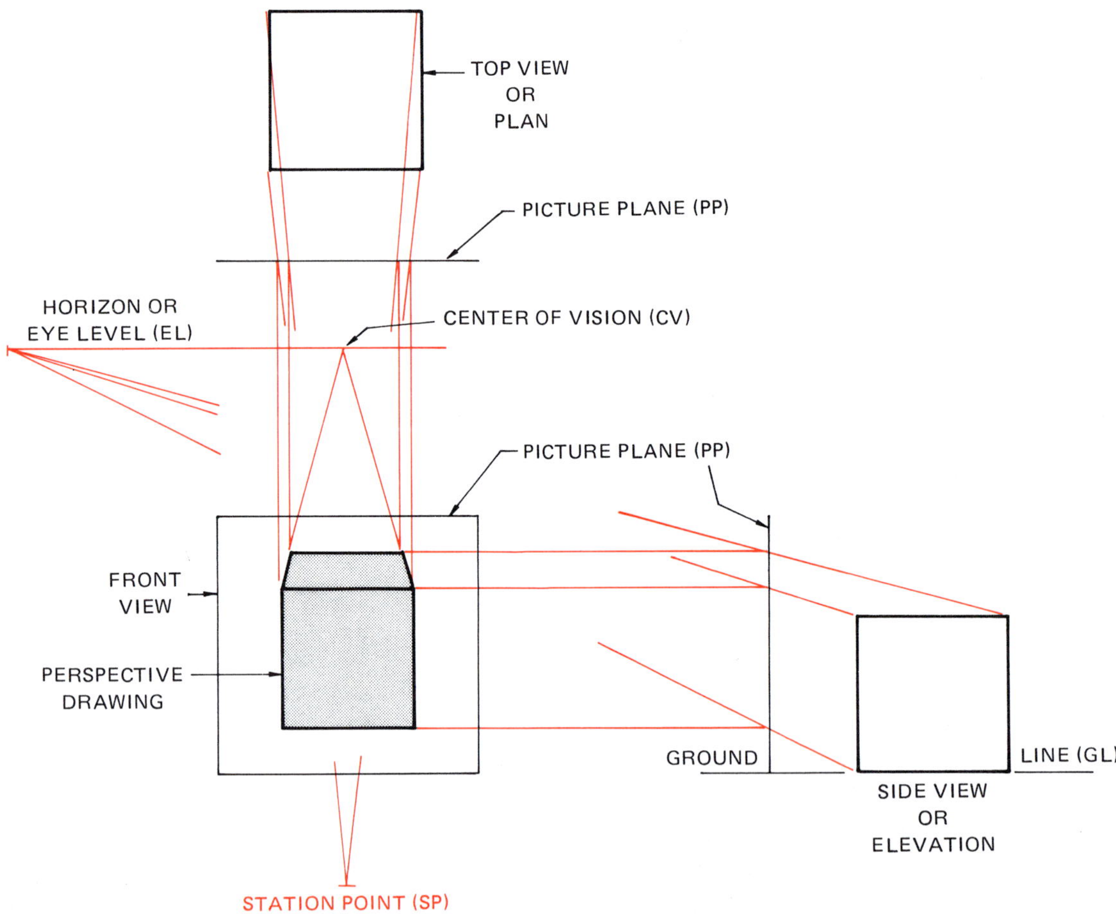

Fig. 10-46 Perspective drawing obtained by orthographic projection

MAKING A PARALLEL PERSPECTIVE DRAWING

The method used in making a parallel perspective drawing of a cube is shown in figure 10-47. The plan is drawn first, then the picture plane and the station point are located. Next, visual rays are drawn from the corners in the plan to the station point. These rays pierce the picture plane. The ground line is located and the front elevation is drawn on the ground line directly beneath the plan. The center of vision is located in line vertically with the station point and at the level of the eye. Receding lines are then drawn from the corners of the front elevation to the center of vision. Lines are not projected downward from the piercing points. These lines intersect the converging receding lines to locate the corners of the perspective view.

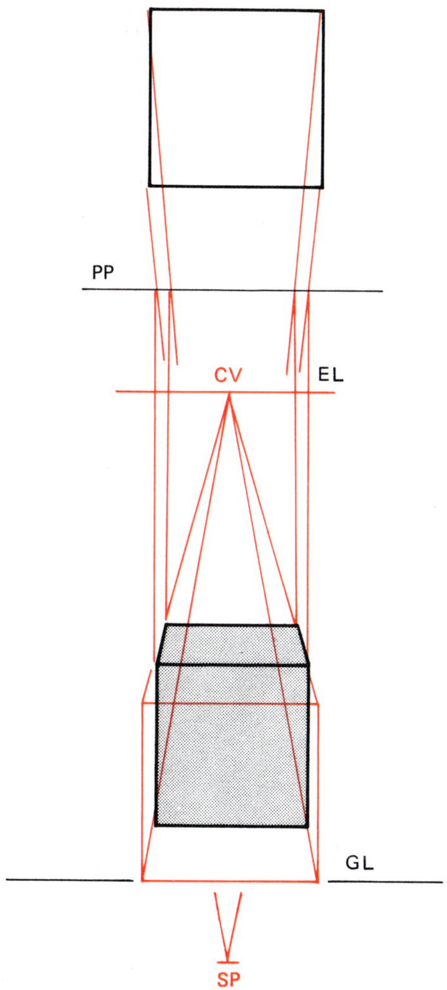

Fig. 10-47 A parallel perspective drawing of a cube

THE POSITION OF THE PICTURE PLANE

Various positions of the picture plane in relation to the object are given in figure 10-48. When the front face of the object is placed against the picture plane, this face is full size in the per-

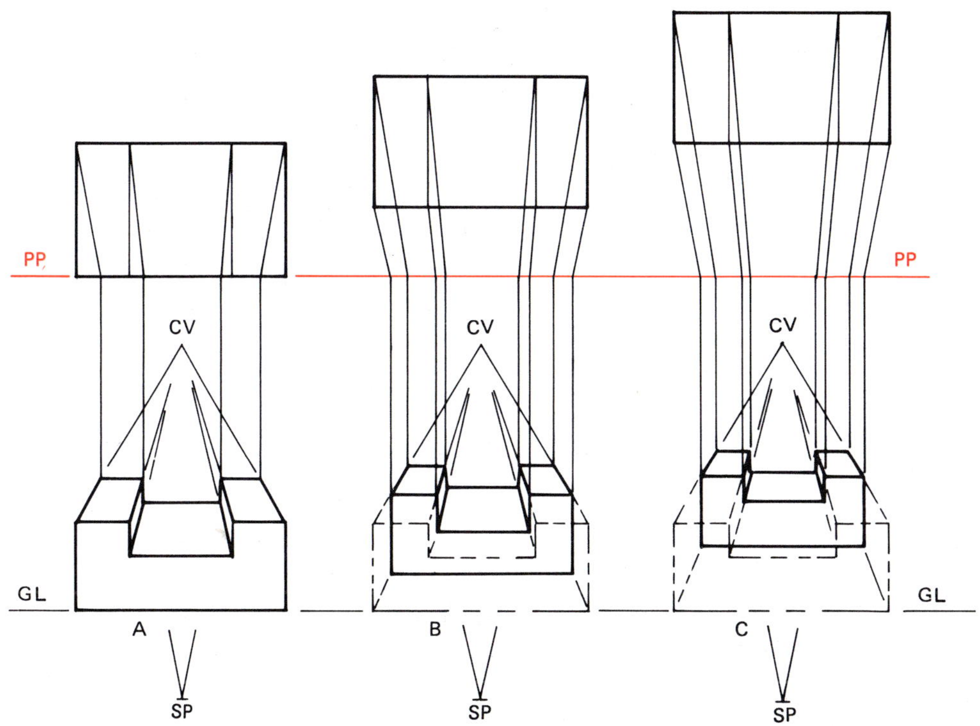

Fig. 10-48 The position of the picture plane

spective view, as shown at A. The perspective view is reduced proportionally as the distance from the picture plane to the object is increased, as shown at B and C.

A study of figure 10-48 reveals that the best results are obtained if the object is placed against the picture plane, as shown at A, since the front face is then shown in its true shape and size, and the construction of the perspective view is simplified.

THE HEIGHT OF THE CENTER OF VISION

The position of the center of vision is on the eye level line directly above the station point. As the height of the center of vision above the ground line increases, more of the top surface is seen in the perspective view. This is illustrated in figure 10-49.

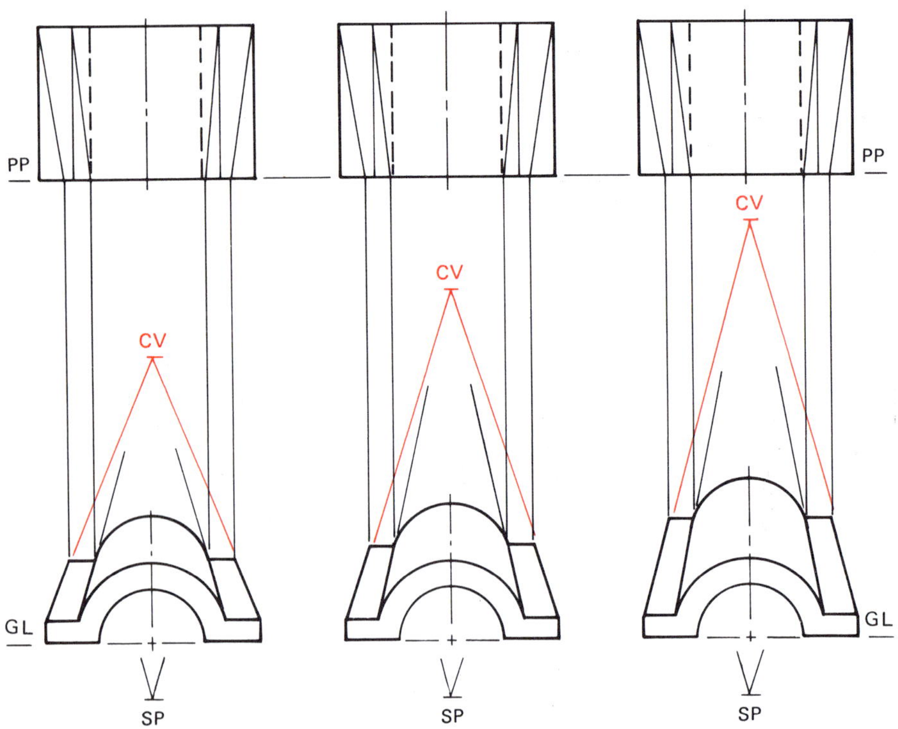

Fig. 10-49 The height of the center of vision above the ground line

POSITION OF THE STATION POINT

In the preceding illustrations the station point is directly in front of the center of the object. When the station point is directly in front of the object, the distance from the station point to the picture plane should not be less than twice the width nor twice the height of the object. Placing the station point nearer to the picture plane creates distortion in the perspective view.

More often, the object can be seen better if the station point is moved to the right or left of the center, figure 10-50. As the distance to the right or left of the center is increased, more of the side of the object is seen in the perspective view. When the station point is away from the center, the distance from the nearest corner of the plan to the station point should be not less than twice the height of the object, nor twice the width of the plan, as seen from the station point.

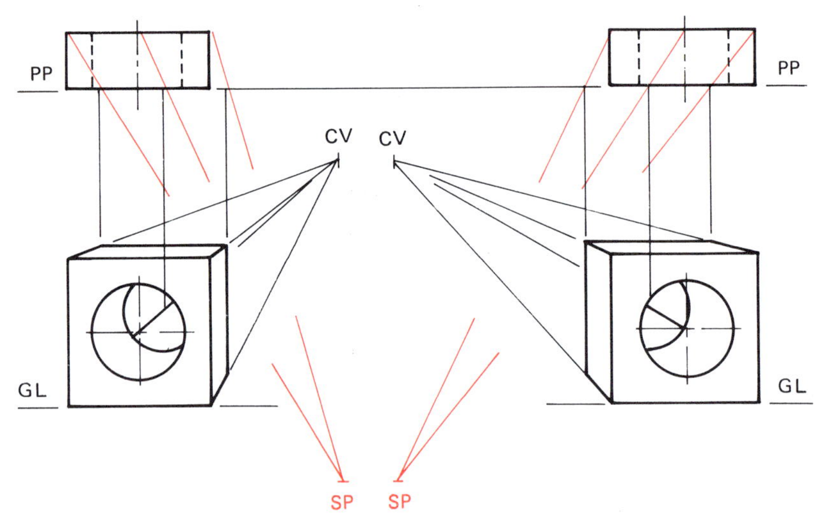

Fig. 10-50 Position of station point at right of left of center

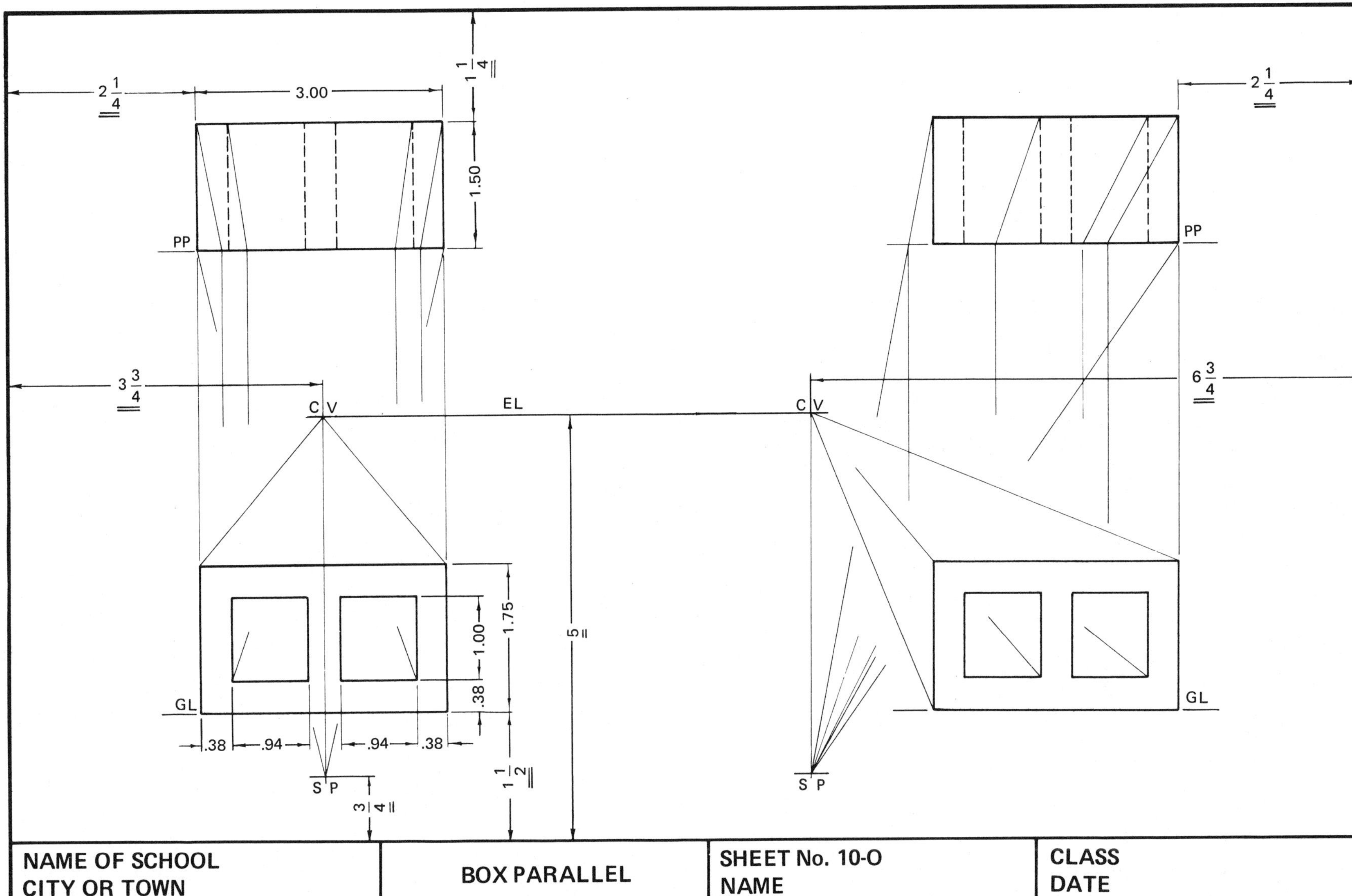
2 1/4
3.00
1 1/4
1.50
PP
2 1/4
PP
3 3/4
C V
EL
C V
6 3/4
1.75
1.00
5
.38
GL
GL
.38
.94
.94
.38
S P
1 1/2
3/4
S P
NAME OF SCHOOL
CITY OR TOWN
BOX PARALLEL
SHEET No. 10-O
NAME
CLASS
DATE

Fig. 10-51

MAKING A PARALLEL PERSPECTIVE DRAWING

Parallel perspective drawings are similar to oblique drawings except that the receding lines converge instead of being parallel. Therefore, the two rules for oblique drawings also apply to parallel perspective drawings. Perspective drawings are dimensioned following the same rules given for isometric drawings.

Two parallel perspective drawings of the box parallel, figure 10-51, are drawn on this next sheet. The picture plane in both drawings is placed against the front of the box parallel as shown in the elevation. Consequently, the front surface of the part is drawn in its true shape and size in the perspective views. Dimensions may be omitted from all of the perspective drawings in this group.

ASSIGNMENT O

Make two parallel perspective drawings of the box parallel. In the first drawing, the station point is located directly in front of the center of the part. In the second drawing, the station point is placed at the left of the center.

Procedure

1. Lay out the margin lines on the drawing sheet and letter the title strip.
2. Locate the station points. Draw the ground line, the eye level line, and the picture plane. Locate the center of vision for both drawings.
3. Draw the elevation and the plan for both drawings.
4. Draw lines (visual rays) from the corners of the plans toward the station points. These rays pierce the picture plane.
5. Draw receding lines from the corners in the elevations. These lines converge at the centers of vision.
6. From the piercing points in the picture plane, project lines downward to intersect the receding converging lines. These projectors determine the location of the vertical edges and locate the end points of the receding horizontal edges. They also determine the position of the horizontal edges which are behind and parallel to the picture plane.
7. Complete the drawing.
8. Check the station points and all lines on both parallel perspective drawings of the box parallel.

REVIEW QUESTIONS

1. What is the outstanding difference between oblique drawings and parallel perspective drawings?
2. In perspective drawing, what is the top view called? Front and side view?
3. What is the picture plane?
4. What is the station point?
5. What is the ground line?
6. What is the eye level?
7. What is the center of vision?
8. What is a parallel perspective drawing?
9. In parallel perspective, what is the best position for the picture plane in relation to the object?
10. What effect does increasing distance between the picture plane and the object have?
11. What effect has increasing the height of center of vision above the ground have?
12. What is the effect when the station point is to the right of center of the object?
13. Give the smallest distance the station point should be from the picture plane?
14. Explain what is meant by (a) station point; (b) picture plane; (c) eye level; and (d) the center of vision.
15. Explain how vertical edges are located in the perspective view.
16. What is meant by parallel or one-point perspective drawing?

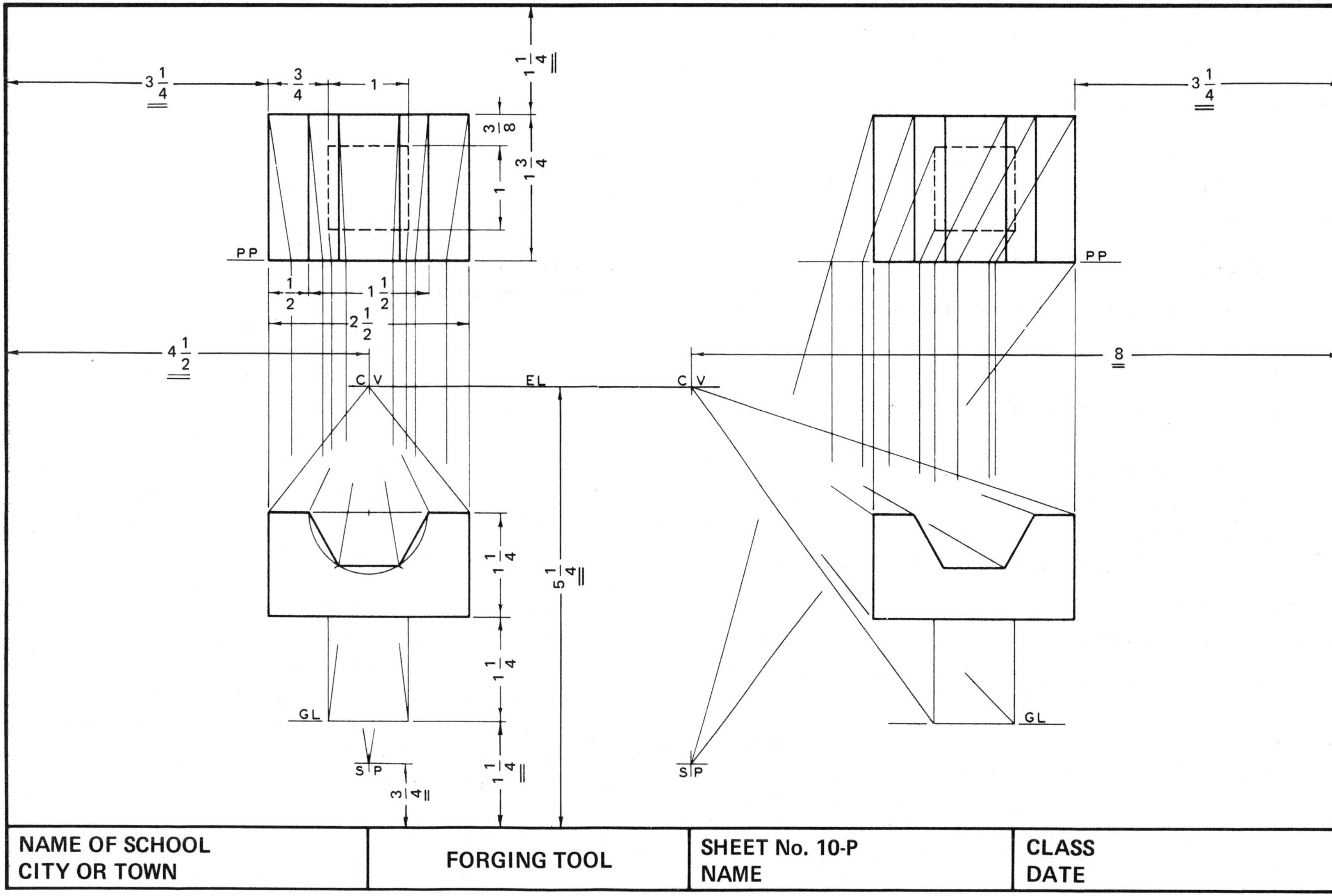
PP
CV
EL
GL
SP
NAME OF SCHOOL
CITY OR TOWN
FORGING TOOL
SHEET No. 10-P
NAME
CLASS
DATE

Fig. 10-52

SLOPING OR OBLIQUE EDGES IN PERSPECTIVE DRAWINGS

In the box parallel drawing, the vertical edges are parallel to the picture plane. Some of the horizontal edges are parallel to the picture plane; others are perpendicular to it. The forging tool in figure 10-52 has several edges which are not vertical or horizontal.

When sloping or oblique edges occur in perspective drawings, the end points must first be located in the perspective view. The end points are then joined with sloping lines.

ASSIGNMENT P

Make two parallel perspective drawings of the forging tool.

Procedure

1. Lay out the margin lines on the drawing sheet and letter the title strip.
2. Locate the station points. Draw the ground line, the eye level line, and the picture plane. Locate the center of vision for both drawings.
3. Draw the elevation for both drawings with light lines.
4. Draw the plans.
5. Draw the visual rays from the plans to the station points.
6. Draw the receding lines from the corners of the elevations to the centers of vision.
7. From the piercing points in the picture plane, project lines downward to intersect the receding lines. These projectors determine the position of the vertical edges and locate the end points of the receding horizontal edges and the sloping edges. They also determine the position of the horizontal edges which are behind and parallel to the picture plane.
8. Complete the drawing, omitting the dimensions.
9. Check all lines and sizes on the parallel perspective drawings of the forging tool.

REVIEW QUESTIONS

1. What is the picture plane?
2. What is the station point?
3. What is the eye level?
4. What is the center of vision?

ECCENTRICS

The eccentric in figure 10-53 contains a number of circular edges that are parallel to the picture plane. Circles parallel to the picture plane are true circles in the perspective view, but only those circles which lie against the picture plane are their true size. The circles lying behind the picture plane are smaller than actual size and have different centers.

The drawing shows that C1 is the center of the front end of the hub and bore. C3 is the center and R3 is the radius for the circle on the other end of the hub. C4 is the center and R4 is the radius for the rear end of the bore. This circle appears only partially in the first drawing and is not seen in the second drawing.

The center of the eccentric circle is at C2 in the front elevation. This circle is behind the picture plane so, in the perspective view, its center is C5 and its radius is R5. Likewise, C6 is the center and R6 is the radius of the rear eccentric circle.

A study of this drawing shows that the radius of each circle is the same as the distance from its piercing point to the center line of the plan. Also, the horizontal parallel center lines are located in the perspective view where the projectors from the piercing points intersect the receding lines.

ASSIGNMENT Q

Make two parallel perspective drawings of the eccentric.

Procedure

1. Lay out the margin lines on the drawing sheet and letter the title strip.

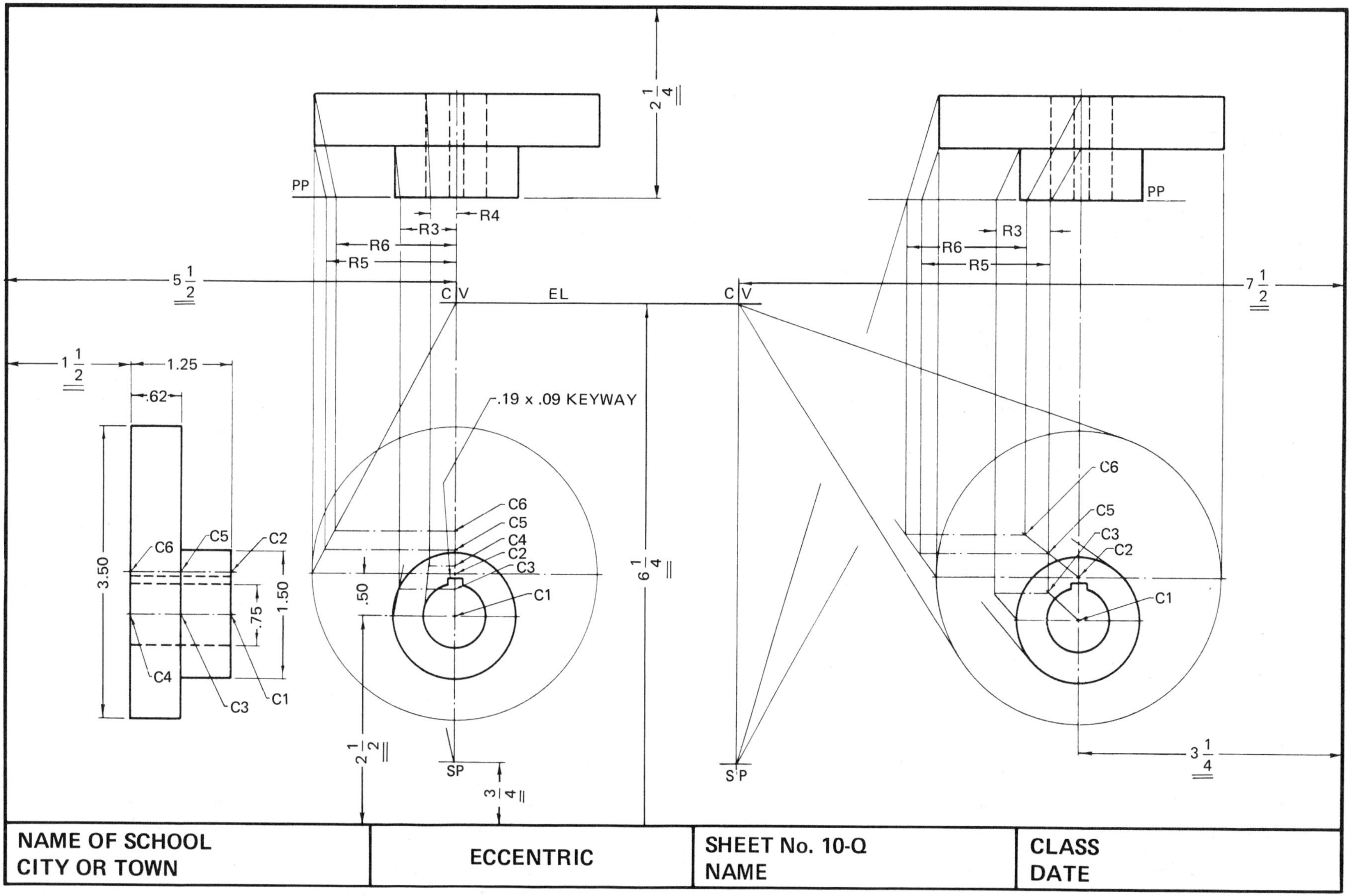
PP
R3
R4
R6
R5
5 1/2"
C V
EL
2 1/4"
1 1/2"
1.25
.62
.19 x .09 KEYWAY
C6
C5
C4
C2
C3
C1
3.50
1.50
.75
.50
6 1/4"
2 1/2"
SP
3/4"
7 1/2"
3 1/4"
NAME OF SCHOOL
CITY OR TOWN
ECCENTRIC
SHEET No. 10-Q
NAME
CLASS
DATE

Fig. 10-53

2. Locate the station points. Draw the ground line, the eye level line, and the picture plane. Locate the center of vision in both drawings.
3. Draw with light lines the front elevation in both drawings.
4. Draw the side elevation.
5. Draw the plan in both drawings.
6. In the first drawing, draw the visual rays from the corners on the left side of the plan.
7. Draw the receding lines from the left end of each radius in the front elevation to the center of vision.
8. Draw projectors downward from the piercing points to locate the horizontal parallel center lines.
9. Complete the drawing using the centers C3, C4, C5, C6, and radii R3, R4, R5, and R6.
10. In the second drawing, draw visual rays from the corners on the left side of the plan and from the two center points.
11. In the front elevation, draw the receding lines from the left end of each radius and from the two centers.
12. Draw projectors downward from the piercing points to locate the horizontal parallel center lines and the center points.
13. Complete the drawing using centers C3, C5, C6, and radii R3, R5, and R6.
14. Check for accuracy.

CIRCLES ON RECEDING PLANES IN PERSPECTIVE DRAWINGS

The intersecting cylinders in figure 10-54 have some circles which will appear on the receding planes in the perspective view.

A circle on a receding plane appears as an ellipse in the perspective view. To draw the ellipse in the perspective view, the circle is first divided into a number of equal parts. The points on the circle are located in the perspective view and are then joined using an irregular curve.

This drawing has a circle on the receding horizontal plane and a circle on each of the receding vertical planes.

ASSIGNMENT R

Make two parallel perspective drawings of the intersecting cylinders.

Procedure

1. Lay out the margin lines on the drawing sheet and letter the title strip.
2. Locate the station points. Draw the ground line, the eye level line, and the picture plane. Locate the center of vision for both drawings.
3. Draw the plan for both drawings. Divide the circle into twelve equal parts.
4. Project these twelve points to the right and left ends of the cylinder.
5. Draw, with light lines, the front elevation for both drawings.
6. At one end of the elevations draw a semicircle and divide it into six equal parts.
7. Project these points to the right and left ends of the cylinder.
8. Draw visual rays from each point on the circles and from the points on the right and left ends of the plans.
9. Draw receding lines from each point on the top edge and from each point on the right and left edges to the center of vision.
10. Locate the points for the ellipses. Each point is located in the perspective view by drawing a vertical projector from its piercing point to the corresponding receding line in the perspective view. For example: The vertical projector from the

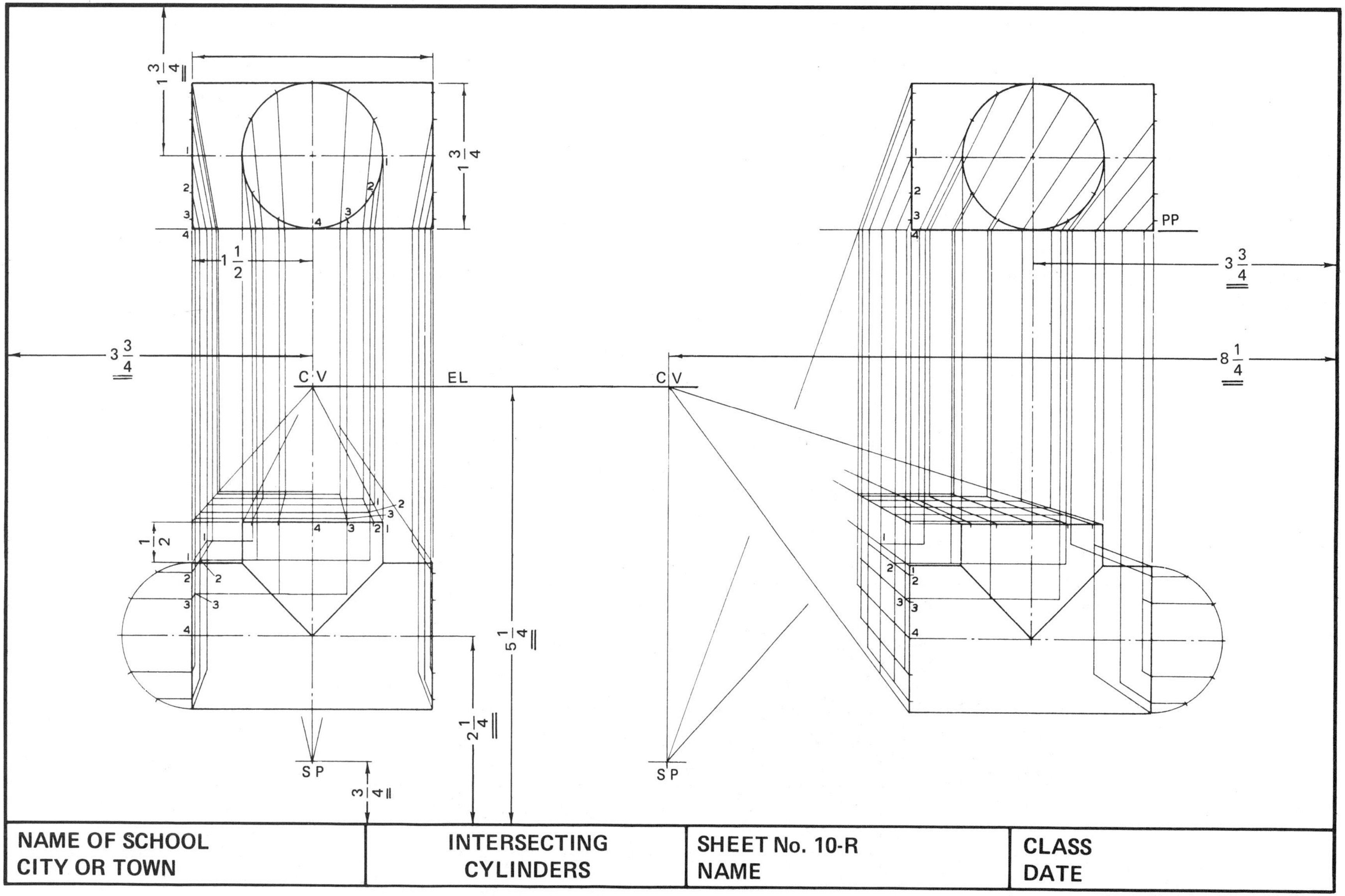
1 3/4
1 3/4
1 1/2
3 3/4
C V
EL
C V
PP
3 3/4
8 1/4
1/2
5 1/4
2 1/4
S P
3/4
S P
NAME OF SCHOOL
CITY OR TOWN
INTERSECTING
CYLINDERS
SHEET No. 10-R
NAME
CLASS
DATE

Fig. 10-54

piercing point of point 1 intersects the receding line from point 1 to locate this point in the perspective view.

11. To check the accuracy of the points on the horizontal receding plane, draw horizontal parallel lines from the intersections at the left end, as shown on the drawing.

12. Draw the ellipses on the horizontal planes. In the first drawing, only the front half of the ellipses on the vertical planes is seen. In the second drawing, the ellipse on the left end is seen in full while only the front half of the ellipse on the right end is seen. Draw these semiellipses.

13. When the ellipses are drawn, locate the points for the line where the two cylinders intersect. Draw the line of intersection.

14. Complete the drawing, omitting the dimensions.

15. Check the accuracy of all lines and ellipses.

ANGULAR PERSPECTIVE

In the preceding perspective drawings, one face of the object is parallel to the picture plane. The vertical edges and one group of horizontal edges are parallel to the picture plane. The other horizontal edges are perpendicular to the picture plane.

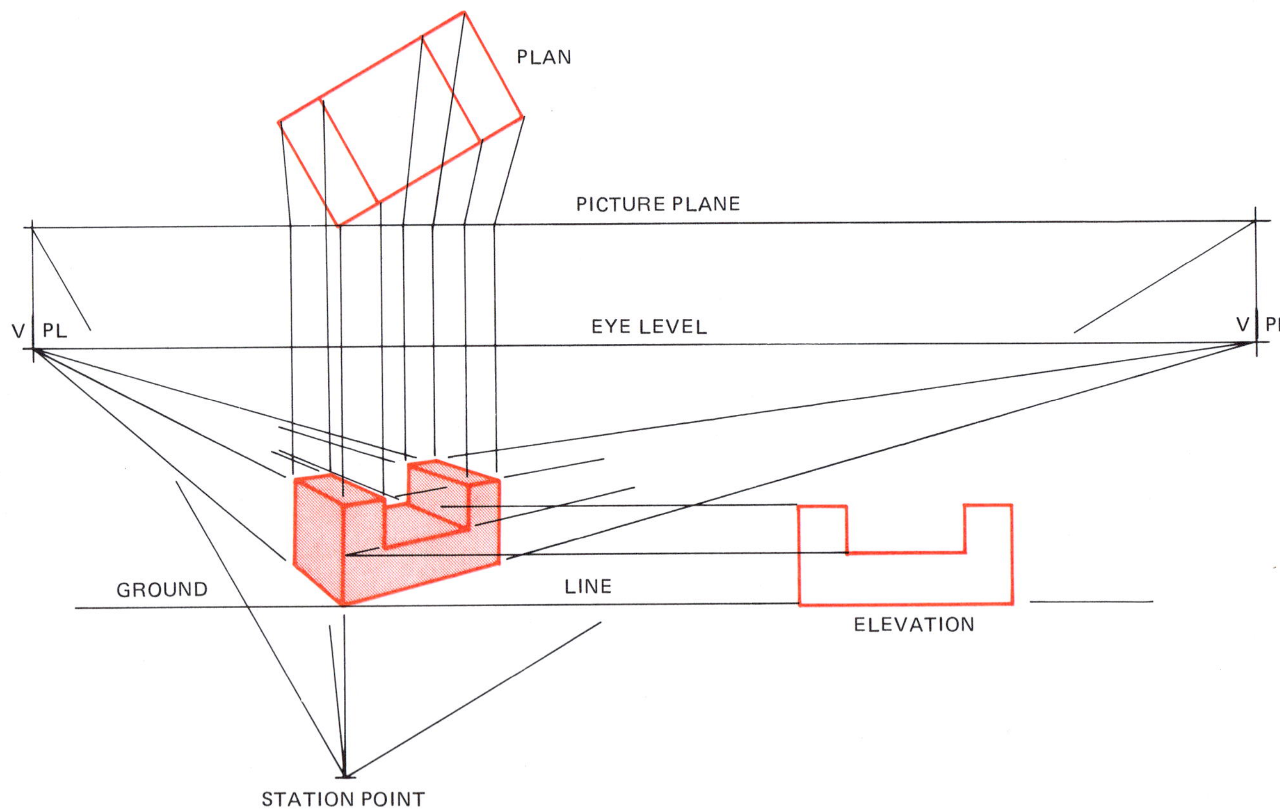

Fig. 10-55 Angular perspective drawing

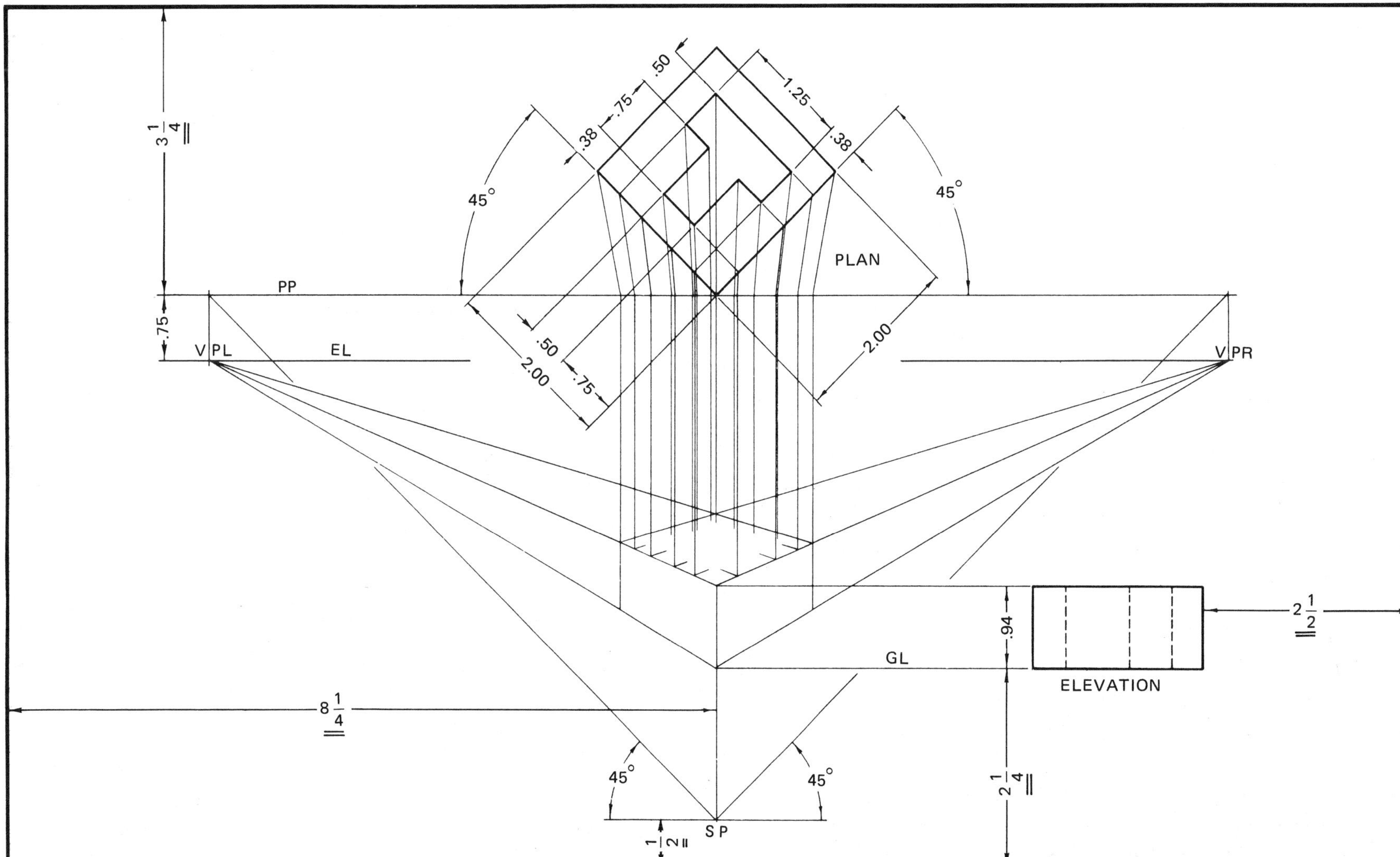

NAME OF SCHOOL CITY OR TOWN	DIE BLOCK	SHEET No. 10-S NAME	CLASS DATE

Fig. 10-56

In the perspective view, these horizontal perpendicular edges converged toward a point in the distance called the center of vision. Since all horizontal perpendiculars vanish at the center of vision, this point is called their vanishing point, and this type of perspective is called *one-point perspective.*

In angular perspective the position of the object is such that all of its horizontal edges are at an angle with the picture plane, figure 10-55. One group of horizontal edges slopes at an angle to the right; the other slopes at an angle to the left.

The parallel lines which slope to the right will vanish at a point in the distance, called the right vanishing point (VPR), and those which slope to the left will vanish at the left vanishing point (VPL). This type of perspective is called *two-point perspective.*

Since the vanishing points are always on a level with the eye, perpendiculars are dropped from the vanishing points on the picture plane to locate the vanishing points on the eye level line.

In any perspective view, only those edges which are against the picture plane are their true lengths. All others are foreshortened. Therefore, to find the foreshortened lengths of the horizontal lines and to determine the position of the vertical lines, visual rays are drawn from the corners of the plan to the picture plane, and lines are projected downward from the piercing points, figure 10-55.

The die block, figure 10-56, has a number of horizontal edges on its upper surface that are foreshortened in the perspective view. The foreshortened lengths of these lines are found by drawing visual rays and projecting downward from the piercing points.

In angular perspective, the object may be placed at any angle in relation to the picture plane. More of the right surface is shown in the perspective view if this surface is placed nearest the picture plane. Likewise, features on the left surface are shown more clearly if this surface is placed nearest the picture plane. As the angle between the picture plane and the surface is reduced, more of the surface is shown.

ASSIGNMENT S

Make an angular perspective drawing of the die block.

Procedure

1. Lay out the margin lines on the drawing sheet and letter the title strip.
2. Locate the station point. Draw the ground line, the picture plane, and the eye level line.
3. From the station point, draw lines at 45 degrees to locate the two vanishing points on the picture plane.
4. From the vanishing points on the picture plane, drop perpendiculars to locate the right and left vanishing points on the eye level line.
5. Draw the plan and the elevation of the die block.
6. The front vertical edge is against the picture plane. It is its true length in the perspective view. Locate and draw this edge.
7. Draw receding lines from each end of this vertical edge to the right and left vanishing points.
8. Draw rays from the corners of the die block to the picture plane. Project downward from the piercing points to locate the right and left vertical edges.
9. In the plan, extend the edges of the hole to the outer edge of the die block.
10. Draw rays from the points on the outer edges to the picture plane.
11. From the piercing points, project lines downward to the upper edges of the perspective view.
12. The points on the upper edges of the perspective view determine the location of the horizontal edges of the upper sur-

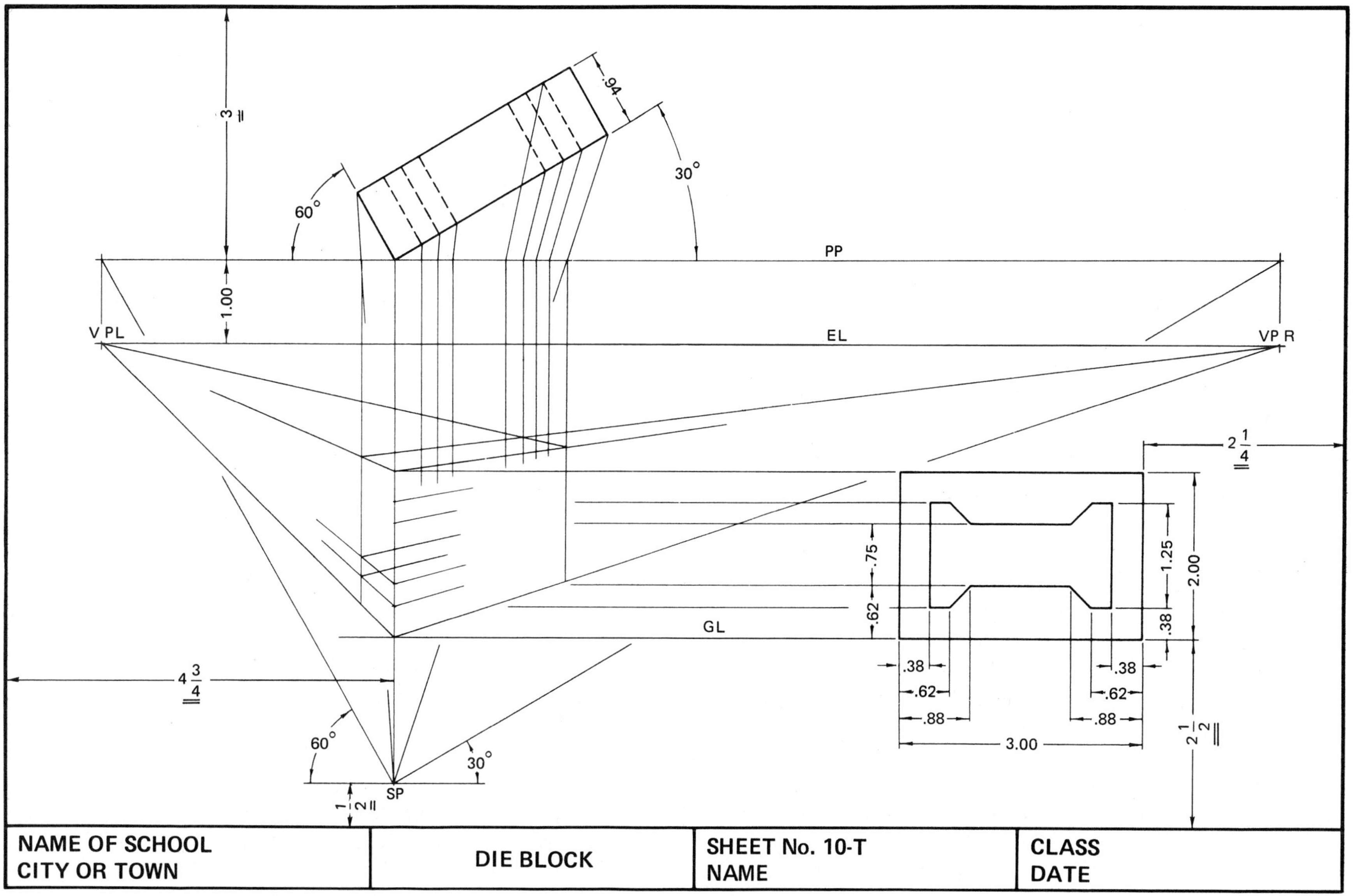
3"
.94
30°
60°
PP
1.00
V PL
EL
VP R
2 1/4"
2.00
1.25
.75
.62
.38
GL
4 3/4"
.38
.62
.88
.38
.62
.88
3.00
2 1/2"
60°
30°
SP
1/2"
NAME OF SCHOOL
CITY OR TOWN
DIE BLOCK
SHEET No. 10-T
NAME
CLASS
DATE

Fig. 10-57

face. Draw receding lines from these points to the vanishing points.

13. Draw rays from each corner of the hole. Project downward from the piercing points to locate the ends of the horizontal and the vertical edges in the hole.
14. Complete the drawing omitting the dimensions.
15. Check all lines and views for this angular perspective drawing of the die block.

REVIEW QUESTIONS

1. How are the right and left vanishing points located?
2. How are the horizontal edges on the upper surface of the die block located?
3. How are the corners of the hole and the vertical edges located?
4. Why is angular perspective called two-point perspective?

CHANGING THE PICTURE PLANE ANGLE

The die block, figure 10-57, has horizontal edges, vertical edges, and sloping edges on its front surface.

The horizontal edges are located in the perspective view by projecting from the corners in the elevation to the front vertical edge. Receding lines are then drawn from the points on the front edge to the right vanishing point.

The location of the vertical edges and the end points of the horizontal and sloping edges are found by projecting downward from the piercing points on the picture plane.

Two horizontal edges on the rear surface are partially seen in the perspective view. They are found by drawing receding lines from the points on the front edge to the left vanishing point. Receding lines are then drawn from the points on the rear vertical edge to the right vanishing point.

ASSIGNMENT T

Make an angular perspective drawing of the die block.

Procedure

1. Lay out the margin lines on the drawing sheet and letter the title strip.
2. Locate the station point. Draw the ground line, the eye level line, and the picture plane.
3. Locate the vanishing points.
4. Draw the elevation and the plan.
5. Draw the outlines of the perspective view.
6. Draw the edges of the hole which are on the front surface.
7. Locate and draw the edges of the hole which are on the rear surface.
8. Complete the drawing, omitting the dimensions.
9. Check the accuracy of all points, lines and edges of the hole.

SLOPING EDGES

The book end in figure 10-58, has edges which are horizontal, vertical, sloping, and oblique.

The vertical edges and the end points of the horizontal edges are found by projecting downward from the piercing points on the picture plane. The sloping edges are drawn in their respective places by joining the end points of the horizontal edges.

The horizontal edges are projected from the elevation to the front edge of the perspective view. Receding lines are then drawn to the right vanishing point. The horizontal edges in the plan are extended to the outer edges. Rays and projectors from the end points of these extended lines will locate their end points in the perspective view. Receding lines are drawn from the end points of the extended lines to the vanishing points. The horizontal edges are on these receding lines.

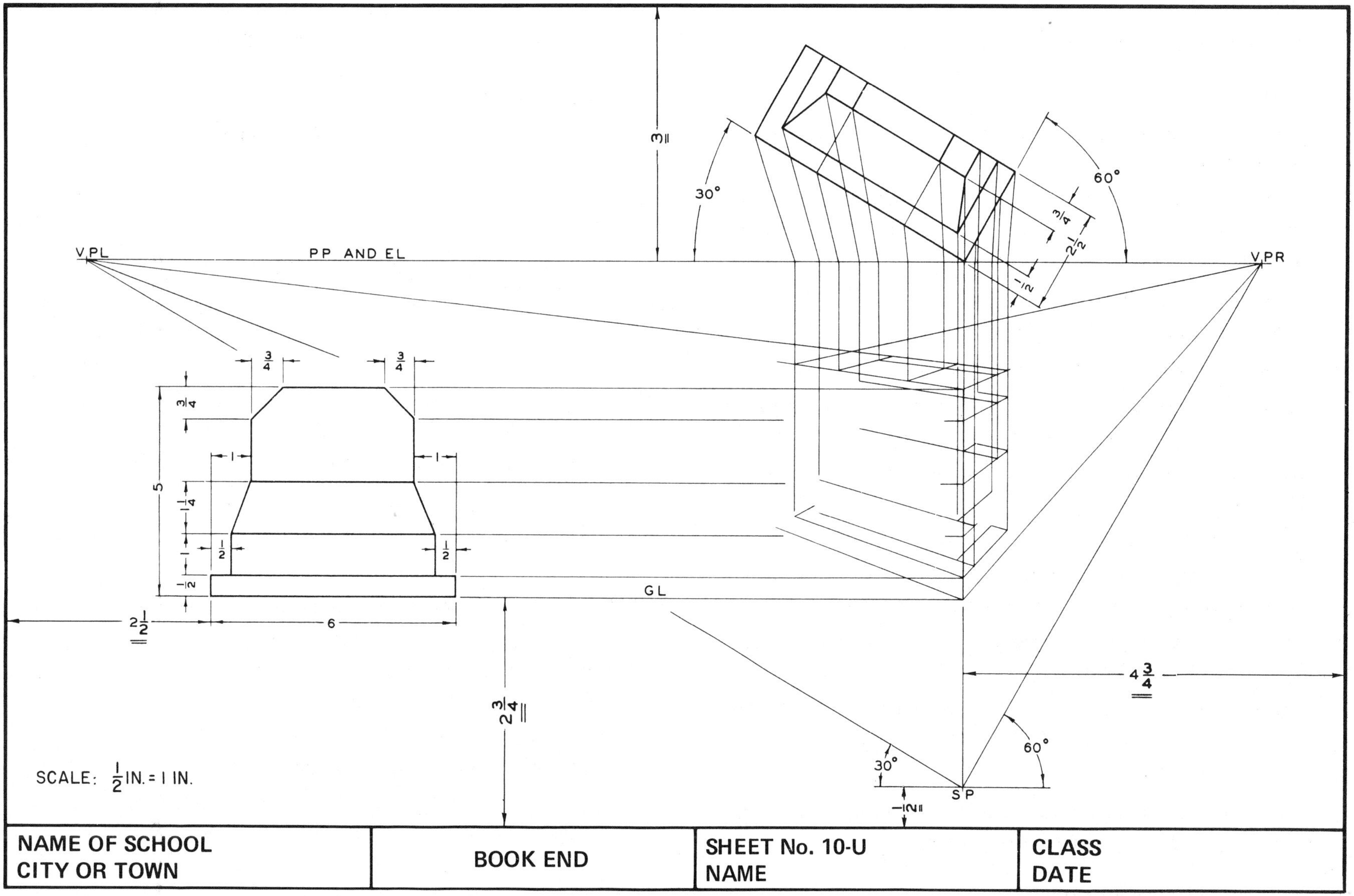
VPL
PP AND EL
VPR
30°
60°
GL
SP
SCALE: 1/2 IN. = 1 IN.
NAME OF SCHOOL
CITY OR TOWN
BOOK END
SHEET No. 10-U
NAME
CLASS
DATE

Fig. 10-58

ASSIGNMENT U

Make an angular perspective drawing of the book end.

Procedure

1. Lay out the margin lines on the drawing sheet and letter the title strip.
2. Locate the station point. Draw the ground line and the picture plane. On this drawing, the picture plane and the eye level are one line.
3. Locate the vanishing points.
4. Draw the elevation and the plan.
5. Draw the outlines of the perspective view.
6. Project lines from the horizontal edges in the elevation to the front edge of the perspective view.
7. Draw receding lines from the front edge to the right vanishing point.
8. Extend the horizontal lines in the plan to the front edges.
9. From the points on the front edges of the plan, draw rays and projectors to locate these points in the perspective view.
10. From these same points in the perspective view, draw receding lines to the vanishing points.
11. Draw the vertical edges. The vertical edges locate the end points of the horizontal and sloping edges.
12. Complete the drawing.
13. Check all horizontal, vertical, sloping, and oblique edges and lines as drawn on the perspective drawing.

CIRCLES IN ANGULAR PERSPECTIVES

Circles on the receding planes appear as ellipses in angular perspective drawings. The circles in the plan and elevation are divided into a number of equal parts, and the points on the circles are located in the perspective view. The points then are joined using an irregular curve to draw the ellipses.

ASSIGNMENT V

Make an angular perspective drawing of the intersecting cylinders, figure 10-59.

Procedure

1. Lay out the margin lines on the drawing sheet and letter the title strip.
2. Locate the station point. Draw the ground line and the picture plane. The eye level and the picture plane are one line on this drawing.
3. Locate the vanishing points.
4. Draw the elevation and the plan.
5. Divide the circles into twelve equal parts.
6. In the plan, project the points on the circle to both ends of the cylinder.
7. Project the points on the circle in the elevation to the front vertical line in the perspective view.
8. Draw receding lines from the points on the front vertical line to the left vanishing point.
9. Draw rays from the points on the left end of the cylinder. Project downward from the piercing points to locate the twelve points for the ellipse. Draw the ellipse.
10. From the points on the front half of the ellipse draw receding lines to the right vanishing point.
11. Draw rays from the points on the front half of the right end of the cylinder.
12. Draw projectors from the piercing points to locate the points for the semiellipse on the right end. Draw this semiellipse.

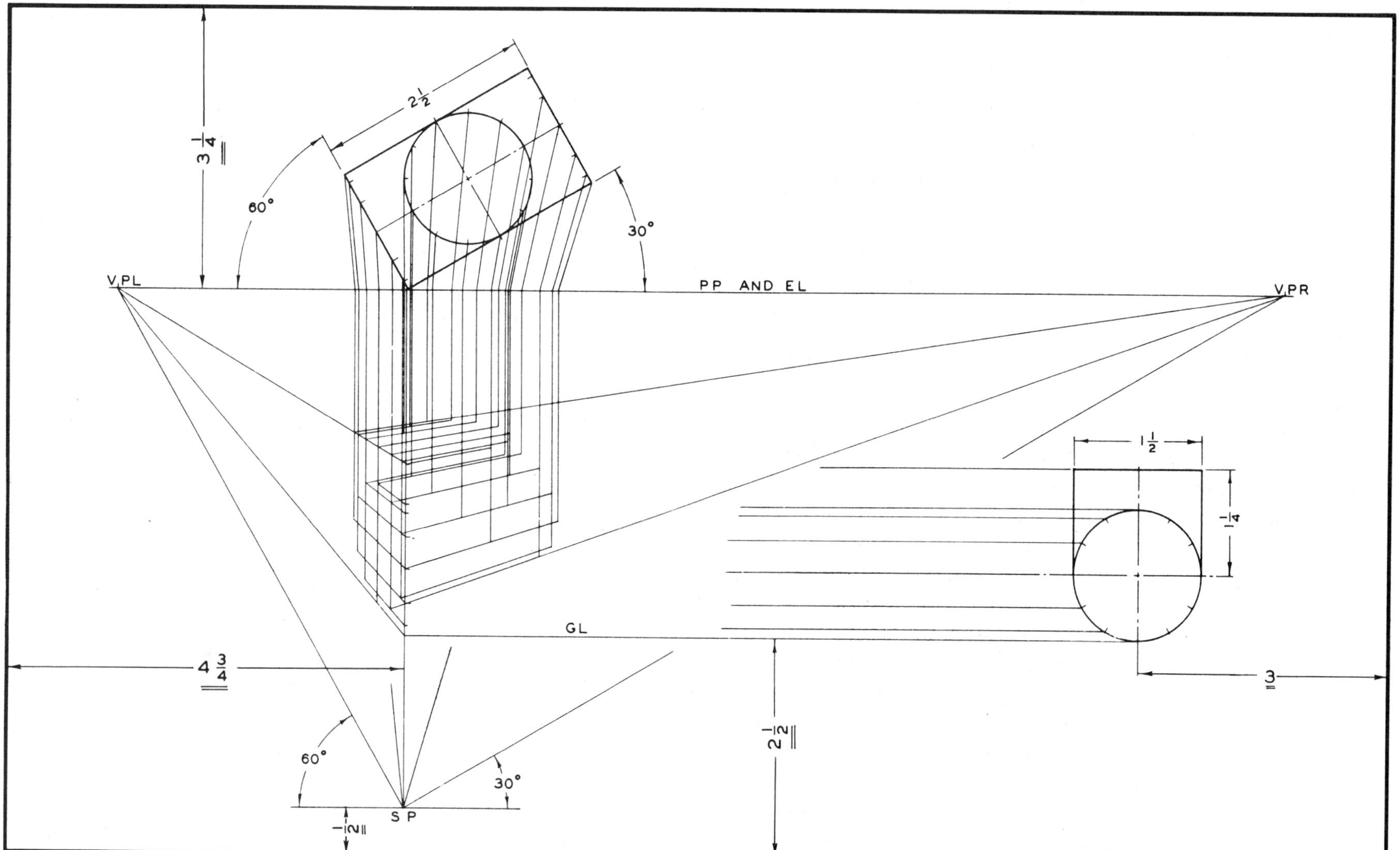

NAME OF SCHOOL CITY OR TOWN	INTERSECTING CYLINDERS	SHEET No. 10-V NAME	CLASS DATE

Fig. 10-59

13. For the ellipse on the horizontal plane, draw receding lines to the right vanishing point as shown on the drawing.
14. Draw rays from the points on the circle in the plan.
15. Project downward from the piercing points to locate the points for this ellipse. Draw the ellipse.
16. Locate the points and draw the line where the two cylinders intersect.
17. Complete the drawing and erase all unnecessary lines.
18. Check the elevation and plan, and all points, lines, and ellipses.

MEASURING LINES IN ANGULAR PERSPECTIVE DRAWING

Angular perspective drawing is used principally by architects to make pictorial drawings of proposed buildings. A plan and elevation fully describing the building is made to a selected scale. Then, the perspective drawing is made.

Measuring lines are used in angular perspective drawings to locate certain points. In figure 10-60, measuring lines locate the peaks and ridges on the roof, and find the height of the edges at the corners of the walls.

Parts which are against the picture plane can be measured in their true heights. The peak of the roof is not against the picture plane. In order to measure its true height, it is assumed that it is moved forward to contact the picture plane. A measuring line is drawn downward from the point of contact. The true height of the peak is marked off on the measuring line.

A receding line is drawn from the true height to the right vanishing point. The peak is located on this receding line by a projector from its piercing point. The peak of the dormer is found in the same manner.

The front vertical edge of the cottage is not against the picture plane. In order to measure its true height, this edge is assumed to be moved forward to the picture plane and a measuring line is drawn where it contacts the picture plane. The true height is measured on this line. Receding lines drawn from the measuring line determine the upper and lower ends of the vertical edges at each corner of the walls.

ASSIGNMENT W

Make an angular perspective drawing of the cottage.

Procedure

1. Lay out the margin lines on the drawing sheet and letter the title strip.
2. Locate the station point. Draw the ground line and the picture plane.
3. Using the scale, 1/8 inch equals one foot, measure and draw the eye level line. The eye level is usually five feet six inches above the ground line.
4. Locate the vanishing points.
5. Using the same scale, draw the elevations and the plan. The partial elevation is drawn to locate the peak of the roof.
6. Assume that the peak of the roof is moved forward to contact the picture plane. A measuring line is drawn from the point where the peak contacts the picture plane. Project the true height of the peak from the elevation to this measuring line.
7. From the true height mark, draw a receding line to the right vanishing point.
8. Locate the peak in the perspective view on this receding line. Draw the outlines of the roof.
9. Find the peak of the dormer in the same manner and draw the dormer.
10. Assume that the front vertical edge of the cottage is moved forward and to the right until it contacts picture plane. Draw measuring line at a point of contact.

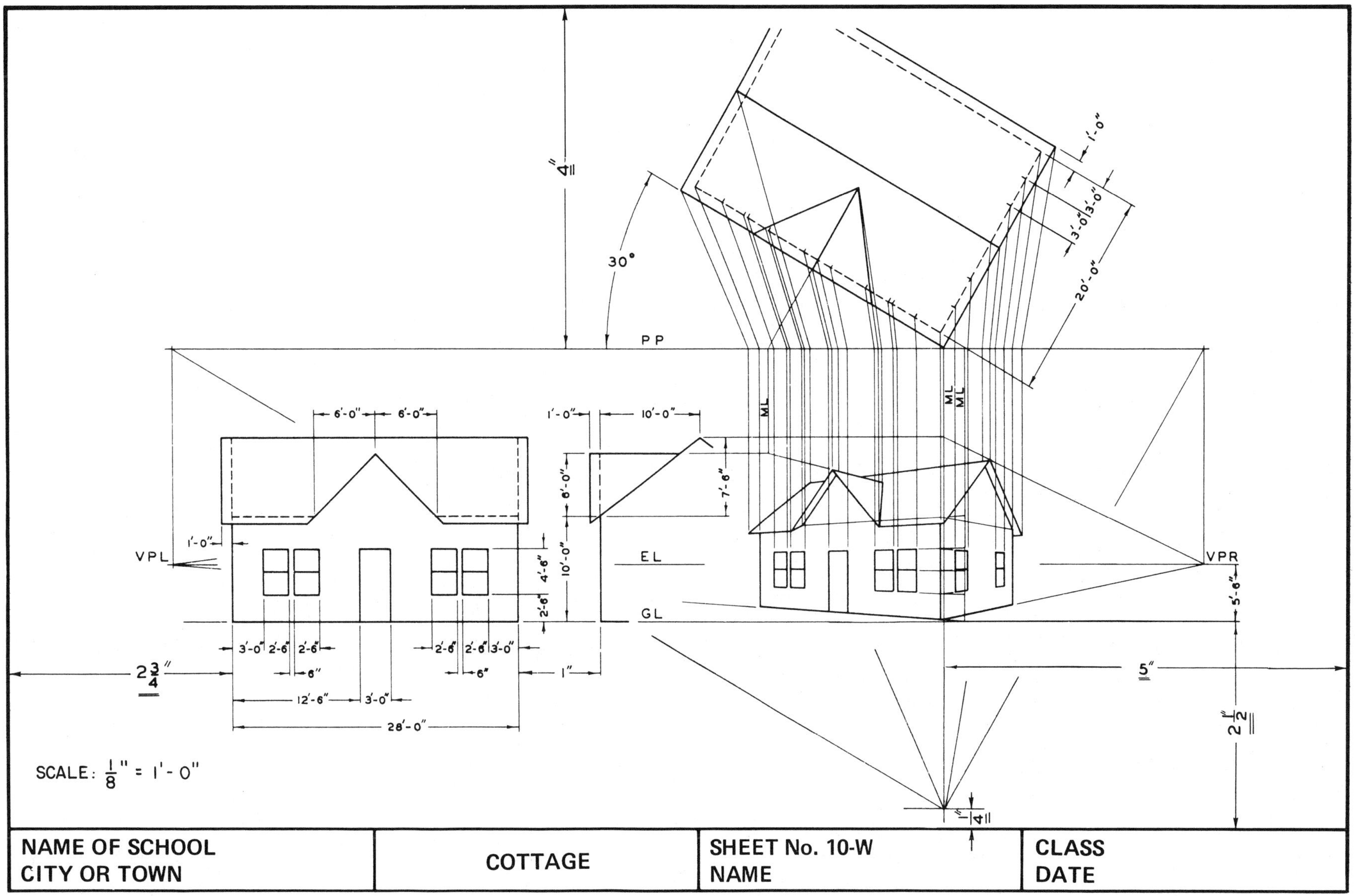
PP
30°
4"
VPL
VPR
EL
GL
ML
ML
ML
6'-0"
6'-0"
1'-0"
10'-0"
8'-0"
7'-6"
10'-0"
4'-6"
2'-6"
3'-0"
2'-6"
2'-6"
6"
12'-6"
3'-0"
28'-0"
1"
2 3/4"
5"
2 1/2"
5'-6"
1/4"
1'-0"
3'-0"
3'-0"
20'-0"
SCALE: 1/8" = 1'-0"
NAME OF SCHOOL
CITY OR TOWN
COTTAGE
SHEET No. 10-W
NAME
CLASS
DATE

Fig. 10-60

11. Mark the true height of the wall and the height of the windows on this measuring line.
12. Draw receding lines from the true height mark and from the lower end of the measuring line to the left vanishing point.
13. Locate and draw the front vertical edge and the left vertical edge of the wall between these two receding lines.
14. From the ends of the front vertical edge of the wall, draw receding lines to the right vanishing point.
15. Locate and draw right vertical edge of wall between these two receding lines.
16. The height of the windows and the door are found in the same manner. Draw the door and the windows and complete the drawing.
17. Check the accuracy of all points, lines, and views.

UNIT 11
SKETCHING

The ability to make rapid, accurate sketches freehand is in great demand today. Time to make accurate detail drawings is an expensive luxury in industry. Every engineer, designer, builder, trades person, and technician needs to know how to express ideas through drawing without instruments.

Designs for machines, tools, and new products begin as rough sketches from which working drawings are made later. Ideas spontaneously communicated in design conferences through sketches can be analyzed, discussed, improved, and redrawn quickly to allow the creative process to continue without interruption for detailed drawings. Sketches are also used to:

- supplement, clarify, and record verbal explanations.
- record technical notes and descriptions to be used in detail drawings.
- explain processes and procedures to others.
- make notes of repairs or improvements needed on equipment.
- substitute for detail drawings when a sketch suffices.

Accuracy in sketching depends on two skills; the ability to see an object accurately and analyze how it is shaped, and the ability to make the hand draw on a two-dimensional surface what the eye sees as three-dimensional. Both skills improve with practice.

PROPORTION IN SKETCHING

Correct proportion is the most important element of a good drawing. *Proportion* refers to the relationship of the features of an object to the whole in both size and location. Two techniques to improve the ability to estimate and record distance accurately are the sight method and the blind contour exercise.

The *sight method* is illustrated in figure 11-1. Any object, such as a pencil, ruler, or thumbnail can help to estimate relative size and location of features. Determine the overall height and width and relate smaller units within this space by enclosing each feature within a rectangle of light construction lines. Keep the elbow straight and stand in the same position each time an object is measured this way. It is also important to work from the larger, general shapes to the smaller, specific details when laying out distances. The final sketch can then be carefully drawn within these spaces.

The second exercise works by forcing the sketcher to depend on eye-hand coordination to judge distances. It is called a *blind contour* study because the sketcher does not look at the paper at all while drawing the object. This exercise is done as follows: Touch the top, bottom, right, and left sides of the paper to feel the maximum size the drawing can be made. Now put the pencil point down at the place on the paper where the top of the object should be and fix your eye on that part of the object. The trick is to feel as though the pencil is touching the object. Then, as the eye wanders around every bend, fold, nut, and bolt on the object, the pencil follows in one continuous line from top to bottom in-

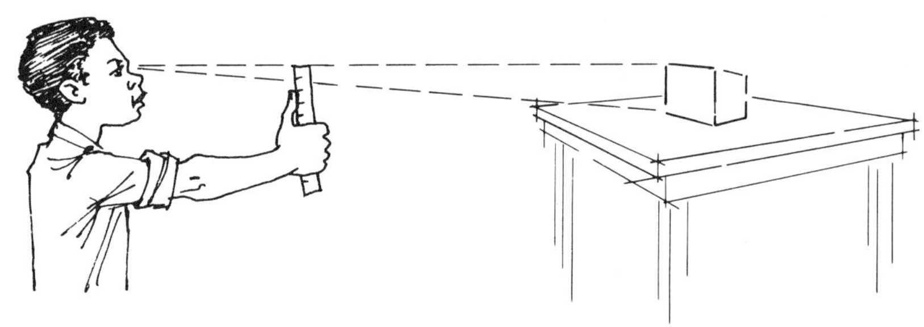

Fig. 11-1 Sighting across a measuring "instrument" to estimate the proportion of an object for sketching

cluding every detail and returns to the starting point. At first the details may be in the wrong places or the wrong size, but that is unimportant. What is important is that the eye is being trained to see accurately.

Select an object that seems too detailed and difficult to draw. Engines, typewriters, bicycles, even stools and chairs with a lot of detail make good subjects for practice sketching, figure 11-2. The drawing should be timed and not take more than ten minutes. This exercise is designed to improve visual concentration, the ability to judge distances by eye, and the ability to make the hand draw what the eye tells it to. The greatest advantage of this exercise is that it can be practiced anywhere without being obvious.

Fig. 11-2 A "blind" contour exercise develops the ability to "see" an object kinesthetically.

ASSIGNMENT A

Make a blind contour drawing of any selected object. Use no more than ten minutes for this sketch. Any size paper will do.

Make a sketch of the same object using the sight measuring and blocking-in techniques explained above. Any size paper will do.

EQUIPMENT

Freehand sketching requires a pencil (F, H, or HB), a plastic or kneaded rubber eraser, and a pad of paper. The paper may be plain or graph (coordinate) paper. There are many types of ruled graph paper designed to aid in more controlling proportions or for special drawing problems such as pictorial solutions in isometric, oblique or perspective (figure 11-3). Sketching in this series is done on plain paper.

When sketches are made on plain paper, the sizes and the proportions of the various parts are estimated as accurately as possible by eye. Freehand sketching should be done with a soft pencil (F, H, or HB). Fine lines are sketched with a sharp conical point, heavier lines with a blunt point. The hand is held in a comfortable position with the pencil grasped firmly, but not tightly enough to cause strain. Most sketching is done using the elbow or the shoulder as a pivot. The larger the radius, the straighter the line will be, figure 11-4.

SKETCHING HORIZONTAL LINES

Horizontal lines are sketched from left to right, using the elbow as a pivot, without moving the wrist or fingers as they hold the pencil in position. It helps if the beginner places a series of dots in a straight line about one inch apart. The dots are then connected with short, clean strokes of the pencil. As the beginner becomes more skilled, the dots may be spaced further apart, until finally only the ends need to be marked. The trick is to keep the eye focused on the dot to which the pencil is going instead of

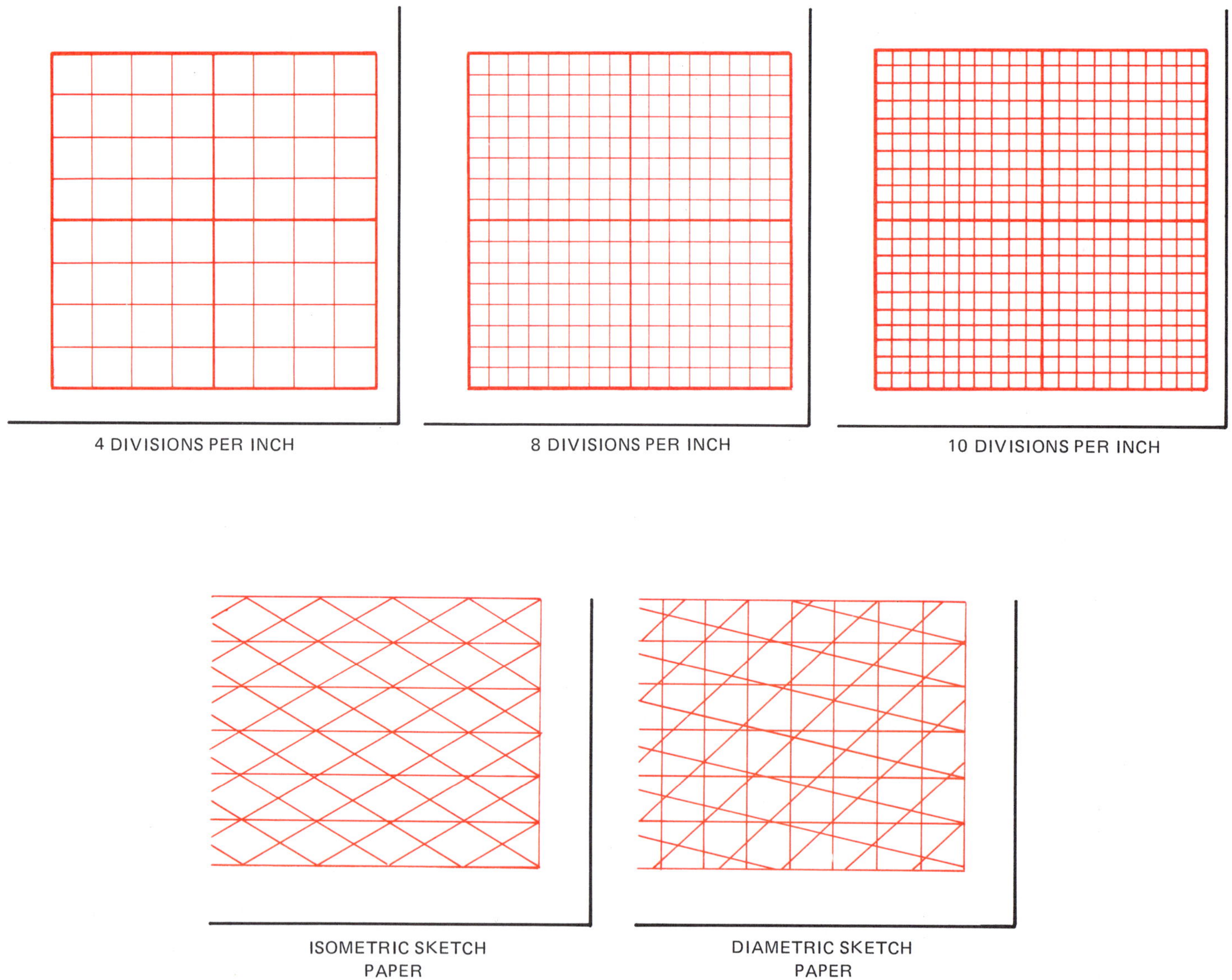

Fig. 11-3 Grid papers aid in freehand sketching.

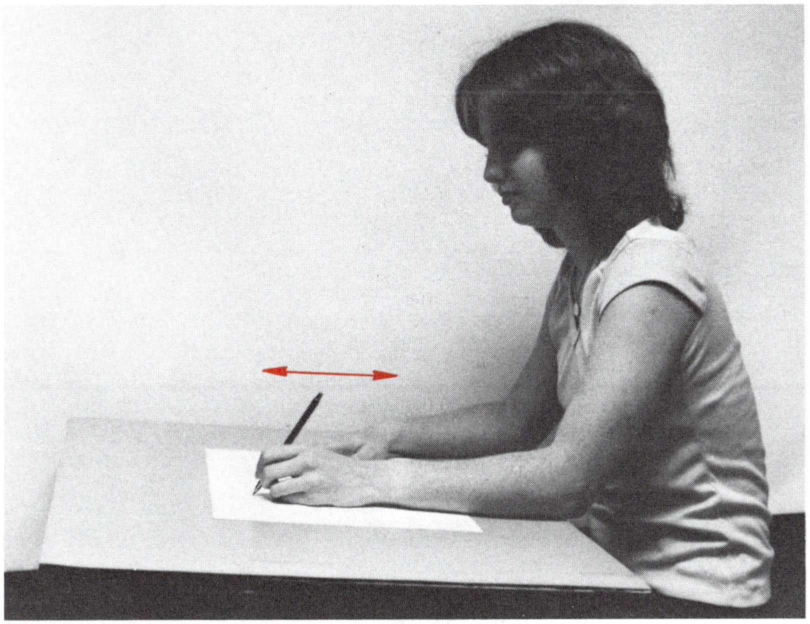

Fig. 11-4 Hold the pencil firmly, keep hand and wrist still, and pull from the elbow when drawing.

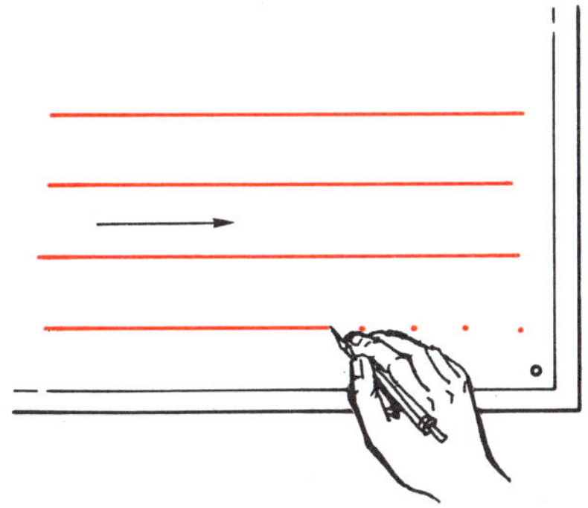

Fig. 11-5

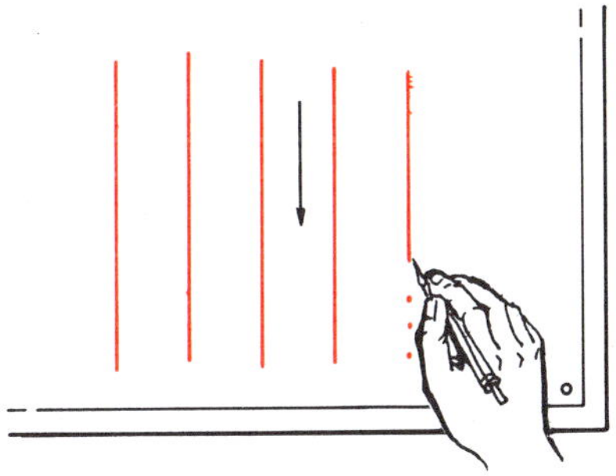

Fig. 11-6

looking at the pencil point. Stop at each point and focus on the next dot before moving on, figure 11-5.

SKETCHING VERTICAL LINES

Vertical lines are sketched from the top downward. The pencil strokes are made by using the shoulder as the pivot and pulling the arm straight toward the body. When sketching vertical lines, use a series of dots in a straight line. Increase the spacing of the dots until a straight line can be made with only the end points located, figure 11-6. A common aid in making a straight vertical line is the edge of the sketch pad or a similar straight edge used as a guide. Rest the fingers against the guide while pulling the pencil, figure 11-7.

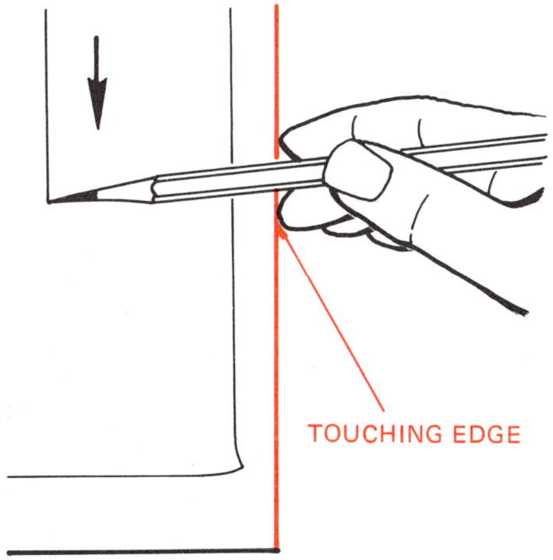

Fig. 11-7 Use the edge of a board or the paper to aid in making a straight, vertical line.

SKETCHING SLOPING LINES

Sloping lines are sketched in the same manner as either horizontal or vertical lines depending on the angle of the incline, figure 11-8. They are frequently sketched by turning the paper so the inclined lines are horizontal or vertical. Remember to pull the pencil toward the body.

SKETCHING ARCS AND CIRCLES

Sketching curved lines requires an understanding of the hand and arm movements. Very small arcs are made with finger movements. Slightly larger arcs are made accurately with a wrist movement holding the fingers still. Larger arcs and circles require the use of the elbow as a pivot with both the wrist and fingers locked comfortably in place. Very large circles are made by holding finger, wrist, and elbow in place and swinging from the shoulder. This movement must be made standing up, figure 11-9.

There are methods for increasing accuracy in making circles and arcs. For example, when drawing an arc the paper is turned to the position which is most convenient. When an arc joins two straight lines, the end points are located on the lines which are drawn first, figure 11-10. Circles are sketched using the enclosing square method, the center line method, the hand pivot method, or the free circle method.

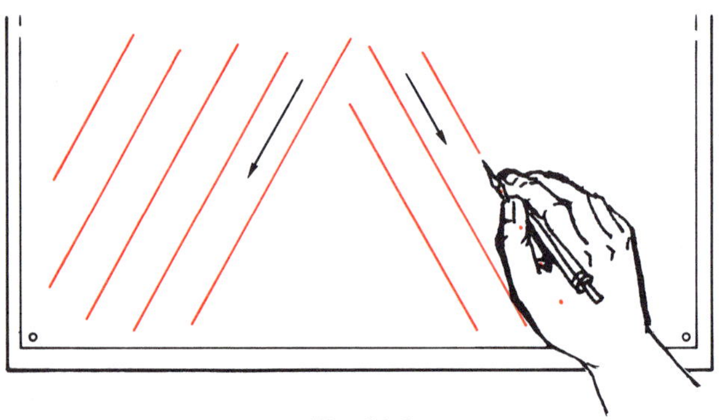

Fig. 11-8

Fig. 11-9 Large circles use the shoulder as a pivot point.

The Enclosing Square Method

Draw a light square the size of the desired diameter. Locate and draw center lines. Draw arcs tangent to the square at midpoints, figure 11-11.

The Center Line Method

Sketch horizontal, vertical, and 45-degree center lines. Mark points approximately the length of the desired radius on each line, figure 11-12. Or use a piece of scrap paper marked to the desired circle, figure 11-13. Connect each dot with a short arc. Be sure to keep the eye on the dot ahead instead of looking at the pencil while drawing.

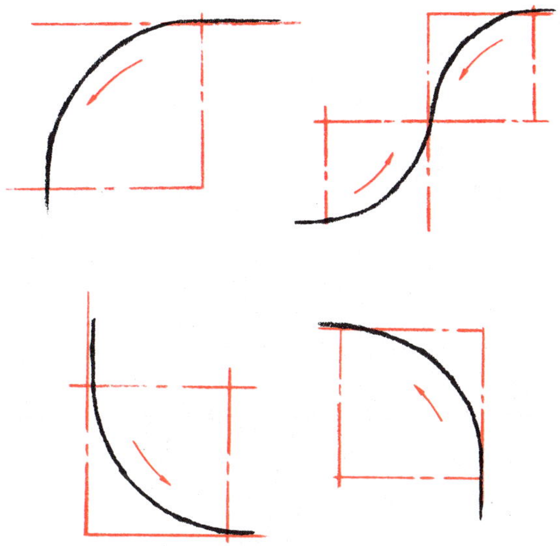

Fig. 11-10 Sketching arcs

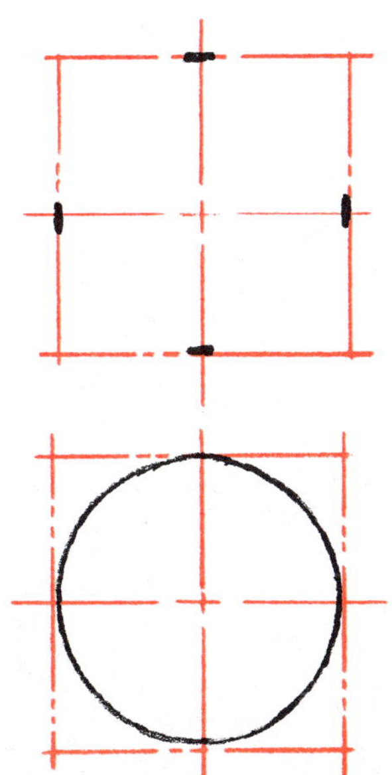

Fig. 11-11 Draw a square for sketching a circle.

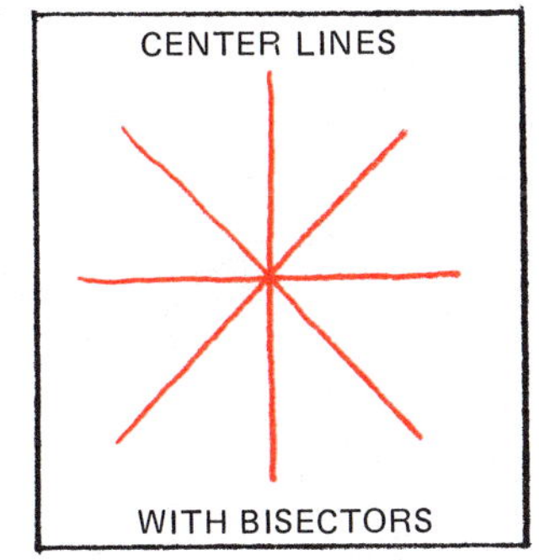

Draw center lines for a circle with bisecting lines.

Mark off the estimated radii on all lines before sketching the circle.

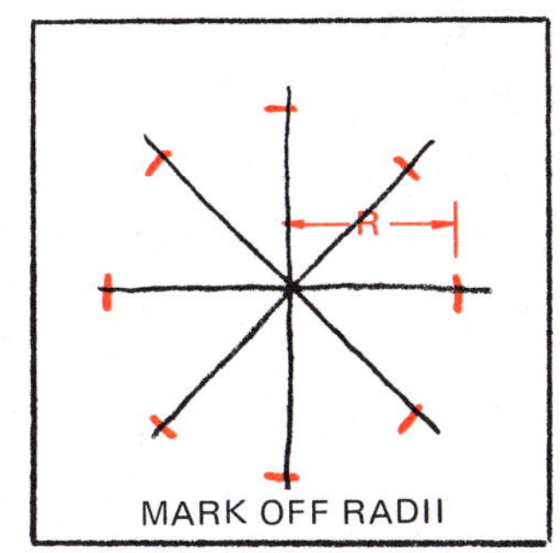

Fig. 11-12

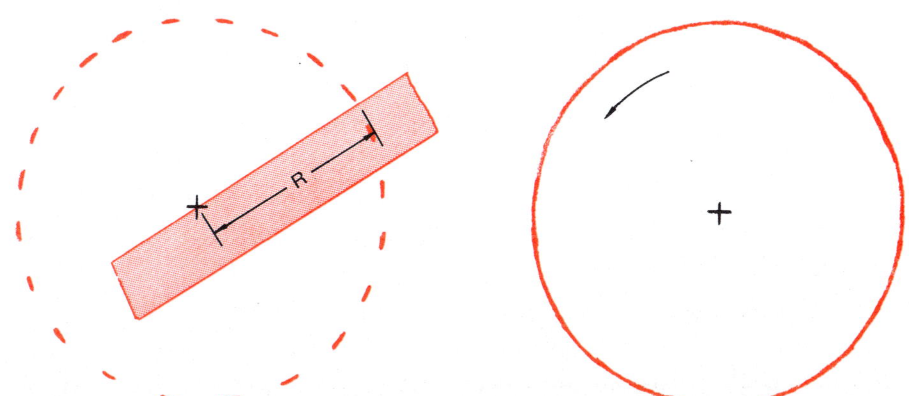

Fig. 11-13 Use a piece of paper to mark off radii for sketching large circles.

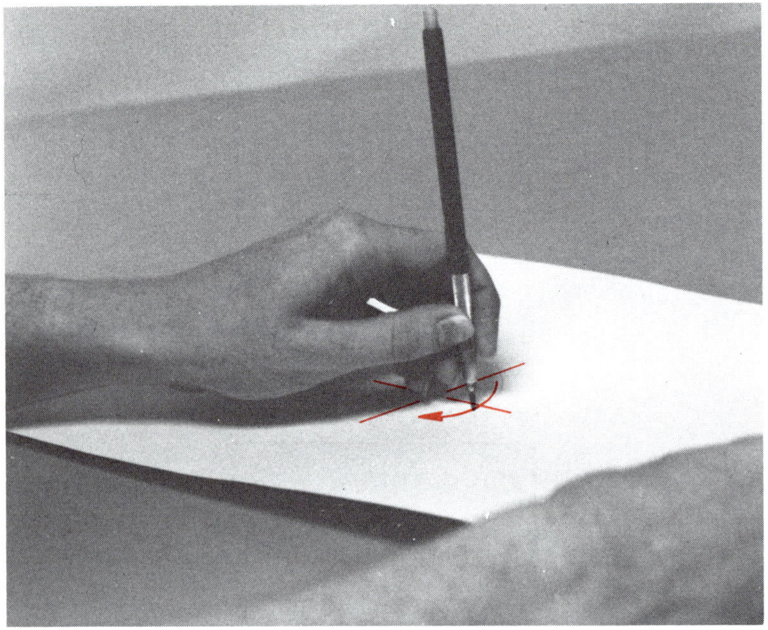

Fig. 11-14 Use the little finger as a pivot point for freehand circle.

Fig. 11-15 Two pencils used as a "compass" for sketching a circle.

The Hand Pivot Method

Use the little finger of the hand holding the pencil as a pivot point while rotating the paper to form a circle, figure 11-14. An alternate method is to use two pencils as a compass and rotate the paper to make the circle, figure 11-15.

The Free Circle Method

This method depends on skill which is only developed through practice. Controlling all movement from the elbow, lightly sketch several full circles of the approximate size in the correct location. Select the best circle and darken the line, figure 11-16.

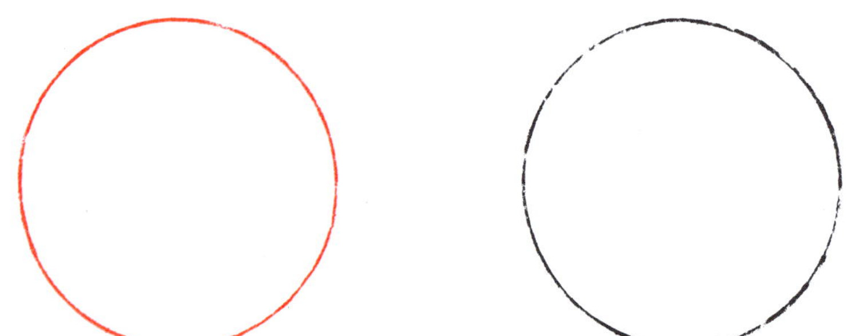

Fig. 11-16 Free circle method of sketching a circle

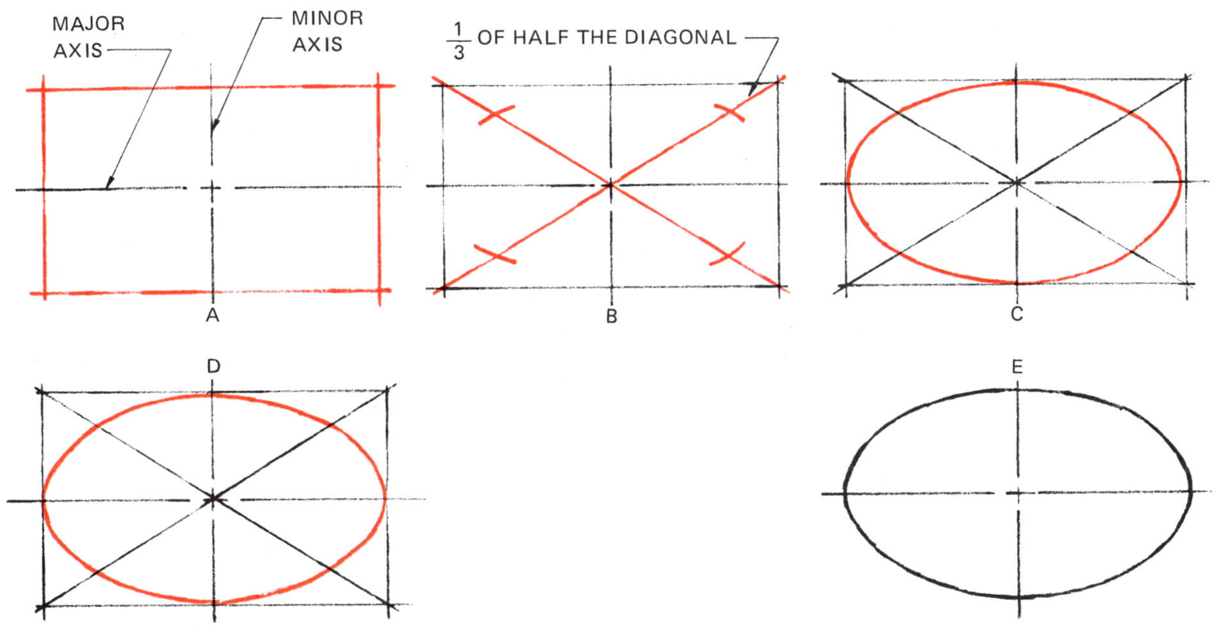

Fig. 11-17 An ellipse can be sketched by drawing a rectangle. One side is equal to the minor axis. The other is equal to the major axis.

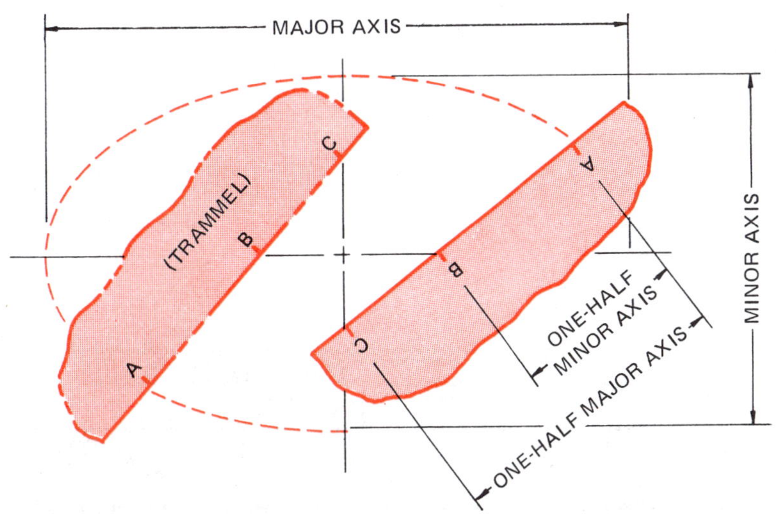

Fig. 11-18 Trammel method of sketching an ellipse

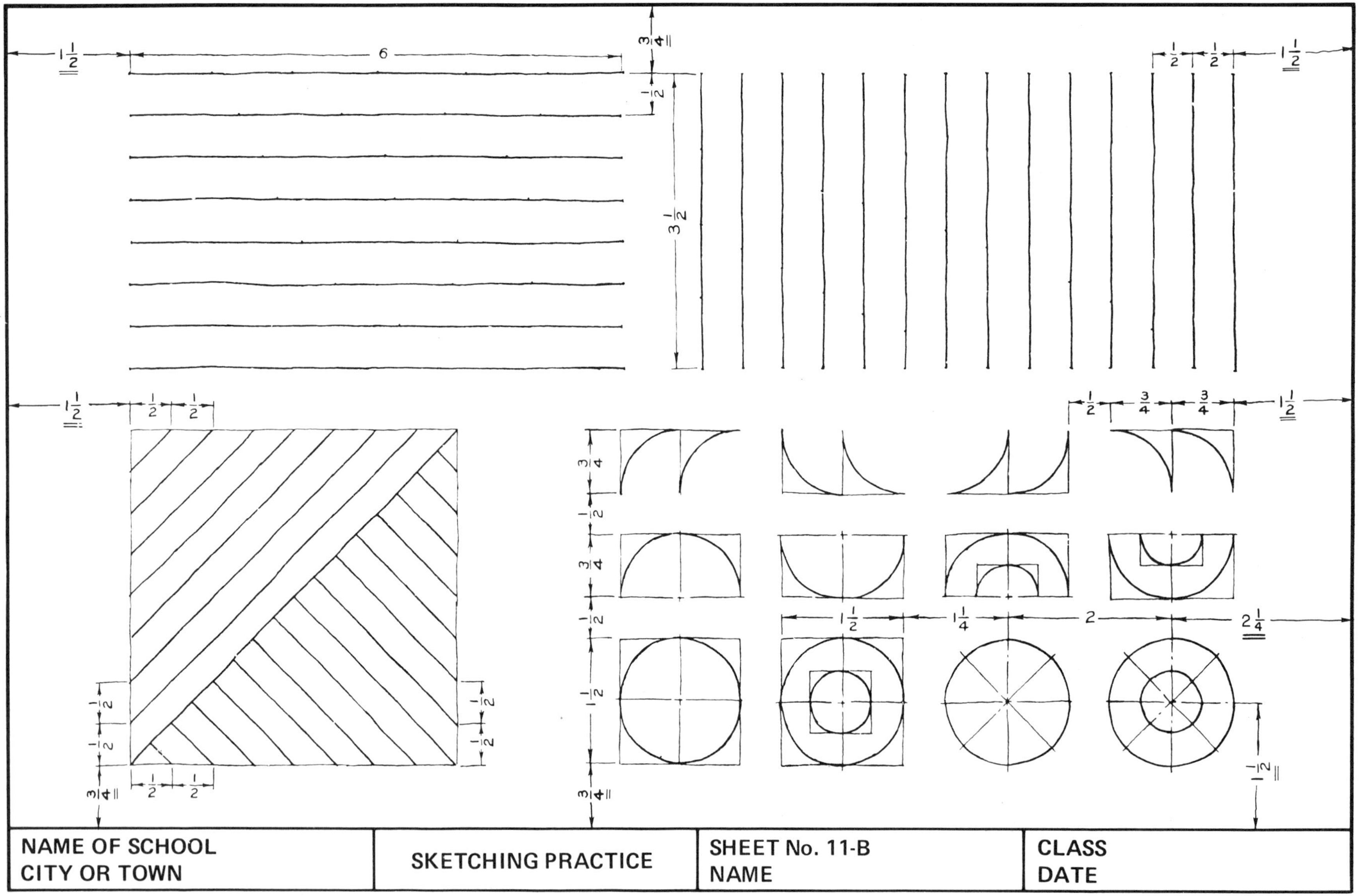
NAME OF SCHOOL
CITY OR TOWN
SKETCHING PRACTICE
SHEET No. 11-B
NAME
CLASS
DATE

Fig. 11-19

SKETCHING ELLIPSES

There are two methods of sketching ellipses. One is to block in a light rectangle the size of the major and minor axes and sketch arcs tangent to the midpoints of the rectangle, figure 11-17.

The other method uses a *trammel,* a scrap of paper on which measurements are marked. First sketch the major and minor axes. Then mark one-half of each of these measurements on the trammel. Sliding these marks along the axes of the ellipse as shown in figure 11-18, plot the arcs. Connect these dots with short strokes.

ASSIGNMENT B

Sketch freehand the lines, arcs, semicircles, and circles shown in figure 11-19.

Procedure

1. Lay out the margin lines on the drawing sheet and letter the title strip.
2. Sketch freehand the horizontal and vertical lines. The length of the lines and spacing may be measured with a scale on this drawing. Omit all dimensions.
3. Sketch the sloping lines.
4. Sketch the first and second rows of blocks. Sketch the arcs and semicircles within these blocks.
5. Sketch the next two blocks and center lines. Sketch circles within these blocks.
6. Sketch the center lines and point off the radius on each line for the last two circles. Sketch the circles.
7. Check all lines.

FREEHAND ORTHOGRAPHIC SKETCHING

When freehand sketches are made in orthographic projection, it is good practice to block-in the views first. The size of the blocks are estimated as accurately as possible by eye. The blocks are spaced approximately the same distance from the side margin lines, and about the same distance above the title strip and below the upper margin line, for a well-balanced sheet, figure 11-20. The correct view is sketched in each block.

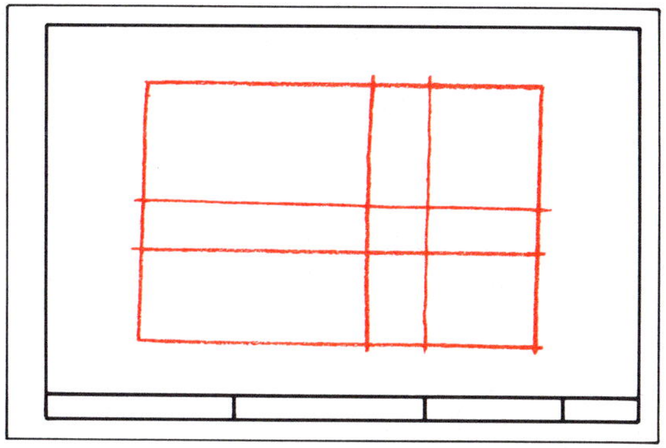

Fig. 11-20 Blocking-in sketched views

The sizes and proportions of the various details should be reasonably accurate. Estimate these dimensions and proportions carefully. Erase as often as necessary. Take time to do a neat sketch; correct in every detail.

ASSIGNMENT C

Make a freehand three-view orthographic sketch of the drill vise base, figure 11-21.

Procedure

1. Lay out the margin lines on the drawing sheet and letter the title strip.
2. Sketch the blocks for the three views. These blocks are about 2 1/2 inches from the side margin lines and about 1 1/4 inches from the upper margin line and the title strip.

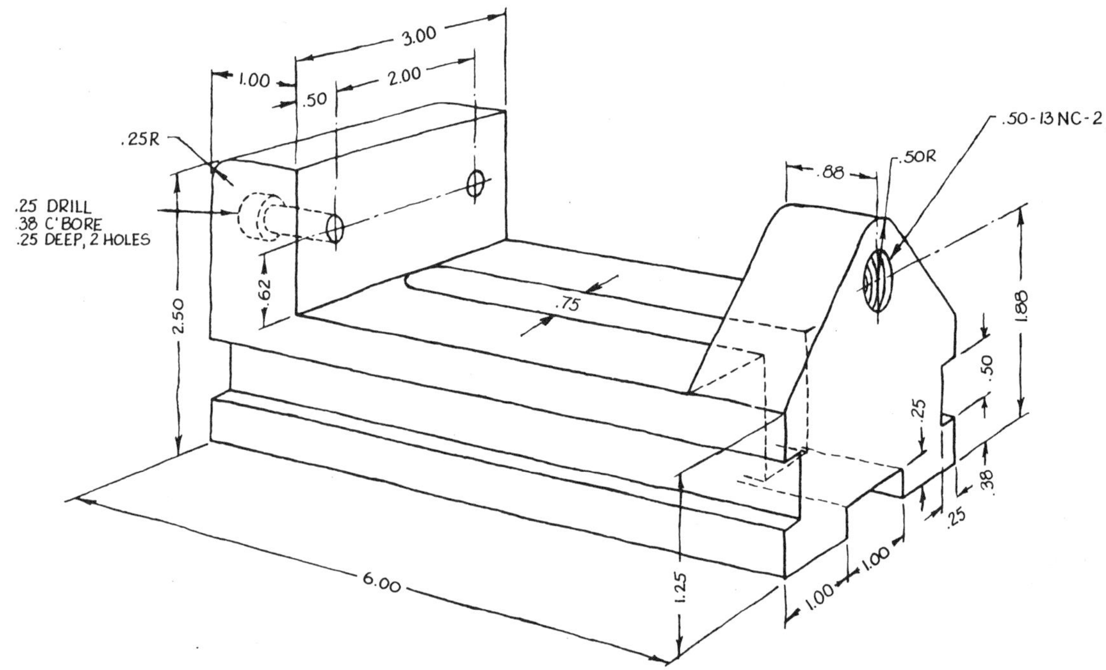

NAME OF SCHOOL CITY OR TOWN	DRILL VISE BASE	SHEET No. 11-C NAME	CLASS DATE

Fig. 11-21

3. Sketch the front, right side, and top views.
4. Dimension the views and erase unnecessary lines.
5. Check each line in the sketched orthographic views.

SKETCHING IN ISOMETRIC

This drawing shows sketches in orthographic projection of the jaw of a drill press vise and a cutter holder. From these orthographic sketches, freehand isometric sketches are made following the principles and methods making isometric drawings.

When a sketch is made in isometric, an isometric box is sketched first. The axes of the box should be approximately equal to the length, width, and height of the part to be sketched. The part is then sketched within the box.

SKETCHING CIRCLES IN ISOMETRIC

Circles in isometric are sketched within an isometric square. The four-center method for drawing isometric circles is followed. Lines are sketched from the midpoint of each side to the opposite obtuse angle. These lines locate the centers and the end points for each arc of the isometric circle. The isometric circle is then sketched tangent to the sides of the isometric square, figure 11-22.

Fig. 11-22 Circles in isometric sketches

ASSIGNMENT D

Make freehand an isometric sketch of the drill vise jaw and the cutter holder, figure 11-23.

Procedure

1. Lay out the margin lines on the drawing sheet and letter the title strip.
2. Sketch an isometric box on the left half of the sheet. The lower corner of this box is about 2 1/2 inches above the title block and about 3 1/2 inches from the left margin line. The length of the vertical axis is the approximate height of the jaw. The receding axis on the right side is the approximate length. The receding axis on the left side is the approximate width of the jaw.
3. Sketch the vise jaw within this box. The sketch shows the front, left side, and top of the jaw. The drilled and counterbored holes cannot be shown clearly in the sketch without using hidden lines, but these holes can be adequately described with a note.
4. Dimension the sketch.
5. Sketch an isometric box on the right half of the sheet for the cutter holder. The lower corner of this box is about 2 inches above the title block, and 4 1/2 inches from the right margin line. The length of the vertical axis is the approximate height of the cutter holder. The receding axis on the right side is the approximate length. The receding axis on the left side is the approximate width of the cutter holder.
6. Sketch the cutter holder within the box. The sketch shows the left side, the front, and the top of the cutter holder.
7. Dimension the sketch.
8. Check all lines and dimensions and erase unnecessary lines.

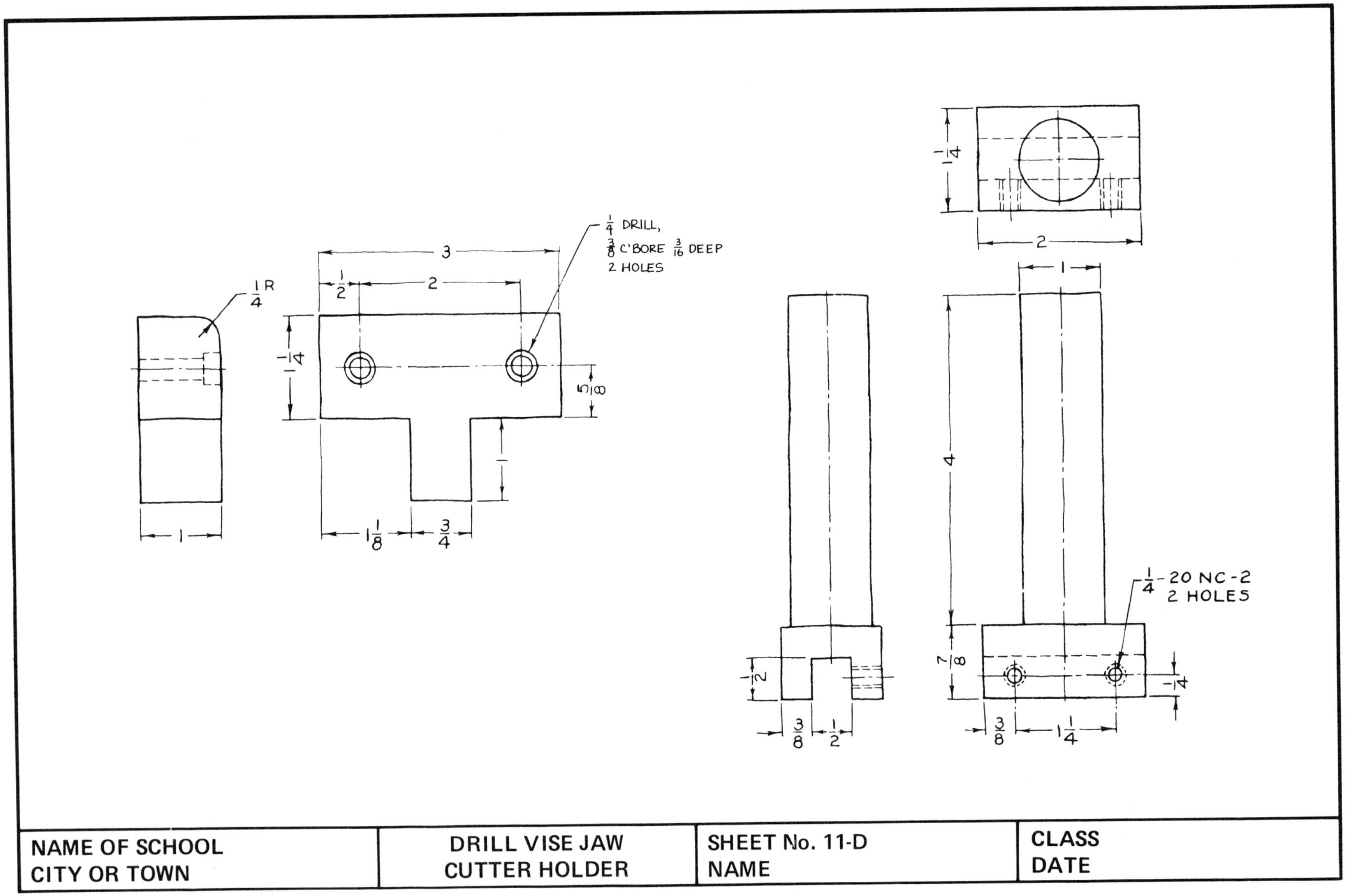

1/4 DRILL,
3/8 C'BORE 3/16 DEEP
2 HOLES
1/4 R
1/4-20 NC-2
2 HOLES
NAME OF SCHOOL
CITY OR TOWN
DRILL VISE JAW
CUTTER HOLDER
SHEET No. 11-D
NAME
CLASS
DATE

Fig. 11-23

OBLIQUE AND CABINET SKETCHING

This drawing shows sketches in orthographic projection of a gear blank and a drill jig base. From these orthographic sketches, freehand sketches are to be made following the principles and methods for making oblique and cabinet drawings.

In oblique drawing, the face of the object which shows the contour or irregular outline of the circular outline, is placed parallel to the plane of the projection. This face is drawn in its true shape and size. The same rule applies when oblique sketches are made.

In cabinet drawings, the effect of distortion is lessened by reducing the length of the receding edges by one-half. This also applies when making cabinet sketches.

In the next assignment, the circular face of the gear blank and the irregular contour of the drill jig base are sketched parallel with the plane of projection. The receding axes are reduced one-half.

ASSIGNMENT E

Make a freehand cabinet sketch of the gear blank and the drill jig base, figure 11-24.

Procedure

1. Lay out the margin lines of the drawing sheet and letter the title strip, or use one sheet of 8 1/2-by-11-inch plain paper or graph paper of any kind for each object.
2. Sketch the vertical center line for the circle on the front of the hub of the gear blank about 3 1/4 inches from the left margin line. Sketch the horizontal center line about 4 1/2 inches above the title block.
3. Sketch the axis of the gear blank at 45 degrees.
4. On this axis, locate the approximate centers for the other three circles on the gear blank. The measurements on this receding axis are approximately one-half of the actual size for cabinet sketches.
5. Mark the approximate radius for the first circle on the edge of a piece of scrap paper and then point off this radius about the center. Sketch the first circle.
6. Repeat this process for the other three circles.
7. Complete the sketch.
8. Dimension the sketch.
9. For the drill jig base, an oblique box is sketched first. The front face of this box is the approximate length and height of the drill jig base. The receding axis is one-half of the depth of the drill jig base. Sketch the front face of this box about 2 3/4 inches above the title block and about 3 inches from the right margin line.
10. Sketch the drill jig base within the box. Sketch the outline of the front face first.
11. Dimension the sketch.
12. Check the accuracy of all lines and curved surfaces and erase unnecessary lines.

SKETCHING IN PERSPECTIVE

Sketches often are made using the principles of perspective drawing. The receding lines are drawn so as to converge toward imaginary vanishing points. Since edges behind the picture plane are foreshortened in perspective drawing, the lengths of these edges are judged by eye so that the sketch presents a natural appearance of the object.

In freehand sketching, the principles remain the same, but it is impossible to use the projection methods with the plan and elevation. Therefore, there are no obsolete measurements to de-

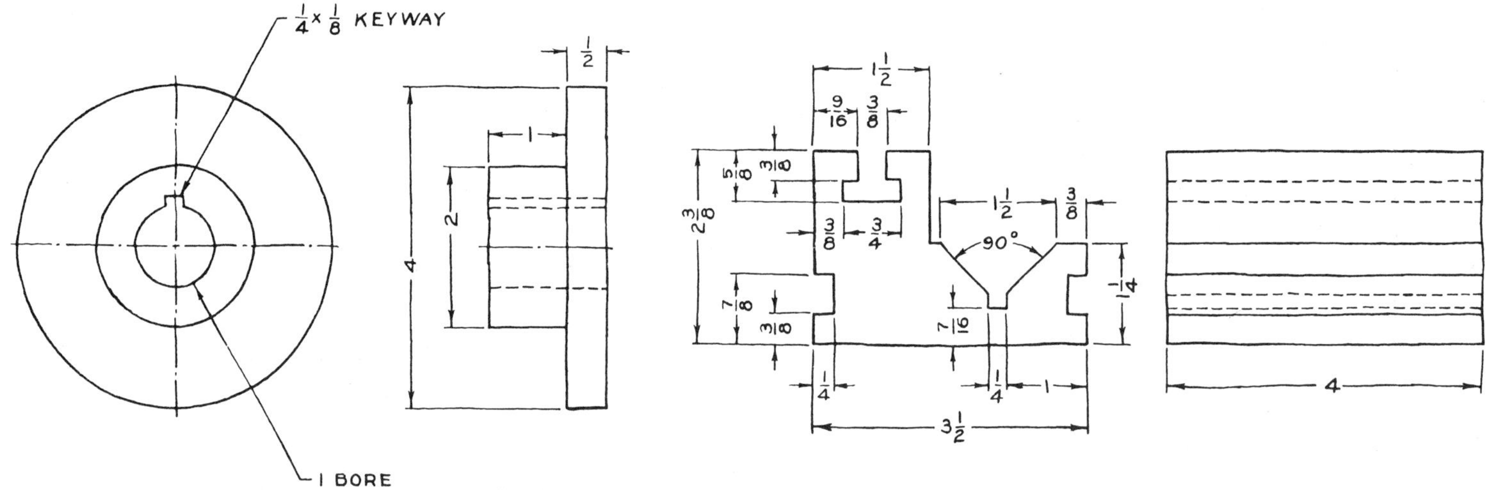

NAME OF SCHOOL CITY OR TOWN	GEAR BLANK DRILL JIG BASE	SHEET No. 11-E NAME	CLASS DATE

Fig. 11-24

pend upon. Many things are left to personal judgment. The appearance of the object can be altered by:

- moving the horizon line up or down.
- moving the vanishing points closer together or farther apart.
- changing the height, width, or depth of the object when making the first decision on size.

The best way to learn is to experiment with placement and the number of vanishing points. Some things to remember are:

- There is always a horizon line or eye level whose position determines whether the object is viewed from above or below. This line is always assumed to be at the level of the eye of the observer.
- The vanishing points for one- and two-point perspective are always on the horizon line.
- In one-point perspective, one face of the object is drawn as it appears with horizontal and vertical lines and only one set of receding lines. It is similar to an oblique drawing without true length measurements.
- In two-point perspective, only one corner of the object touches the picture plane. So it is similar to isometric drawings with two sets of receding lines. There are vertical lines, but not horizontal lines.
- Three-point perspective is used only for special effects in special relations. There are no horizontal or vertical lines, but all lines converge on one of the three vanishing points.

The procedures to follow when sketching in perspective are:

1. Decide whether to draw the object with a surface or a corner toward the viewer. A surface requires only one vanishing point. A corner requires two.

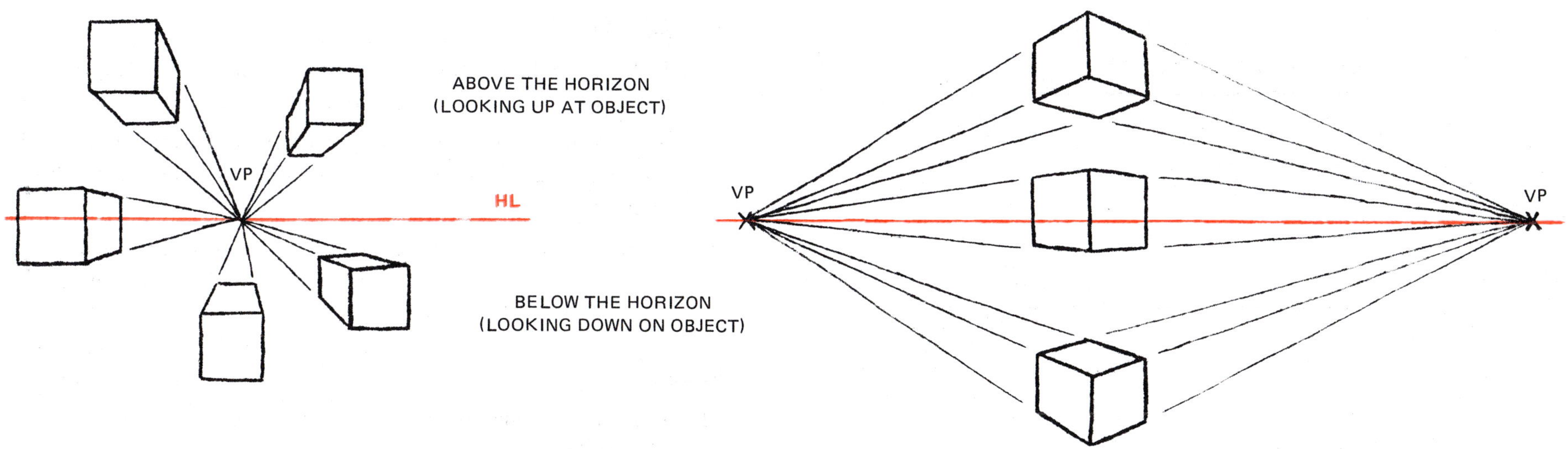

Fig. 11-25 Decide on the location of the horizon line.

Fig. 11-26 Locate the vanishing point or points on the horizon line.

2. Decide on the location of the horizon line, figure 11-25. A horizon line below the object indicates the viewer is looking up at it. The horizon line above the object indicates the viewer is looking down on it. If the object is on the horizon line, it is at eye level.
3. Locate the vanishing point or points on the horizon line, figure 11-26.
4. Establish the height of the object and its placement in relation to the vanishing points and the horizon line, figure 11-27.
5. Sketch the face of the object to be drawn in one-point perspective, figure 11-28.

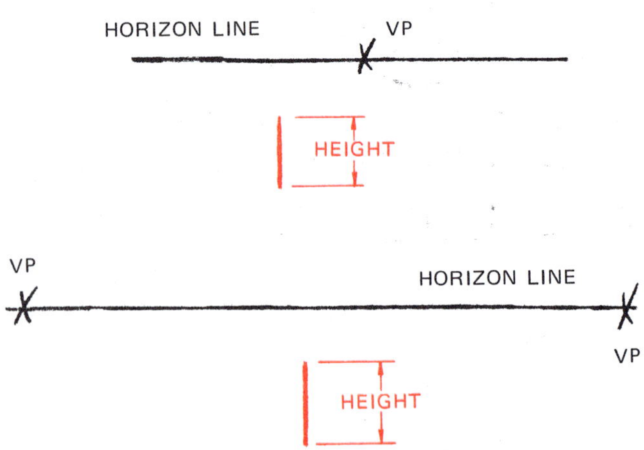

Fig. 11-27 Establish the height of the object and its placement in relation to the vanishing points and horizon line.

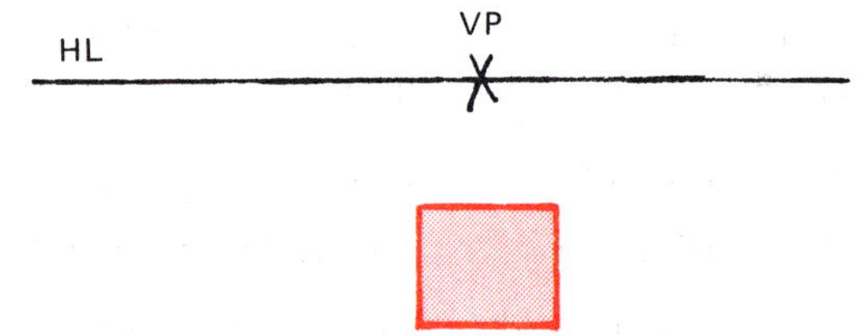

Fig. 11-28 Sketch the face of the object to be drawn in one-point perspective.

6. Project lines for the receding axes to the correct vanishing points, figure 11-29.
7. Estimate relative depth and sketch vertical and horizontal lines, figure 11-30.
8. Project from these intersections to the vanishing points, figure 11-31.
9. Darken in the outlines of the object. Erase all projection lines, figure 11-32.

ASSIGNMENT F

A. Make a freehand one-point perspective drawing of the drill jig base, figure 11-24. Place the horizon line above the object.

B. Make a freehand two-point perspective drawing of the drill vise base, figure 11-21. Decide on the placement of the horizon line.

C. Make a freehand one- or two-point perspective drawing of the drill vise jaw, figure 11-23. Place the horizon line below the object.

Procedure

1. Lay out the margin lines and letter the title strip on each drawing sheet.
2. Locate the horizon line and vanishing points for each drawing.

3. Locate each object in relation to the horizon line and vanishing points.
4. Sketch the object. Omit hidden lines and dimensions.
5. Check drawings for accuracy and clean up the sheets.

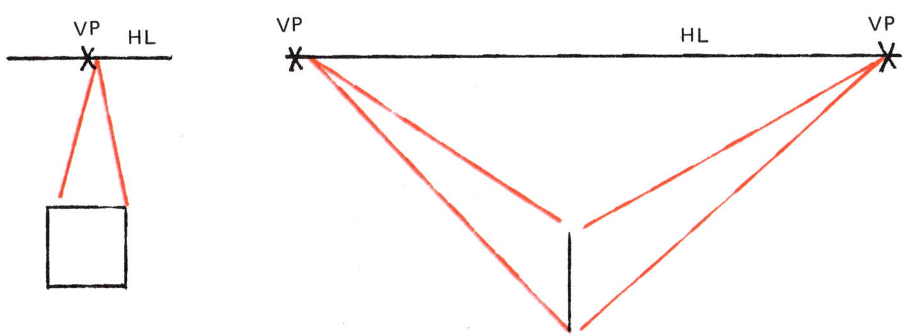

Fig. 11-29 Project lines for receding axes to the correct vanishing points.

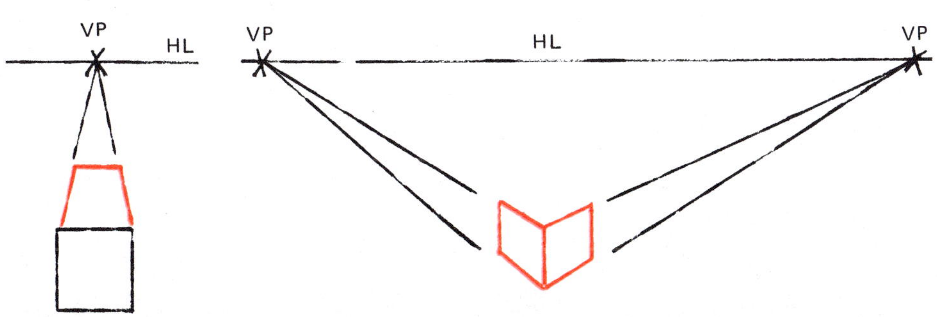

Fig. 11-30 Estimate relative depth and sketch vertical and horizontal lines.

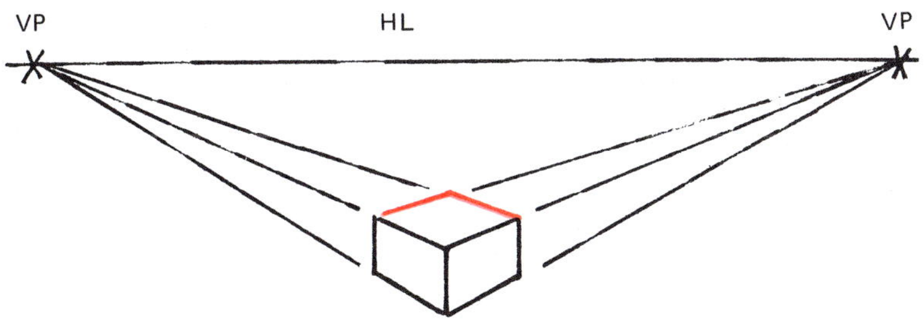

Fig. 11-31. Project from these intersections to the vanishing points.

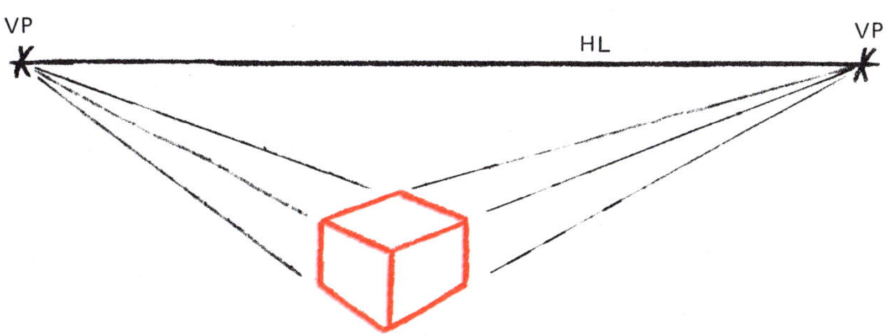

Fig. 11-32 Darken in the outlines of the object. Erase all projection lines.

UNIT 12 DEVELOPMENTS AND INTERSECTIONS

Surface development is used extensively in industries that fabricate three-dimensional objects from two-dimensional sheets of metal, paper, or fabric. Pattern developments and intersections must be projected in solving problems in the sheet metal industries for constructing heating, cooling, and air-conditioning units; in clothing design and manufacture; in piping transition pieces; and in package design and construction. Machine designers use pattern developments for machinery guards, feed hoppers, and chutes, and to design jig and fixture bodies and machine parts made from heavy plate that must be joined by welding.

DEVELOPMENT

Development is the drawing of all of the surfaces of an object on a plane to make a pattern which, if folded or rolled, forms the required object.

The development of a cube is illustrated in figure 12-1. A pictorial drawing is shown at A. At B the cube is opened at the seams. The development on a plane is shown at C.

The surface patterns are bent or rolled to form the required shape and then fastened with suitable seams. Allowance for the various types of seams must be computed and added to the patterns. However, the surface patterns in this book do not show the seam allowance because the seam allowance differs according to the type of seam used. Also, the allowance varies with different thicknesses of material.

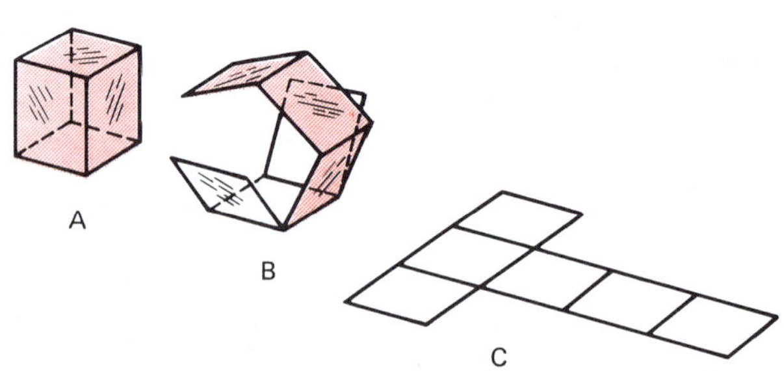

Fig. 12-1 The development of a cube

TERMS

- A *prism* is a solid with bases or ends that are equal parallel polygons and faces that are parallelograms.
- A *parallelogram* is a four sided figure whose opposite sides are equal and parallel.
- A *right prism* is one in which all of its sides are rectangles.
- A *truncated prism* is the part left when one end is cut off by a cutting plane.

The lines where the pattern is to be folded or bent are called *bend lines.* They are marked with a small, freehand cross at each end of the line, figure 12-2.

There are three basic types of development problem:

- those having parallel lines that are projected by using parallel line development.
- those, such as cones or pyramid shapes, having elements that extend to meet at an apex that is used in radial line development.
- those transition pieces whose form changes from one type of basic geometric shape to another and are developed through a *triangulation* method.

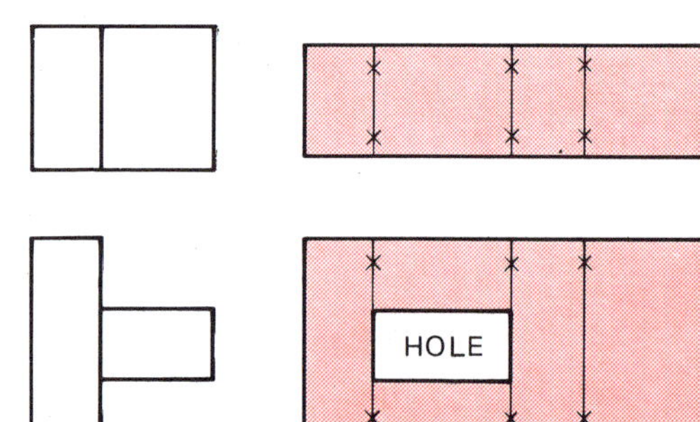

Fig. 12-2 The development of intersecting rectangular pipes

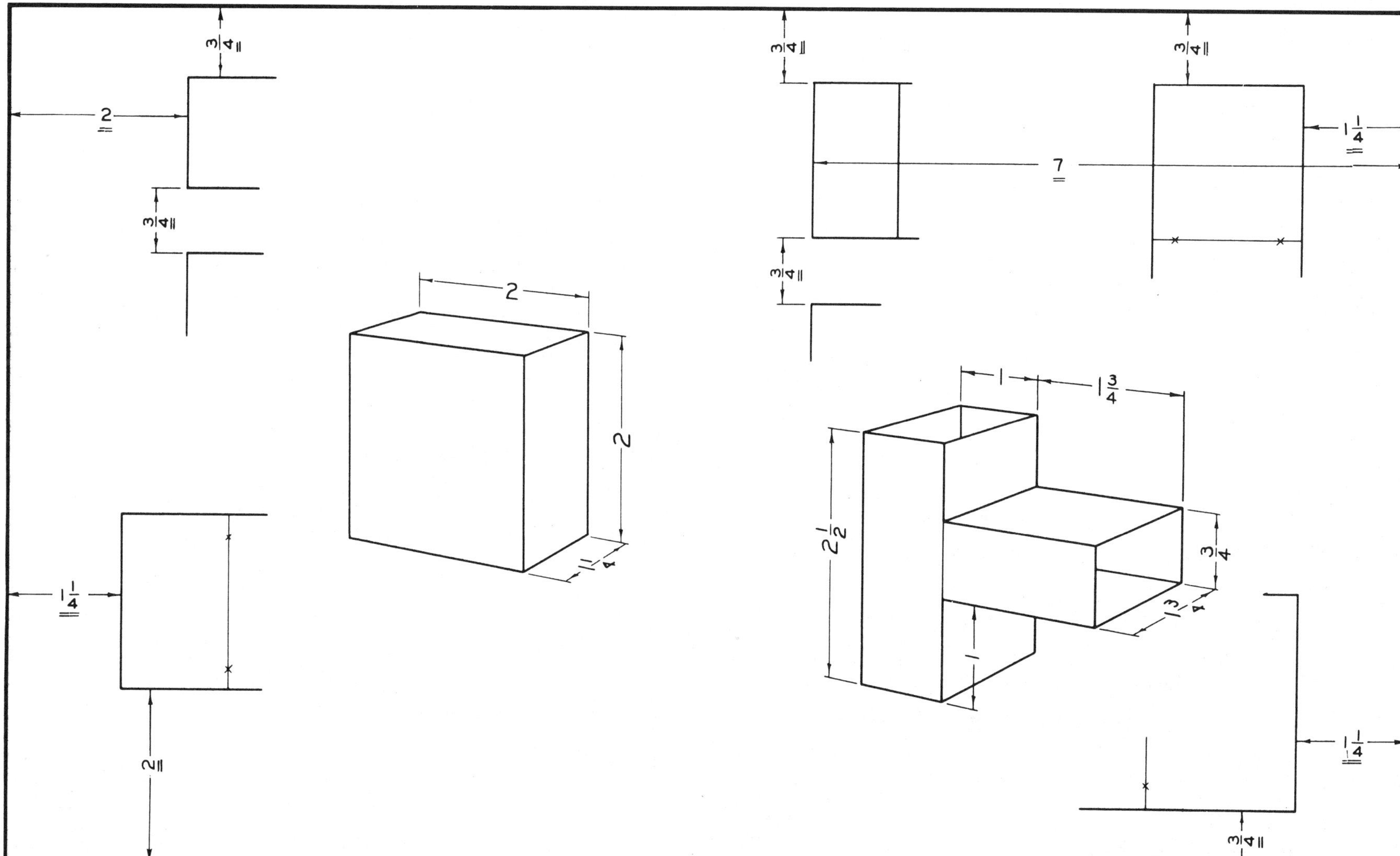

NAME OF SCHOOL CITY OR TOWN	RECTANGULAR PRISM RECT. PIPE INTERSECTION	SHEET No. 12-A NAME	CLASS DATE

Fig. 12-3

INTERSECTIONS

Intersections are the points and edges located where the surfaces on one part join or cut the surfaces of another part. The lines of intersections must be determined before the surfaces of intersecting parts are developed.

Figure 12-2 pictures two intersecting rectangular pipes, a top and a front view of the pipes, and the development of each pipe.

ASSIGNMENT A

Make a two-view drawing of the prism shown in figure 12-3. Make a development drawing of its surfaces.

Draw a top and front view of the intersecting rectangular pipes in figure 12-3. Make the development drawings of each pipe.

Procedure

1. Lay out the margin lines on the drawing sheet.
2. Draw the top view and the front view of the rectangular prism. The underscored dimensions show the location of these views on the drawing. These underscored locating dimensions should not be on the completed drawing.
3. Draw the development of the surfaces of the rectangular prism. The patterns for the ends may be attached to any one of the sides. The bend lines are marked at each end with a small cross.
4. Draw the top view and the front view of the intersecting rectangular pipes.
5. Draw the development of each pipe.
6. Check the drawing for the development of the rectangular prism and intersecting rectangular pipes.

REVIEW QUESTIONS

1. What is meant by the term "development"?
2. What are intersections?
3. How are bend lines indicated on the pattern?

TRUNCATED PRISM

The development of a truncated square prism is shown pictorially in figure 12-4. A *truncated prism* is the part left when one end is cut off. A drawing and the development of the same truncated prism is shown in figure 12-5. The drawing, shows a bottom view, a front view, and a frontal auxiliary view. The auxiliary view is drawn to show the true shape and size of the upper sloping surface. The development of the truncated prism is obtained by projecting lines from the front view.

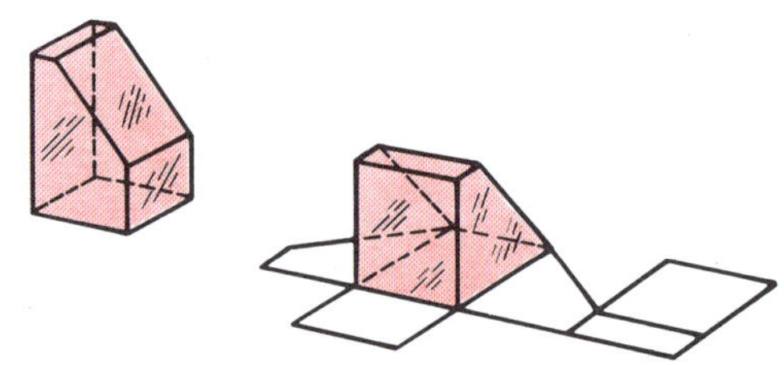

Fig. 12-4 Pictorial development of a truncated square prism

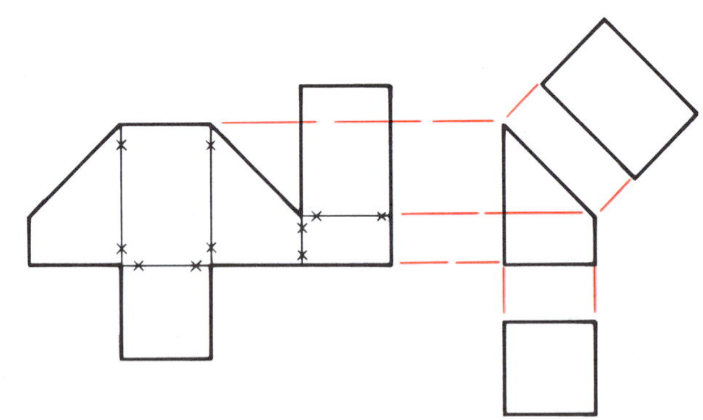

Fig. 12-5 Drawing and development of a truncated square prism

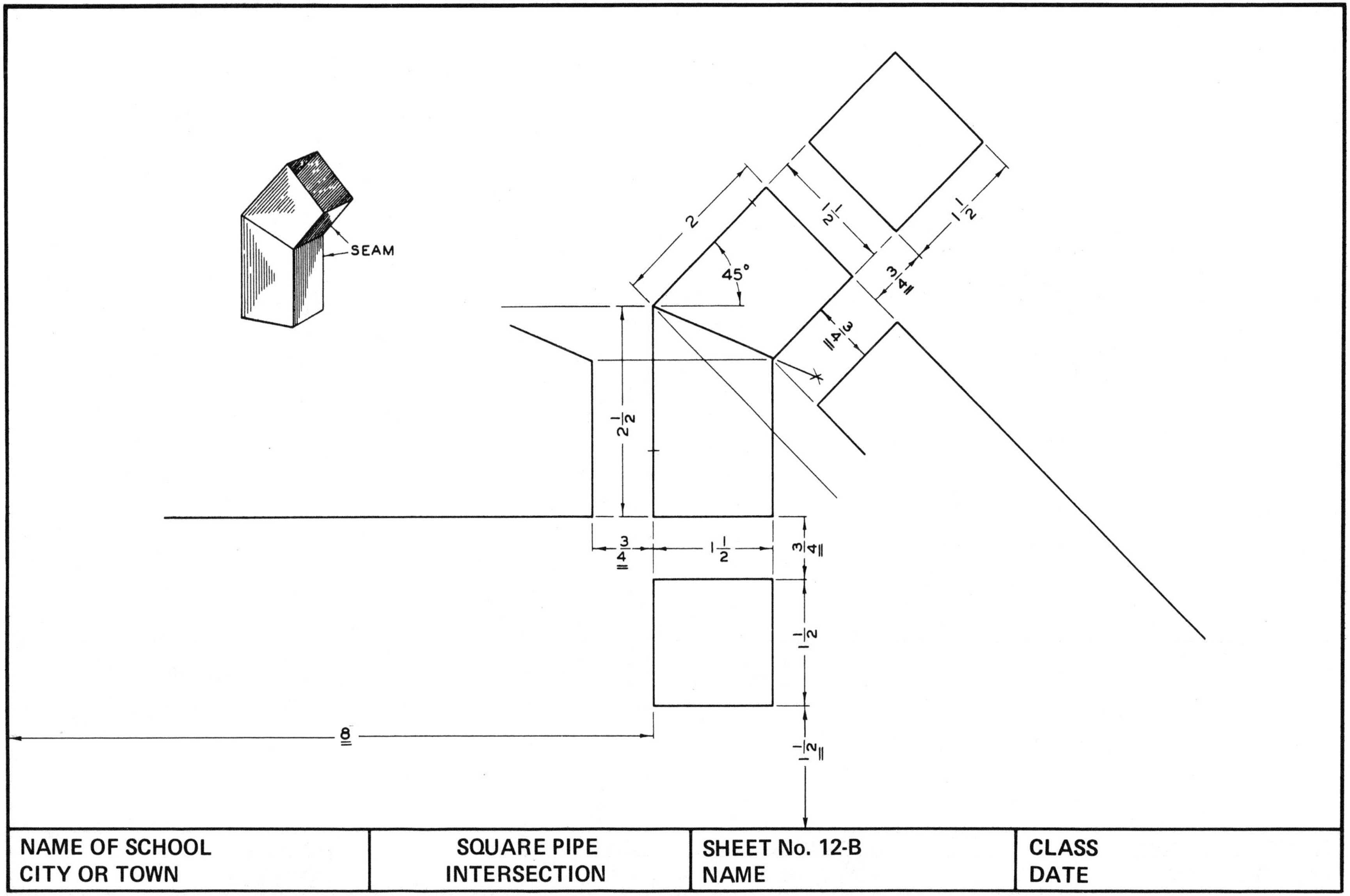
SEAM
2
45°
$1\frac{1}{2}$
$1\frac{1}{2}$
$\frac{3}{4}''$
$\frac{3}{4}''$
$2\frac{1}{2}$
$\frac{3}{4}''$
$1\frac{1}{2}$
$\frac{3}{4}''$
$1\frac{1}{2}$
$8''$
$1\frac{1}{2}''$
NAME OF SCHOOL
CITY OR TOWN
SQUARE PIPE
INTERSECTION
SHEET No. 12-B
NAME
CLASS
DATE

Fig. 12-6

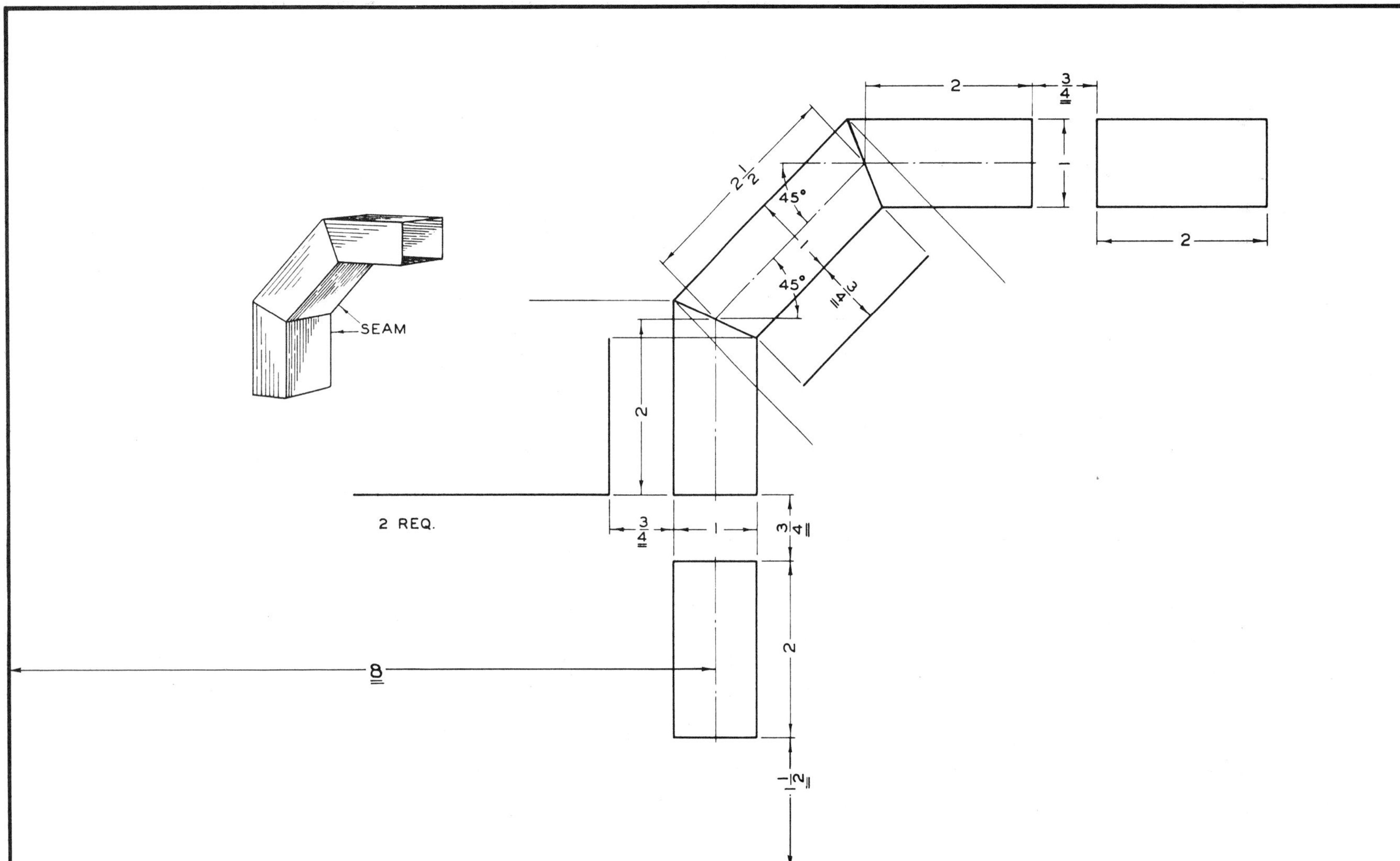

NAME OF SCHOOL CITY OR TOWN	RECTANGULAR PIPE ELBOW	SHEET No. 12-C NAME	CLASS DATE

Fig. 12-7

Two intersecting square pipes are shown in figure 12-6. Each pipe is a truncated square prism. The development of each pipe is drawn in the same manner as pictured in figure 12-5.

ASSIGNMENT B

Draw the bottom view, the front view, and the frontal auxiliary view of the intersecting square pipes.

Lay out the development or surface pattern of each pipe.

Procedure

1. Lay out the margin lines on the drawing sheet and letter the title strip.
2. Draw the bottom view of the upright pipe, locating it on the drawing as shown by the underscored dimensions. When orthographic drawings are used to develop surface patterns, it is not always necessary to draw complete views. Therefore, the entire pipe intersection is not shown in the bottom view. Only the shape of the upright pipe and its length and width are shown.
3. Draw the front view of both pipes. The line of intersection is obtained by bisecting the angle of bend.
4. Draw the frontal auxiliary view of the slanting pipe. This view shows the shape, length, and width of the pipe.
5. Draw the surface pattern of the upright pipe to the left of the front view.
6. Draw the surface pattern of the slanting pipe to the right and perpendicular to the front view. Lay out the patterns so that the seams on both pipes are at the same corner when the pipes are joined.
7. Complete the drawing. Check each view and the development of the surface pattern.

REVIEW QUESTIONS

1. What is a truncated prism?
2. Must complete views for orthographic drawings be drawn to develop surface pattern? Justify the answer.

RECTANGULAR PIPE ELBOW

The rectangular pipe elbow, figure 12-7, is made in three sections. The two end sections are rectangular prisms truncated at one end. The center section is a rectangular prism truncated at both ends.

The surface patterns of each pipe section are developed in the same manner as on the preceding drawing. Because both end sections have the same dimensions, only one surface pattern is required for them.

ASSIGNMENT C

Draw a view of the bottom end, the front view, and the right end view of the rectangular pipe elbow.

Develop the surface pattern of an end section and the center section.

Procedure

1. Lay out the margin lines on the drawing sheet and letter the title strip.
2. Draw a view of the bottom end. Draw the front view and the right end view.
3. Draw the surface pattern of the lower end section to the left of the front view. Lay out the patterns so that the seams are at the corners designated in the picture on the drawing.
4. Draw a surface pattern of the center section to the right and perpendicular to the front view of this section.
5. Complete and check the drawing.

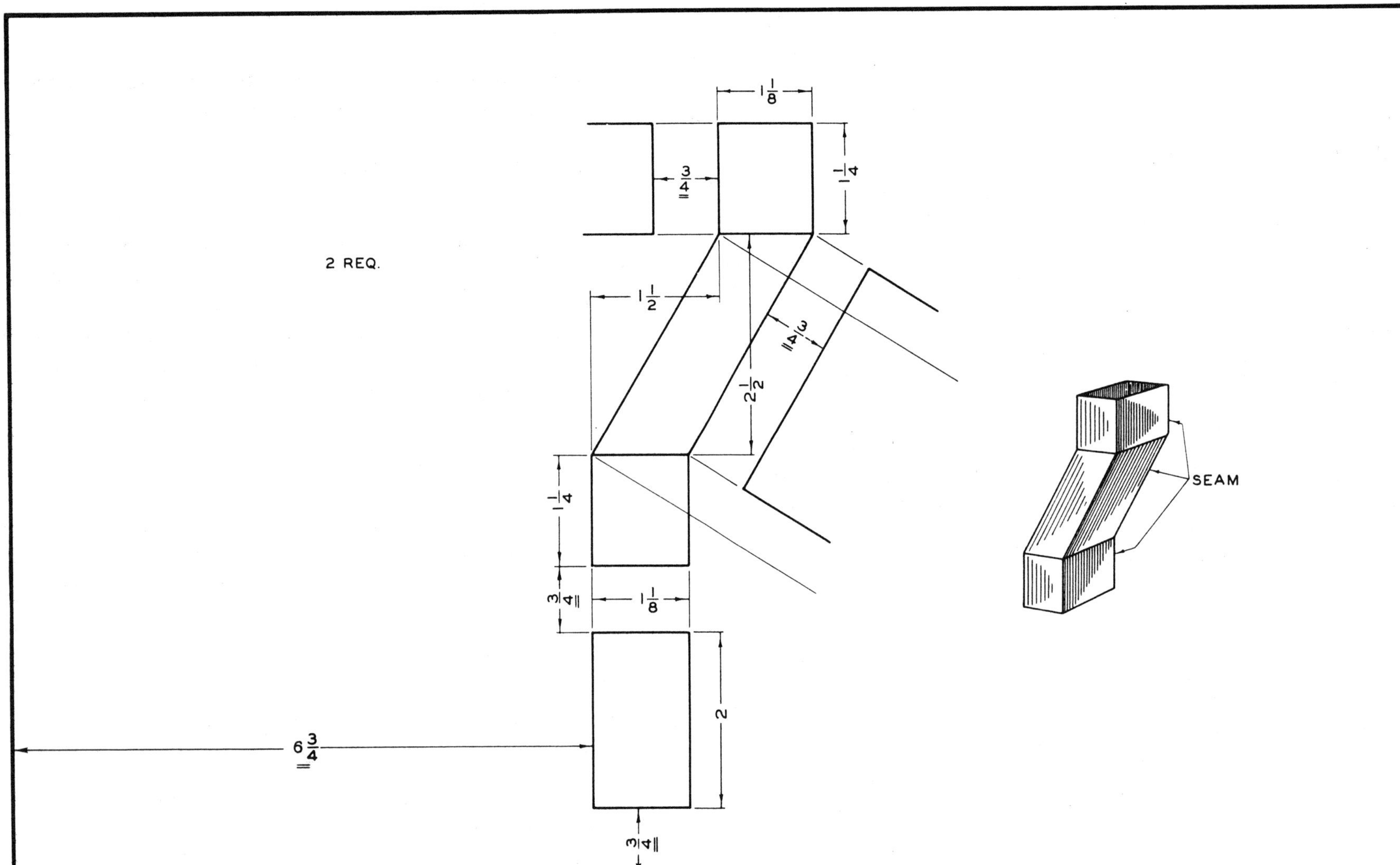

NAME OF SCHOOL CITY OR TOWN	RECTANGULAR PIPE OFFSET	SHEET No. 12-D NAME	CLASS DATE

Fig. 12-8

REVIEW QUESTIONS

1. What is a prism?
2. What is a right prism?
3. What is a truncated prism?

OBLIQUE PRISM

The rectangular pipe offset in figure 12-8 is made in three sections. The upper and lower sections are right rectangular prisms because the ends are equal rectangles and all of the sides are rectangles.

The center section is an oblique rectangular prism. An *oblique prism* is one whose lateral edges are not perpendicular to its base. All of its sides are not rectangles.

An oblique square prism and its development is pictured in figure 12-9. A front view, a view of the bottom end, and the development of the same oblique square prism are drawn in figure 12-10. In the development of the oblique square prism, all of the upper and lower edges are of equal length. Notice also that the two rectangles and the other two parallelograms are not the same width. This difference is pointed out so that care is taken when drawing the surface patterns of oblique prisms to make the edges the correct length.

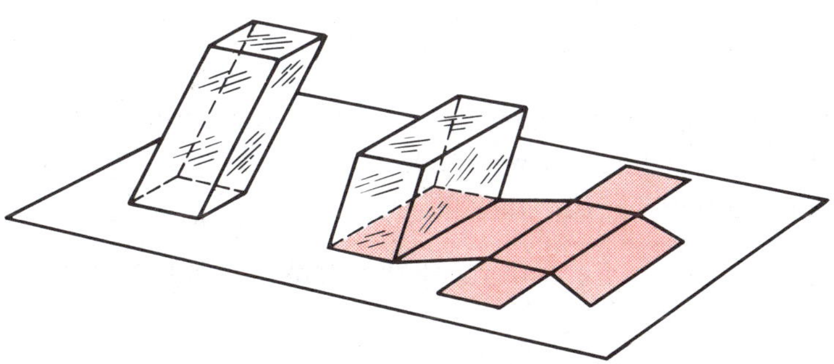

Fig. 12-9 Pictorial development of an oblique prism

ASSIGNMENT D

Draw a view of the bottom end and front of the rectangular pipe offset.

Develop the surface patterns of the upper section and the center section.

Procedure

1. Lay out the margin lines on the drawing sheet.
2. Draw a view of the bottom end. Draw the front view.
3. Draw the surface pattern of the upper section to the left of the front view. Lay out the patterns to have the seams at the corners designated in the picture on the drawing.
4. Draw the surface pattern of the center section to the right and perpendicular to the front view of this section. Be sure that the upper and lower edges of the surface pattern are the correct lengths.
5. Complete the drawing. Check the accuracy of the working drawing and the development of the surface pattern.

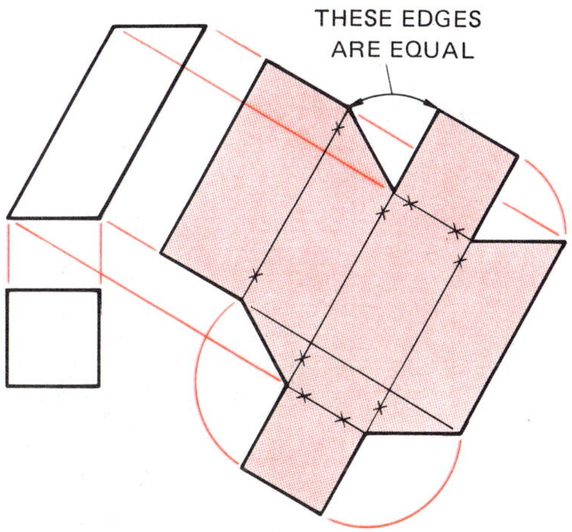

Fig. 12-10 Drawing and development of an oblique prism

HEXAGONAL PRISMS

The surface patterns of hexagonal prisms are developed as shown in figure 12-11. A drawing and development of a truncated hexagonal prism is illustrated in figure 12-12. A frontal auxiliary view is required to obtain the pattern of the upper sloping surface. The reference plane method is used to obtain the width of the auxiliary view.

An octagonal stack or pipe intersecting a roof flange is described in figure 12-13. The stack is a truncated octagonal prism and its surface pattern is developed the same as figure 12-12. The surface pattern of the roof flange is shown when the frontal auxiliary view is drawn. The horizontal center line in the top view and the center line of the auxiliary view are used as reference lines to obtain the width of the opening for the stack.

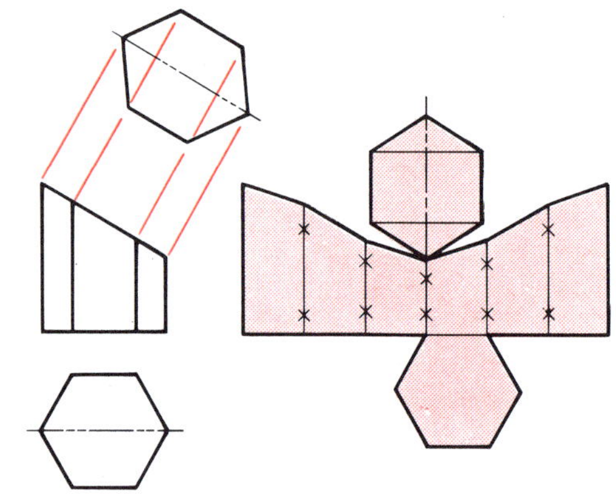

Fig. 12-12 Drawing and development of a truncated hexagonal prism

ASSIGNMENT E

Draw a top view and a front view of the octagonal stack and roof flange.

Develop the surface patterns of each piece.

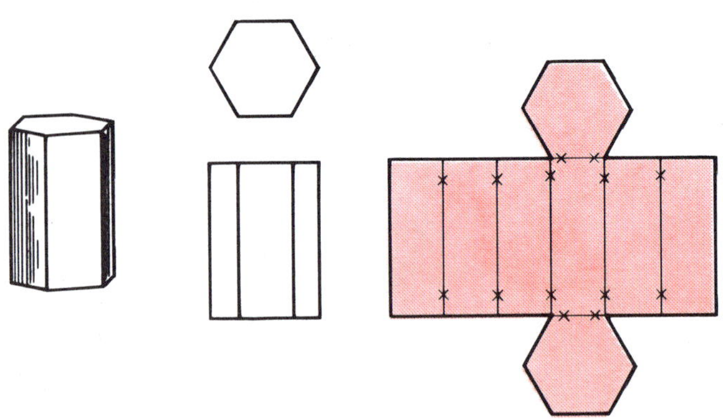

Fig. 12-11 Drawing and development of a hexagonal prism

Procedure

1. Lay out the margin lines on the drawing sheet and letter the title strip.
2. Draw the center lines for the top view.
3. Draw a circle. Draw the octagon about the circle.
4. Draw the front view of the stack and roof flange.
5. Draw the roof flange in the top view.
6. Draw the frontal auxiliary view of the roof flange. Use the center lines as reference lines to lay out the width of the opening for the stack.
7. Draw the surface pattern of the stack to the left of the front view.
8. Complete the views. Check the accuracy of the top and front view and the development of the surface patterns.

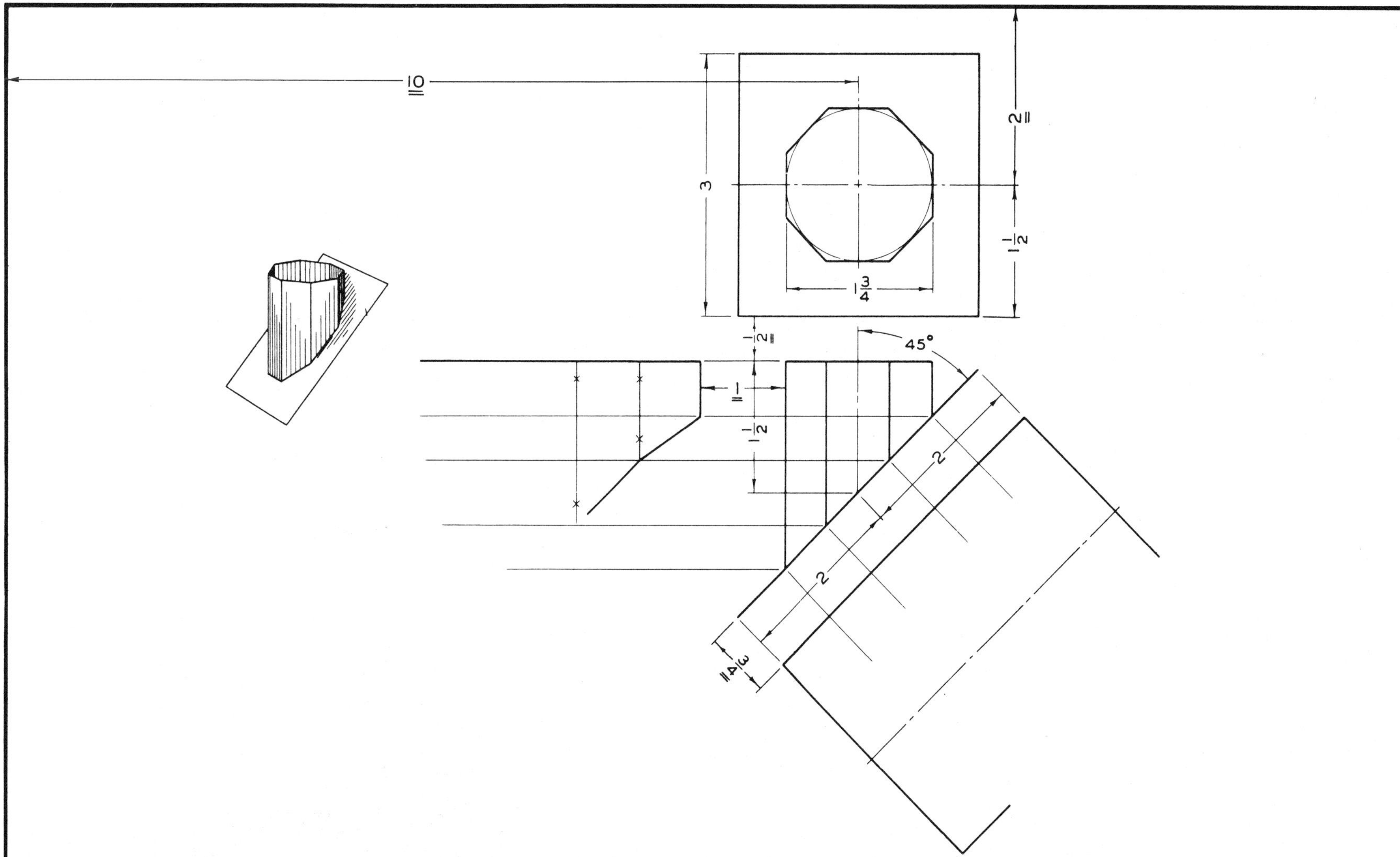

NAME OF SCHOOL CITY OR TOWN	OCTAGONAL STACK INTER- SECTING A ROOF FLANGE	SHEET No. 12-E NAME	CLASS DATE

Fig. 12-13

CYLINDERS

The development of a cylinder is shown pictorially in figure 12-14. The drawing and the development of the same cylinder is covered in figure 12-15.

To develop the surface pattern of a cylinder, consider the cylinder as being a prism having a large number of sides. In figure 12-15, the curved surface is divided into sixteen equal parts. This curved surface is then stretched out on a plane and the sixteen parts are accurately pointed off to obtain the length of the pattern. The length of the pattern should be approximately equal to the circumference of the circle (3.1416xD).

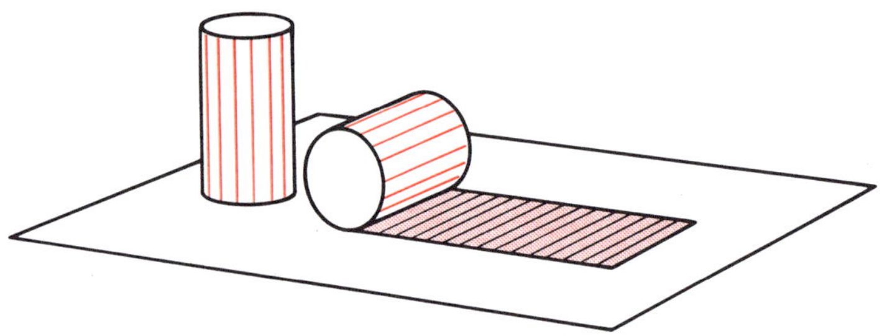

Fig. 12-14 Pictorial development of a cylinder

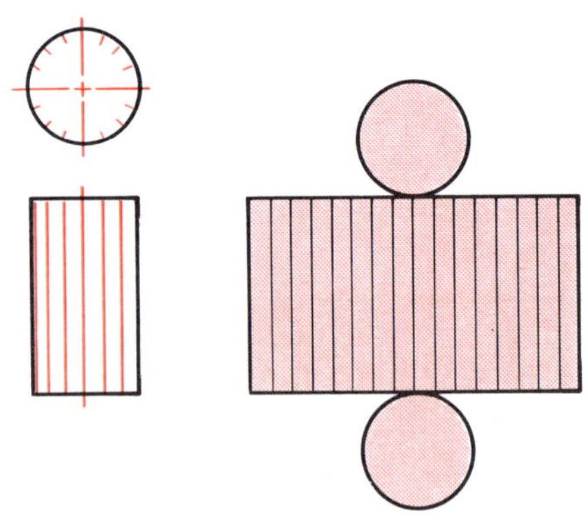

Fig. 12-15 Drawing and development of a cylinder

A truncated cylinder is developed in figure 12-16. The lines on the sides of the cylinder are called *elements.* The length of each element in the pattern is obtained by projecting from the corresponding element in the front view. A frontal auxiliary view shows the pattern of the sloping surface.

The four-piece elbow in figure 12-17 is made up of sections which are truncated cylinders. The surface patterns of each section are developed as in figure 12-16.

TERMS

- A *cylinder* is a solid of uniform diameter having bases or ends which are equal, parallel circles.
- A *right cylinder* is one whose elements are perpendicular to its base.
- An *oblique cylinder* is one whose elements are not perpendicular to its base.
- A *truncated cylinder* is the part left when an end is cut off a cylinder by a cutting plane.

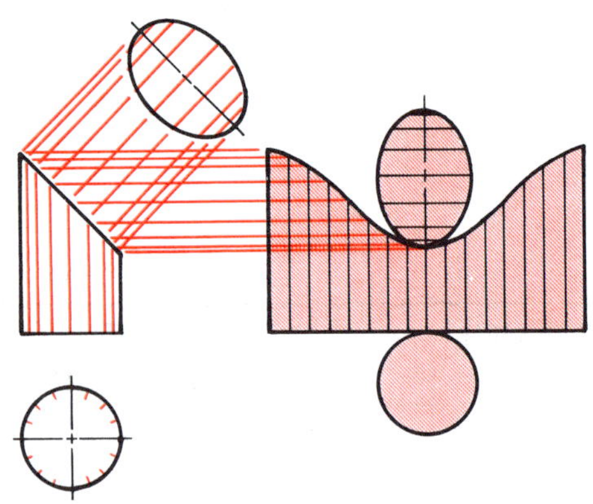

Fig. 12-16 Drawing and development of a truncated cylinder

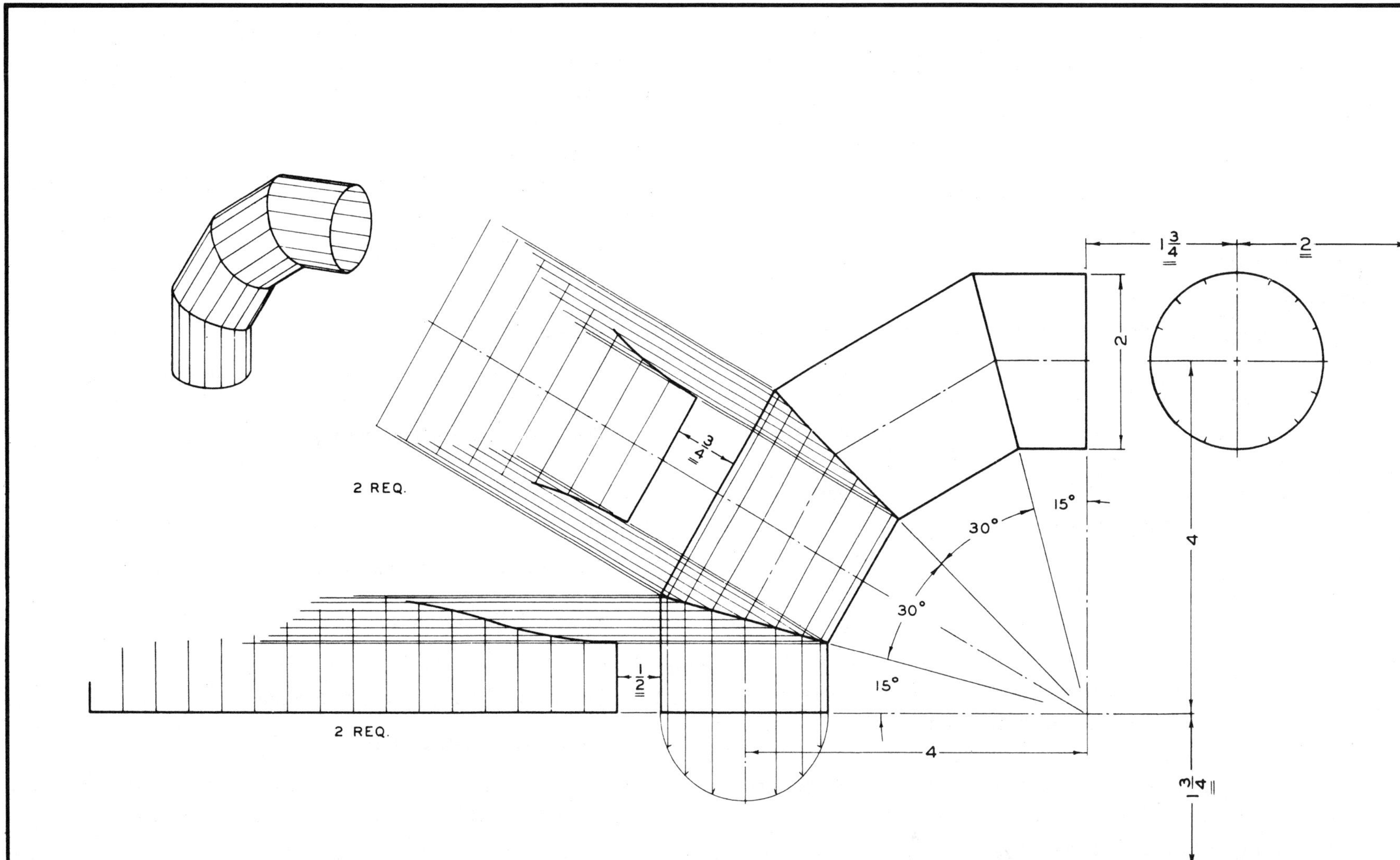

NAME OF SCHOOL CITY OR TOWN	FOUR-PIECE ELBOW	SHEET No. 12-F NAME	CLASS DATE

Fig. 12-17

ASSIGNMENT F

Draw a right end view and a front view of the four-piece elbow.

Develop a surface pattern of one end section and one center section.

Procedure

1. Lay out the margin lines on the drawing sheet.
2. Draw the right end view.
3. Draw the front view.
4. Divide the circle into sixteen equal parts.
5. Draw a semicircle at the bottom end and divide it into eight equal parts.
6. Draw the elements on the two lower sections of the elbow.
7. Divide the circle into sixteen equal parts. With the dividers then point off these sixteen spaces to the left of the front view to obtain the length of the pattern. Check the length of the pattern using formula 3.1416 x D.
8. Point off these sixteen equal spaces to the left and perpendicular to the next section of the elbow.
9. Draw the elements for both patterns.
10. Project from the ends of each element in the front view to find the ends of the corresponding elements in the patterns.
11. Using the irregular curve, connect the ends of the elements with smooth curves. Only one pattern is required for the end sections because they are identical. Likewise, the center sections are identical so only one pattern is developed.
12. Complete the drawing. Check the accuracy of the right end view, front view, and surface pattern.

ROUND PIPE INTERSECTING A ROOF FLANGE

The round pipe in figure 12-18 is a variation of a truncated cylinder. Its surface pattern is developed in the same manner as described on the last drawing.

The roof flange in the front view is drawn with two straight lines at a 90-degree angle. Imagine that one-half of the flange is bent back to make a flat surface as indicated by the two arcs. The front view of the flange would then be a straight, slanting line. A frontal auxiliary view projected from this line is the surface pattern.

A semicircle divided into eight equal parts is used to obtain the width of the pipe opening. The diameter of this semicircle is the same as that of the pipe.

ASSIGNMENT G

Draw a top and front view of the round pipe intersecting a roof flange.

Develop the surface pattern of each piece.

Procedure

1. Lay out the margin lines on the drawing sheet.
2. Draw the top view.
3. Draw the front view.
4. Draw the surface pattern of the pipe.
5. Draw the surface pattern of the flange.
6. Complete the drawing. Check accuracy of all lines and dimensions on both views and the surface patterns.

REVIEW QUESTIONS

1. What is a cylinder?
2. What is a right cylinder?
3. What is an oblique cylinder?

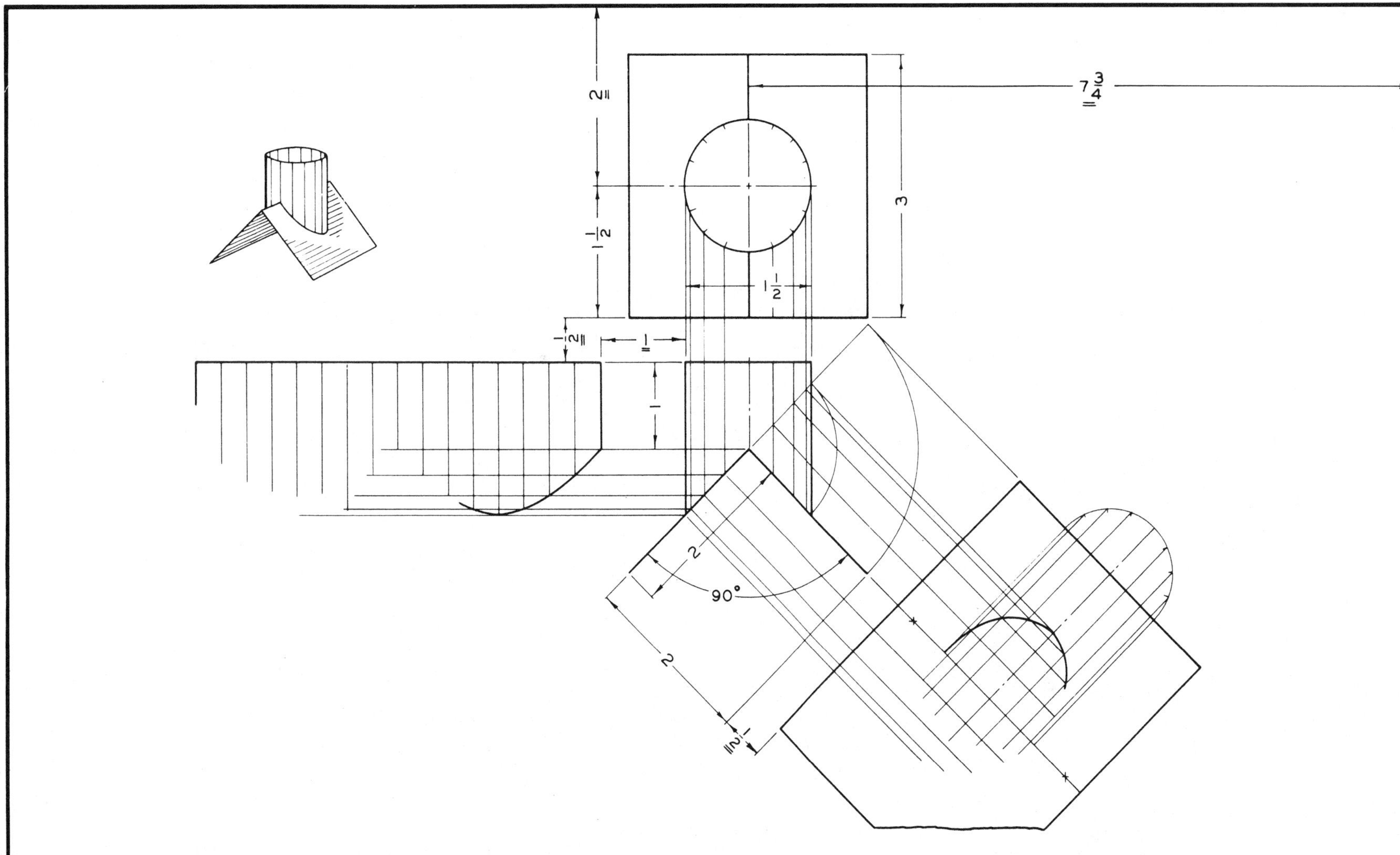

NAME OF SCHOOL CITY OR TOWN	ROUND PIPE INTERSECTING A ROOF FLANGE	SHEET No. 12-G NAME	CLASS DATE

Fig. 12-18

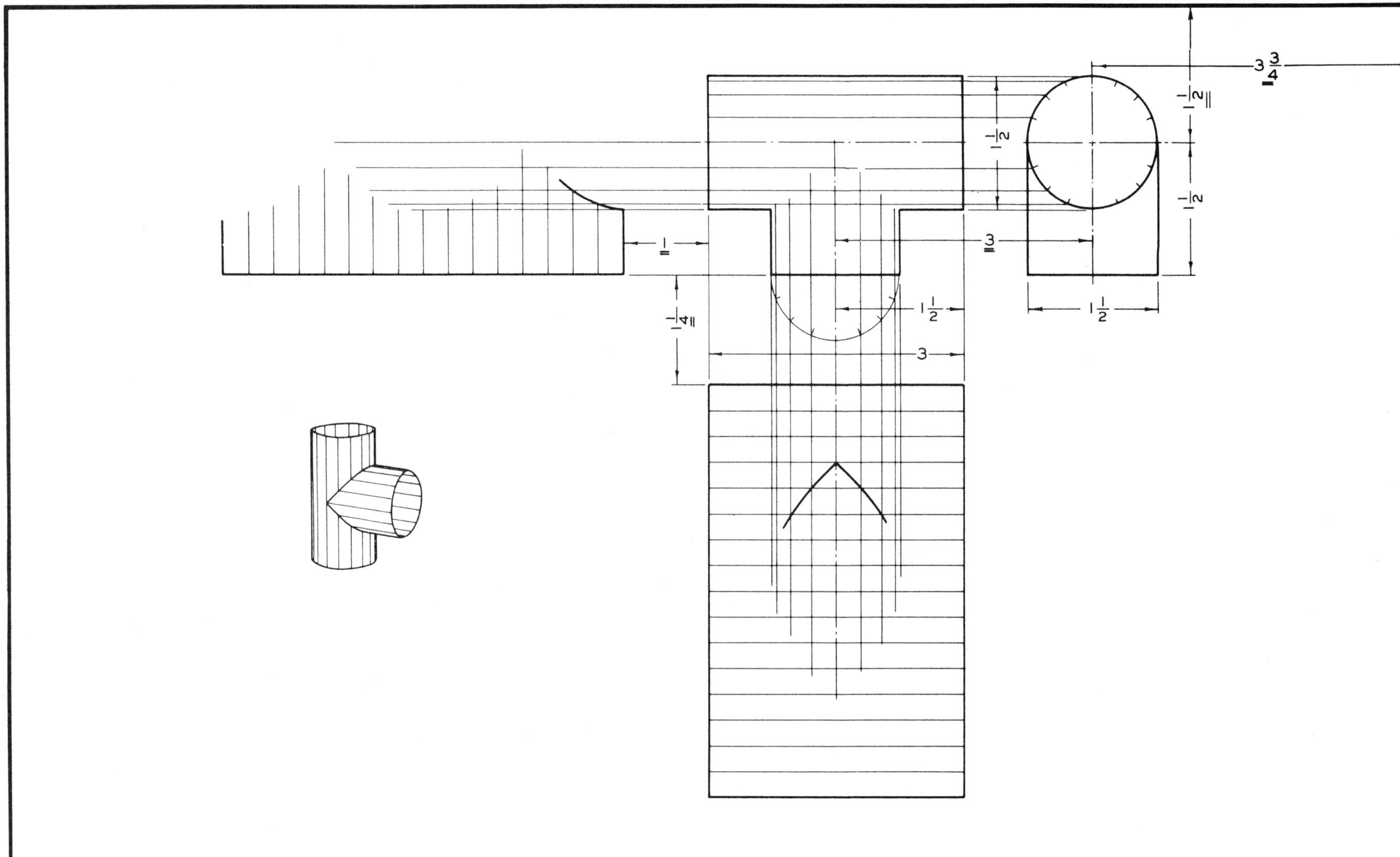

NAME OF SCHOOL CITY OR TOWN	T-JOINT	SHEET No. 12-H NAME	CLASS DATE

Fig. 12-19

T-JOINT

The T-joint pictured in figure 12-19 is made of two cylindrical pipes intersecting at a right angle. Before the surface patterns can be developed, the line of intersection must be determined.

When cylinders of equal diameters intersect, the equally spaced elements on both cylinders intersect. The points where the elements intersect are connected to find the line of intersection.

The surface pattern of the vertical pipe is drawn to the left of the front view. The length of each element in the pattern is found by projecting from the corresponding element in the front view. The ends of the elements are then connected with a smooth curve.

ASSIGNMENT H

Draw a right-side view and a front view of the T-joint.
Develop the surface pattern of each cylindrical pipe.

Procedure

1. Lay out the margin lines on the drawing sheet.
2. Draw the right-side view. Divide the circle into sixteen equal parts.
3. Draw the front view. Draw a semicircle at the end of the vertical pipe. Divide this semicircle into eight equal parts. This semicircle is used in place of the end view of the vertical pipe.
4. Draw the elements on both pipes in the front view. Draw the line of intersection.
5. Draw the surface pattern of the vertical pipe.
6. Draw the surface pattern of the horizontal pipe. Locate the necessary points and draw the opening in this pattern.
7. Complete the drawing. Check the accuracy of each regular view and the surface pattern of the cylindrical pipe.

REVIEW QUESTIONS

1. What is a T-joint?
2. How is the intersection of the pipes found in the front view?

ROUND PIPE JOINT

Figure 12-20 involves two cylindrical pipes of equal diameters intersecting at other than a right angle. When cylinders of equal diameter intersect, the equally spaced elements on both cylinders intersect. The points where the elements intersect are connected to find the line of intersection in the front view.

ASSIGNMENT I

Draw a top and front view of the round pipe joint.
Develop the surface pattern of both pipes.

Procedure

1. Lay out the margin line on the drawing sheet.
2. Draw the top view of vertical pipe. Divide the circle into sixteen equal parts.
3. Draw the front view. Draw a semicircle at the end of the branch pipe and divide it into eight equal parts. This semicircle replaces an auxiliary view of the branch pipe.
4. Draw the elements on both pipes in the front view; then, the line of intersection.
5. Draw the branch pipe in the top view.
6. Draw the surface pattern of the vertical pipe to the left of the front view.
7. Draw the surface pattern of the branch pipe to the right and perpendicular to the front view of this pipe.
8. Dimension and complete the drawing. Check the accuracy of each regular view and the surface pattern.

REVIEW QUESTIONS

1. What is meant by intersecting cylinders?
2. What shape is the cut-out section of the vertical pipe?

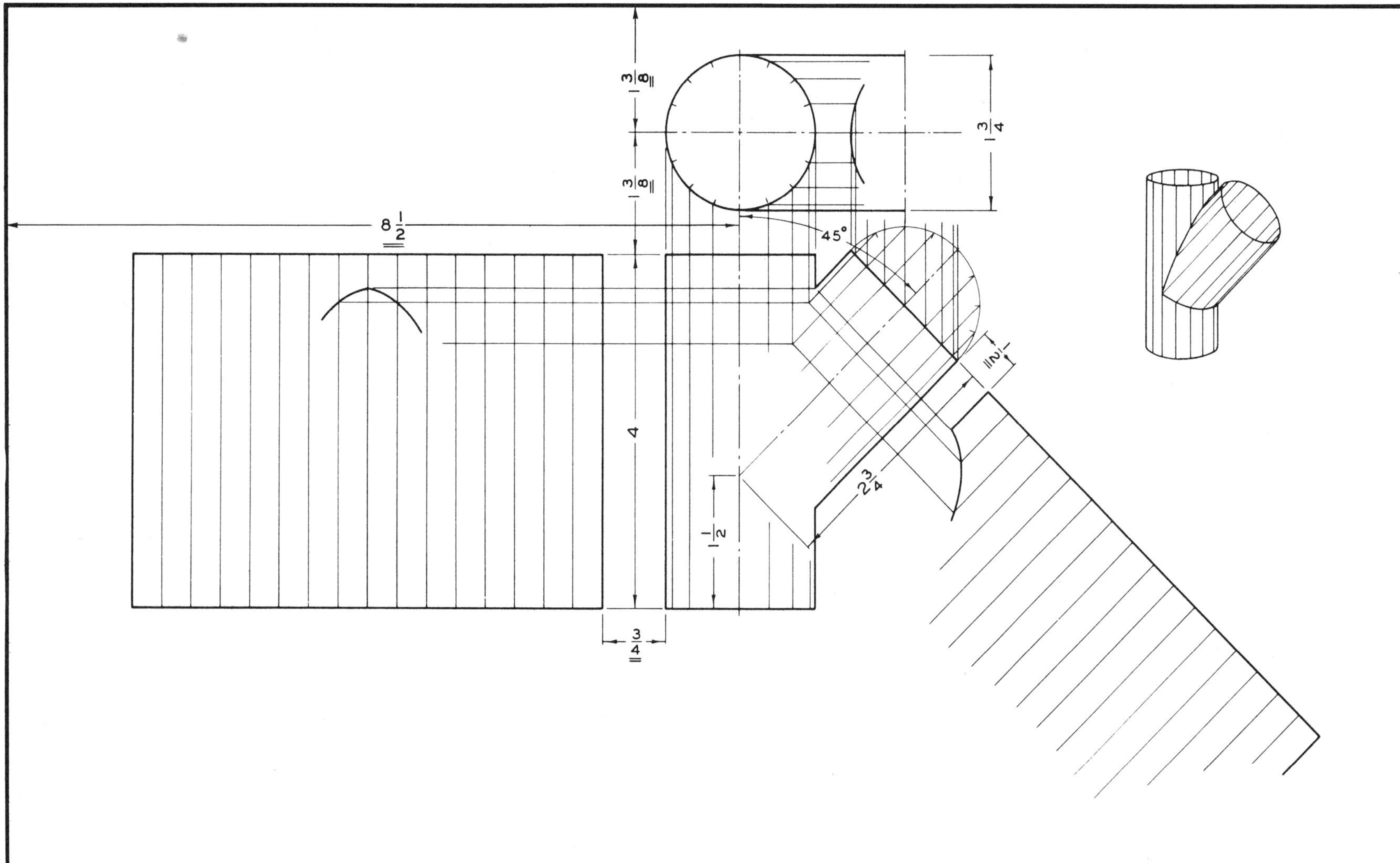

NAME OF SCHOOL CITY OR TOWN	ROUND PIPE JOINT	SHEET No. 12-I NAME	CLASS DATE

Fig. 12-20

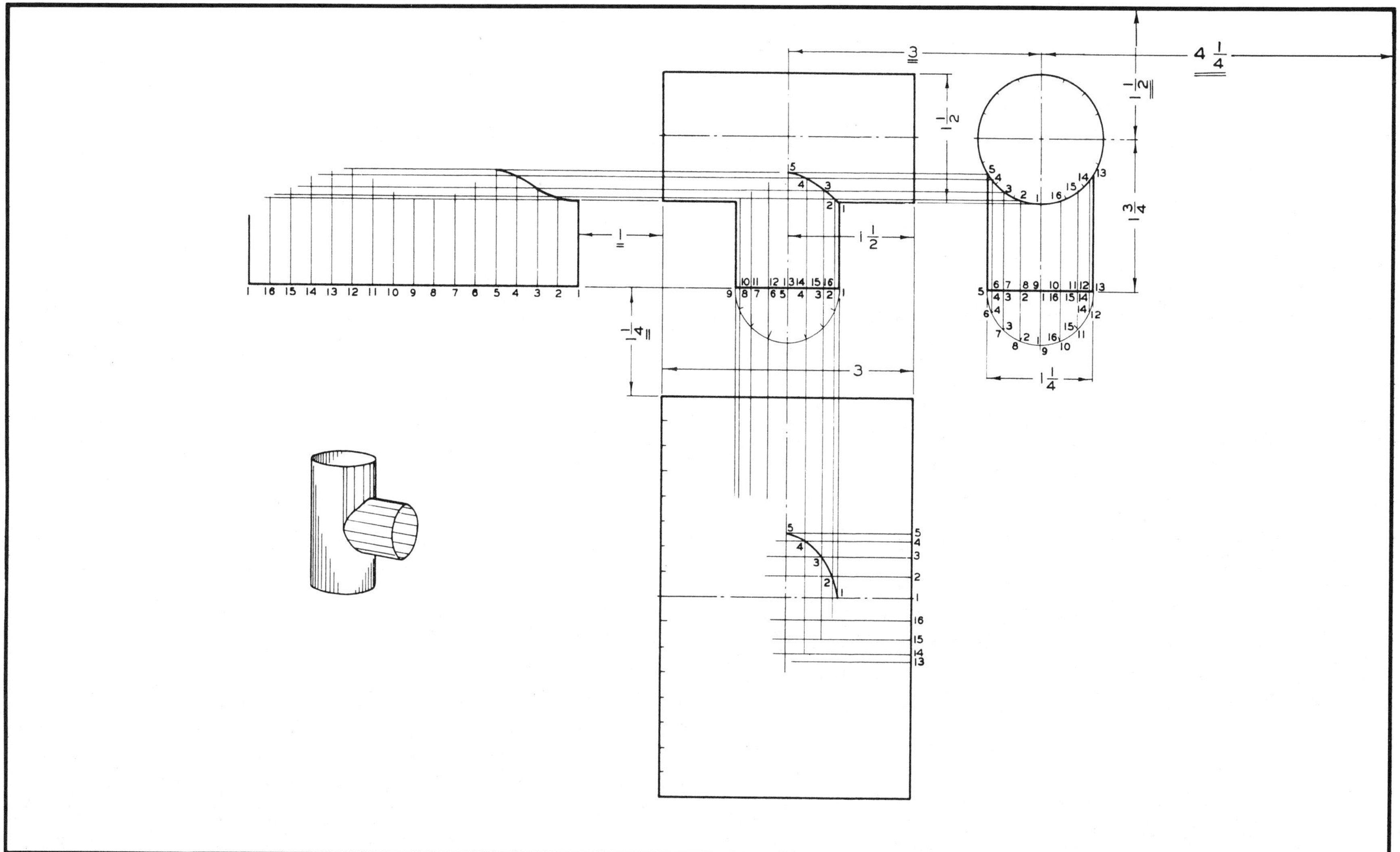

NAME OF SCHOOL CITY OR TOWN	T-JOINT	SHEET No. 12-J NAME	CLASS DATE

Fig. 12-21

T JOINT USING UNEQUAL DIAMETER PIPES

This T joint, figure 12-21, is made of two cylindrical pipes of unequal diameter intersecting at a right angle.

The smaller diameter branch pipe is divided into sixteen equal parts with elements drawn in both views. The length of each element is shown in the right-side view. Lines are projected from the upper end of each element in the side view to locate the upper end of each corresponding element in the front view. When the upper end of each element is located in the front view, the line of intersection is drawn.

The projection lines are then continued to the left to locate the upper end of each corresponding element in the pattern. When the upper end of each element in the pattern is determined, these ends are connected with a smooth curve and the pattern is completed.

The pattern of the larger diameter main pipe is drawn beneath the front view. This pattern has a hole or opening in it for the branch pipe. To develop this opening, lines are projected down from the points of intersection in the front view. The chord 1-2 on the circle in the side view is measured with the dividers. A *chord* is a straight line between two points on a circle. This distance is marked off on each side of the horizontal center line in the pattern. The chord 2-3 is then marked off on each side, followed by chords 3-4 and 4-5. Horizontal lines drawn from these points intersect the vertical lines from the front view. The points of intersection are connected with a smooth curve to complete the opening in the pattern.

ASSIGNMENT J

Draw a right-side view and a front view of the T joint.
Develop the surface pattern of each pipe.

Procedure

1. Lay out the margin lines on the drawing sheet.
2. Draw the right-side view.
3. Draw a semicircle on the end of the branch pipe and divide it into eight equal parts. This semicircle is used in place of the end view of the branch pipe.
4. Draw the elements on the branch pipe. The elements are numbered from 1 to 16 to make it easier to find the line of intersection and the pattern.
5. Draw the outline of the front view.
6. Draw the elements on the branch pipe in this view. Then locate and draw the line of intersection.
7. Draw the surface pattern of the branch pipe to the left of the front view.
8. Draw the surface pattern of the larger diameter pipe beneath the front view.
9. Locate the necessary points and draw the opening in this pattern.
10. Dimension and complete the drawing. Check the accuracy of the two regular views and the development of the patterns.

ROUND PIPE JOINT USING UNEQUAL DIAMETERS

The round pipe joint in figure 12-22 is made of two cylindrical pipes of unequal diameter intersecting at an angle other than 90 degrees. The elements are drawn on the branch pipe in both views.

The points of intersection in the front view are located by projecting lines down from the left end of each element in the top view to the corresponding element in the front view. The points of intersection are then connected with a smooth curve.

The end of the branch pipe in the top view is an ellipse. Points for this ellipse are located by projecting lines up from the upper end of each element in the front view to the corresponding element in the top view. The points are then connected to complete the ellipse.

The pattern of the branch pipe is drawn to the right and perpendicular to the front view of this pipe.

The pattern of the main pipe is drawn to the left of the front view. The opening in this pattern is found in the same manner as on the last drawing. The chord 1-2 in the circle in the top view is

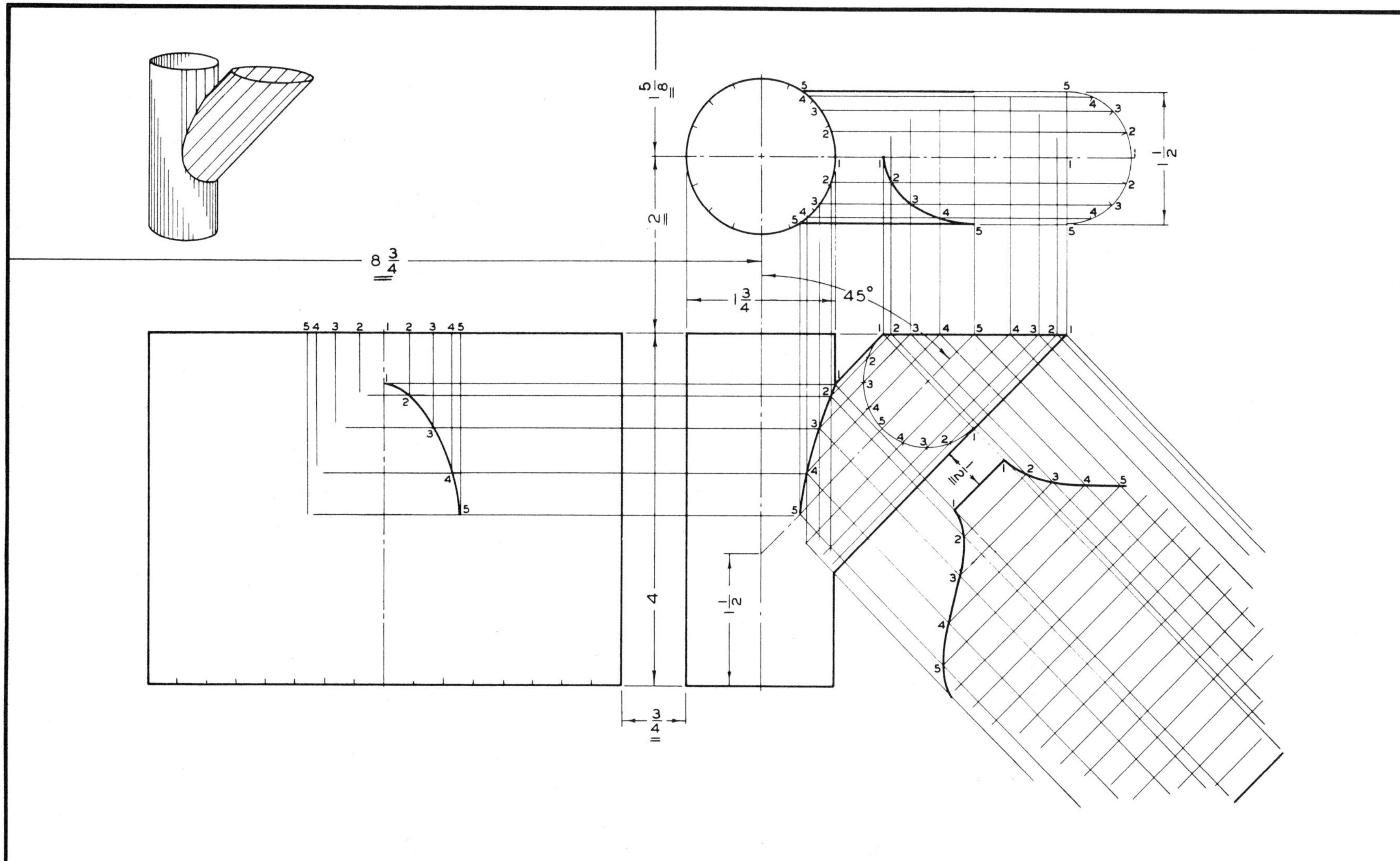

NAME OF SCHOOL CITY OR TOWN	ROUND PIPE JOINT	SHEET No. 12-K NAME	CLASS DATE

Fig. 12-22

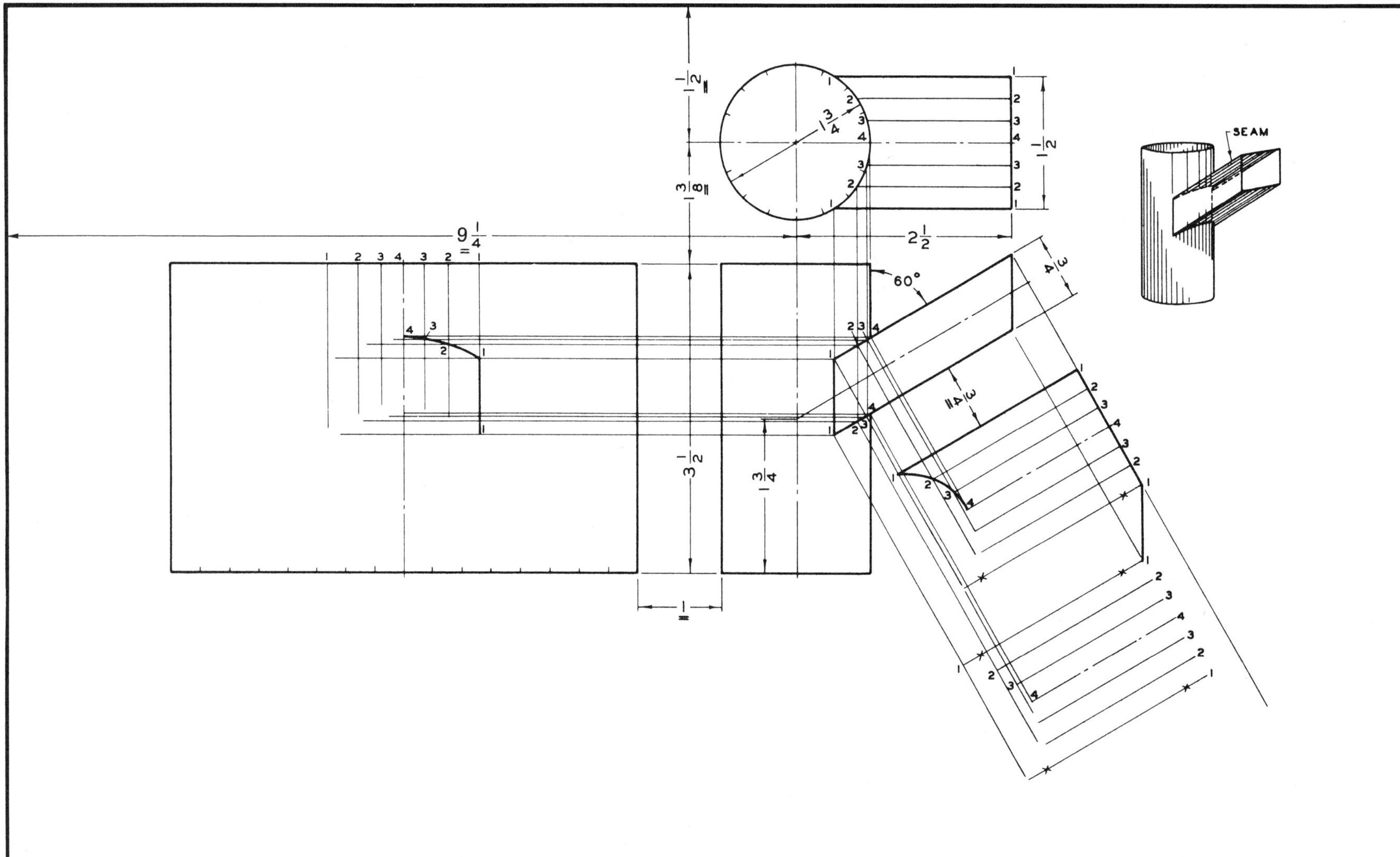

NAME OF SCHOOL CITY OR TOWN	ROUND AND RECTANGULAR PIPE INTERSECTION	SHEET No. 12-L NAME	CLASS DATE

Fig. 12-23

measured with the dividers and marked off on each side of the vertical center line in the pattern. The chord 2-3 is then marked off on each side, followed by chords 3-4 and 4-5. Lines are drawn down from these points. Next, horizontal lines are drawn to the left from the points of intersection in the front view to intersect the vertical lines. The points of intersection are connected with a smooth curve to complete the opening in the pattern.

ASSIGNMENT K

Draw a top view and a front view of the round pipe joint. Develop the surface pattern of each pipe.

Procedure

1. Lay out the margin lines on the drawing sheet.
2. Draw the circle in the top view.
3. Draw the outline of the front view.
4. Draw the elements on the branch pipe in both views. The semicircles are used to locate the elements.
5. Draw the line of intersection in the front view.
6. Complete the top view.
7. Draw the pattern of the branch pipe.
8. Draw the pattern of the main pipe.
9. Locate the necessary points and draw the opening in this pattern.
10. Complete the drawing. Check the accuracy of the regular top and front views, and the surface patterns of each pipe.

ROUND AND RECTANGULAR PIPE INTERSECTION

The pipe joint in figure 12-23 consists of a rectangular pipe intersecting a round pipe at an angle other than 90 degrees.

In the top view, the upper surface of the branch pipe is divided into six equal parts. Lines intersect the circle. These points of intersection are then projected down to the upper and lower surfaces of the branch pipe in the front view. The patterns are developed by projecting lines from the front view.

ASSIGNMENT L

Draw the top and front view. Develop the surface pattern of each pipe.

Procedure

1. Lay out the margin lines on the drawing sheet.
2. Draw the top view.
3. Draw the front view.
4. Draw the surface pattern of the branch pipe.
5. Draw the surface pattern of the round pipe. Measure the chords 4-3, 3-2, and 2-1 and mark off these distances on each side of the center line to obtain the length of the opening.
6. Complete the drawing. Check all elements for accuracy in surface patterns of the round and rectangular pipes.

REVIEW QUESTIONS

1. What is a prism?
2. What is a truncated prism?
3. What is a right cylinder?
4. What is an oblique cylinder?
5. What is a truncated cylinder?

RADIAL LINE DEVELOPMENT

The last group of drawings included a number of different forms of prisms and cylinders. In those drawings, the lateral edges of the prisms and the elements of the cylinders are parallel. The lines representing these edges and elements are also parallel in the surface patterns. Because these lines are parallel, the method used to develop the surface patterns is called *parallel line development.*

The drawings in this new group deal with various forms of pyramids and cones and the development of their surfaces. The edges on the sides of a pyramid and the elements of a cone meet at a point called the *apex.* When their surface patterns are developed, these lines also meet at a point. These lines are said to radiate from the point, and the method used to develop the surface of pyramids and cones is called *radial line development.* The pictorial development of a pyramid with lines radiating from the apex is shown in figure 12-24.

TRUE LENGTH OF LINES

A pictorial drawing of a square pyramid is given in figure 12-24. The edges meeting at the apex are oblique edges; that is, they are not parallel to the horizontal or vertical planes of projection. Consequently, these edges are foreshortened in the top and front views as shown in figure 12-25 at A.

To find the true length of the oblique foreshortened edges, the pyramid may be rotated to the position shown at B. In these views, the edges 2-5 and 4-5 are inclined to the horizontal plane

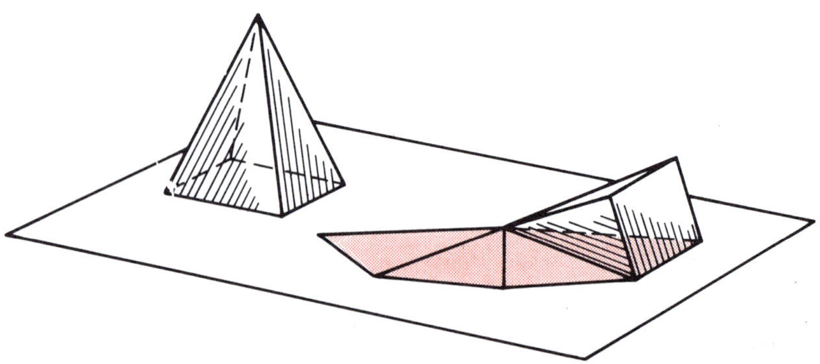

Fig. 12-24 Pictorial development of a pyramid

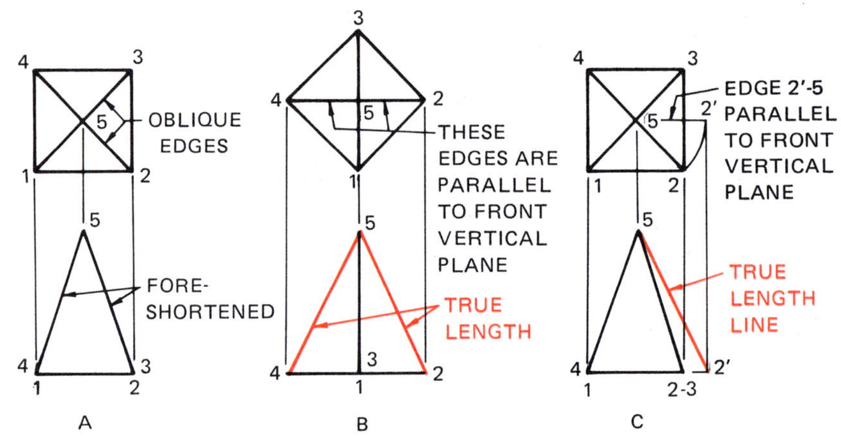

Fig. 12-25 The true length of a line

but parallel to the front vertical plane. Therefore, they are true lengths in the front view.

Another method commonly used to find the true length of the oblique edges is to assume that only one oblique edge is rotated parallel to the front vertical plane, as in the top view at C. This edge 2-5 is then shown in its true length, 2′-5, in the front view.

The true lengths of inclined or oblique edges must always be used when surface patterns are drawn. The drawing should be checked against the following statements to visualize and distinguish the true length lines from the foreshortened lines.

- When a line is horizontal in the top view, the corresponding line in the front view is the true length.
- When a line is horizontal in the front view, the corresponding line in the top view is the true length.

Conversely,

- When a line is slanting in the top view, the corresponding line in the front view is foreshortened.
- When a line is slanting in the front view, the corresponding line in the top view is foreshortened.

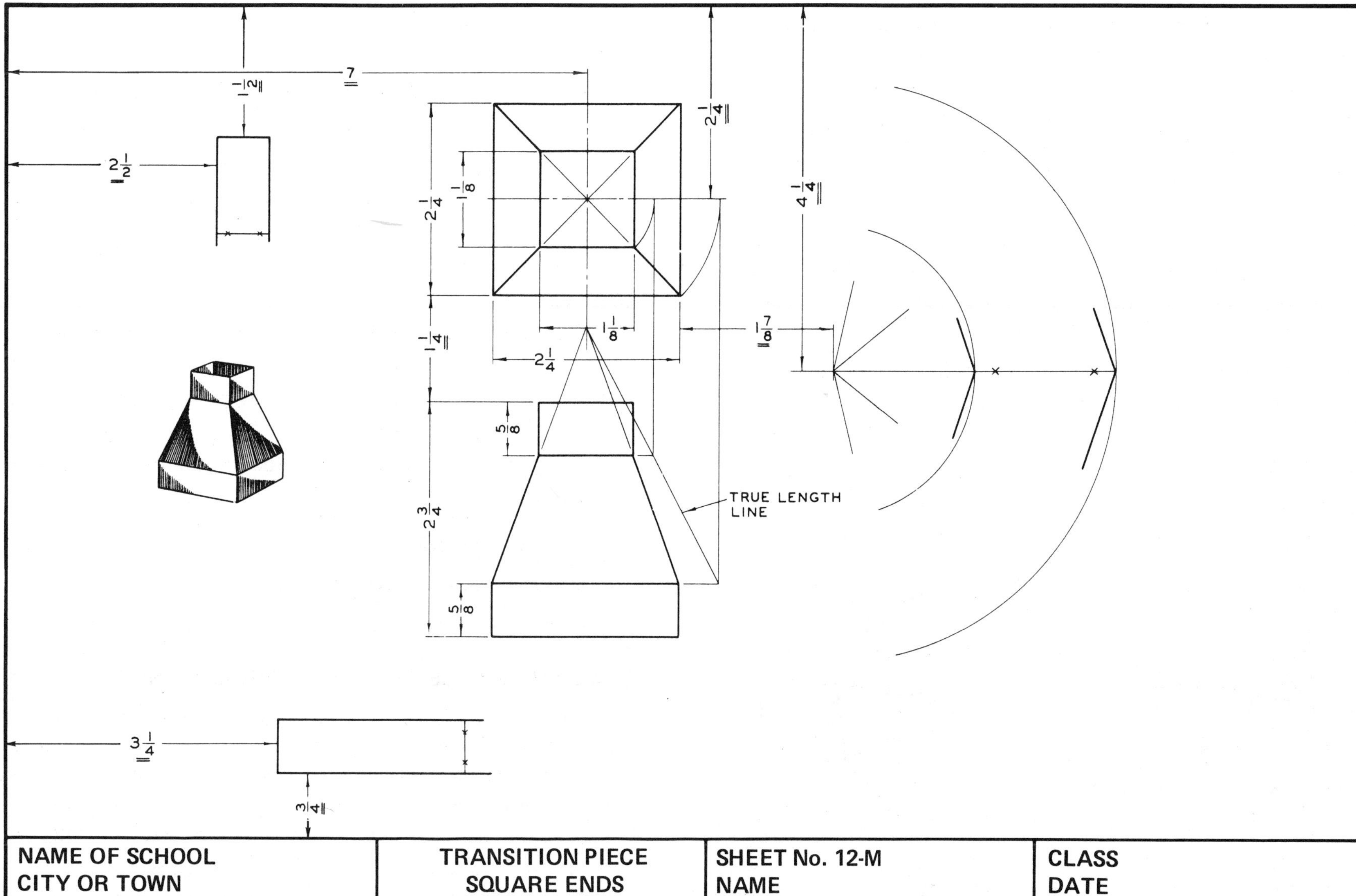
7
1 1/2
2 1/2
2 1/4
1 1/8
2 1/4
1 1/4
1 1/8
2 1/4
1 7/8
4 1/4
5/8
2 3/4
5/8
TRUE LENGTH LINE
3 1/4
3/4
NAME OF SCHOOL
CITY OR TOWN
TRANSITION PIECE
SQUARE ENDS
SHEET No. 12-M
NAME
CLASS
DATE

Fig. 12-26

The part to be drawn in figure 12-26 is a transition piece or a transformer. A transition piece is a fitting which makes a change from one size to another, or from one shape to another.

This transition piece connects a large square pipe with a smaller one. The center section has the form of a truncated pyramid. The drawing and the development of a pyramid and of a truncated square pyramid are represented in figure 12-27.

TERMS

- A *pyramid* is a solid whose base is a polygon and whose sides are triangles which meet at a common point called the apex, or vertex.
- A *right pyramid* is one whose axis is perpendicular to its base.
- An *oblique pyramid* is one whose axis is inclined to its base.

ASSIGNMENT M

Draw a top and front view of the transition piece. Develop the surface pattern of each part.

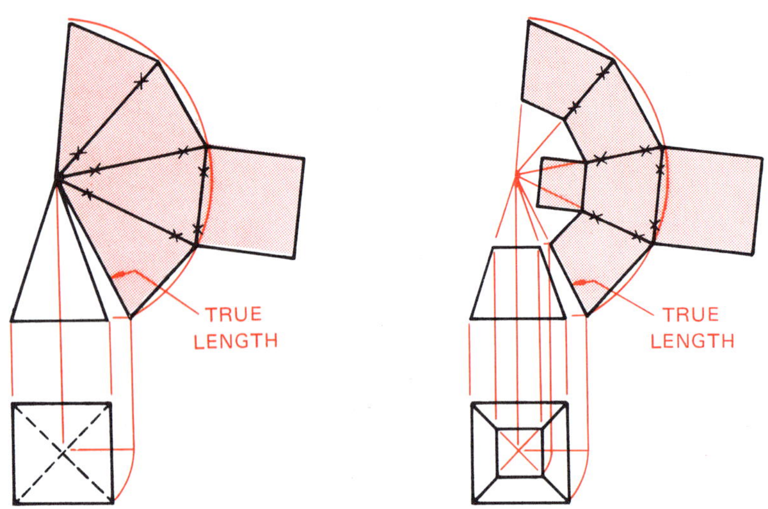

Fig. 12-27 Drawing and development of pyramids

Figure 12-26 shows a practical arrangement of the views and the surface patterns.

Procedure

1. Lay out the margin lines on the drawing sheet and letter the title strip.
2. Draw the top view. Draw the front view.
3. Draw the true length line using the method shown in figure 12-25 at C.
4. Draw the surface pattern of the truncated pyramid to the right of the views.
5. Draw the surface pattern of the lower collar below the front view.
6. Draw the surface pattern of the upper collar to the left of the views.
7. Complete the drawing. Check the top and front views of the transition piece.

REVIEW QUESTIONS

1. Define a transition piece.
2. Explain an oblique edge.

TRANSITION PIECE WITH RECTANGULAR ENDS

The transition piece with rectangular ends, figure 12-28, connects a large rectangular pipe with a smaller one. The center section is a truncated rectangular pyramid. The oblique edges on the pyramid are foreshortened in both views. The true length of these oblique edges must be determined before the surface pattern is developed.

ASSIGNMENT N

Draw a top and front view. Develop the surface pattern of each section.

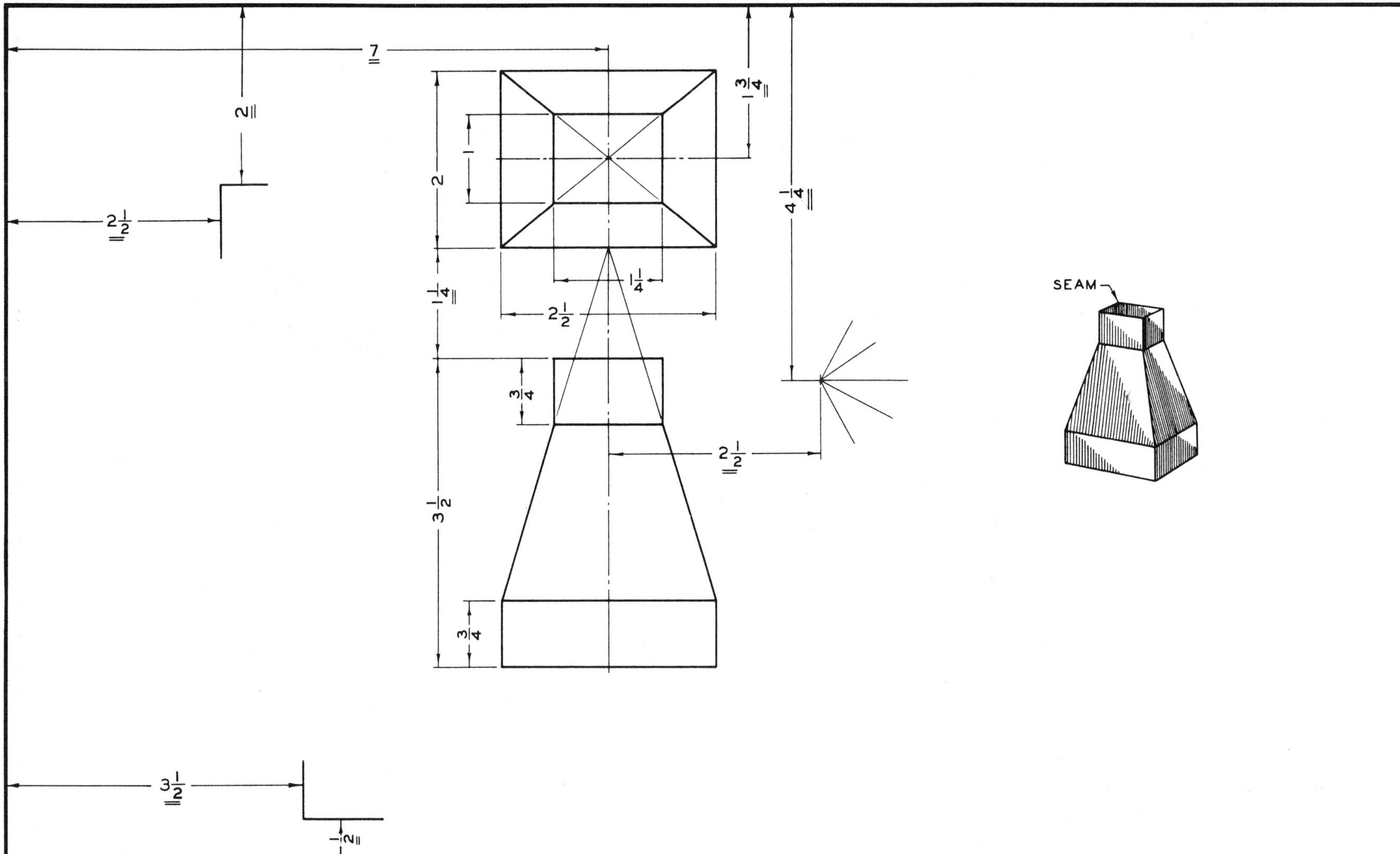

NAME OF SCHOOL CITY OR TOWN	TRANSITION PIECE RECTANGULAR ENDS	SHEET No. 12-N NAME	CLASS DATE

Fig. 12-28

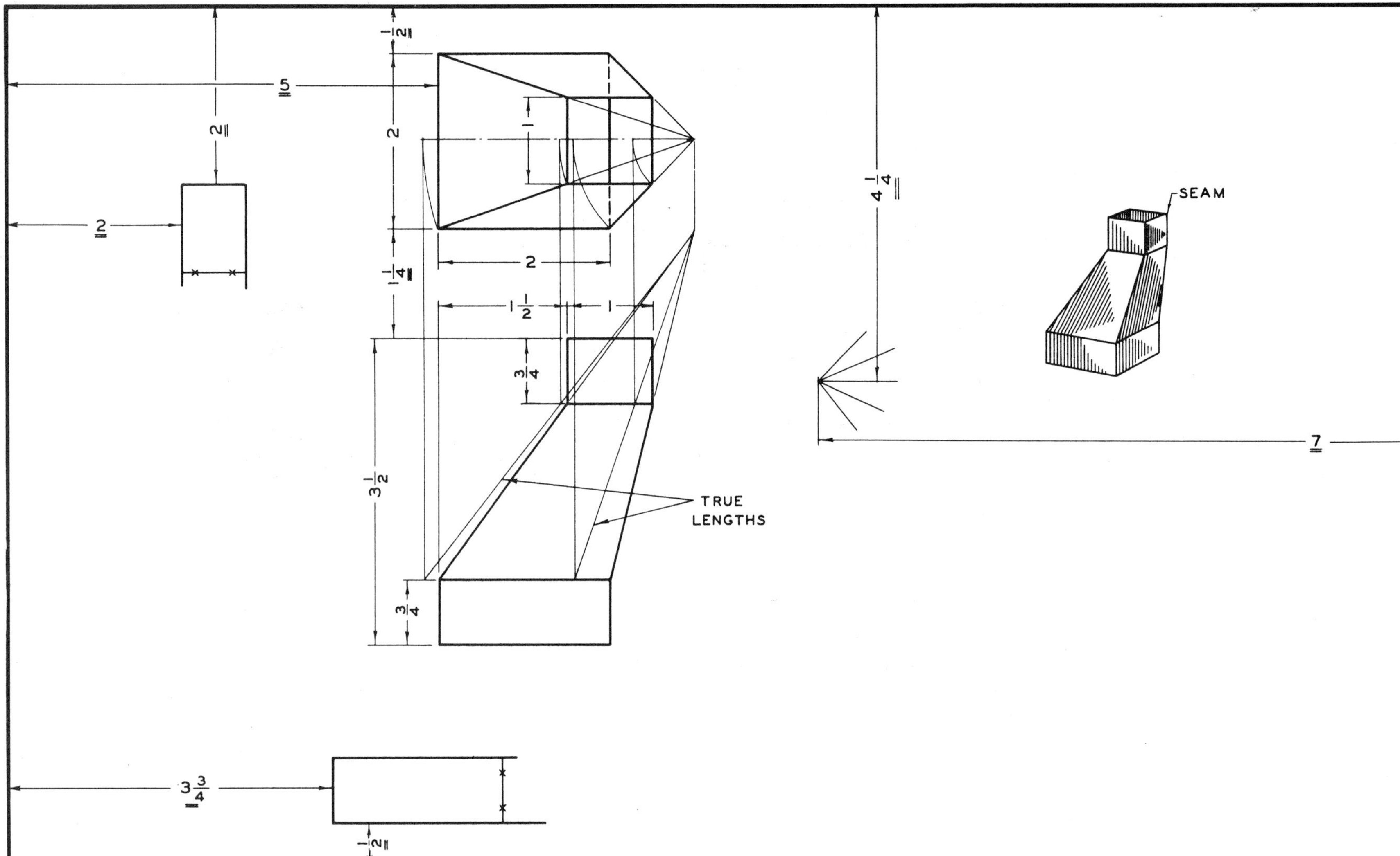

NAME OF SCHOOL CITY OR TOWN	OFFSET TRANSITION PIECE SQUARE ENDS	SHEET No. 12-O NAME	CLASS DATE

Fig. 12-29

Procedure

1. Lay out the margin lines on the drawing sheet and letter the title strip.
2. Draw the top view. Draw the front view.
3. Locate and draw the line representing the true length of the oblique edges of the pyramid. Use the method previously covered in figure 12-25 at C.
4. Draw the surface pattern of the truncated pyramid to the right of the views.
5. Draw the surface pattern of the lower collar below the front view.
6. Draw the surface pattern of the upper collar to the left of the views.
7. Complete the drawing. Check the accuracy of the two views and the surface patterns.

REVIEW QUESTIONS

1. Explain in detail how to distinguish the true length edges from the foreshortened edges in the views.
2. Explain how to find the true length of an oblique foreshortened edge.

OFFSET TRANSITION PIECE

The fitting pictured in figure 12-29 is called offset because the center of one end is not in line with the center of the opposite end. Since it is offset, the center section of the fitting has the form of a truncated oblique square pyramid.

An oblique pyramid is one in which the axis is not perpendicular to its base. The two-view drawing of an oblique square pyramid, figure 12-30 illustrates the method used to find the true lengths of the oblique edges of the pyramid and the development of its surfaces.

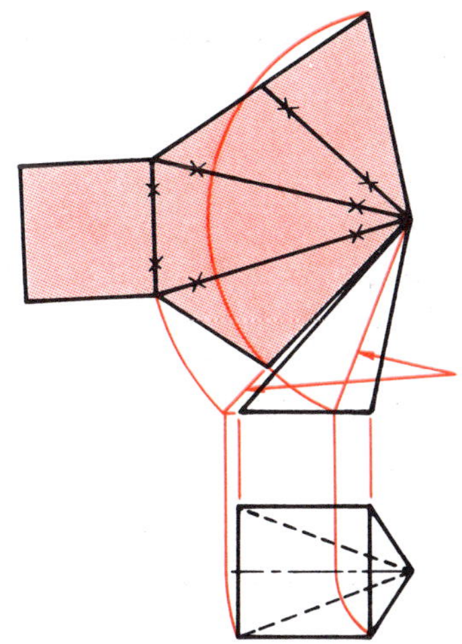

Fig. 12-30 Drawing and development of an oblique pyramid

In the assignment, the oblique edges of the pyramid are rotated to the horizontal position. Refer to figure 12-30. Then, the true lengths of these edges can be found in the front view.

ASSIGNMENT O

Draw the top and front views of the offset transition piece. Develop the surface pattern of each section.

Procedure

1. Lay out the margin lines on the drawing sheet and letter the title strip.
2. Draw the top view. Extend oblique lines to find the apex of the oblique pyramid.
3. Draw the front view. Again extend oblique lines to find the apex of the pyramid.
4. Draw the true length lines in the front view.

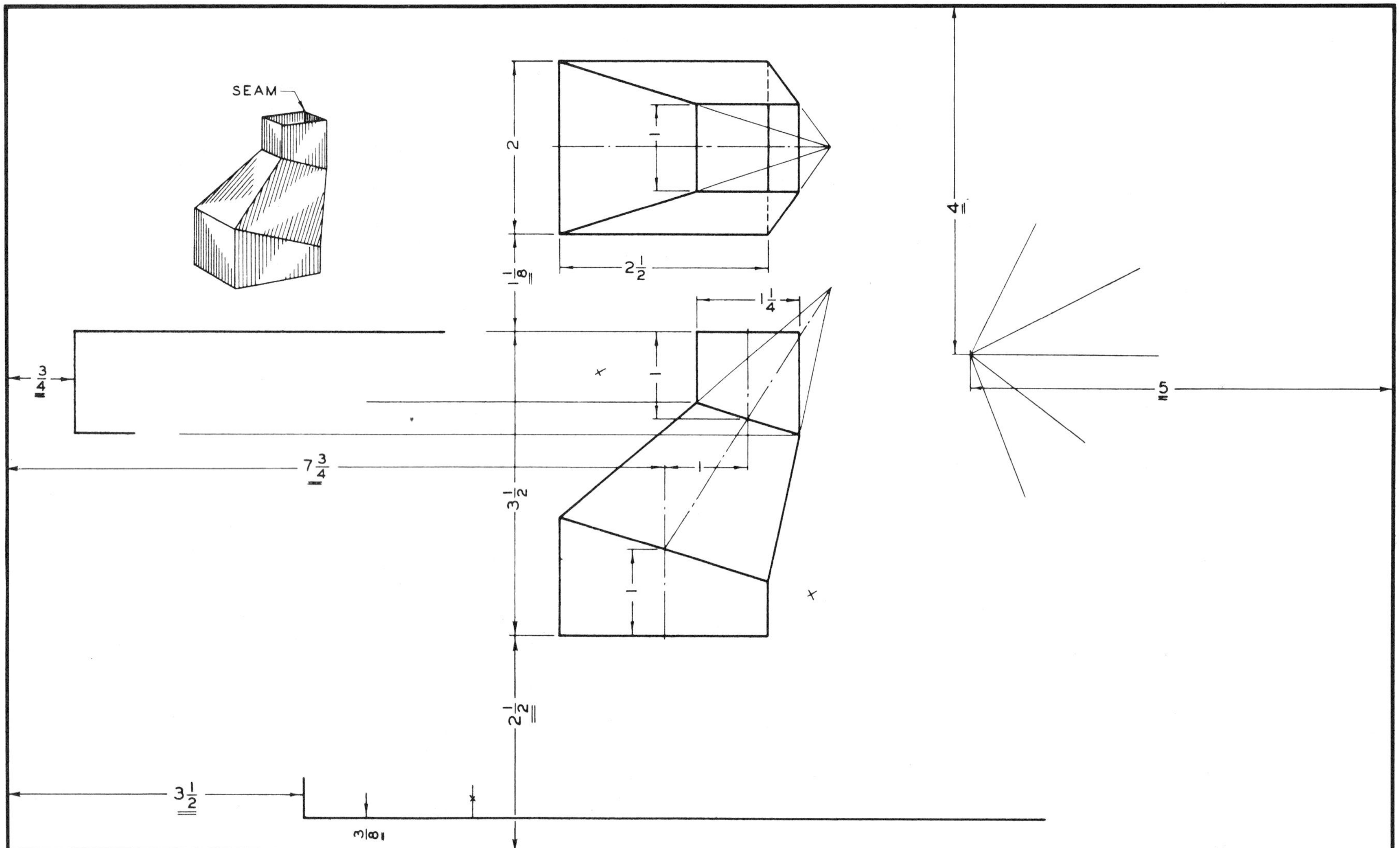

NAME OF SCHOOL CITY OR TOWN	OFFSET TRANSITION PIECE RECTANGULAR ENDS	SHEET No. 12-P NAME	CLASS DATE

Fig. 12-31

5. At the right of the views develop the surface pattern of the whole oblique square pyramid. Each triangle must be constructed individually. The length of the three sides of each triangle is taken from the views.
6. Locate and draw the upper edges to complete the pattern of the truncated pyramid.
7. Draw the surface pattern of the upper collar to the left of the views.
8. Draw the surface pattern of the lower collar below the front view. The accuracy of surface patterns can easily be proven. The length of the edges in one section must be equal to the corresponding edges in the section to which it is joined.
9. Check the surface patterns as follows:
   - The bottom edges in the pattern of the truncated pyramid must be equal to the edges of the lower collar.
   - The top edges in the pattern of the truncated pyramid must be equal to the edges of the upper collar.
10. Complete drawing and erase all unnecessary lines. Check the regular views and the development of the surface pattern.

OFFSET TRANSITION PIECE WITH RECTANGULAR ENDS

The center section of the offset transition piece in figure 12-31 is an oblique rectangular pyramid that is truncated at both ends. The upper and lower sections are truncated rectangular prisms.

The true lengths of the oblique edges must be determined before the pattern of the oblique pyramid can be developed. The pattern of the truncated oblique pyramid is then developed in the same manner used on the last drawing.

The pattern of the upper section is drawn to the left of the front view. The pattern of the lower section is drawn beneath the front view because there is insufficient space for it at the right or left of the front view.

ASSIGNMENT P

Draw a front view and a top view of the offset rectangular transition piece. Develop the pattern of each section.

Procedure

1. Lay out the margin lines on the drawing sheet and letter the title strip.
2. Draw the center lines for the front view.
3. Bisect the angles between the vertical and sloping center lines to obtain the lines of intersection.
4. Complete the front view. Draw the top view.
5. Extend the oblique lines to find the apex of the pyramid in the top and front views.
6. Find the true lengths of the oblique lines.
7. At the right of the views develop the surface pattern of the whole oblique pyramid. Each triangle must be drawn individually. The length of the three sides of each triangle is taken from the views.
8. Locate and draw the upper edges to complete the pattern of the truncated pyramid.
9. Draw the surface pattern of the upper section of the transition piece to the left of the front view.
10. Draw the surface pattern of the lower section of the transition piece beneath the front view.
11. Check the patterns as follows:
    - The top edges on the pattern of the truncated pyramid must be equal in length to the corresponding edges on the pattern of the upper section.
    - The bottom edges on the pattern of the truncated pyramid must be equal in length to the corresponding edges on the pattern of the lower section.

12. Complete the drawing and erase all unnecessary lines. Check the front and top views and the development of the surface patterns.

REVIEW QUESTIONS

1. What is a pyramid?
2. What is a right pyramid?
3. What is an oblique pyramid?
4. When is a line foreshortened in the views on a drawing?
5. When is a line the true length in the views on a drawing?

CONES

To review some basic principles, the elements of a cone meet at a point called the *apex.* When the surface pattern of a cone is developed, the elements radiate from a point. The method used to develop the pattern is called *radial line development.* All of these statements are illustrated in the pictorial development of a cone, figure 12-32, and shown in figure 12-33.

All of the elements of a right cone are the same length. The true length of the elements is shown by the elements at the extreme right and left in the front view. These two elements are horizontal lines in the top view. Therefore, they are parallel to the front vertical plane. Consequently they are their true lengths in the front view.

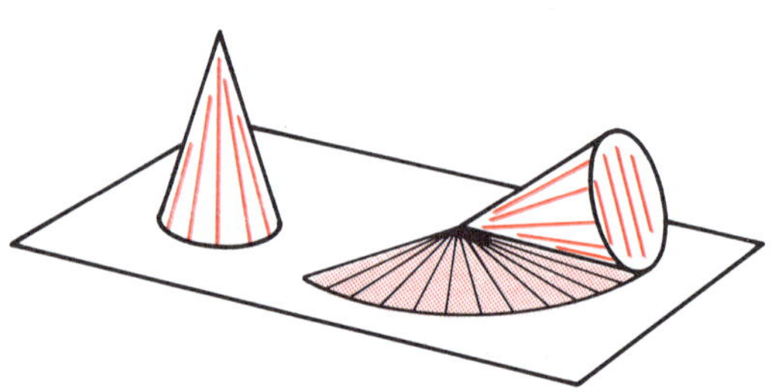

Fig. 12-32 Pictorial development of a cone

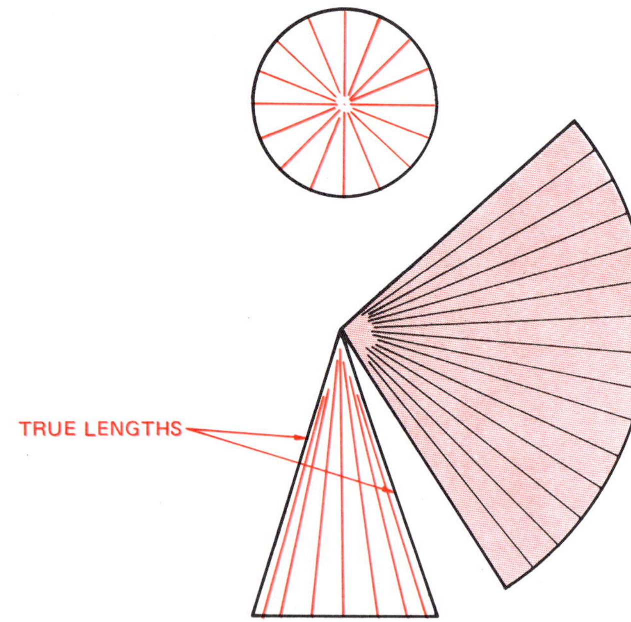

Fig. 12-33 Drawing and development of a cone

The funnel in figure 12-34 consists of two truncated right cones. The pattern of each cone is developed as shown in figure 12-33.

TERMS

- A *cone* is a solid whose base is a circle and whose sides taper uniformly to the apex. A cone may be considered as a pyramid having an infinite number of sides.
- A *right cone* is one whose axis is perpendicular to its base.

ASSIGNMENT Q

Draw one-half of the top view, and the front view of the funnel. Develop the surface pattern of each section.

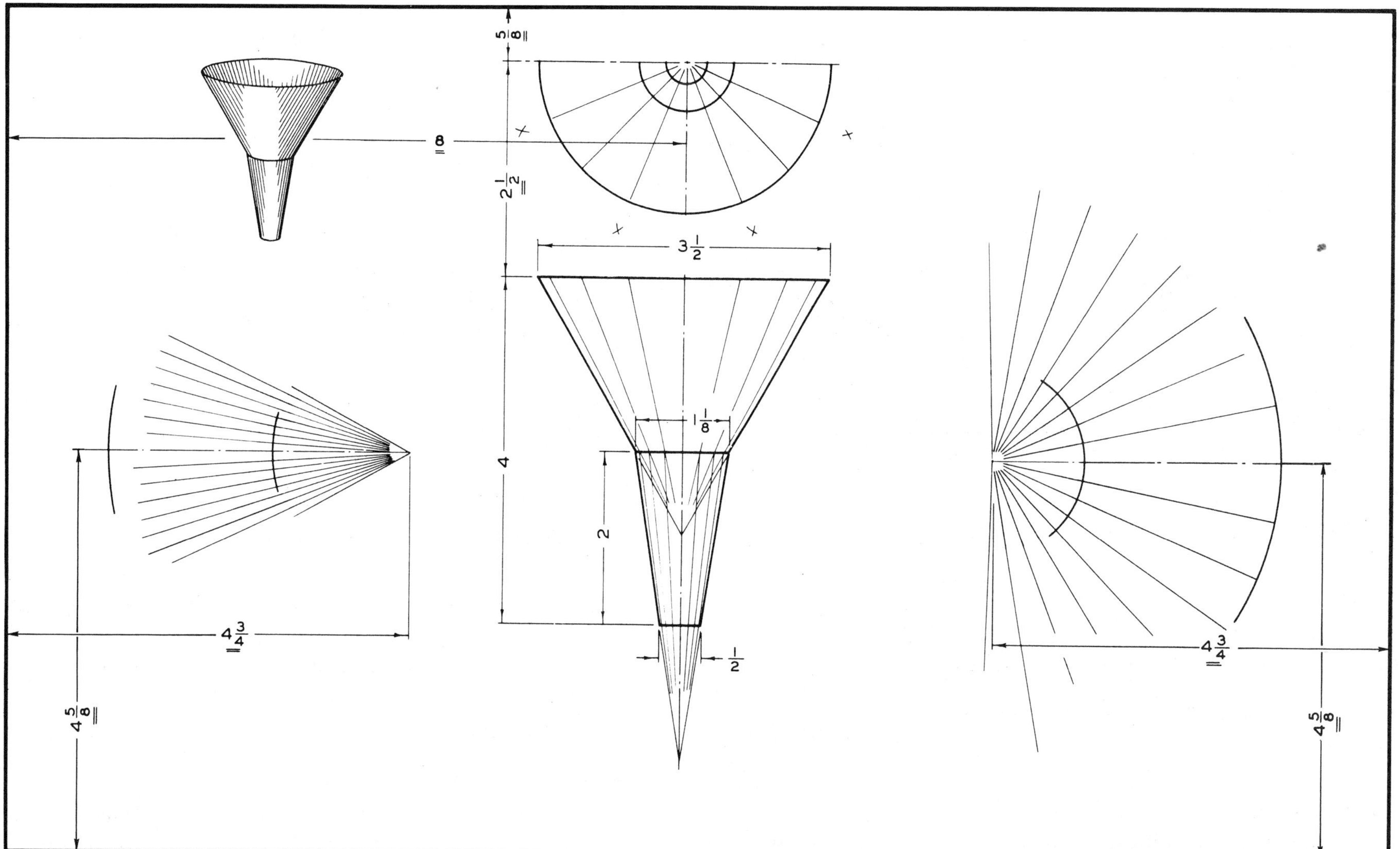

NAME OF SCHOOL CITY OR TOWN	FUNNEL	SHEET No. 12-Q NAME	CLASS DATE

Fig. 12-34

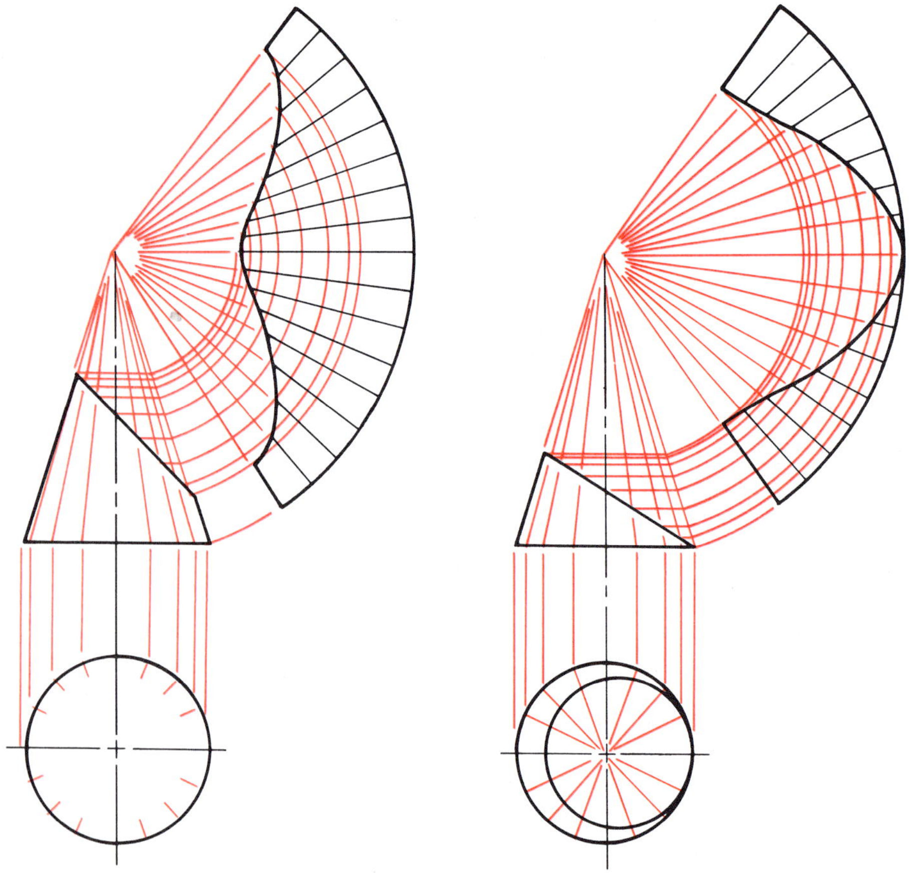

Fig. 12-35 Development of truncated cones

Procedure

1. Lay out the margin lines on the drawing sheet and letter the title strip.
2. Draw one-half of the top view.
3. Divide larger semicircle into eight equal parts. Draw the elements.
4. Draw the front view.
5. Extend the elements at the extreme right and left to find the apex of each cone. Since all of the elements of a right cone are the same length, it is unnecessary to draw the other elements in the front view.
6. At the right of the views draw an arc using a radius equal to the true length of the elements of the upper cone.
7. Adjust the dividers to divide the largest semicircle into eight equal spaces. Then, with the dividers, point off on the arc eight of these spaces above the horizontal center line, and eight below the center line.
8. Draw lines from each point to complete the pattern of the whole cone.
9. Draw an arc using a radius equal to the true length of the elements from the line of intersection to the apex. This completes the pattern of the truncated cone.
10. Draw the pattern of the smaller cone at the left of the views.
11. Check the accuracy of the patterns with the dividers. The length of the outer arc on the pattern of the small cone must be equal to the length of the inner arc on the pattern of the larger one.
12. Complete the drawing. Check the accuracy of both views and the development of the surface patterns.

TRUNCATED CONES

The funnel on the last drawing consists of two truncated right cones. These truncated cones are cut by planes which are parallel to the base of the cones. When a right cone is cut by a plane parallel to its base, all of the elements on the truncated portion are of equal length.

Two more truncated right cones are contained in figure 12-35. These cones are cut by planes not parallel to the base of the cones. The first drawing shows the development of the section below the

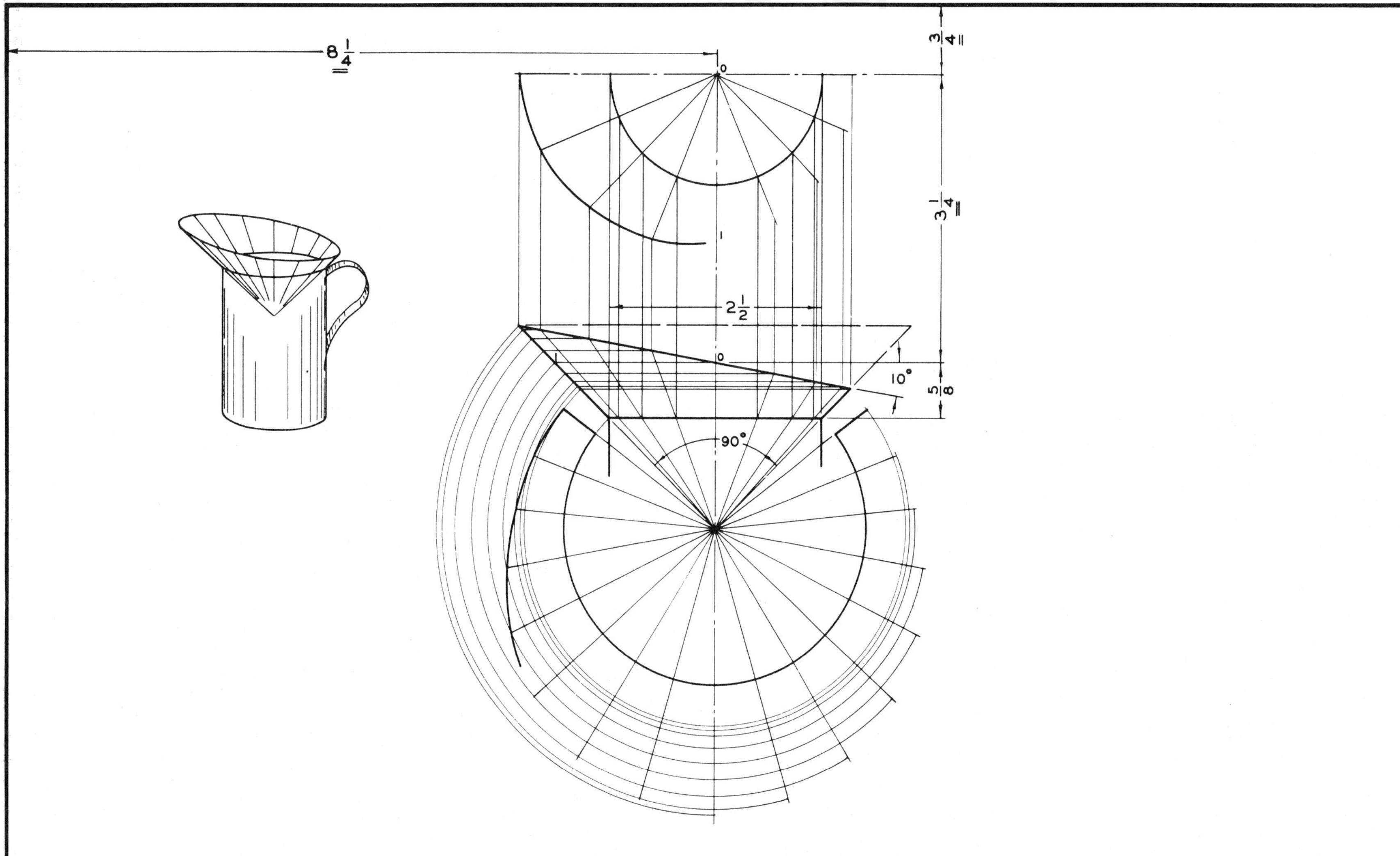

NAME OF SCHOOL CITY OR TOWN	DEVELOPMENT OF A POURING LIP	SHEET No. 12-R NAME	CLASS DATE

Fig. 12-36

inclined cutting plane; the second, the development of the section above the inclined cutting plane.

When a right cone is truncated by an inclined cutting plane, the elements from left to right on the truncated portion differ in length. The elements on the extreme right and left in the front view of figure 12-35 are their true lengths, while the other oblique elements are foreshortened. To find the true lengths of the oblique elements, the ends of these elements are projected to the true length line. The true lengths are then transferred with the compass to the corresponding element in the pattern.

The pouring lip on the round measure shown in figure 12-36 is a part of an inverted right cone. The lower end is cut by a plane parallel to the base and the upper end is cut by a plane inclined to the base.

Again, the elements on the extreme right and left are their true lengths, while the other oblique elements are foreshortened. The ends of the foreshortened elements are projected to the true length line at the left to determine their true length. The true length of each element is then transferred to the corresponding element in the pattern. The outer ends of the elements are connected with a smooth curve to complete the pattern.

ASSIGNMENT R

Draw one-half of the top, and front view of the pouring lip. Develop the surface pattern.

Procedure

1. Lay out the margin lines on the drawing sheet and letter the title strip.
2. Draw the center lines and the semicircle in the top view.
3. Divide the semicircle into eight equal spaces. Draw the elements in this view.
4. Draw the upper and lower edges of the lip in the front view.
5. From each point on the semicircle, project lines down to the lower edge of the lip (the line of intersection).
6. Locate the apex of the cone in the front view. Draw the elements.
7. Project a line up from the upper end of each element in the front view to the corresponding element in the top view. Draw the outer edge of the lip. The distance 0-1 in the top view is equal to 0-1 in the front view.
8. Draw an arc. Use the apex of the cone for a center and a radius, which is the true length from the apex to the line of intersection.
9. The dividers should be adjusted to divide the semicircle in the top view into eight equal spaces. With the dividers, point off on the arc in the pattern, eight of these spaces on each side of the vertical center line.
10. Draw the elements through the points on the arc.
11. From the upper end of each element in the front view draw a line to the left to the true length line. The true length of each oblique element is then found on this line.
12. With the compass, transfer the true length of each element to the corresponding element in the pattern.
13. Connect the outer ends of elements with a smooth curve to the complete pattern.
14. Complete the drawing. Check the top and front views, and the development of the surface pattern.

REVIEW QUESTIONS

1. What is a cone?
2. What is a right cone?
3. What is a truncated cone?

SCALENE CONES

The center section in this new offset transition, figure 12-38, is part of an oblique or scalene cone. An *oblique* or *scalene cone* is one whose axis is inclined to its base. The development of scalene cones is shown in figure 12-37. The elements on a scalene cone differ in length. In the front view of the first illustration, the elements on each side are their true lengths because these elements are horizontal lines in the top view. Consequently, they are parallel to the front vertical plane of projection.

To find the true lengths of the oblique elements, rotate each element in the top view about the apex to the horizontal line position. The end of each element is projected to the front view, and a line representing the true length of each element, is drawn. These true lengths are transferred to the pattern. Triangles representing the spaces between the elements are then constructed to complete the pattern. The base of each triangle is equal to the chord between the elements in the top view.

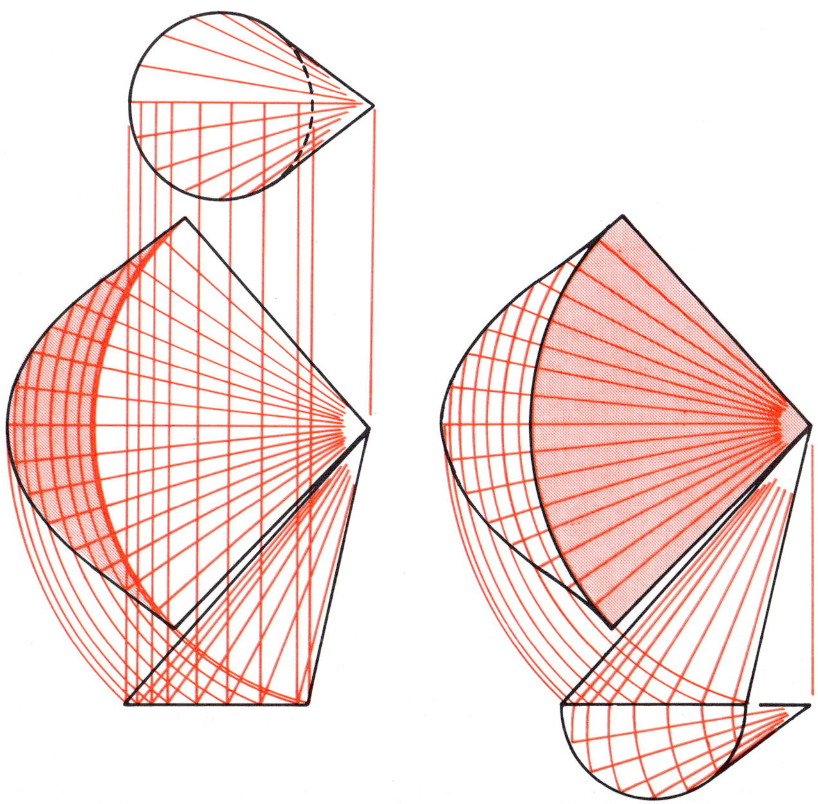

Fig. 12-37 The development of oblique or scalene cones

The second illustration, figure 12-37, shows a shorter method of developing a scalene cone. In this illustration, one-half of the bottom view (called the *plan*) is attached to the front view (called the *elevation*). Each element in the plan is rotated about the apex to the horizontal position and lines representing the true lengths of the elements are drawn in the elevation. These true lengths are transferred to the pattern. The second method is used to develop the pattern of the truncated scalene cone on this drawing.

ASSIGNMENT S

Draw one-half of the plan, and the elevation of the offset transition piece. Develop one-half of the surface pattern of the truncated scalene cone. The surface patterns of the collars are not made on this drawing.

Procedure

1. Lay out the margin lines on the drawing sheet and letter the title strip.
2. Draw the center lines and the two semicircles in the plan.
3. Divide the larger semicircle into eight equal spaces and draw the elements.
4. Draw the elevation. Using point 0 in the plan as a center, rotate each oblique element to the horizontal position.
5. Project lines upward from the ends of the arcs in the plan to the base of the cone. Draw the lines representing the true length of each element.

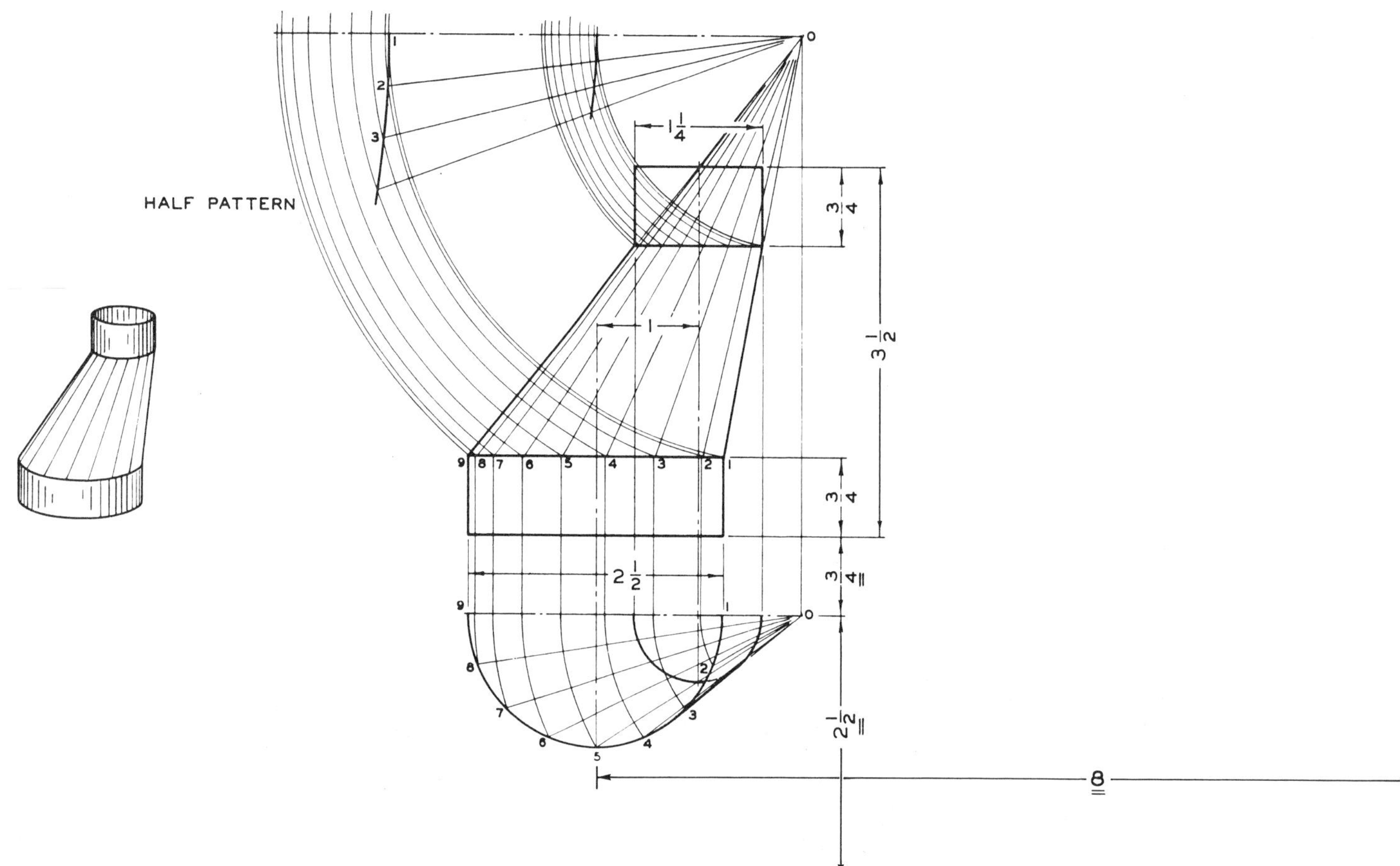

NAME OF SCHOOL CITY OR TOWN	OFFSET TRANSITION PIECE ROUND ENDS	SHEET No. 12-S NAME	CLASS DATE

Fig. 12-38

6. Using point 0 in the elevation as a center, draw arcs from points 1, 2, 3, etc.
7. The dividers are adjusted to divide the larger semicircle into eight equal spaces. With the dividers, point off spaces 1-2, 2-3, etc., to locate the bases of the triangles in the half pattern.
8. Draw the triangles in the half pattern.
9. Using point 0 in the elevation as a center, draw arcs from the upper end of each true length line on the truncated cone to the corresponding line in the half pattern.
10. Connect the ends of the true length lines to complete the half pattern.
11. Complete the drawing. Check the plan and elevation views, and the development of the surface pattern.

REVIEW QUESTIONS

1. What is a plan?
2. What is an elevation?

ROUND OFFSET TRANSITION PIECE

The round transition piece, figure 12-39, differs from the last drawing which had horizontal lines of intersection with the ends of the truncated cone parallel to the horizontal plane of projection. The ends of the cone were true circles in the plan.

The lines of intersection on this next transition piece are slanting, and an auxiliary view of each end of the truncated cone shows that the ends are ellipses. This is illustrated in figure 12-40.

In this illustration, the base of the cone appears as a circle in the plan. However, this is not the true shape because the base of the cone is not parallel to the horizontal plane of projection. The auxiliary view shows that the true shape of the base is an ellipse. In this ellipse the distances 1′ - 2′, 2′ - 3′, etc., are the true lengths of the spaces between the ends of the elements.

The true lengths of distances 1-2, 2-3, etc., may be determined by a second method. Points 1, 2, 3, 4, and 5 in the elevation are rotated to the horizontal line. Lines are then projected to the plan. These lines intersect horizontal lines extending outward from the same points in the plan. When the points of intersection are connected, one quarter of the ellipse is drawn, and the distances 1′ - 2′, 2′ - 3′, etc., are true lengths. Only one quarter of the ellipse is required because the true lengths are the same in each quarter.

The true lengths on this drawing are found by the second method. The oblique elements on the cone are foreshortened. The true lengths are found by constructing a *true-length diagram*. The end of each element in the plan is rotated about the apex to the horizontal line. Lines are then projected downward. The lines intersect horizontal lines extending to the right from the end of each element in the elevation. The lines drawn from each point of intersection to the apex are the true lengths of the elements.

The pattern is developed by constructing adjoining triangles, one at a time. The lengths of the sides of each triangle are taken from the true-length diagram, and the length of each base is taken from the partial ellipse.

ASSIGNMENT T

Draw the plan and the elevation of the round offset transition piece. Develop the surface pattern of each section. Use the same method described in figure 12-40 to develop the pattern and locate true lengths.

Procedure

1. Lay out the margin lines on the drawing sheet and letter the title strip.
2. Draw the plan.
3. Divide the base circle into sixteen equal spaces and draw the elements.

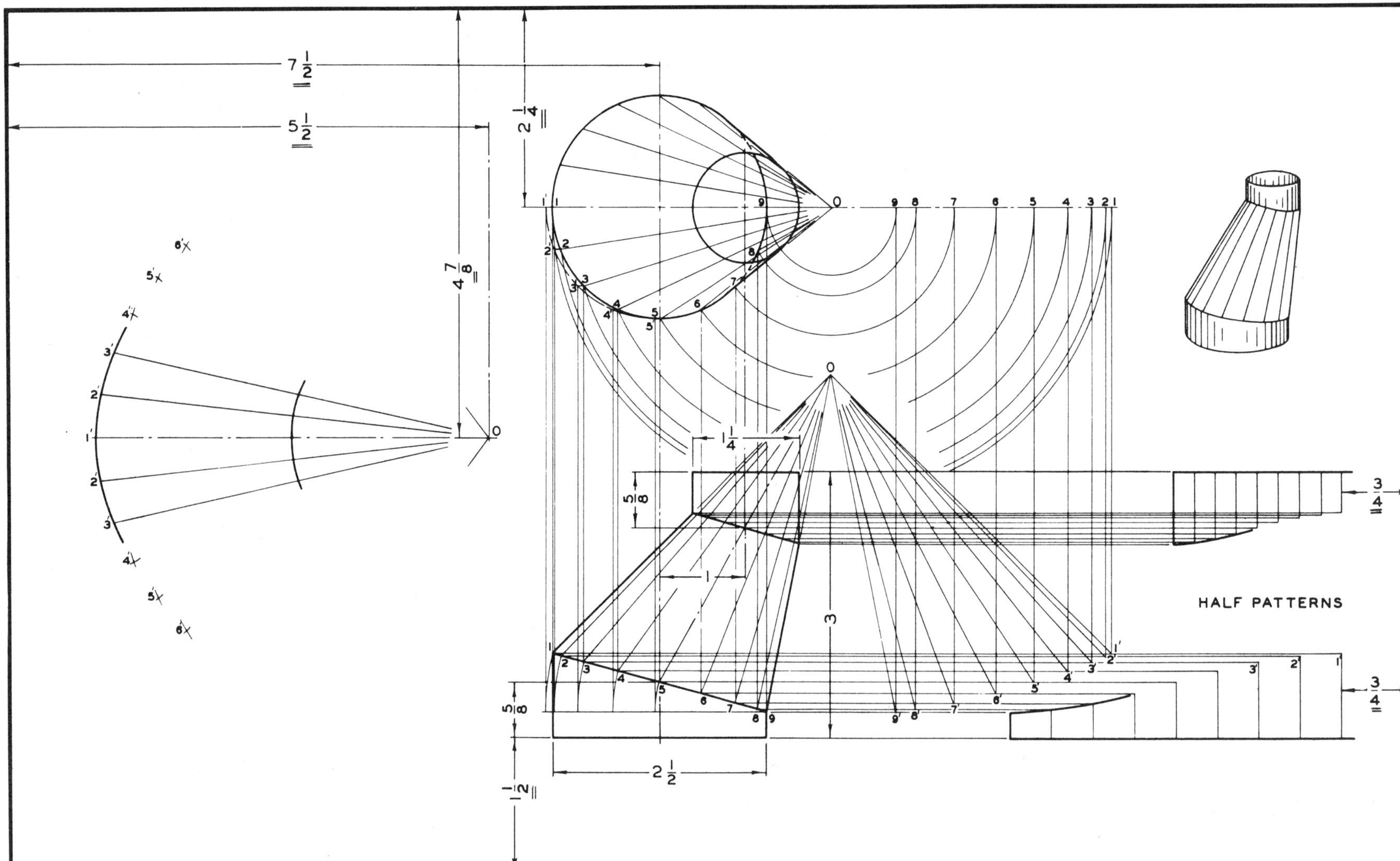

NAME OF SCHOOL CITY OR TOWN	OFFSET TRANSITION PIECE ROUND ENDS	SHEET No. 12-T NAME	CLASS DATE

Fig. 12-39

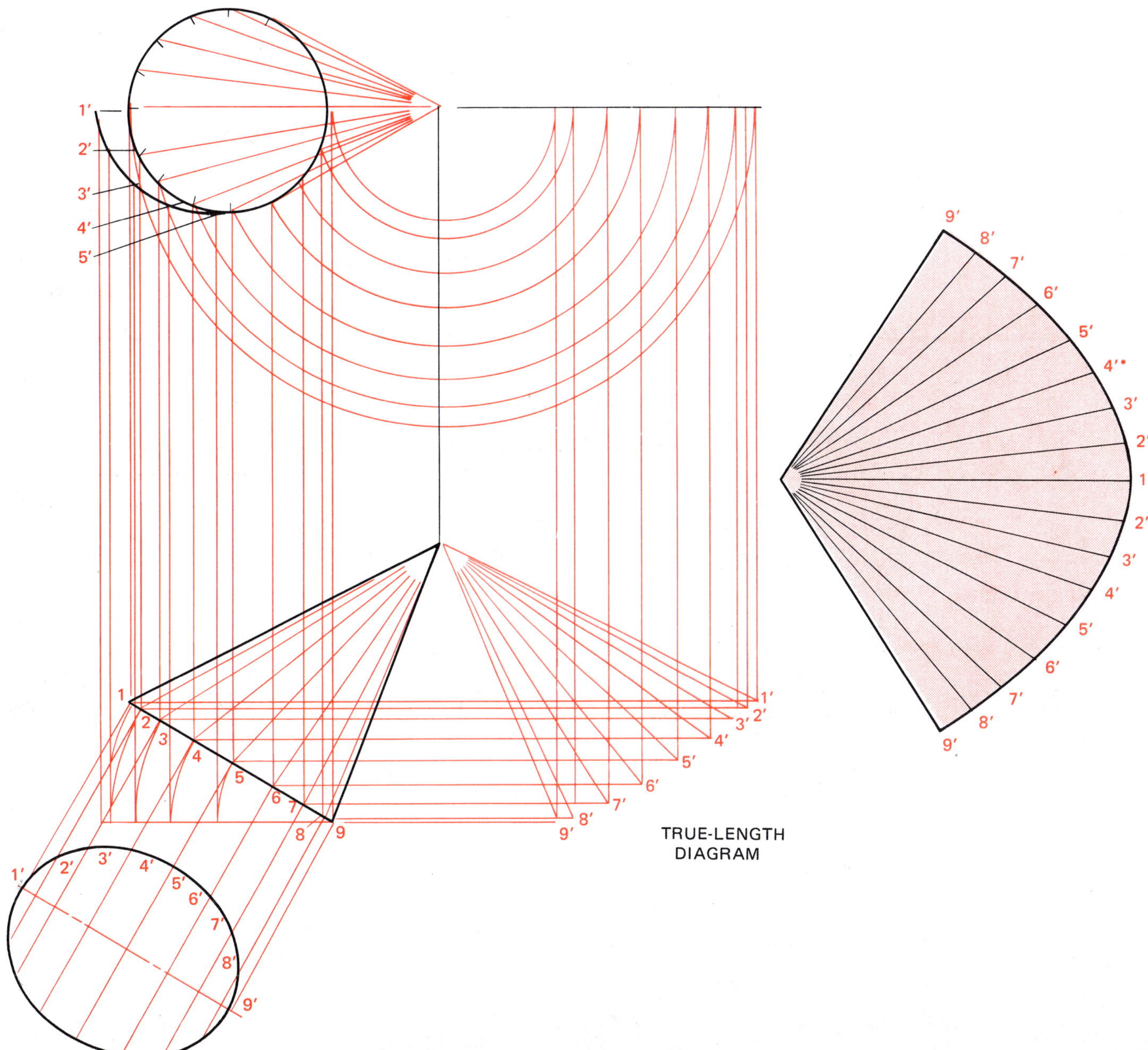

Fig. 12-40 Development of a scalene cone by triangulation

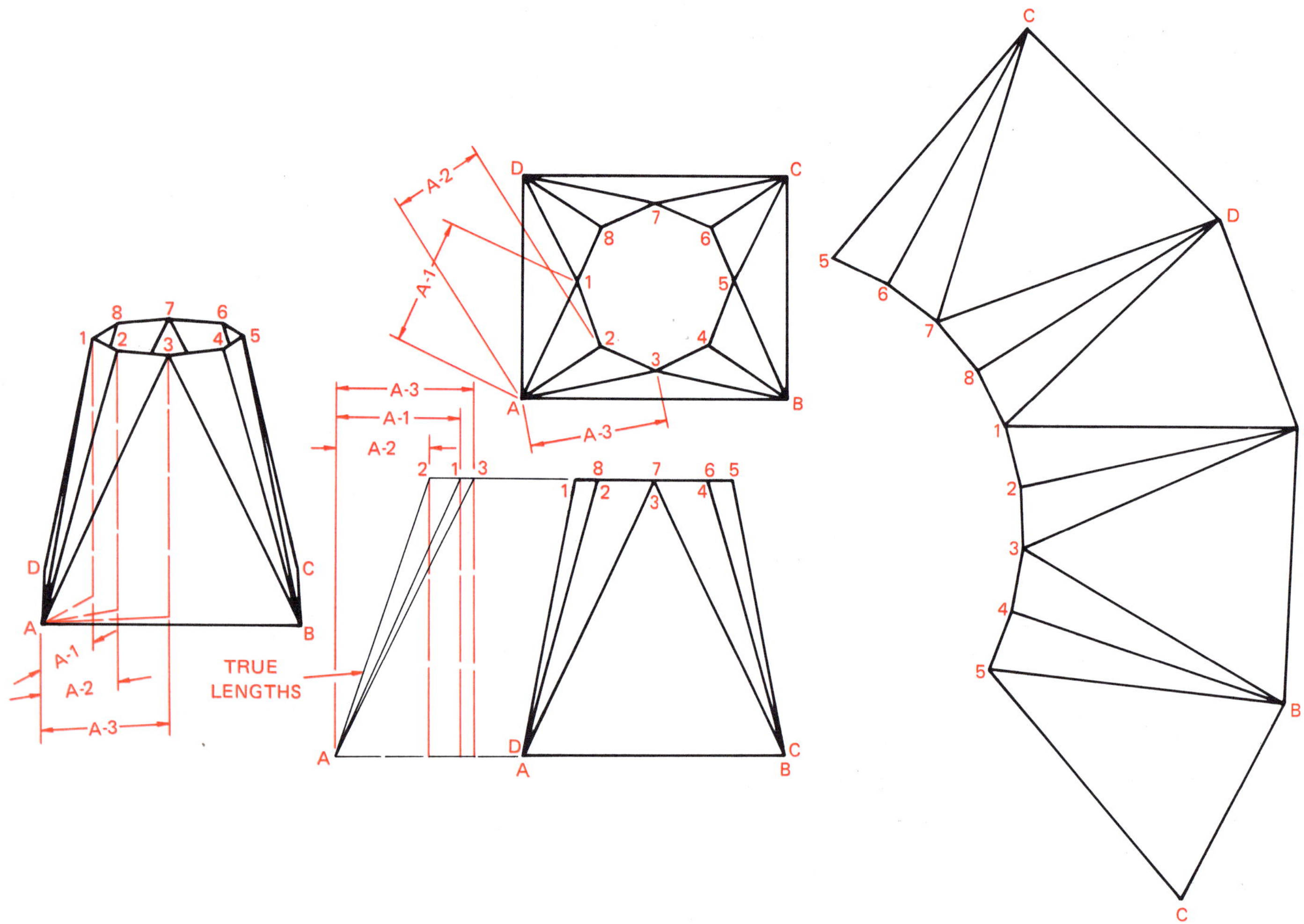

Fig. 12-41 Development of surface patterns by triangulation

4. Draw the elevation. Bisect the angles between the center lines to locate the lines of intersection.
5. Draw the elements in the elevation.
6. In the plan, draw a horizontal line to the left from points 1, 2, 3, 4, and 5.

DEVELOPMENT BY TRIANGULATION

Surfaces which cannot readily be developed either by the parallel line method or the radial line method can be developed approximately by the *triangulation method*.

Development by triangulation consists of three steps:

1. The surface to be developed is divided into triangles.
2. The true lengths of the sides of the triangles are found.
3. Adjoining triangles are constructed, one at a time, until the pattern is completed.

These three steps are illustrated in figure 12-41. This pictorial drawing shows a top view and a front view, a true length diagram, and the pattern of a transition piece. The lower end of this transition piece is a rectangle. The upper end is an octagon. The upper and lower corners are connected with edges which divide the surface to be developed into twelve triangles (step 1). The sides of the triangles are oblique lines and they are foreshortened in both views. The true lengths of these oblique lines are found (step 2), and then adjoining triangles are drawn to complete the pattern (step 3).

The true lengths of the oblique lines (step 2) are found by constructing right angle triangles. Each oblique line may be considered as being the hypotenuse of one of the right angle triangles shown in the pictorial drawing, figure 12-41. When these three right triangles are constructed, the hypotenuse of each triangle is the true length of one of the oblique lines. The altitude of these right triangles is taken from the front view. The length of each base is taken from the top view. Only three right triangles are constructed because this transition piece is symmetrical on both axes and the four quarters are alike.

TRANSITION PIECE SQUARE TO ROUND

The transition piece in figure 12-42 has a square base and a round top. The circle is divided into sixteen equal parts. Lines drawn from the points on the circle to the four corners of the base divide the surface into twenty triangles (step 1). Some of these triangles are slightly curved, but they are considered to be flat when the pattern is developed.

Right triangles are drawn to find the true lengths of the oblique line (step 2). To conserve space these right triangles are drawn one on another. The pattern is developed by constructing the adjoining triangles (step 3).

ASSIGNMENT U

1. Draw a top view and a front view of the square to round transition piece.
2. Find the true lengths of the oblique lines.
3. Develop the surface pattern by triangulation.

PROCEDURE FOR DEVELOPING BY TRIANGULATION

1. Lay out the margin lines on the drawing sheet and letter the title strip.
2. Draw the top view. Draw the front view.
3. Draw the true length diagram to find the true lengths of the first five oblique lines. The right triangles are drawn one on another to conserve space on the drawing. Triangles are drawn for the first five oblique lines only because this transition piece is symmetrical on both axes and the four quarters are alike.
4. Construct the triangle A - 1 - D using the true length A - 1 for the sides and distance A - D for the base.

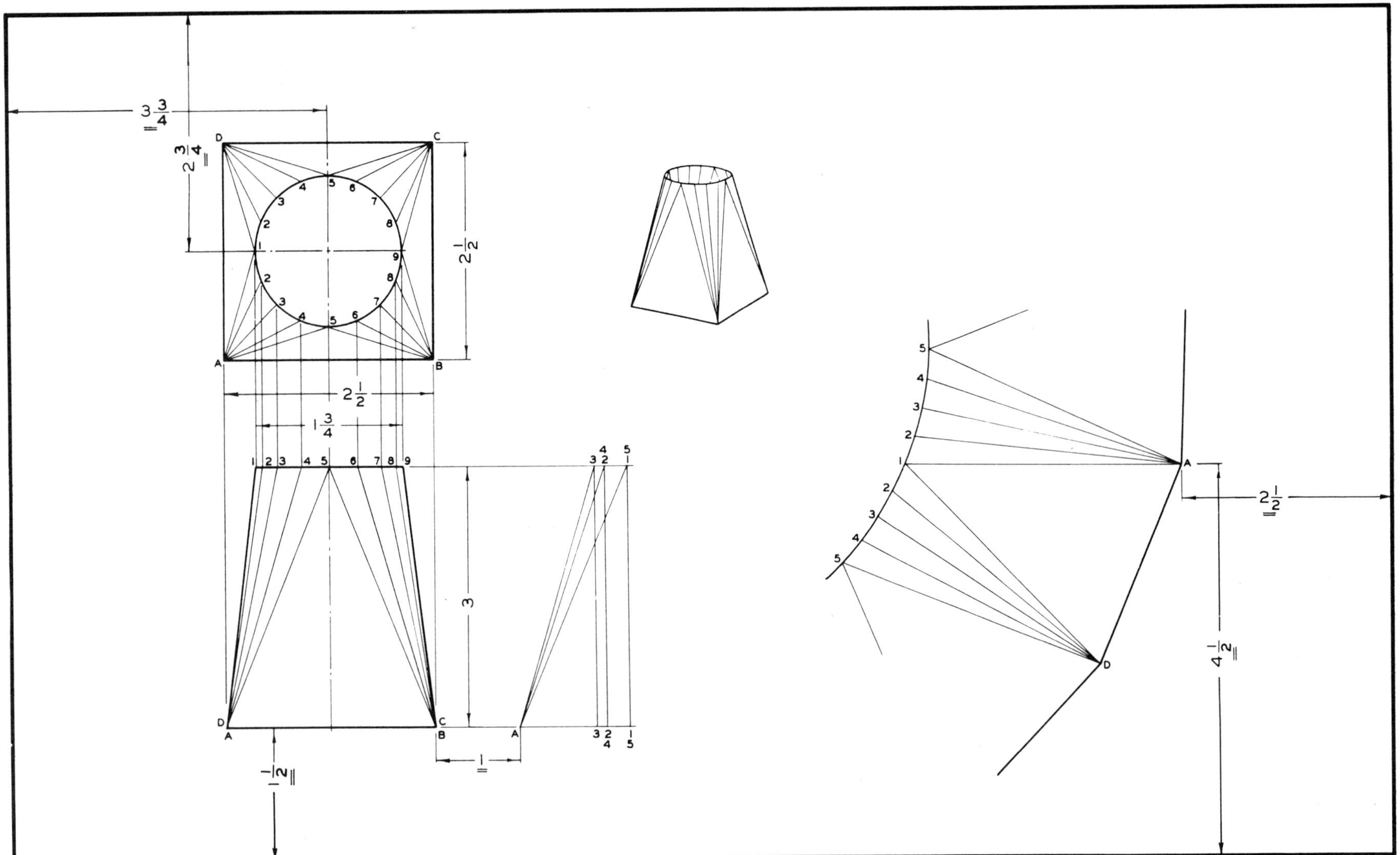

NAME OF SCHOOL CITY OR TOWN	TRANSITION PIECE SQUARE TO ROUND	SHEET No. 12-U NAME	CLASS DATE

Fig. 12-42

5. Construct the remaining triangles one at a time. Join them on their common sides until the twenty triangles are drawn.
6. Complete the drawing and erase unnecessary lines.
7. Check the top and front view, the true length of the oblique lines, and the development of the surface pattern by triangulation.

REVIEW QUESTIONS

1. What three steps are essential when patterns are developed by the triangulation method?
2. How are the dimensions obtained for the right triangles in the true length diagram?
3. How are the dimensions obtained for triangle A - 1 - D in the pattern?

TRANSITION PIECE ROUND TO RECTANGULAR

The lower end of the transition piece in figure 12-43 is round; the upper end, rectangular. This transition piece is symmetrical on one axis and offset on the other. Therefore, only the front and rear halves are alike. The ends of the transition piece are horizontal and parallel.

The circle is divided into sixteen equal spaces. When the points on the circle are connected with the upper corners, the surface is divided into twenty triangles.

A true length diagram is drawn to find the true lengths of the oblique sides of these triangles. Because the ends are horizontal and parallel, the altitude of all the right triangles in the true length diagram is the same.

The surface pattern is constructed by triangulation.

ASSIGNMENT V

Draw a top view and a front view of the round to rectangular transition piece. Draw a true length diagram. Develop the surface pattern of the center section by triangulation.

Procedure

1. Lay out the margin lines on the drawing sheet and letter the title strip.
2. Draw the top view.
3. Draw the front view.
4. Draw the true length diagram to find the true lengths of the ten oblique lines on the front half.
5. Construct triangle A - 1 - D using the true lengths A - 1 and D - 1 for the sides, and distance A - D for the base.
6. Construct triangles 1 - A - 2 and 1 - D - 2.
7. Construct the remaining triangles. Join them on their common sides until the twenty triangles are completed. The patterns for the round and rectangular collars are not drawn on this drawing.
8. Complete the drawing and erase unnecessary lines.
9. Check the top and front view, the true length diagrams, and the surface patterns.

REVIEW QUESTIONS

1. What three steps are used to develop the surface pattern by triangulation?
2. How are the dimensions obtained for the right triangles in the true length diagram?
3. This transition piece is symmetrical on one axis. What does this mean?
4. Why is this transition piece unsymmetrical in the front view?

TRANSITION PIECE RECTANGULAR TO ROUND

The lower end of the transition piece in figure 12-44 is rectangular; the upper end, round. The transition piece is symmetrical on one axis and offset on the other, so the front and rear halves are alike.

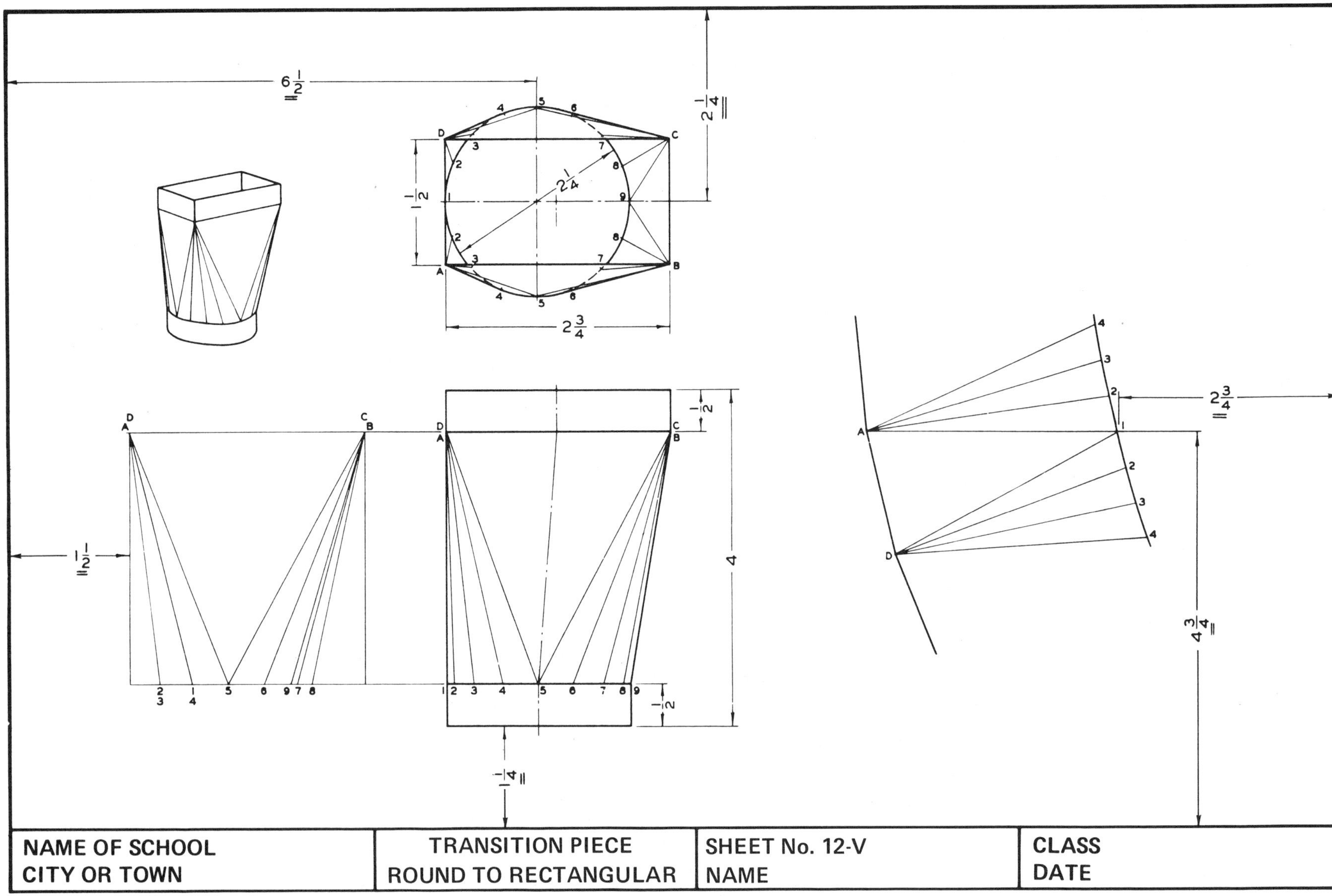

NAME OF SCHOOL
CITY OR TOWN
TRANSITION PIECE
ROUND TO RECTANGULAR
SHEET No. 12-V
NAME
CLASS
DATE

Fig. 12-43

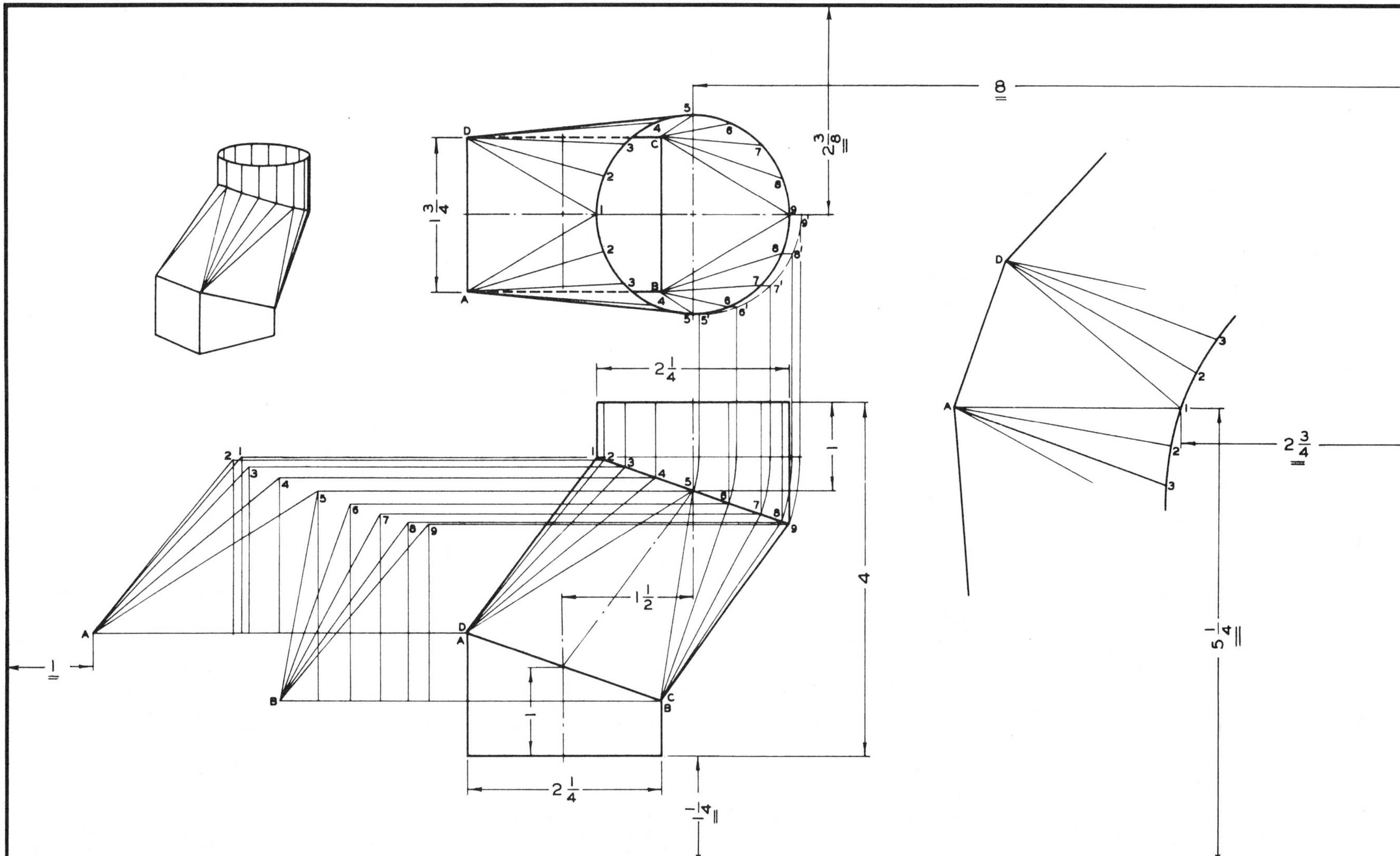

NAME OF SCHOOL CITY OR TOWN	TRANSITION PIECE RECTANGULAR TO ROUND	SHEET No. 12-W NAME	CLASS DATE

Fig. 12-44

The lines of intersection on this transition piece are inclined to the horizontal plane. If an auxiliary view is projected from the upper line of intersection, it shows that the top end of the center section is elliptical. Instead of drawing the elliptical auxiliary view, points 5, 6, 7, 8, and 9 are rotated to a horizontal line and then projected to the top view. These lines intersect horizontal lines extending to the right from points 5, 6, 7, 8, and 9 in the top view. When the points of intersection are connected, one-fourth of the ellipse is drawn. The distances 5′ - 6′, 6′ - 7′, etc., are the true lengths and these distances are used when the triangles are constructed in the pattern.

The points on the upper line of intersection are connected with the lower corners to divide the center section into twenty triangles. Because the lines of intersection are inclined to the horizontal plane, the altitude of each right triangle in the true-length diagram is different.

ASSIGNMENT W

Draw a top view and a front view of the rectangular to round transition piece. Draw a true length diagram. Develop the surface pattern of the center section by triangulation.

Procedure

1. Lay out the margin lines on the drawing sheet and letter the title strip.
2. Draw the top view. Draw the front view.
3. Locate points 5′, 6′, 7′, and 9′ in the top view.
4. Draw the true length diagram. Draw triangle A - 1 - D.
5. Construct triangles 1 - A - 2 and 1 - D - 2. The base of 1 - 2 is the same distance as 8′-9′ .
6. Construct the remainder of the triangles, joining them on their common sides. When triangles A - 5 - B and C - 5 - D are constructed, the true length of their bases is taken from line A - B in the front view. The rectangular and round collars are not drawn on this drawing.
7. Complete the drawing and erase all unnecessary lines.
8. Check the top and front views, the true length diagrams and the development of the surface pattern.

REVIEW QUESTIONS

1. How is the surface to be developed divided into triangles?
2. How is the true length diagram drawn?
3. How is the pattern constructed?

ELBOW

This elbow, figure 12-45, is a rectangular to round transition piece. Because it is symmetrical on one axis, the front and rear halves are alike.

A side view normally is drawn to show the shape of the round opening. However, the surface pattern can be developed without a side view, so it is omitted. The semicircles attached to the top and front views are used in place of the side view.

The lines of intersection are not parallel to the horizontal or profile planes. An auxiliary view, projected from the upper line of intersection, shows that the upper end of the center section is an ellipse. Instead of drawing the elliptical auxiliary view, points 1′, 2′, 3′, 4′, and 5′ are located by the same method used on the last drawing. When these points are connected, one-fourth of the ellipse is shown. The distances 1′ - 2′, 2′ - 3′, etc., are used when the triangles are constructed in the pattern.

ASSIGNMENT X

Draw a front view and a top view of the elbow. Draw the true length diagram. Develop the surface pattern of the center section by triangulation.

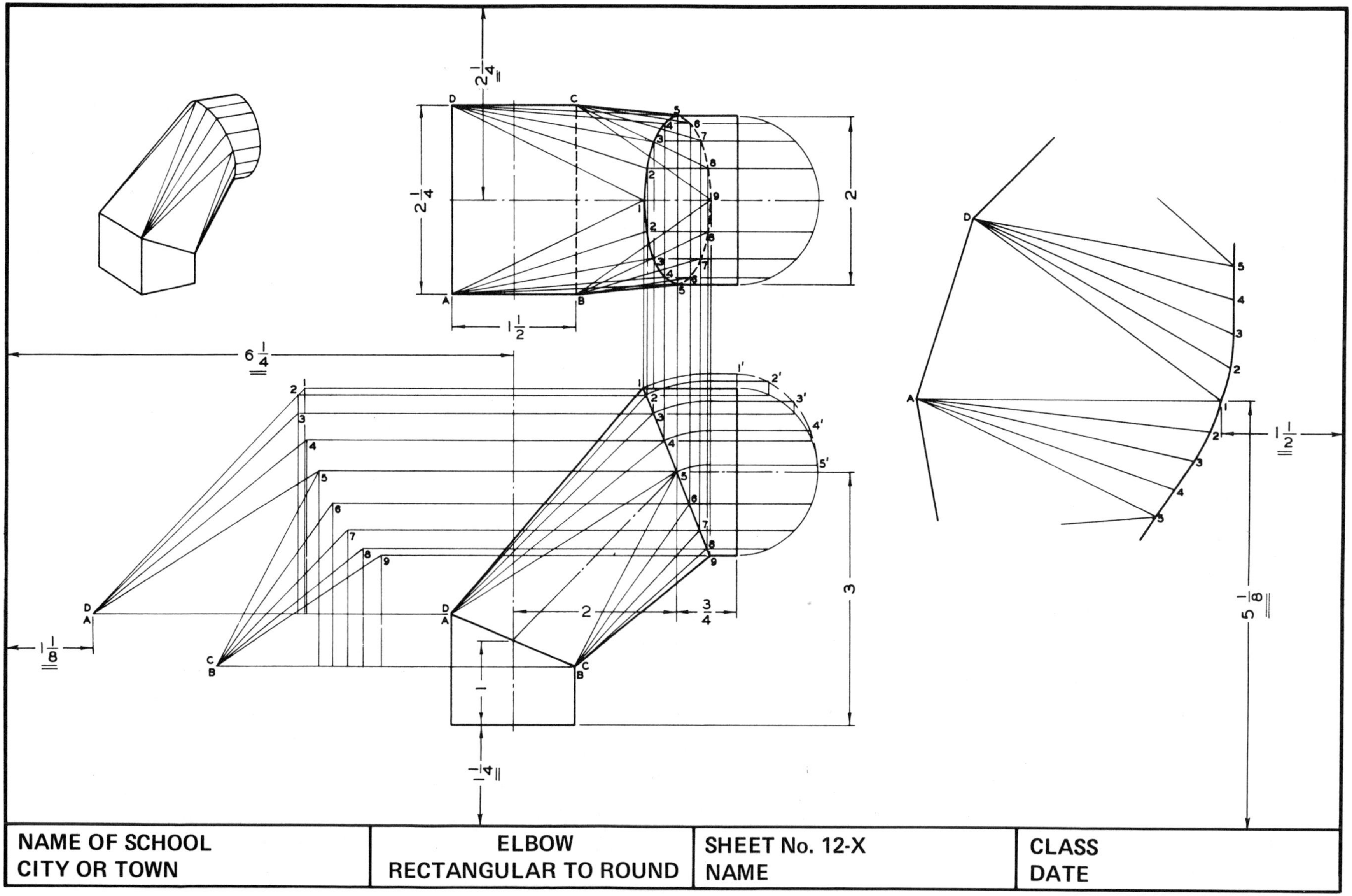
NAME OF SCHOOL
CITY OR TOWN
ELBOW
RECTANGULAR TO ROUND
SHEET No. 12-X
NAME
CLASS
DATE

Fig. 12-45

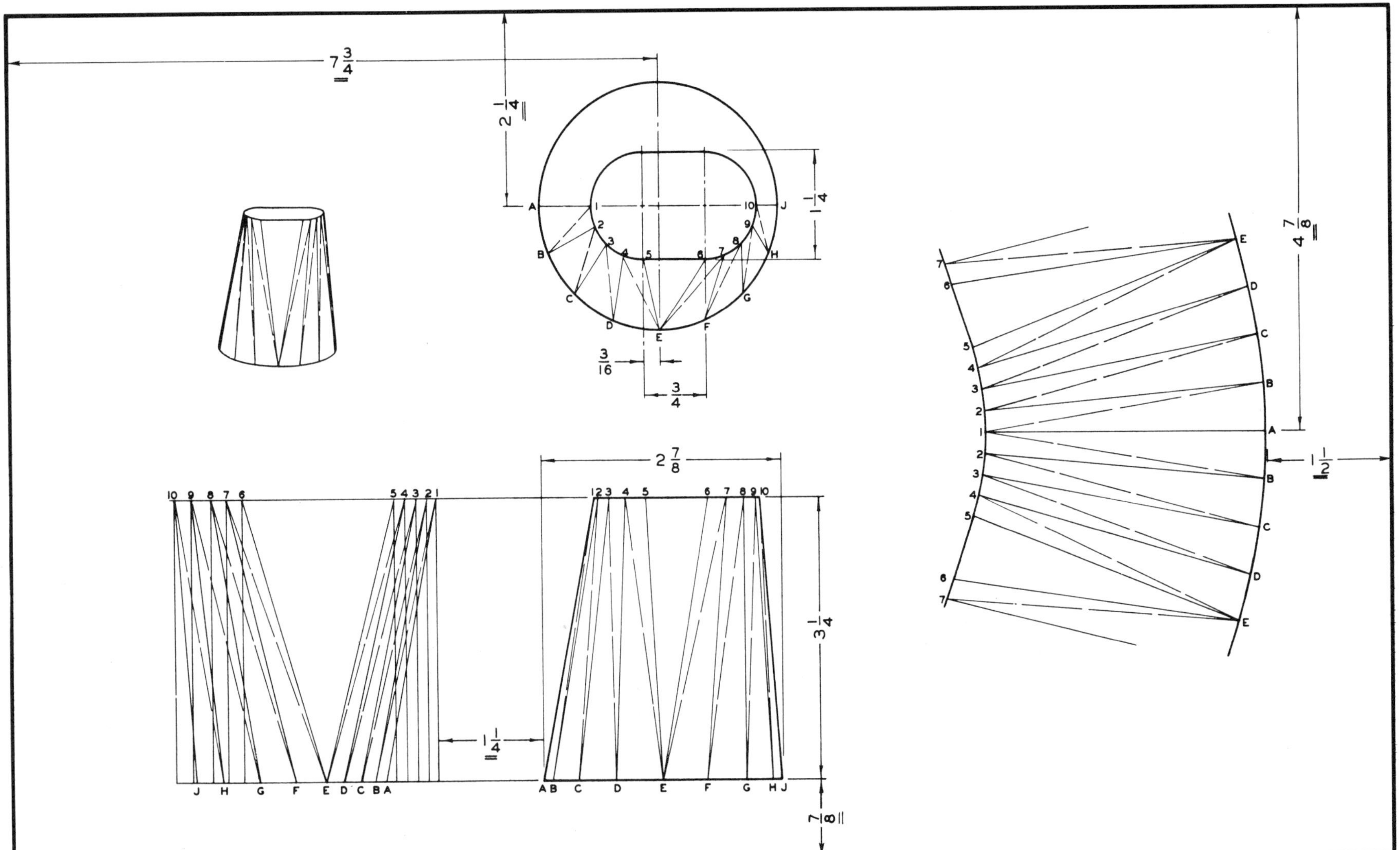

NAME OF SCHOOL CITY OR TOWN	ROUND TO OBLONG FITTING	SHEET No. 12-Y NAME	CLASS DATE

Fig. 12-46

Procedure

1. Lay out the margin lines on the drawing sheet and letter the title strip.
2. Draw the front view.
3. Locate points 1′, 2′, 3′, 4′, and 5′ in the front view.
4. Draw the top view. Draw the true length diagram.
5. Draw triangle A - 1 - D.
6. Construct triangles 1 - A - 2, and 1 - D - 2. The base of these two triangles is equal to distance 1′ - 2′.
7. Construct the remaining triangles, joining them on common sides. When triangles A - 5 - B and C - 5 - D are constructed, the length of their bases is taken from line A - B in the front view.
8. Complete the drawing and erase unnecessary lines.
9. Check the top and front view, the true length diagram, and the development of the surface pattern.

REVIEW QUESTIONS

1. Explain the three steps necessary to develop patterns by the triangulation method.
2. Name two places where rectangular to round elbows are used.

ROUND TO OBLONG FITTING

The method by which the surface on this round to oblong, figure 12-46, fitting is divided into triangles differs from the method used on the preceding drawings.

Because this fitting is symmetrical on one axis, and the front and rear halves are alike, only the front half is divided into triangles. In the front half, the base circle is divided into eight equal spaces, and the ends of the oblong are each divided into four equal spaces. Points A - 1, B - 2, C - 3, etc., are connected with lines. Points B - 1, C - 2, D - 3, etc., are connected with dashed lines to divide the surface into triangles. Dashed lines are an aid in laying out the triangles in the pattern.

The true-length diagram is drawn to find the true lengths of the oblique lines. The surface pattern is then drawn by triangulation.

ASSIGNMENT Y

Draw the top view and the front view of the round to oblong fitting. Draw the true-length diagram. Develop the surface pattern by triangulation.

Procedure

1. Lay out the margin lines on the drawing sheet and letter the title strip.
2. Draw the top view. Draw the front view.
3. Divide the front half of both views into triangles.
4. Draw the true length diagram.
5. Use the true lengths and construct the triangles one at a time. Join them on their common sides until the thirty-four triangles are drawn.
6. Complete the drawing and erase unnecessary lines.
7. Check the top and front view, the true length diagram, and the development of the surface pattern.

REVIEW QUESTIONS

1. Explain how the method used to divide the surface of this fitting into triangles differs from the method used on the preceding drawings.
2. Give two examples of where the round to oblong fitting is used.

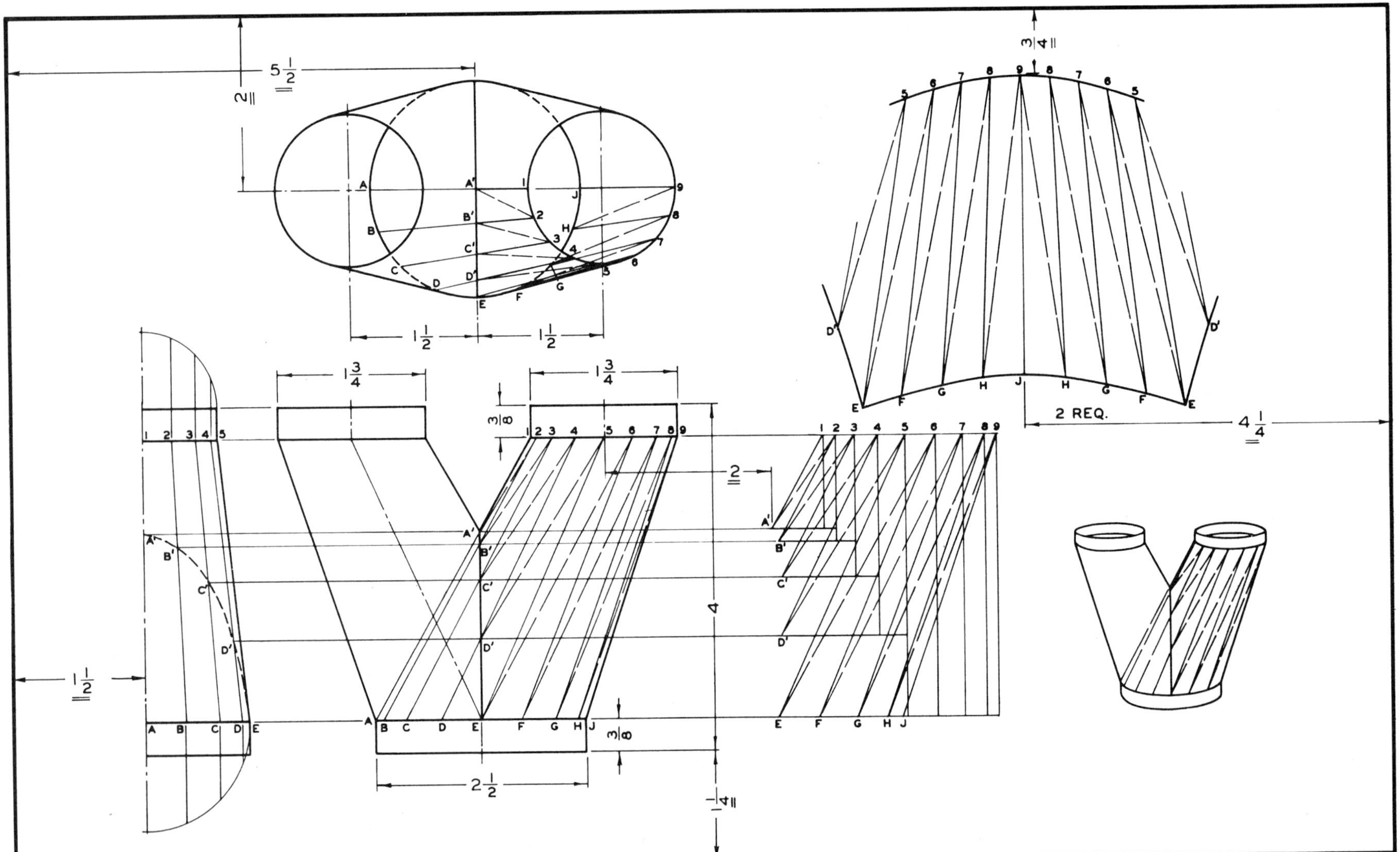

NAME OF SCHOOL CITY OR TOWN	TWO-PRONGED FORK	SHEET No. 12-Z NAME	CLASS DATE

Fig. 12-47

TWO-PRONGED FORK

The two-pronged fork, figure 12-47, consists of two truncated oblique cones joined to cylindrical collars.

This fitting is symmetrical on both axes, and the right and left forks are alike. Because both forks are alike, only one surface pattern is developed. Since the fitting is symmetrical on both axes, the front and rear halves also are alike. Therefore, only the front half of one fork is divided into triangles.

The line of intersection where the two forks join in the center is an elliptical curve. One-half of the side view is drawn to show one-half of this elliptical curve and to determine the distances between points A′ - B′, B′ - C′, C′ - D′, and D′ - E′.

ASSIGNMENT Z

Draw the top view, the front view, and one-half of the side view of the two-pronged fork. Draw the true-length diagram. Develop the surface pattern of one fork.

Procedure

1. Lay out the margin lines on the drawing sheet and letter the title strip.
2. Draw the top view. Draw the front view.
3. Draw one-half of the side view. In the side view, draw one-quarter of a circle at each end.
4. In the top view, divide the front half of the large circle into eight equal spaces. Divide the lower quarter circle in the side view into four equal spaces.
5. Project these points to the base of the cone in the front view and in the side view. Place letters on these points in the three views.
6. In the top view, divide one-half of the small circle into eight equal spaces. Divide the upper quarter circle in the side view into four equal spaces.
7. Project these points to the upper edge of the cone in the front view and in the side view. Place numbers on these points in the three views.
8. Connect A - 1, B - 2, C - 3, etc., in all three views.
9. Lines A - 1, B - 2, C - 3, and D - 4 cross the line of intersection in the top and front views to locate A′, B′, C′ and D′. Project these points from the front view to the side view and draw the elliptical curve in the side view.
10. Draw the dashed lines A′ - 2, B′ - 3, C′ - 4, etc., in the top and front views to divide the front half of one fork into triangles.
11. Draw the true-length diagram to find the true length of the oblique lines.
12. Begin with line J - 9 and construct the triangles, one at a time. Join the triangles on their common sides until the thirty-two triangles are drawn.
13. Complete the drawing and erase unnecessary lines.
14. Check the top, front, and side view; the true-length diagram; and the surface pattern.
15. On another sheet of drawing paper, construct the patterns for the two forks and the three collars.
16. Cut out these patterns and join them together. Use drafting tape or cellulose tape for fastening.

UNIT 13 RELATED MACHINE PROCESSES AND INFORMATION

CASTINGS

Many irregular shaped parts of machines are made from *castings.* A wood, metal, plastic, or wax *pattern* is needed to make a casting. A patternmaker casts the model in the desired material from the drawings of the part.

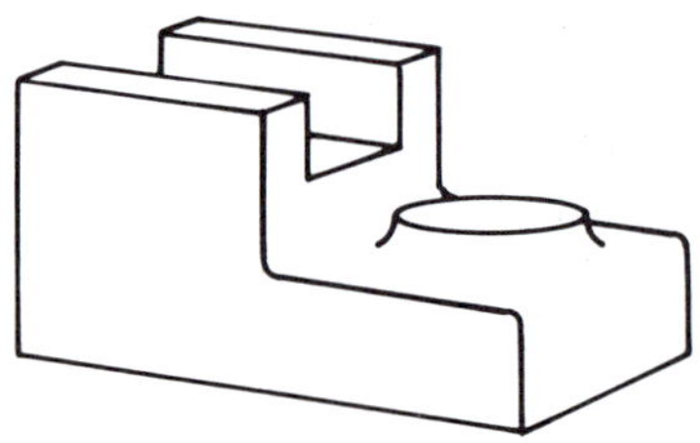

Fig. 13-1 Rough casting

In the foundry where castings are made, the pattern is placed in a box-like arrangement called a *flask.* Molding sand is then packed firmly around the pattern until the flask is completely filled. The construction of both the pattern and the flask is such that the flask can be taken apart to permit the pattern to be removed. The empty mold or cavity left in the sand is the same shape as the pattern used to make it.

After the pattern has been removed, the sections of the flask are put back in place and molten metal is poured into the mold. The casting is allowed to cool before removal from the sand. Castings made in sand molds are called *sand castings.*

The surfaces of sand castings are comparatively rough and pebbly. Consequently, some of the surfaces which touch other parts, commonly called *contact* or *working surface,* may have to be machined to a smooth finish or accurate dimensions. Such surfaces are called *finished surfaces,* figure 13-2.

FINISHED SURFACES

A surface to be machined or finished on a casting is marked with a 60-degree "V". The bottom of the "V" touches the line

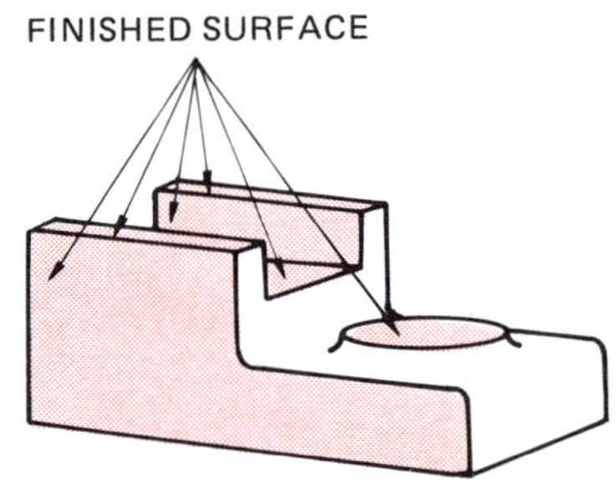

Fig. 13-2 Finished casting

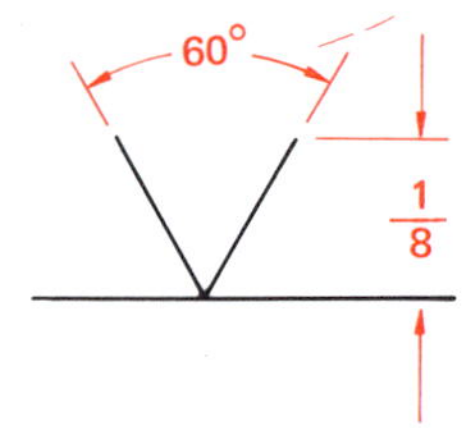

Fig. 13-3 The standard finish mark is drawn with a 30-60° triangle

representing the surface to be machined or finished, figure 13-3. If there is no numerical value placed on the "V", it is a noncontrol finish mark.

To illustrate the use of the finish marks, a picture of a rough casting is shown in figure 13-1. The casting with some finished surfaces is illustrated in figure 13-2. The working drawing in figure 13-4 shows how these finished surfaces are indicated by placing noncontrol finish marks on the lines of the drawing.

Placing Finish Marks

In the working drawing each line with a finish mark represents the edge of a finished surface on the casting. For example, the finished front surface of the casting is represented by the lower line of the top view, and by the line at the left of the side view.

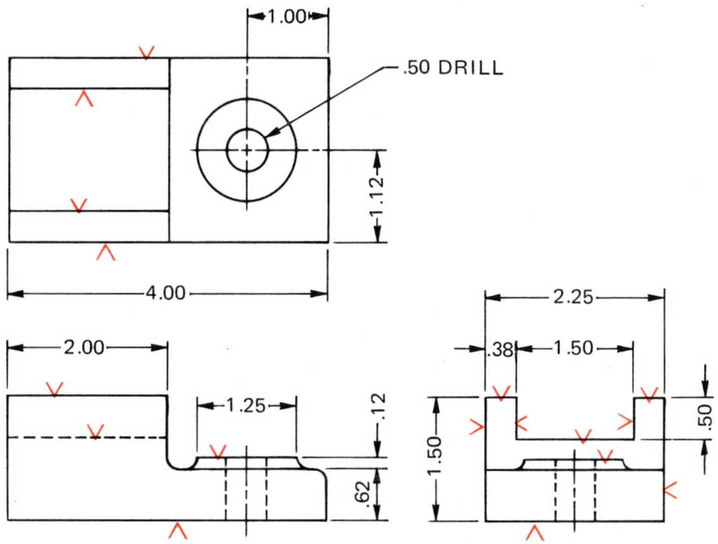

Fig. 13-4 A working drawing with finish marks

The two finished top surfaces are represented by the upper line of the front view, and by the two upper lines of the side view. The lines are often called the "edge view" of a surface.

Finish marks are placed on hidden lines as well as on visible edge lines. For example, the casting has a finished surface in the bottom of the slot. This finished surface is represented with a finish mark in the side view of the working drawing by a visible edge line and in the front view by a hidden line.

The finish marks are not required on holes that are drilled into a casting, but large holes in a casting (cored holes) that are to be finished to an accurate size should have a finish mark.

Formerly, an italic *f* indicated a surface to be finished. Although this symbol is still found on many drawings, its use is gradually being discontinued, figure 13-5.

Fig. 13-5 Old style finish mark

FAO And FAO Outside

If a casting is to be finished all over, a note is placed on the drawing to avoid putting finish marks on every surface. The note is "FINISH ALL OVER" or the abbreviation FAO.

If the casting is to be finished only on the outside, the note states FAO OUTSIDE. No indication of the quality of smoothness is given in either case.

Control Finish Marks

The standard finish developed for use on drawings when it is necessary to specify the quality of a finished surface is the check mark with a horizontal bar, figure 13-6. Numerical values may be added to the symbol to limit the roughness height, figure 13-7. The long line of the check mark is always on the right as the drawing is read. Control indicators are placed on only one view for each surface. If the control must apply to all surfaces, the check mark with its numerical control value follows the note FINISH ALL OVER. No other finish marks need to be placed on the drawing then. It can also be used for exceptions, such as "ALL OVER EXCEPT AS NOTED." This means that some surfaces have individual symbols for control other than the general control limit. Other symbols may be added to specify the direction of lay or the predominant surface pattern required.

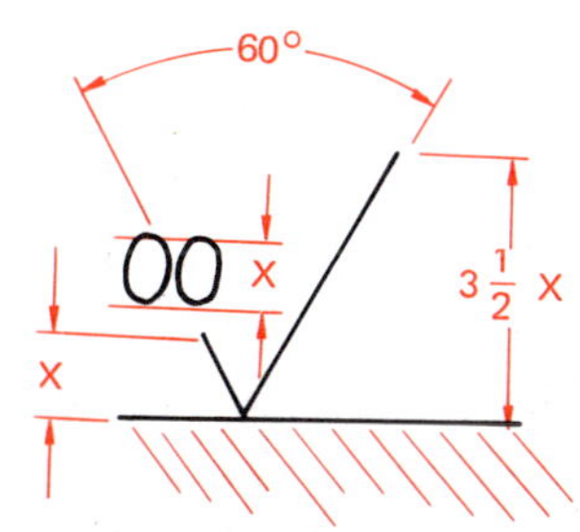

Fig. 13-6 Proportions of the roughness height symbol

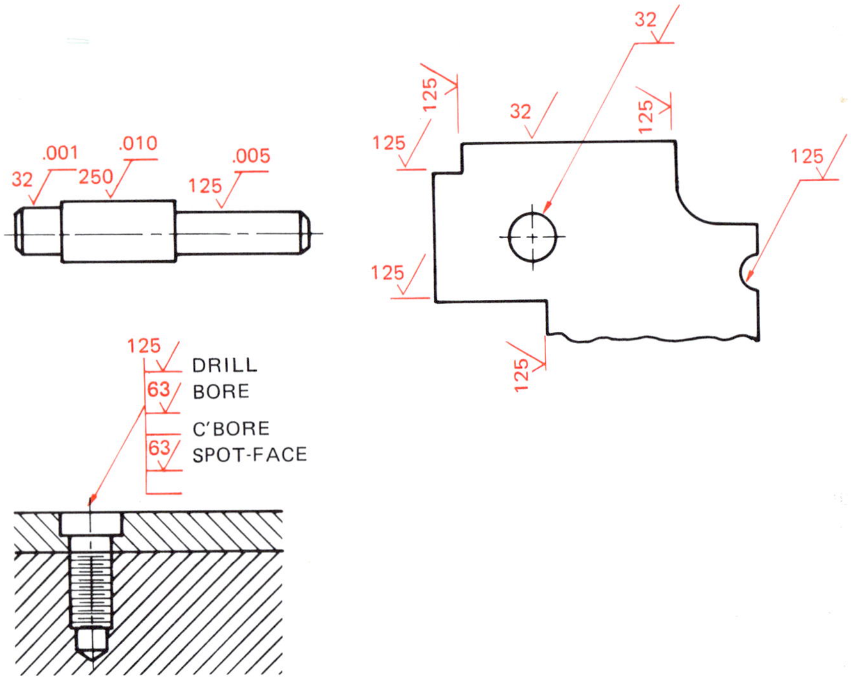

Fig. 13-7 Examples of roughness height symbols applied to objects

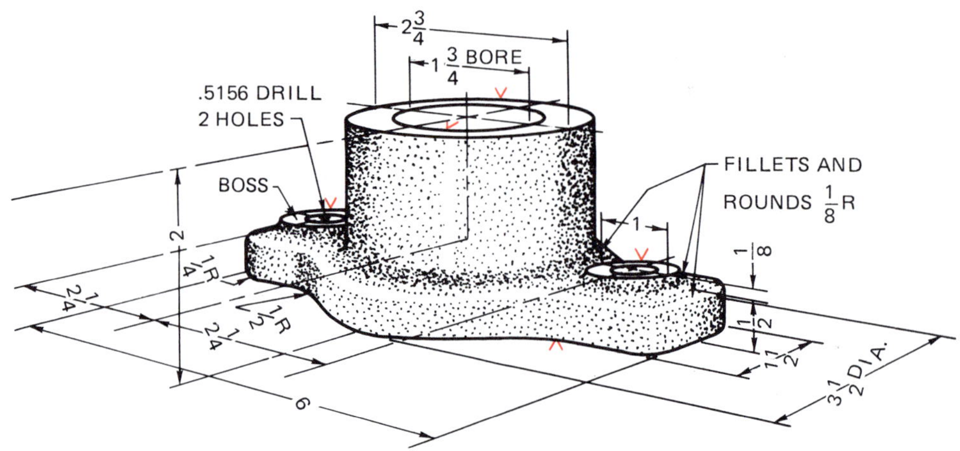

Fig. 13-8

ROUNDS AND FILLETS

Parts that are formed by casting are designed with rounded inside and outside corners, figure 13-9. The inside corners are called *fillets.* The fillet fills in the space making the part stronger and less likely to crack, due to stresses set up in the metal as it cools, than if the two surfaces met at a sharp 90-degree angle.

Finish Allowance

The finish marks on working drawings alert not only the machinist, but the patternmaker who must add material to the pattern before the castings are made. When a casting is finished to the dimensions specified on the working drawing, some metal is cut away. The mental which is removed is called the finish *allowance.* The pattern for the shaft support, figure 13-8, must have bosses approximately 1/4 inch high to allow for finishing the top surfaces to 1/8 inch. This difference provides approximately 1/8 inch finish allowance. Similarly, the finished casting is 2 inches high. Allowing 1/8 inch for finish allowance on the top surface of the pattern and 1/8 inch on the bottom surface, makes the pattern 2 1/4 inches high.

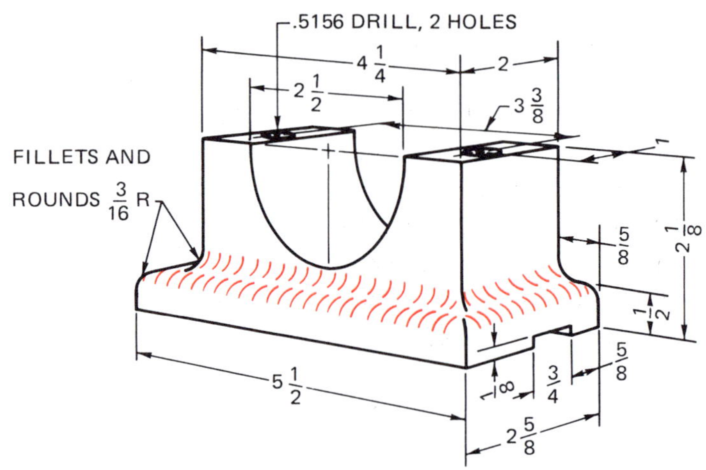

Fig. 13-9 Cast forms are designed with rounds and fillets.

The intersection of two outer surfaces normally forms an edge. When the edge is rounded off, it is called a *round.* An edge may be rounded slightly as a safeguard to prevent cutting the hands of the worker. Rounds may also be used to improve the appearance of a machine part.

Intersecting Rounds and Fillets.

When two rounds meet at corners, a curved edge is formed. In the top view, these curved edges appear as straight lines that slope at 45 degrees, figure 13-10. When fillets meet to form curved edges, these edges appear as straight lines on the top view, figure 13-11.

Fillets and rounds are represented in one of the views by tangent arcs. The arcs may be tangent to the sides of a right angle, tangent to the sides of an acute or obtuse angle, tangent to a straight line and an arc, or tangent to two arcs. There are several methods to locate the centers of tangent arcs as illustrated in figures 13-12 to 13-15.

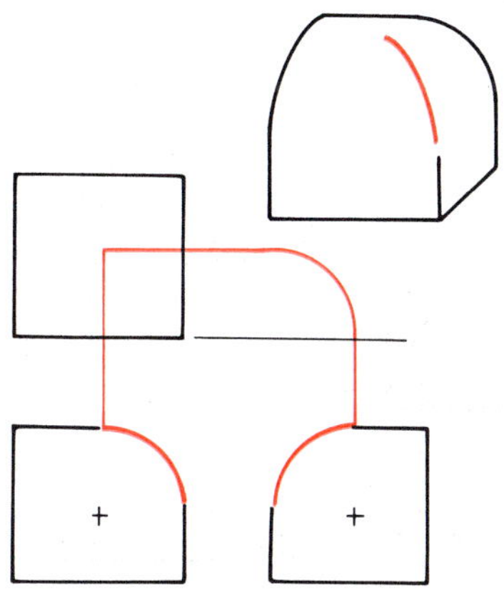

Fig. 13-10 Intersecting rounds

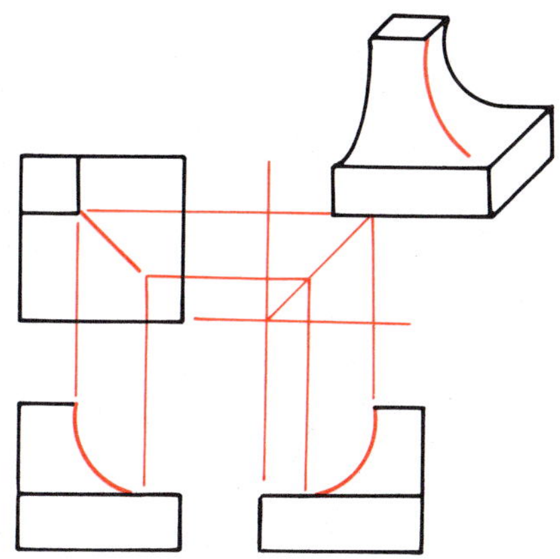

Fig. 13-11 Intersecting fillets

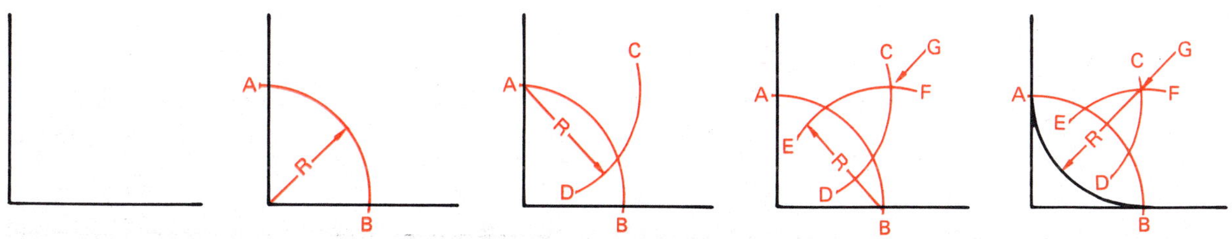

Fig. 13-12 Arc tangent to a right angle

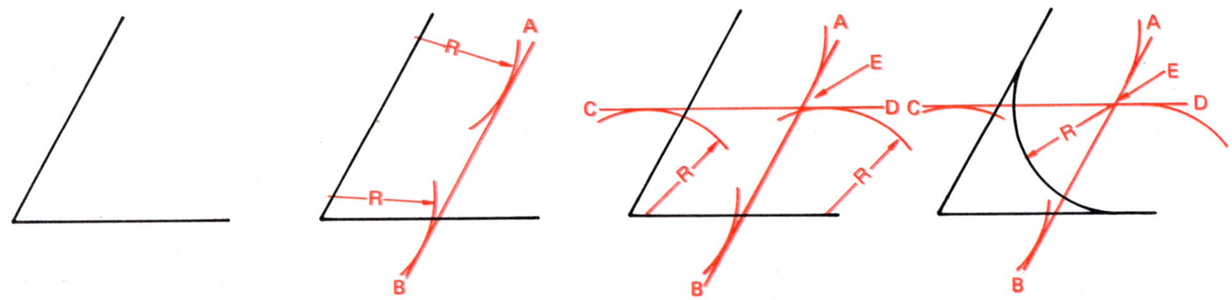

Fig. 13-13 Arc tangent to an acute angle

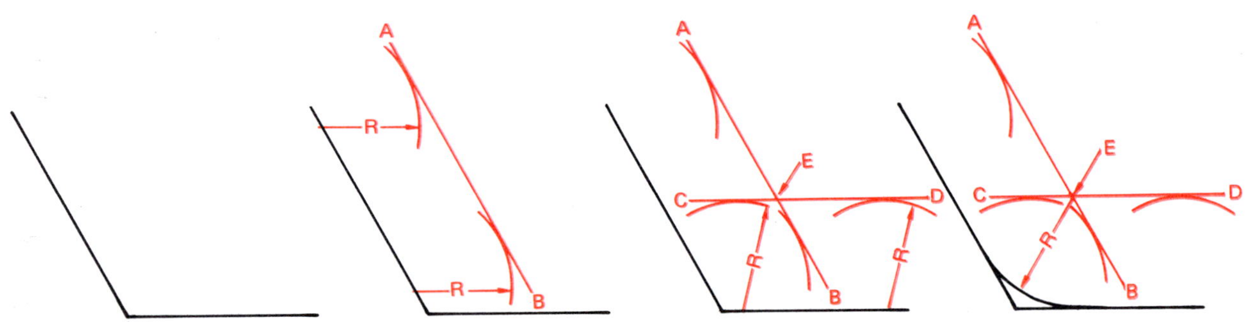

Fig. 13-14 Arc tangent to an obtuse angle

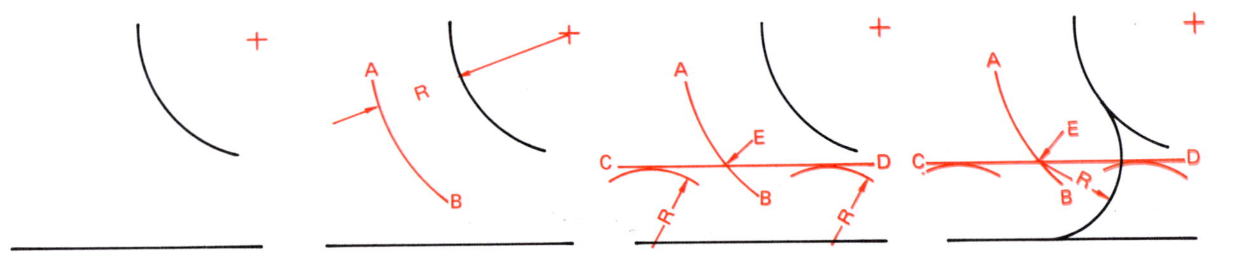

Fig. 13-15 Arc tangent to a straight line and an arc

Fillet Runouts

Intersecting fillets run out or disappear at the tangent points. The method of drawing the runouts of fillets is illustrated in figure 13-16. Lines are projected from the tangent points in the top view to mark the ends of the runouts in the front view. The same radius is used for the runouts and the fillets. However, for the runouts, only one-eighth of a circle is drawn.

How a runout is drawn depends on the shape of the adjoining parts. Figure 13-17 shows how runouts are drawn for intersections of various shapes.

BOSS AND PAD

When parts of a casting are raised to provide a contact or work surface, the raised portion must be finished to a controlled tolerance. The raised portion is called a *boss* if it is circular or cylindrical, figure 13-18 or a *pad* if it is rectangular.

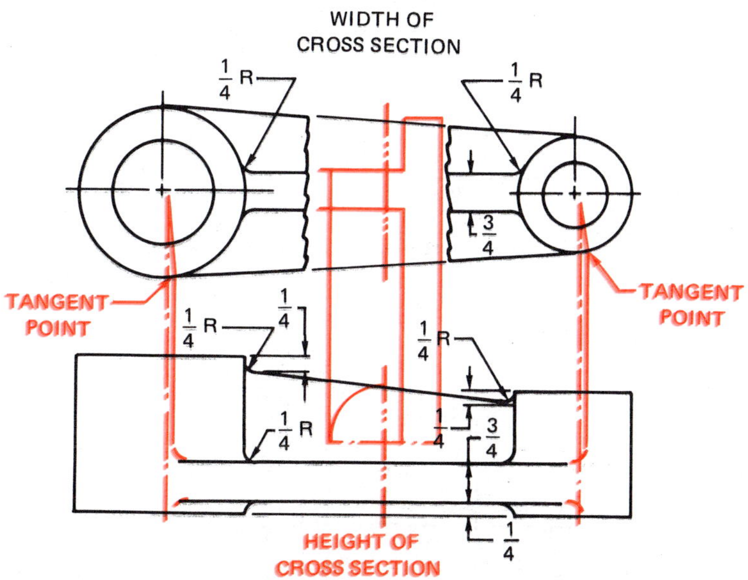

Fig. 13-16 Construction of a revolved section and runouts of fillets

SERRATE

When surfaces must hold a workpiece securely, they are sometimes cut with a series of 45 degree grooves called *serrations*. The serrated faces are hardened to resist wear.

CHAMFERS

The rough or sharp edges left on a machined part at the intersection of two surfaces are beveled or *chamfered* at a given

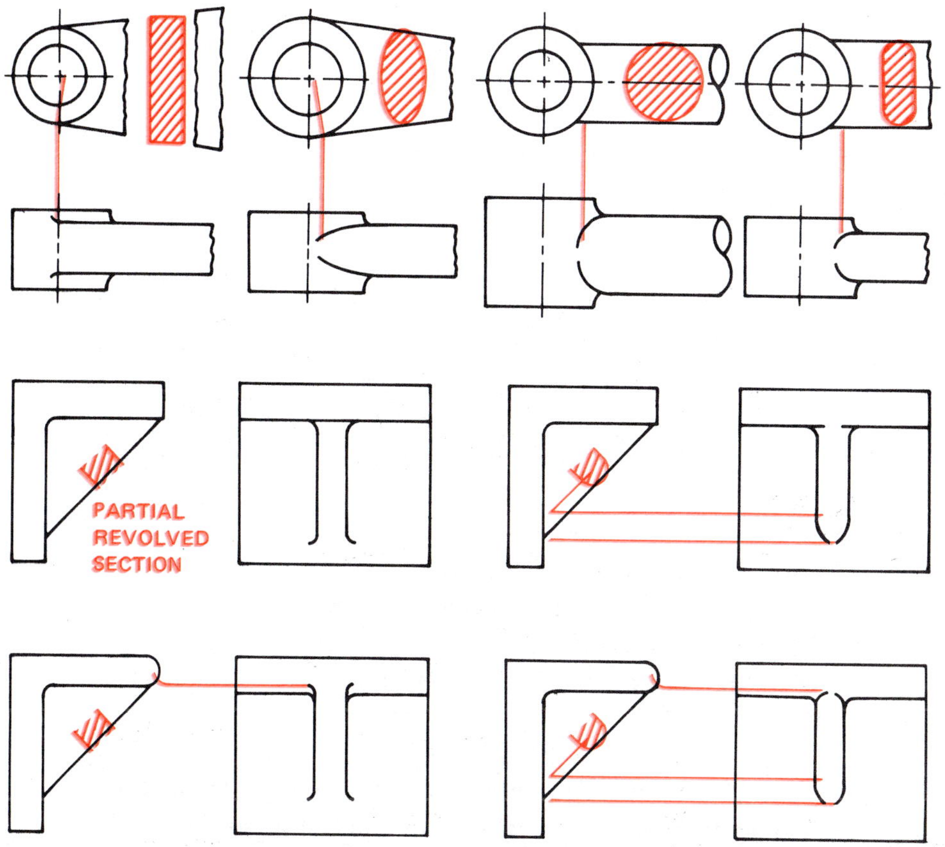

Fig. 13-17 Treatment of fillets, rounds, and runouts

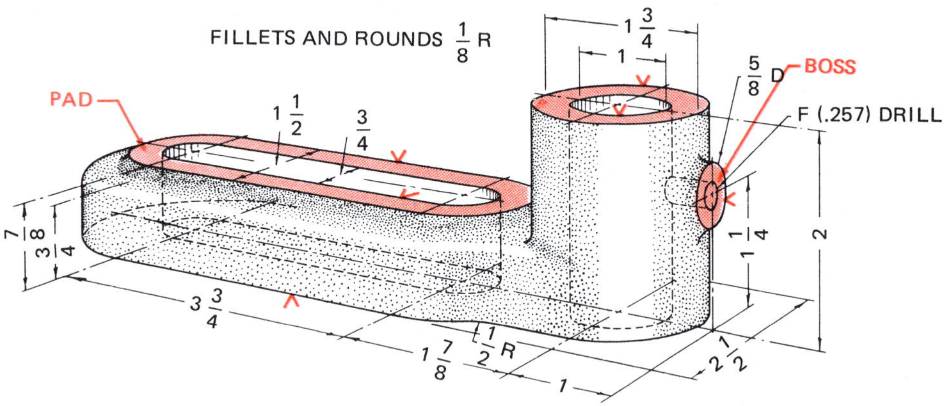

Fig. 13-18 Gear arm

angle to make them safe to handle and better looking. A chamfer on the end of a cylinder is shown in figure 13-19. A chamfer on rectangular surfaces is shown in figure 13-20. Chamfers are dimensioned by size and angle and the word, "CHAMFER," if the angle is 45 degrees. Otherwise, only the angle and the length are given.

KNURLING

Knurling is a machine operation performed on cylindrical parts in a lathe. Knurled surfaces can be gripped and turned more readily with the fingers of the hand than smoothly finished surfaces. Knurling is also used for a press fit between mating parts, figure 13-21.

The tool used for knurling has two rollers called knurls. The knurls have many V-shaped teeth cut across the face. These teeth are cut straight across on straight knurls and at an approximate 30-degree angle for diamond knurls. The angle of the teeth on a right-hand diamond knurl slopes in a direction opposite to that

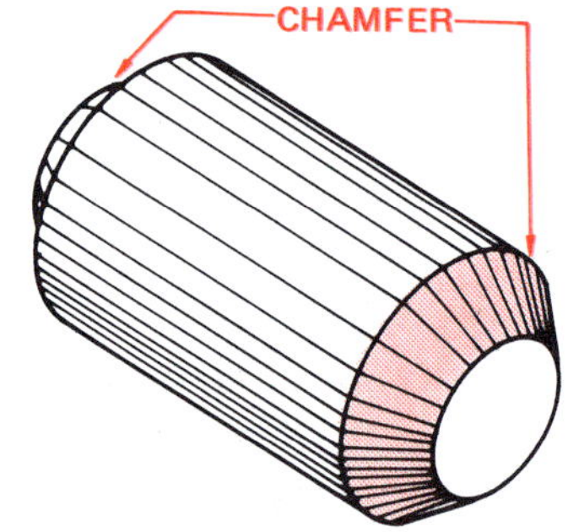

Fig. 13-19 Chamfer on a cylindrical surface

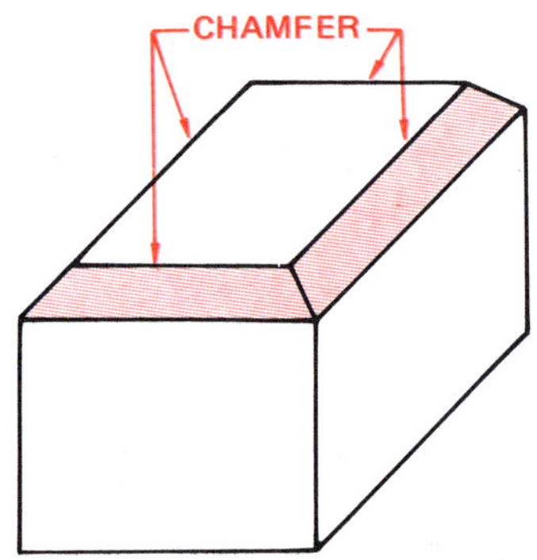

Fig. 13-20 Chamfer on plane surfaces

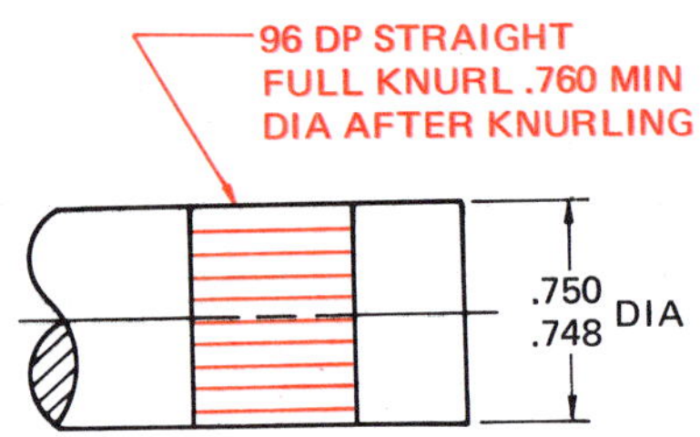

Fig. 13-21 A knurl for a press fit

on a left-hand knurl. As the work revolves in the lathe, the teeth of the knurl are forced into the work causing a series of grooves to be raised in two directions. Small diamond-shaped points are raised between the grooves, figure 13-22. Straight knurls make straight grooves and ridges on the knurled surface, figure 13-23.

Knurled parts are dimensioned with a note stating the pitch and the kind of knurl, figure 13-22 and 13-23. The pitch of a knurl is the number of cuts per inch measured parallel to the axis of the part.

- Fine knurling has 20 cuts per inch or 20 pitch.
- Medium knurling has 12 cuts per inch or 12 pitch.
- Coarse knurling has 8 cuts per inch or 8 pitch.

The knurls used for fine knurling have 80 teeth, medium knurls have 50 teeth, and coarse knurls have 34 teeth.

Diamond knurling is represented on drawings with two series of lines sloping in opposite directions. These lines are drawn at 30 degrees to the axis of the part. The lines are spaced by eye and need not be drawn to the actual pitch. Straight knurling is drawn with straight lines parallel to the axis of the piece.

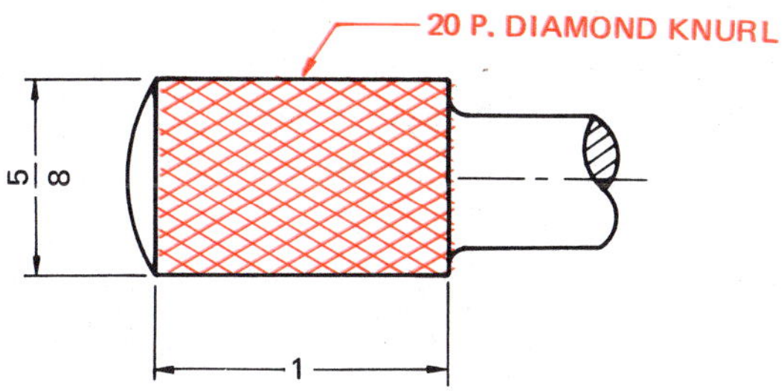

Fig. 13-22 Dimensioning a diamond knurl

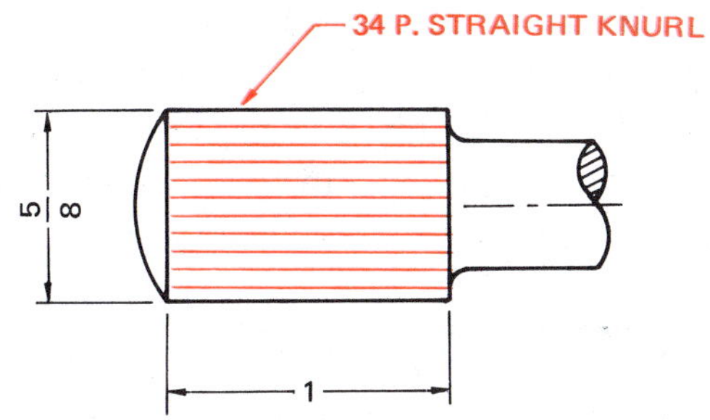

Fig. 13-23 Dimensioning a straight knurl

DIE FORGINGS

Steel machine parts having irregular shapes are made by casting or forging. *Forging* is a method of forming steel to the required shape by hammering or by pressure forming dies.

The forming *dies* are made in two halves. Both halves have a cavity or impression which, when the die is closed, makes a mold having the required form. Pieces of white hot steel, placed on the lower die are pressed into shape between the halves of the die. Space must be provided between the striking surfaces for the excess metal to escape. The excess metal squeezed out is called "the flash" or "the fin". It is removed from the forging in trimming dies. Some forgings may be a single die, others may require several dies to form the part by successive operations. Die forgings made in drop hammers are called *drop forgings.*

A small finish allowance must be provided on die forgings where surfaces are to be finished.

COLD DRAWN STEEL

Cold drawn steel is a steel that is finished while cold by drawing it through a hole in a hardened steel die. The die controls the dimensions of the steel rod or bar within close limits and imparts a smooth, bright finish to its surfaces.

CASE HARDENING

Case hardening is a heat treatment involving two steps: carburizing and hardening.

Carburizing is a process in which low carbon steel (not over .20 percent carbon) is heated to about 1700 degrees F in contact with material that is rich in carbon, such as charcoal. Some of the carbon is absorbed into the steel. The depth that the carbon penetrates the steel depends on the length of time the heat treatment is continued. When the carbon has penetrated to the desired depth, the steel is quenched in oil, water, or brine (a solution containing salt).

The steel is case *hardened* after it has been thoroughly carburized by heating it again to between 1400 and 1450 degrees F and then quenching in oil or water.

This carburizing and hardening treatment causes that portion of the steel which has a higher carbon content due to carburizing to harden. This results in a piece of steel that has a hard surface or case, and an unhardened interior or core.

HARDEN AND TEMPER

Spring steel is hardened by heating it to the correct temperature specified for that particular steel and then quenching in oil. Hardened steel is brittle. Therefore, the steel is again heated to a lower temperature and quenched to remove the brittleness. This second treatment is called *tempering* or *drawing.*

CYANIDE HARDEN

Low carbon steel (not over .20 percent carbon) is surface hardened by heating to 1500-1560 degrees F in a bath of molten sodium cyanide followed by quenching in oil, brine, or water. This heat treatment produces a hard, thin case.

TOOL STEEL

In general terms, tool steel usually contains a sufficient quantity of carbon so that it can be hardened throughout by heating to the correct temperature and then quenching. There are many types of tool steels made for many purposes. In addition to carbon, they may contain other elements such as nickel, chromium, and tungsten, in various quantities and combinations. Each type requires its own specific heat treatment.

GENERATING AND REPRESENTING ROUND HOLES

Holes are generated in many different ways. Some, such as drills, reams, bores, and cores are simple cylindrical shapes that only differ in size or the method of making them. Others, such as the counterbore, countersink, and spot face are designed to provide a special surface for fasteners.

Drilled Holes

Drilling is making a hole with a rotating tool called a *drill.* The sizes of drills are standardized. See the appendix for standard drill size equivalents of both metric and inch drills.

Blind drilled holes are drilled into a part, but not through it. When blind drilled holes are shown on a drawing, an angle is drawn in the bottom of the holes to correspond approximately to the angle on the point of the drill, figure 13-24. The angle is drawn 120 degrees, or 60 degrees on each side of the center line, figure 13-25.

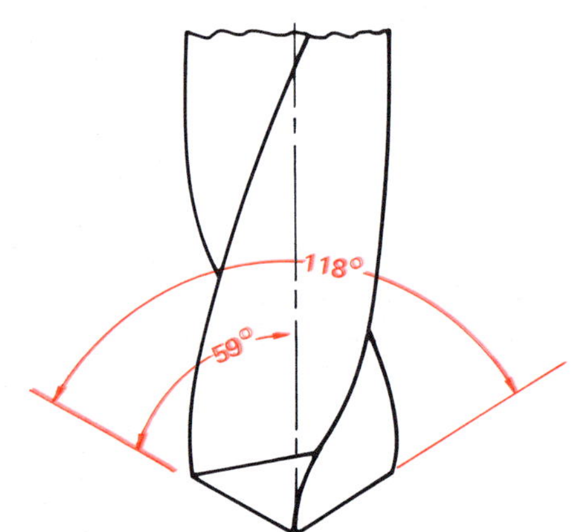

Fig. 13-24 Twist drill

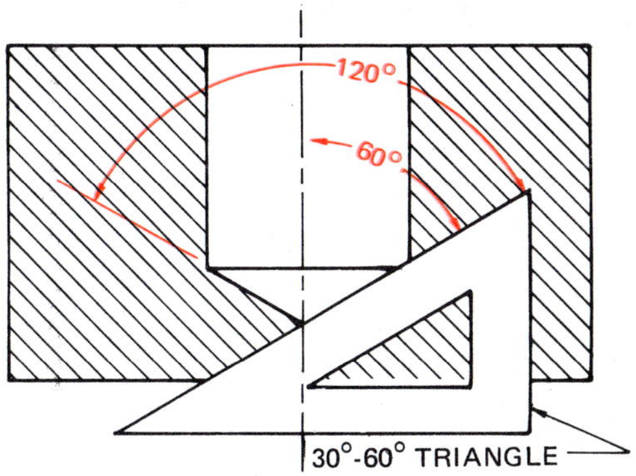

Fig. 13-25 30°-60° triangle

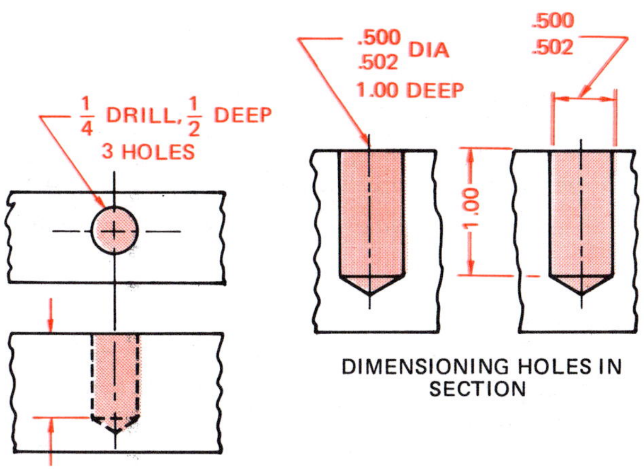

Fig. 13-26 Blind hole measurement

The depth of a blind drilled hole is measured to the depth of the full diameter, figure 13-26.

Reamed Holes

These holes are drilled with a drill whose diameter is slightly less than the finished size of the holes. The holes are then accurately machined to an exact size with a fluted tool called a *reamer*. The reamer produces a smooth finish in the hole.

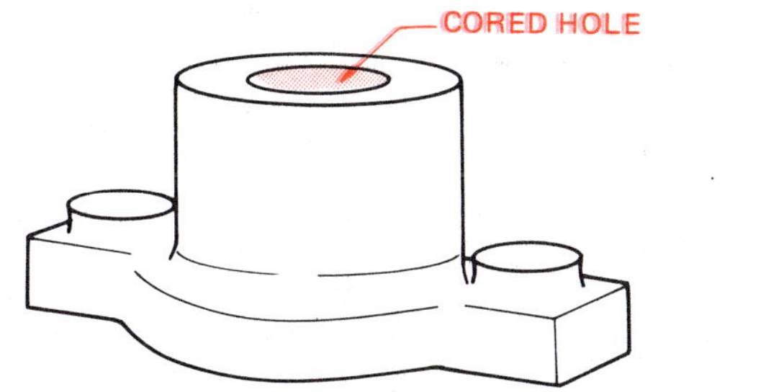

Fig. 13-27 Rough casting for a shaft support

Cored Holes

The large hole in the casting, figure 13-27, is a cored hole. It is made by inserting a baked sand core in the mold. When the metal is poured into the mold, it flows around the sand core. After cooling, the core is broken out, leaving a hole in the casting the size and shape of the core.

Cored holes in castings save metal. Machining time is also saved because less metal must be removed from the casting. Cored holes make a cast piece lighter in weight.

Cored holes that are to be finished to accurate dimensions by boring are marked with a finish mark on the drawing.

Boring

Boring is the process of finishing a cored hole. The cored hole is enlarged to size by a rotating tool bit. Bored holes are specified as in figure 13-28. When a cored hole is to be bored, material is added to the pattern to provide a finish allowance all around the hole. The hole in the rough casting, therefore, is made smaller.

Counterbored Holes

A cylindrical cut on the end of a drilled or bored hole that provides a flat shoulder for the head of a recessed fastener is called a *counterbore*. The diameter and depth of the counterbored hole

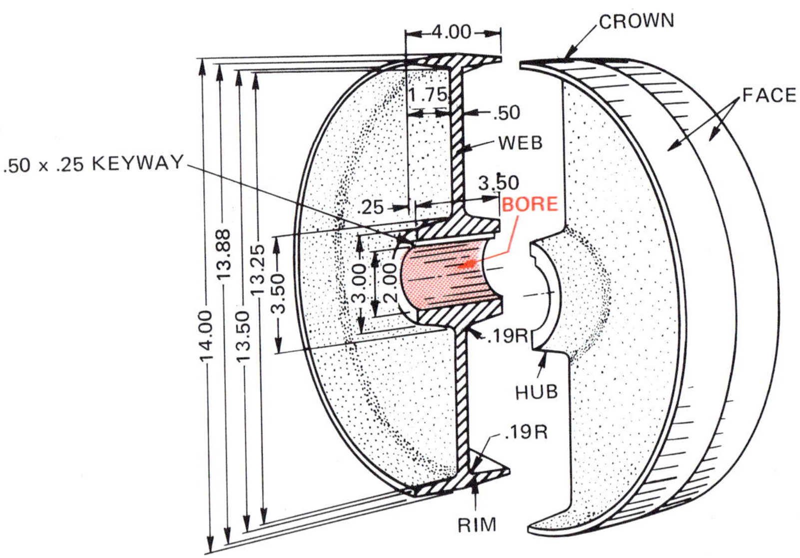

Fig. 13-28

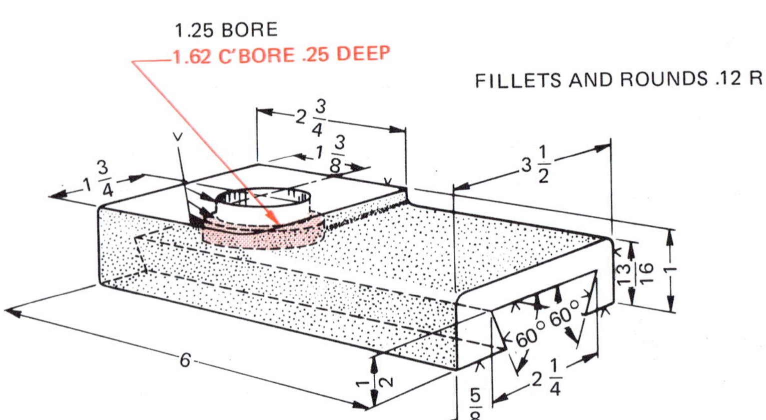

Fig. 13-29 Notation for a counterbored hole

is indicated along with the hole size in one single note, figure 13-29. The note is usually placed on the circular view of the hole with the arrowhead on the leader touching the outside circle. The abbreviation for counterbore is CBORE.

Countersinking

A cone-shaped tool called a countersink (CSK) is used to cut an angle at the end of a drilled or bored hole. The angle of the point may vary depending on the requirements of the job. Countersinking allows the heads of screws or rivets to be flush with the surface so they do not interfere with moving parts. Countersunk holes are specified in two ways, figure 13-30.

Spot-faced Holes

Spot-faced holes are similar to counterbored holes except the finished surface is only 1/32 to 1/16 inch below the rough surface. It is a smooth bearing surface for a bolt head or other fastener that does not have to be flush with the surface, but cannot function

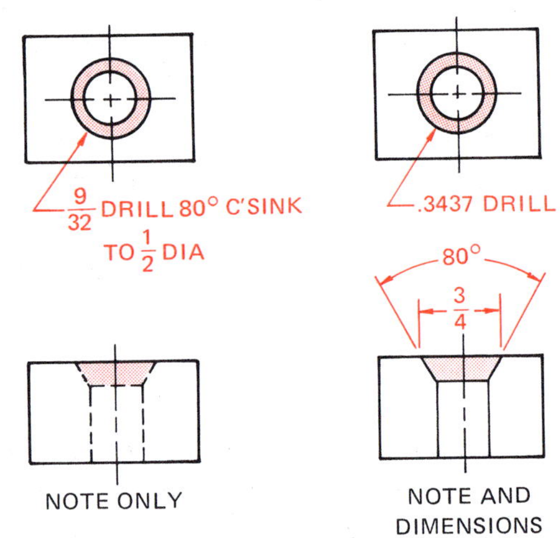

Fig. 13-30 Dimensioning countersinks

effectively against a rough casting surface. It is not necessary to place finish marks on these surfaces because a finish allowance is not required for spot-facing. A spot face is dimensioned with a note as shown in figure 13-31. Only the diameter is indicated. The depth of the spot face is understood.

KEYS, KEYSEATS, AND KEYWAYS

Various shapes of keys are used for locking pulleys and shafts together. A keyway is cut lengthwise in the hub of a pulley. A keyseat of the same width and depth is cut in a shaft that fits into the hole. When the shaft is in place and aligned, a square piece of steel called a *key* is inserted into the keyway and keyseat to prevent slippage, figure 13-32.

Keyways for square and flat keys are dimensioned with a shop note. The width of the keyway is given first, followed by the depth, figure 13-33.

The American Standard woodruff key is a semicircular piece of metal that is inserted in the hub, figure 13-34. The recess in the shaft that holds this key is the *keyseat.*

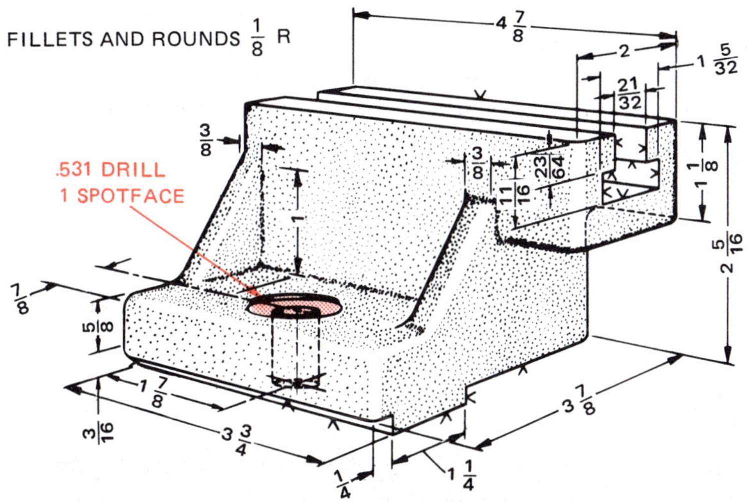

Fig. 13-31

MACHINE TAPERS

Machine tapers are used extensively in the design of machinery as a rapid method of connecting and aligning two mating parts that can still be readily separated when necessary.

A machine taper shank is really a conical-shaped wedge which, when rammed with a little force into a corresponding tapered socket, becomes tightly wedged in the socket. Such tapers are called self-holding tapers or slow tapers.

Several series of tapers are commonly used in the design of

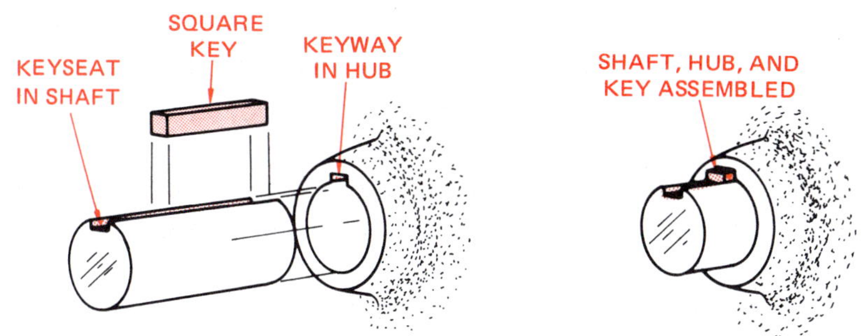

Fig. 13-32 Key and keyway locking pulley and shaft together

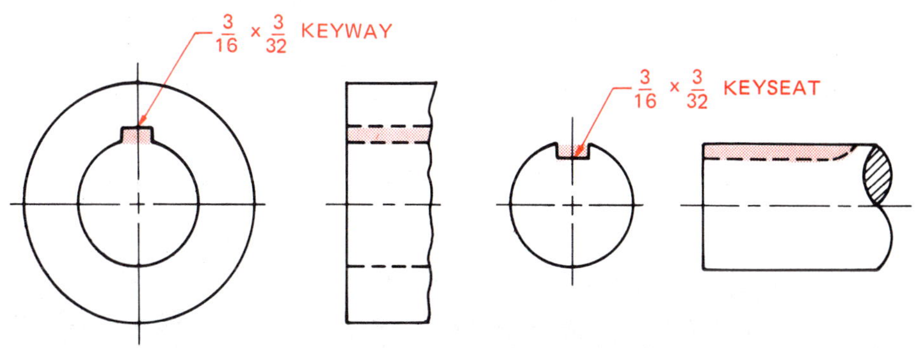

Fig. 13-33 Dimensioning keyways

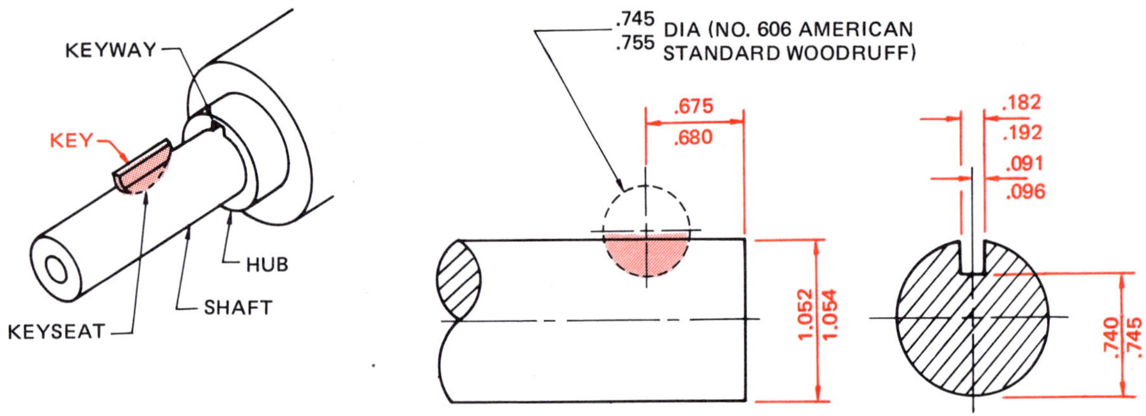

Fig. 13-34 Dimensioning woodruff keyslots

machines. The Morse Taper Series has approximately 5/8-inch taper per foot and is used on the tapered shanks of twist drills. The taper on the shank of the Magic Chuck Body is No. 2 in the Morse Taper Series.

The Brown and Sharpe Taper Series has approximately 1/2-inch taper per foot and is used on a variety of tools and machines.

The American Standard Machine Tapers is a series which aims to reduce the number of sizes of tapers and to combine the best features of the various taper systems. The first three sizes in this system are taken from the Brown and Sharpe Taper Series. The next seven sizes are taken from the Morse Taper Series. These are followed by ten larger sizes having a taper of 3/4 inch per foot. The dimensions of standard tapers are given in tables in the Appendix.

Taper Sockets

A taper socket is finished with a taper reamer. It is tested for size with a taper plug gauge which must enter the hole to a line on the gauge called the gauge line.

Dimensioning Standard Tapers

Standard tapers are best dimensioned by giving the diameter at the gauge line, the length of the taper, and a note stating the number and the standard taper series. For taper shanks, the amount of taper per foot is often given.

Dimensioning Special Tapers

Special tapers are dimensioned by giving the amount of taper per foot, the diameter at one end, and the length of taper. Sometimes the amount of taper per foot and the diameter at both ends are given, with the length of the taper omitted, figure 13-35.

Tang

The *tang* fits into a slot in the bottom of the tapered socket. It prevents the shank from turning in the socket if the friction grip of the taper does not hold.

SCREW THREADS

Screw threads have an important part in the construction and operation of machines. Screw threads are used for fastening parts

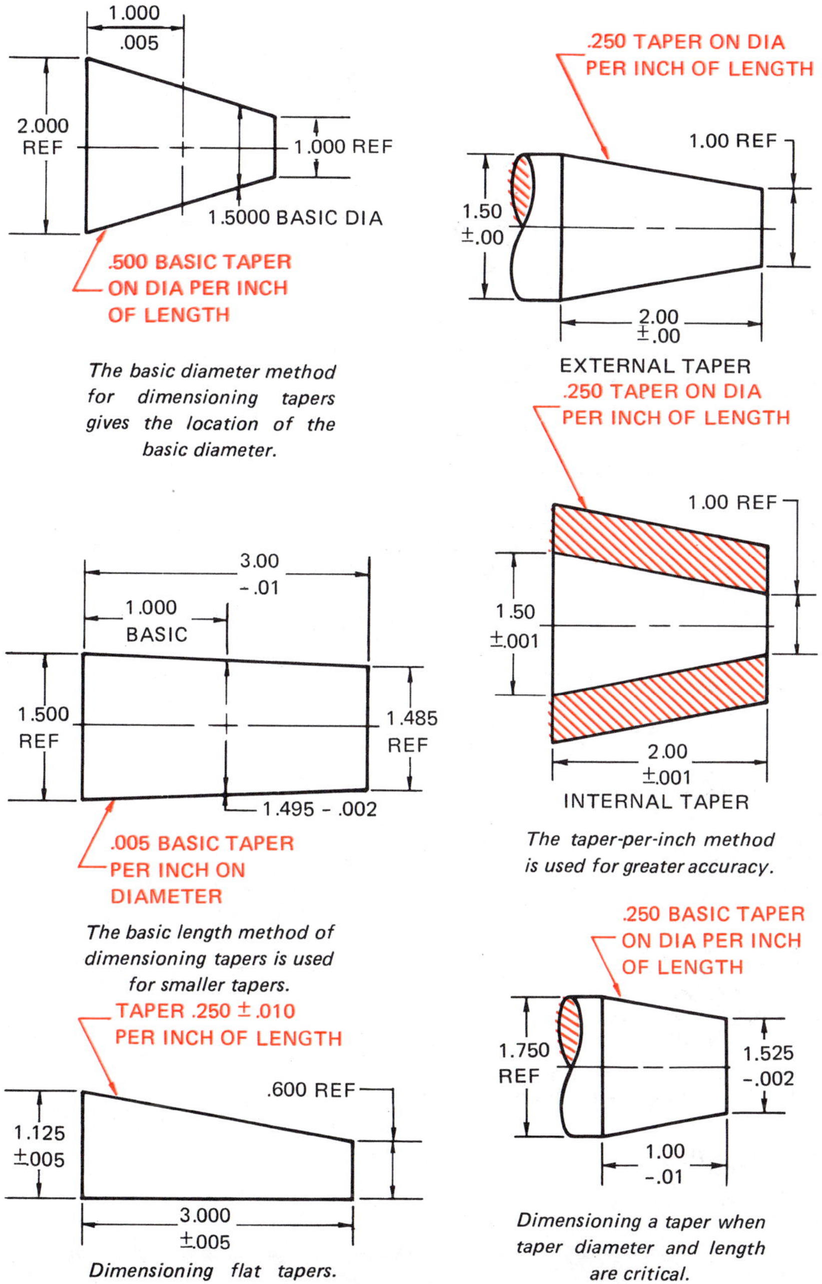

Fig. 13-35 Methods of dimensioning tapers

together, for making adjustments, and for transmitting power. The purpose of this unit is to describe the several forms of screw threads, their various applications, and the conventional methods of representation. The various standard screw-thread forms are shown in figure 13-36.

Screw-Thread Forms

- The Sharp V thread is not commonly used except for some special purposes.
- The American National Form of thread, commonly called the American Standard Thread, is most widely used in the United States for fastening parts together.
- The Whitworth Standard thread is used principally in Great Britain.
- The Acme thread, the 29. Degree Stub thread and the 60-Degree Stub thread (not shown), and the Modified Square thread are used primarily for transmitting power.
- The Standard Worm thread is used to transmit power to worm wheels.
- The Buttress thread is designed to take exceptionally high stresses in one direction only, parallel to the thread axis. They are used on the breech-block mechanisms of large guns and airplane propeller hubs.
- The American National Rolled threads are used for screw shells of electric sockets and lamp bases.
- The American National Taper Pipe threads are used on pipe and pipe fittings. For some applications, the American National Straight Pipe thread (not shown) are used.

Terms

The definitions of terms relating to screw-threads are illustrated in figure 13-37.

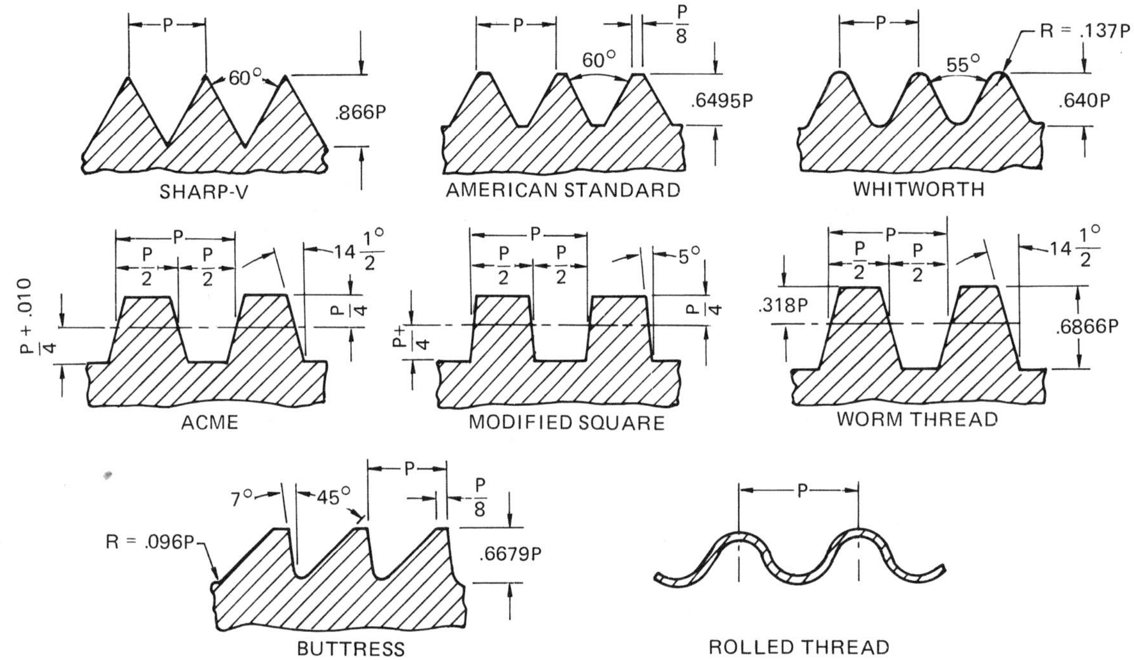

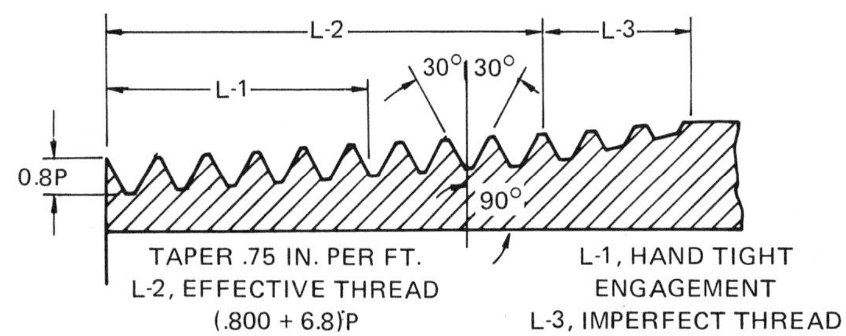

Fig. 13-36 American National taper pipe thread

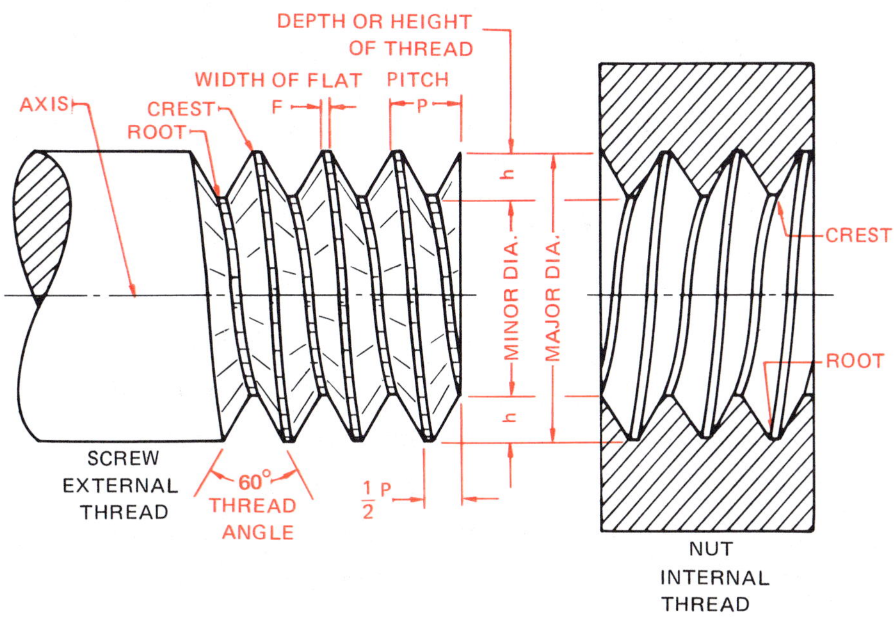

Fig. 13-37 Screw-thread notation

- *Screw thread* is a ridge of uniform section in the form of a helix on the external or internal surface of a cylinder.
- *External thread* is a thread on the outside of a member.
- *Internal thread* is a thread on the inside of a member. Example: A thread in a nut.
- *Major diameter* is the largest diameter of the thread of a screw or nut.
- *Minor diameter* is the smallest diameter of the thread of a screw or nut.
- *Pitch* is the distance from a point on a screw thread to a corresponding point on the next thread measured parallel to the axis.
- *Lead* is the distance a screw thread advances axially in one turn. On a single-thread screw, the lead and the pitch are the same. On a double-thread screw, the lead is twice the pitch. On a triple-thread screw, the lead is three times the pitch.
- *Angle of thread* is the angle included between the sides of the thread measured in an axial plane.
- *Crest* is the surface of a thread joining two sides at the major diameter of a screw and the minor diameter of a nut.
- *Root* is the surface of a thread joining two sides at the minor diameter of a screw and the major diameter of a nut.
- *Depth* is the distance from the root of a thread to a line between two adjacent crests measured perpendicular to the axis.
- *Axis* is the center line running lengthwise through a screw.

American National Thread Series

The American National threads have been classified into seven standard thread series as follows: American National coarse, fine, extra-fine, 8-pitch, 12-pitch, 16-pitch thread series, and special.

- The *American National coarse thread series* (NC) is recommended for general use in machine construction. Sizes range from a No. 1 (.073-inch diameter) to a 4-inch diameter.
- The *Fine thread series* (NF) is recommended where conditions require a fine thread. Sizes range from a No. 0 (.060-inch diameter) to a 1 1/2-inch diameter.
- The *Extra fine series* (NEF) is recommended where thin walled material is to be threaded, and where depth of thread must be held to a minimum. Sizes range from a 1/4-inch diameter to a 2-inch diameter.
- The *8-pitch thread series* (8N) has eight threads per inch for all diameters. It is generally used on bolts for high pressure pipe flanges, cylinder head studs, and similar fastening. Sizes range from a 1-inch diameter to a 6-inch diameter.
- The *12-Pitch thread series* (12N) has twelve threads per inch for all diameters. It is used in boiler practice and for thin

nuts and threaded collars. Sizes range from a 1/2-inch diameter to a 6-inch diameter.

- The *16-Pitch thread series* (16N) has sixteen threads per inch for all diameters. It is used on threaded adjusting collars. Sizes range from a 3/4-inch diameter to a 4-inch diameter.
- *Special threads* (UNS, NS, UN) cover special combinations of diameter, pitch and length of engagement.

The thread diameters and the number of threads per inch for the six standard thread series are given in the Appendix at the end of the book.

Classification of Fits

Five distinct classes of screw-thread fits have been established as a guide to insure interchangeable manufacture of screw-thread parts through the country.

- *Class 1* recommended only for screw-thread work where clearance between mating parts is essential for rapid assembly and where shake or play is not objectionable.
- *Class 2* represents a high quality of commercial screw-thread product and is recommended for the great bulk of interchangeable screw-thread work.
- *Class 3* represents an exceptionally high grade of commercially threaded product and is recommended only in cases where the high cost of precision tools and continual checking of tools and product is warranted.
- *Class 4* intended to meet very unusual requirements more exacting than those for which class 3 is intended. It is a selective fit if initial assembly by hand is required. It is not, as yet, adaptable to quantity production.
- *Class 5* includes interchangeable screw-thread work consisting of steel studs set in hard materials (cast iron, steel, bronze, etc.) where a wrench-tight fit is required.

Symbols for Identifying American Standard Threads

American standard threads are specified by stating the diameter; the number of threads per inch; the thread series, NC, NF, NEF, 8N, 12N, 16N; and the class of fit. For a left-hand thread, the letters LH are placed after the class of fit. The letter A following the class of fit indicates an external thread. An internal thread is signified by using a B after the number for class of fit.

REVIEW QUESTIONS

1. For what three purposes are screw threads used?
2. What is the principal use of the American Standard thread?
3. Which threads are used for transmitting power?
4. What are the names of the seven standard thread series?
5. How many classes of fits are there for screw threads?
6. Define the meaning of each term listed.

a. Screw thread	d. Pitch	g. Depth
b. Major diameter	e. Crest	h. Class of fit
c. Minor diameter	f. Root	

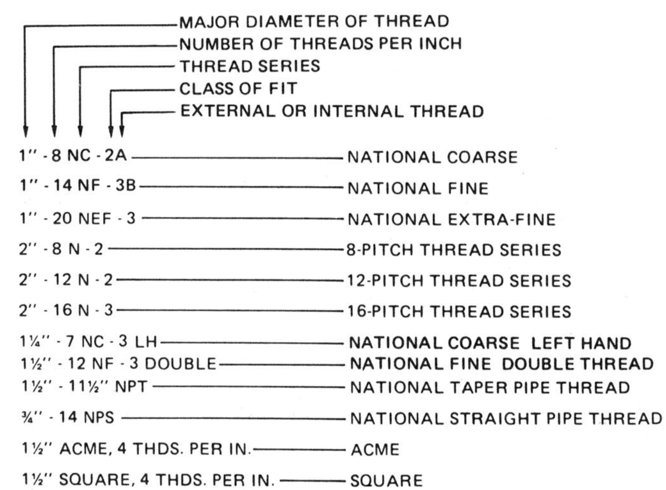

Fig. 13-38

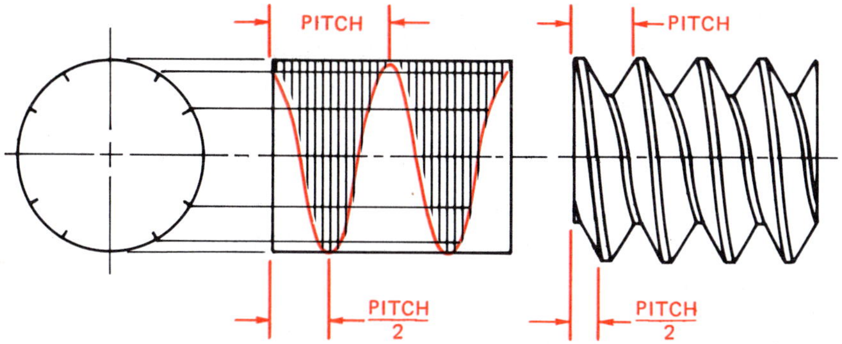

Fig. 13-39 A helix and a screw thread

REPRESENTING SCREW THREADS ON DRAWINGS

A screw thread is defined as a ridge of uniform section in the form of a helix on the external or internal surface of a cylinder. A helix is shown in figure 13-39. It is a curve on the surface of a cylinder that advances lengthwise as it winds around the cylinder.

The screw thread advances lengthwise as it winds around in the same manner as the helix. The amount that the helix and the screw thread advances lengthwise in one turn is called the *lead.* To review, on a single thread the *lead* and the *pitch* are the same. The *pitch* is the distance from a point on a thread to the corresponding point on the next thread measured parallel to the axis. The pitch is always found by dividing one inch by the number of threads per inch.

Semiconventional Screw Thread Representation

In actual practice, it would take too much of the drafter's time to draw American Standard screw threads as in figure 13-39. Instead, semiconventional threads or "thread pictures" are drawn as in figure 13-40.

In place of drawing the crests and roots as flat surfaces, they are drawn with single lines. The outline of the threads is drawn like the 60-degree Sharp-V thread. The curved lines of the

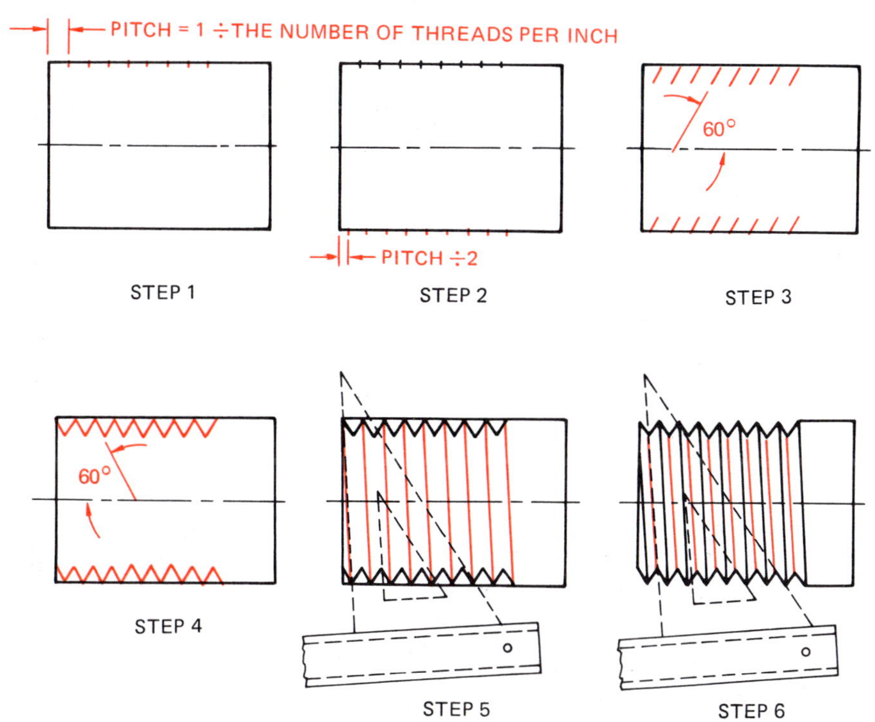

Fig. 13-40 Construction of semiconventional external threads

helix are drawn as straight lines. The drawing shows semiconventional external threads, both right-hand and left-hand.

Figure 13-41 shows semiconventional internal threads. In the end view of the internal threads, the major diameter circle is invisible while the minor diameter circle is visible. When looking into a threaded hole, only the minor diameter circle is seen.

In the exterior view and the full section of semiconventional internal threads, crest and root lines are omitted, but they are drawn to the center in the half section.

ASSIGNMENT A

Draw external screw threads, figure 13-40.

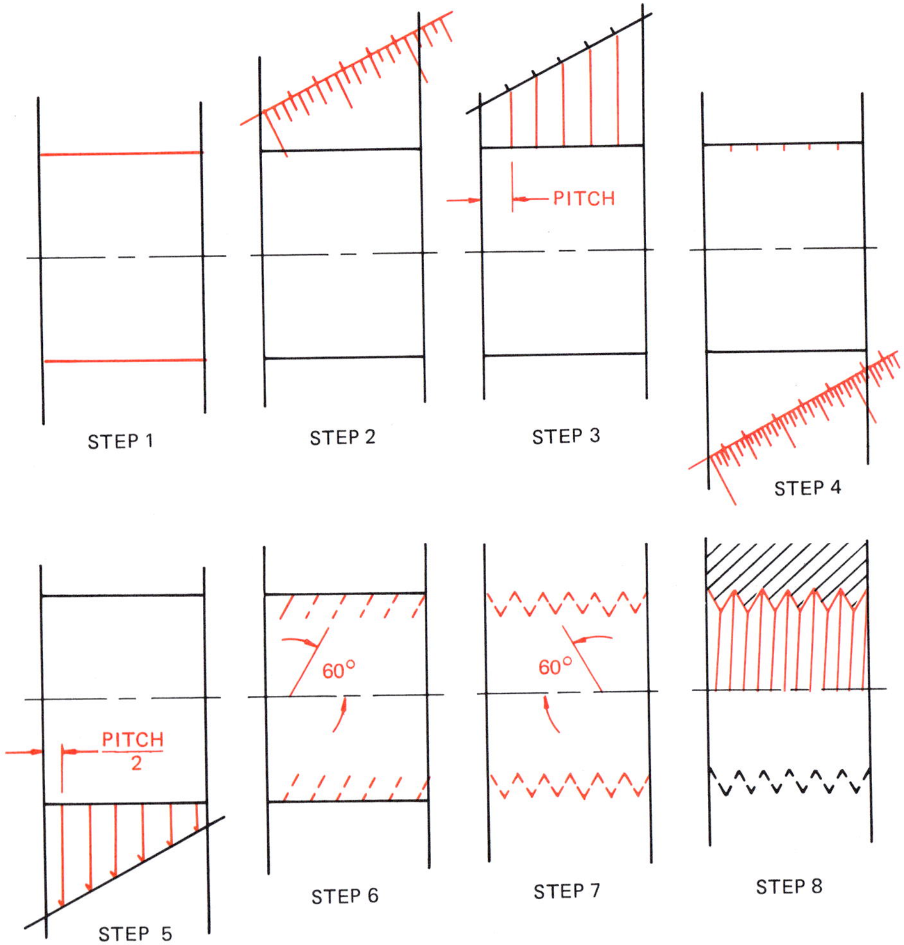

Fig. 13-41 Construction of semiconventional internal threads

Procedure

1. On the upper line, mark off spaces equal to the pitch. The length of the threads is 2 inches. Note: To find the pitch, divide 1 inch by the number of threads per inch.
2. On the lower line, mark off one space which is one-half the pitch. Then continue to mark off the pitch for 2 inches.
3. From each point on the upper and lower line, draw lines sloping 60 degrees to the right for the sides of the threads.
4. Again, from each point on the upper and lower line, draw lines sloping 60 degrees to the left for the other side of the threads.
5. Connect the crests of the threads with slanting lines. The T square is placed against the base of the triangle as a guide so that the crest lines are parallel.
6. Connect the roots of the threads in the same manner. The root lines are not parallel with the crest lines. The root and crest lines are drawn with lines of the same weight.

ASSIGNMENT B

Draw internal screw threads, figure 13-41.

Procedure

1. Draw the two lines for the major diameter of the threads.
2. The pitch of these threads is 1 inch divided by six threads per inch which is 1/6 inch. To divide the 1-inch line into six equal parts, place the scale on a slant so that 1 1/8 inches (18/16th) on the scale lies between the two upper vertical lines. Then, point off six 3/16-inch spaces along the edge of the scale.
3. Draw a vertical line downward from each point to the upper 1-inch line, and it is divided into six equal parts.
4. Place the scale on a slant so that 1 1/8 inches lies between the two lower vertical lines. The first space for the threads is one-half of the pitch so 3/32 inch is pointed off first along the edge of the scale. Continue to point off 3/16 inch spaces.
5. Draw vertical lines up to the lower 1 inch line, and this line is divided for the threads.

6. From each point on both 1 inch lines, draw lines sloping 60 degrees to the right for the sides of the threads.
7. From each point on both 1 inch lines, draw lines sloping 60 degrees to the left for the other side of the threads. The crest and root lines are omitted from the exterior view and from the full section of internal threads.
8. For the half section, the crest and root lines are drawn to the center line only. These lines match the crest and root lines on the back of a bolt. Therefore, these lines slant in a direction opposite to the crest and root lines on the front of a bolt.

REVIEW QUESTIONS

1. What is the pitch of a thread?
2. How is the pitch of a thread found?
3. How is the amount which is measured for the slant of the threads found?
4. What is meant by 2 1/4 - 4NC - 2LH?
5. What is meant by 1 1/2 - 6NC - 2?

REGULAR AND SIMPLIFIED EXTERNAL THREAD SYMBOLS

The regular thread symbol for external threads does not have the 60-degree thread outline in the exterior view. Instead, two parallel lines are drawn at the major diameter. The crest and root lines are drawn at right angles to the thread axis instead of sloping. The root lines are drawn thicker than the crest lines. In actual drafting practice, the crest and root lines are spaced by eye to the approximate pitch. They may be of equal width if preferred, figure 13-42. Figure 13-43 shows the construction of external thread in an exterior view.

The regular thread symbol for threads in section does have the 60-degree thread outline, and on one edge the thread outline is advanced one-half of the pitch.

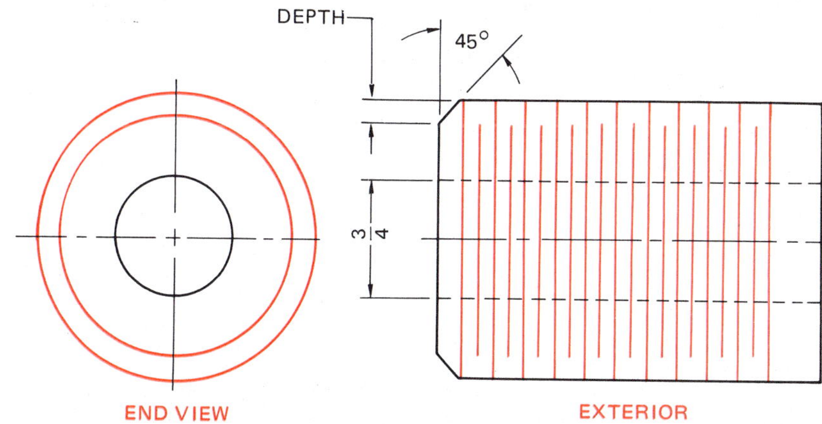

Fig. 13-42 Regular thread symbol

The ends of the screw threads on figure 13-44 are chamfered at 45 degrees to the thread depth to protect the starting thread and permit the engaging nut to start easily. Figure 13-45 shows the construction of external threads in section. Simplified thread symbols are used to further reduce drafting time. The thread outline and the crest and root lines are not drawn. Two dotted lines parallel to the axis are drawn to indicate the depth and the length of the threads, figure 13-46. In the exterior view, a full line is drawn at right angles to the axis at the end of the thread.

REVIEW QUESTIONS

1. What is meant by 1 3/4 - 5NC - 3?
2. What is the main difference between the semiconventional external thread and the regular thread symbol for external threads in an exterior view?
3. Why are regular and simplified thread symbols used on drawings?
4. Which lines are drawn thicker in the regular thread symbol?
5. How is the pitch of a thread found? How is the depth of a thread found?

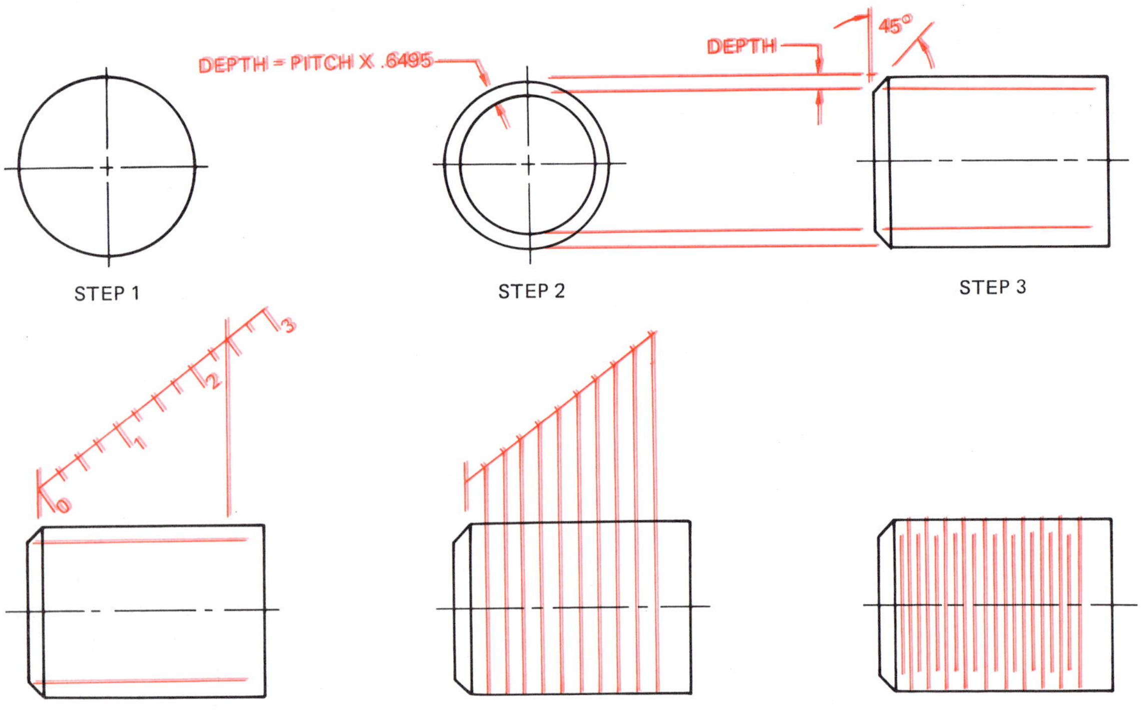

Fig. 13-43 Construction of regular thread symbol for external threads in exterior view

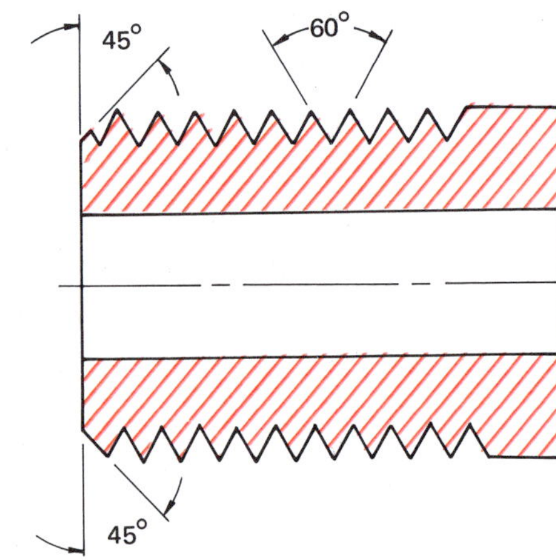

Fig. 13-44 Regular thread symbols

REGULAR AND SIMPLIFIED INTERNAL THREAD SYMBOLS

The crest and root lines are shown in the regular thread symbol in the sectional views of internal threads. In the simplified thread symbol, the crest and root lines are omitted in the sectional views. Instead, lines are drawn parallel to the thread axis to represent the major and minor diameters. The lines representing the major diameter are made with short dashes.

The regular thread symbol and the simplified thread symbol are the same in the exterior views of internal threads.

TAPPED HOLES

Internal threads are usually cut with a tool called a tap. The threaded holes are called *tapped holes.* Figure 13-47 shows three types of tapped holes.

In the first sketch of figure 13-48, the hole is tapped clear through (A). The second hole (B) is tapped part way, and a part of the drilled hole is shown. The third hole (C) is tapped to the bottom.

The three types of tapped holes are produced as shown. In

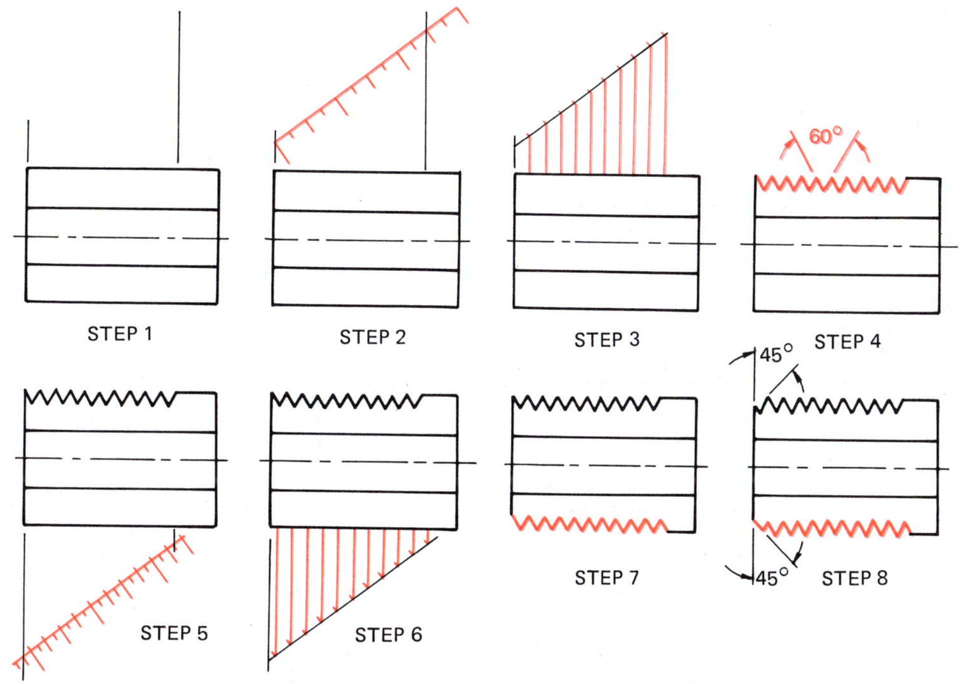

Fig. 13-45 Construction of regular thread symbol for external threads in section

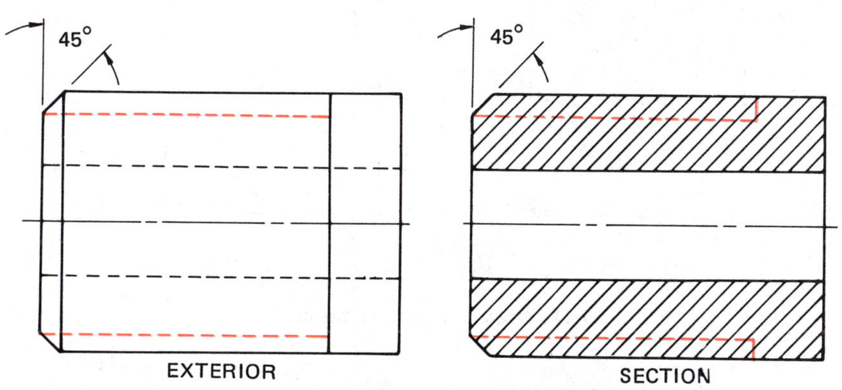

Fig. 13-46 Simple thread symbols

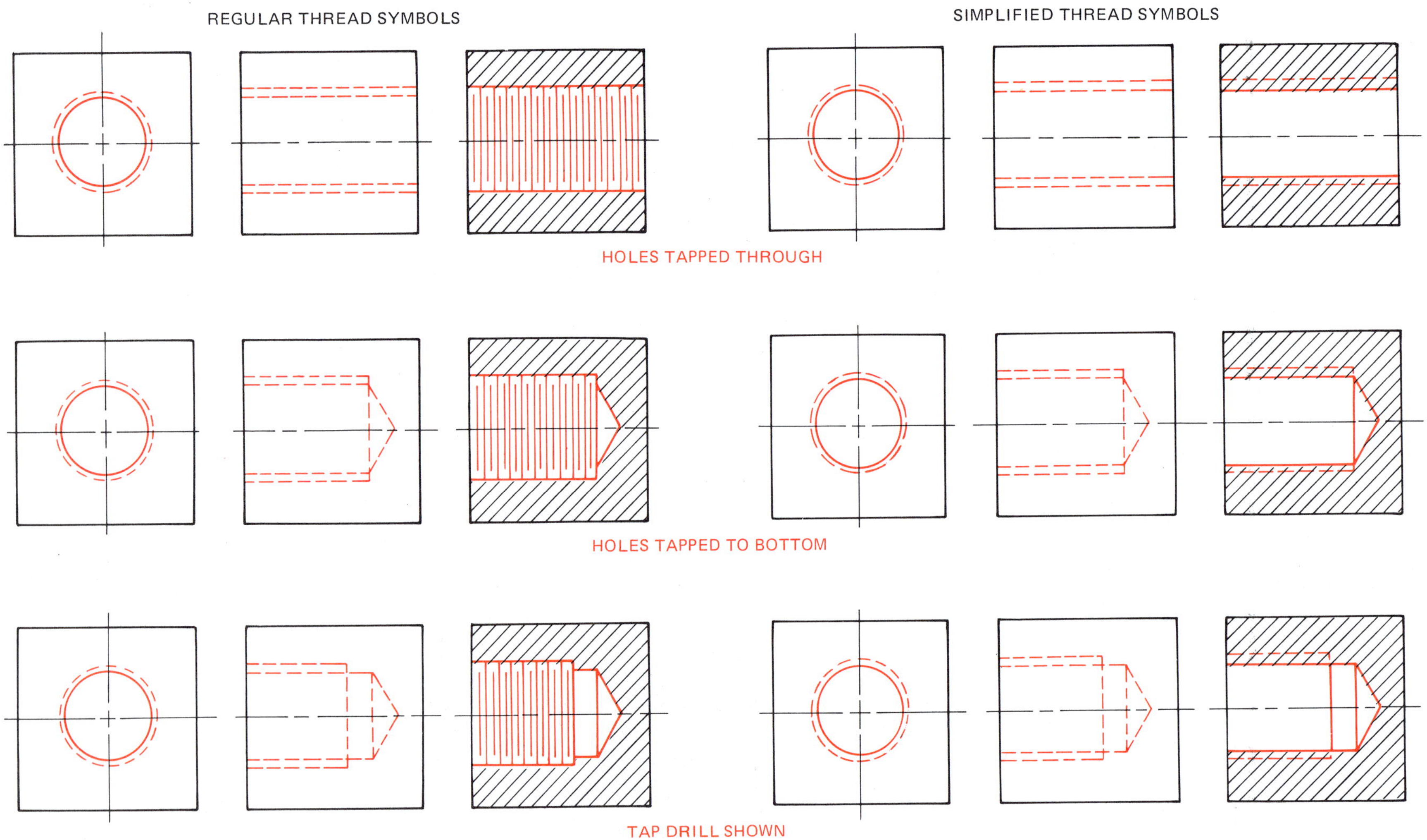

Fig. 13-47 Internal thread symbols

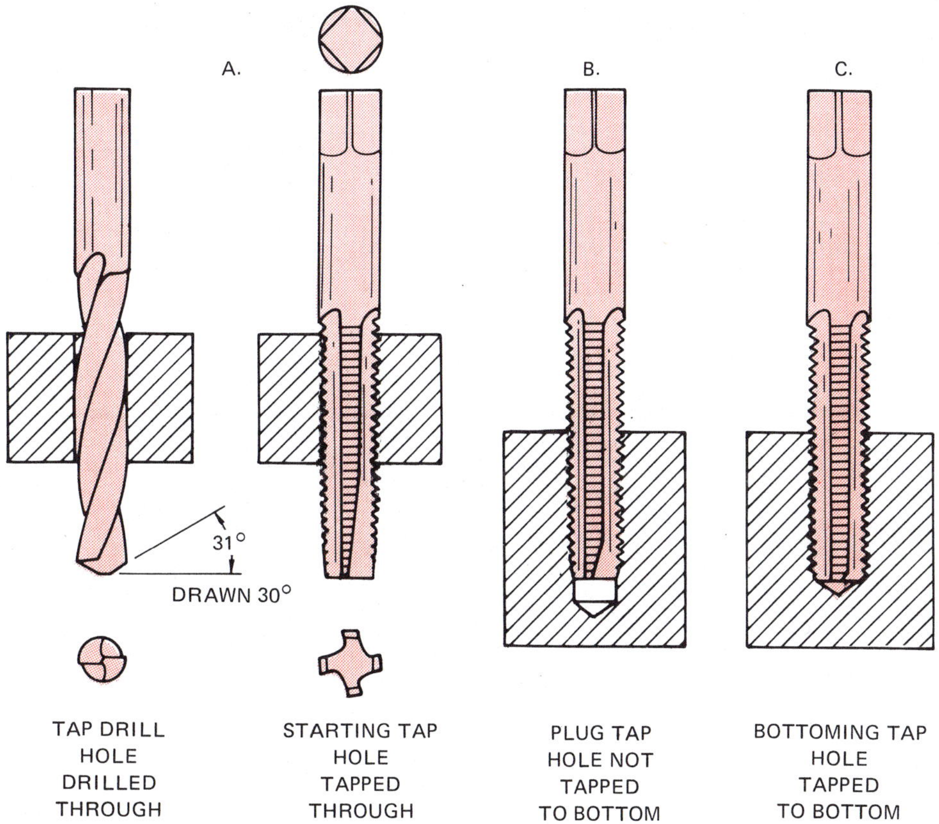

Fig. 13-48 Tap drill, taps and tapped holes

the first illustration (A), a hole is drilled through with a drill called the *tapped drill.* The next hole is tapped with a *starting tap.* The starting tap is also called a *taper tap* because the end of the tap is tapered.

In the next illustration (B), the hole is not drilled through and the threads are tapped part way with a *plug tap.* The end of the plug tap has less taper than a taper tap. When the end of the drilled hole is shown, an angle is drawn in the bottom of the hole corresponding to the angle of the drill point. This angle is drawn 120 degrees, 60 degrees on each side of the center line.

The last sketch shows the hole tapped to the bottom with a *bottoming tap.* On a bottoming tap, only the last thread is chamfered.

Tap Drill Sizes

Holes to be tapped are first drilled with a tap drill. Tap drill diameters are given in the Appendix. The tap drill diameters recommended in this table provide for threads which are approximately 75 percent of full thread depth. These threads are strong enough for most purposes. If smaller tap drills are used, more power is required to tap the holes, and more taps would be broken.

Dimensioning Tapped Holes

Tapped holes are dimensioned with a note which states the tap drill diameter and the depth of the drilled hole, plus the thread notation and the depth of the thread. Example: .4219 DRILL, 7/8 DEEP, 1/2 - 13NC - 2, 3/4 DEEP.

REVIEW QUESTIONS

1. In sectional views of internal threads, what is the main difference between the regular thread symbol and the simplified thread symbol?
2. In the simplified internal thread symbol, which lines are drawn with short dashes in the sectional view?
3. What angle is drawn in the bottom of a drilled hole?
4. What is a tap drill?
5. How are internal threads cut in the holes?

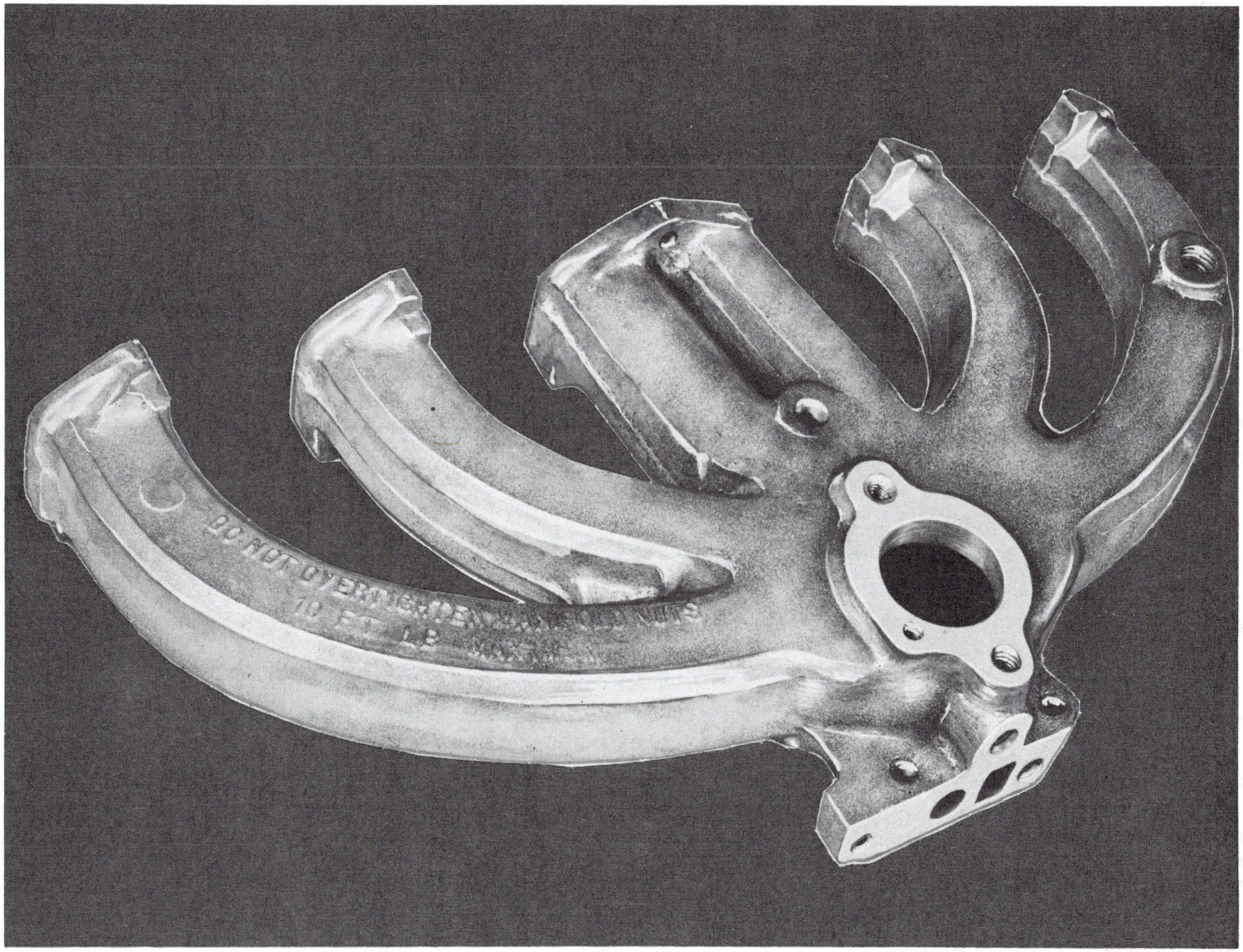

UNIT 14 DRAWING FOR PRODUCTION

WORKING DRAWINGS

A complete set of working drawings includes detail and assembly drawings which supply every item of information for the manufacture and assembly of the product represented.

Lines and symbols are used to make exterior views, sectional views, and auxiliary views of working drawings. The views describe completely the form or shape of a part, a machine, or a structure.

Dimensions and shop notes give the exact size and location of every detail and feature of the object represented. Shop notes describe machine shop processes such as drilling, reaming, boring, countersinking, etc., required to manufacture the part.

The working drawing specifies the kind of material to be used for each part and, if required, the kind of heat treatment.

Ordinary machine finished surfaces on castings, forgings, and welded structures are indicated by finish marks. Special finishes such as grinding, honing, polishing, plating, or painting are specified with shop notes.

Working drawings also include assembly drawings that show how the parts of a machine or structure are put together or assembled.

The first project to be drawn in this group is a toolmaker's clamp. The details are drawn on one sheet and the assembly on another. A pictorial drawing of each detail is given in figure 14-1. The assembly is shown on figure 14-2.

The two clamping members are made of cold-drawn steel. The material is abbreviated as C.D. Steel or CDS. Both clamping members are case hardened to protect the surfaces against wear. The retaining clip is made of spring steel. It is hardened and tempered. The heads of the adjusting screw and the clamping screw are knurled.

To save time, only a few threads are drawn on the clamping screw and the adjusting screw. They are connected with *ditto lines* which indicate that the threads are continued throughout the length of the ditto lines. Ditto lines are a series of short, double dashes, that are evenly spaced and medium width.

ASSIGNMENT A

Make a two-view drawing of each part of the toolmaker's clamp.

Procedure

1. Lay out the margin lines and letter the title strip.
2. Draw the front view of the drilled clamp 1 3/4 inches from the left margin line and 1 inch below the top margin line.
3. Draw the right side view 1 1/8 inches from the front view. Dimension the views and letter the notes.
4. Draw the front view of the tapped clamp 6 1/4 inches from the left margin line and 7/8 inch below the top margin line.
5. Draw the right-side view 1 1/8 inches from the front view.
6. Dimension the views and letter the note.
7. Draw a horizontal center line 1 1/4 inches below the top margin line. Draw a vertical center line 4 3/4 inches from the right margin line.
8. Draw the top view of the adjusting screw.
9. Draw the front view beginning 1 1/4 inches below the center of the top view. Do not draw the ditto lines. Instead, draw the simplified thread symbol for external threads. Dimension the views and letter the notes.
10. Draw a horizontal center line 1 1/4 inches below the top margin line and a vertical center line 2 1/4 inches from the right margin line.
11. Draw the top view of the clamping screw.
12. Draw the front view beginning 1 1/4 inches below the center of the top view.

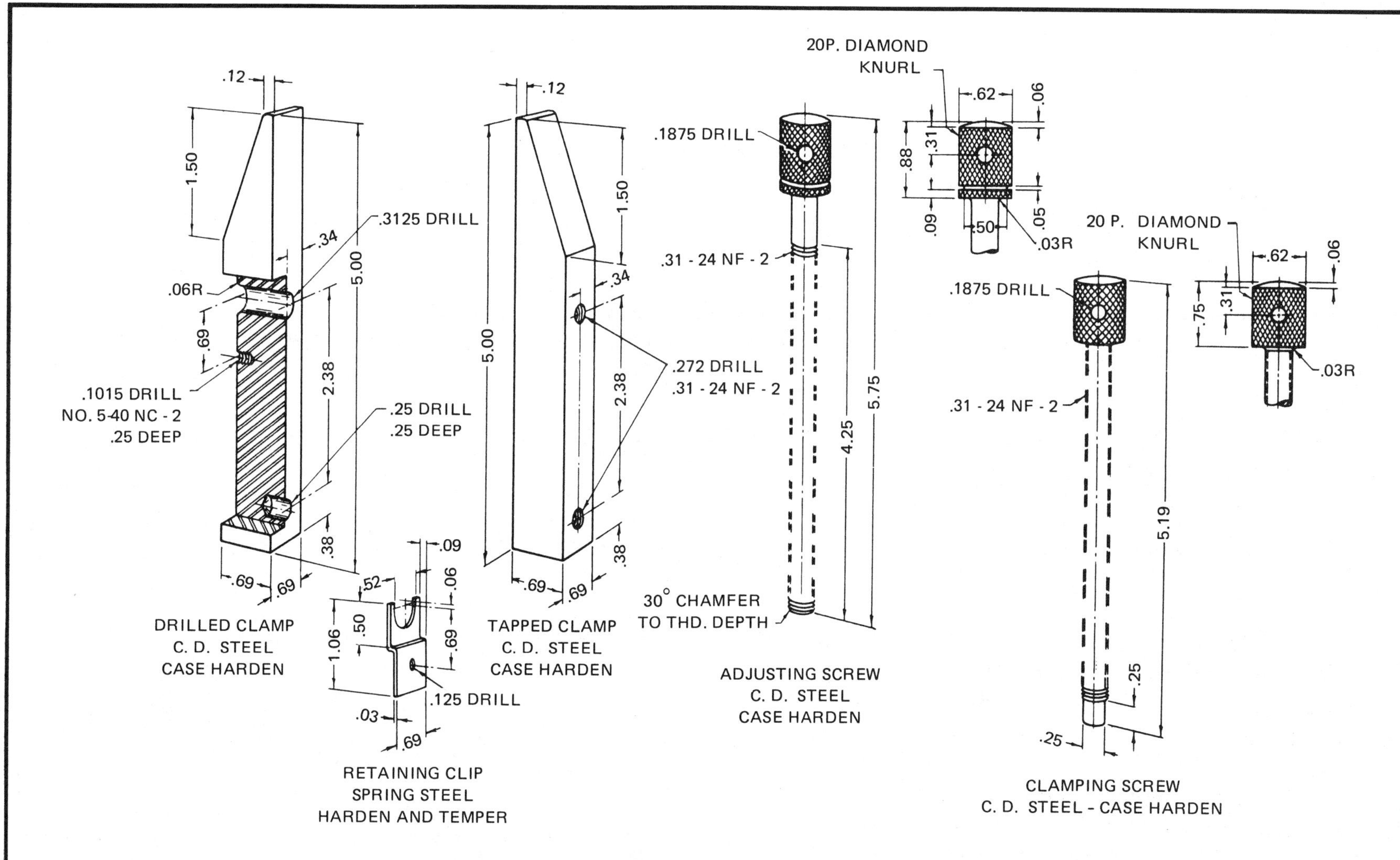

NAME OF SCHOOL CITY OR TOWN	TOOLMAKERS' CLAMP DETAIL	SHEET No. 14-A NAME	CLASS DATE

Fig. 14-1

13. Dimension the views and letter the notes.
14. Draw the front view of the retaining clip 4 1/4 inches from the left margin line and 1 3/8 inches above the title strip.
15. Draw the right-side view 1 1/8 inches from the front view.
16. Dimension the views and letter the note.
17. Below each part, letter the name of the part, the kind of material and the heat treatment required. Check the accuracy of each view of each part, the dimensions and notes.

REVIEW QUESTIONS

1. Give the working definition of drafting.
2. What are working drawings?
3. What is meant by the note CDS? Case hardening? Carburizing?
4. What is meant by the note tempering or drawing?
5. What is the purpose of knurling?
6. For what purpose are ditto lines used?

ASSEMBLY DRAWING

New machines, devices, products, etc., are the outgrowth of ideas or thoughts in the mind of some individual. The natural procedure for designers of new products is to make freehand sketches which are used to illustrate, explain, and clearly present the new ideas to other persons.

From the freehand sketches, design assembly drawings are made. The design assembly drawings usually are made full size in pencil. The design drawing is checked to make sure that the machine functions and operates or produces the work for which it is designed. A check is made to determine the clearance of moving parts, the method of lubrication, the strength of individual parts, and how the machine can be assembled.

When the design is satisfactory, a fully-dimensioned detail drawing is made of each separate part. This is followed by an accurate assembly drawing made from the dimensioned details. Making the assembly drawing from the detail drawing serves as a check on the accuracy of the details.

The assembly drawing shows how the parts of the machine are assembled or placed together. It may consist of exterior views or sectional views. Hidden lines are omitted except when necessary to make the drawing clearer. Sectional views are preferred to show how interior details are assembled. Dimensions are omitted from assembly drawings, except those which aid in the assembly of the machine.

Complex machines may require some unit assembly or subassembly drawings which show how groups of parts are assembled. The unit assembly drawings are combined to make the complete assembly drawing.

ASSIGNMENT B

Make a two-view assembly drawing of the toolmaker's clamp. The assembly drawing is to be made from your detail drawings of the toolmaker's clamp.

Procedure

1. Lay out the margin lines on the drawing sheet and letter the title strip.
2. Draw a horizontal center line 4 1/2 inches below the top margin line.
3. Draw a vertical center line 4 inches from the left margin line.
4. The intersection of these two lines is the center of the adjusting screw. Beginning with this point, draw the left-side view of the clamp assembly.
5. Draw the front view of the clamp assembly. Start with the drilled clamp, drawing it 3 1/4 inches from the center of the side view. Complete the front view. The tapped clamp may be placed anywhere on the threads of the adjusting screw. Draw the simplified thread symbol for the threads.

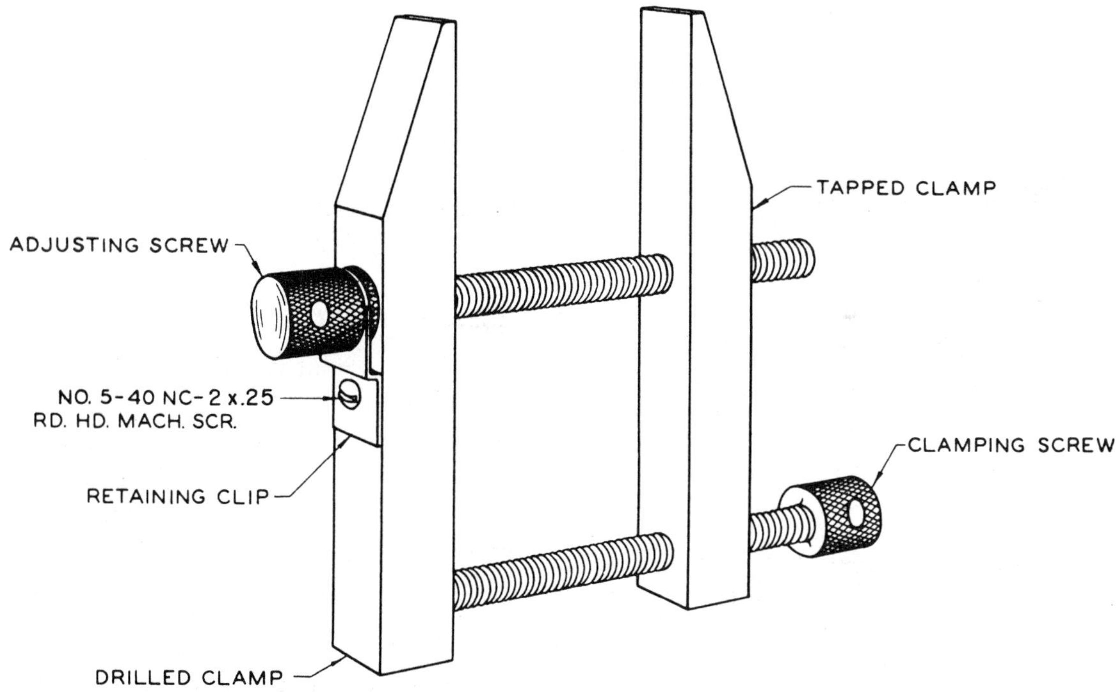

NAME OF SCHOOL CITY OR TOWN	TOOLMAKERS' CLAMP ASSEMBLY	SHEET No. 14-B NAME	CLASS DATE

Fig. 14-2

6. Letter the name of each part in a convenient space. Draw a leader with an arrowhead to connect the name with the part.
7. Letter the notation for the machine screw and attach it to the screw with a leader and an arrowhead.
8. Check the accuracy of every part of the complete toolmaker's clamp.

REVIEW QUESTIONS

1. What is the purpose of design assembly drawings?
2. What are detail drawings?
3. What is the purpose of assembly drawings?
4. Should hidden lines be drawn in assembly drawings?
5. Why are sectional views sometimes used for assembly drawings?
6. Should assembly drawings be dimensioned to show the size of the parts?
7. What is meant by No. 5 - 40 NC - 2 x .25 RD HD MACH SCR?

DETAIL DRAWING

A tap wrench, figure 14-3, is a tool used to hold and drive the square end of a tap to cut threads in a hole by hand. Tap wrenches also are used to hold the square ends of reamers when holes are reamed by hand.

This tap wrench, figure 14-4, has four parts: frame, clamping handle, and two jaws. The frame has a handle on one end. The end of this handle is rounded. The other end is drilled, counterbored, and tapped for the clamping handle. The frame has an opening in the center. The right end of this opening is large enough to permit each jaw to be inserted. The remainder of the opening is beveled on each side at an angle of 90 degrees to form a slide for the jaws. The outside of the frame is usually polished.

The clamping handle is threaded to fit the hole in the frame. A knurled band is provided to make it easier to turn the handle. A pin can be inserted in the drilled hole to turn the handle when clamping the tap. The handle is polished and the right end is rounded or crowned. The left end is cyanide hardened to resist wear.

The jaws have a 90-degree groove at the top and bottom to fit the slide in the frame. They also have a 90-degree groove on one side which clamps the square end of the tap. The jaws are made of tool steel.

ASSIGNMENT C

Draw a front view, top view, right-side view of the frame, and a full section view on cutting-plane line A-A of the tap wrench. Draw a front view and a left end view of the clamping handle. Draw a front view and a side view of the jaw.

Procedure

1. Lay out the margin lines on the drawing sheet and letter the title strip.
2. Draw a horizontal center line for the top view of the frame 1 1/4 inches below the top margin line.
3. Draw a horizontal center line for the front, side, and full section views, 3 3/4 inches below the top margin line.
4. Draw the vertical center line for the front and top views 8 inches from the left margin line.
5. Draw a vertical center line for the side view 5 inches from the right margin line.
6. Draw a vertical center line for the full section view 3 1/2 inches from the right margin line.

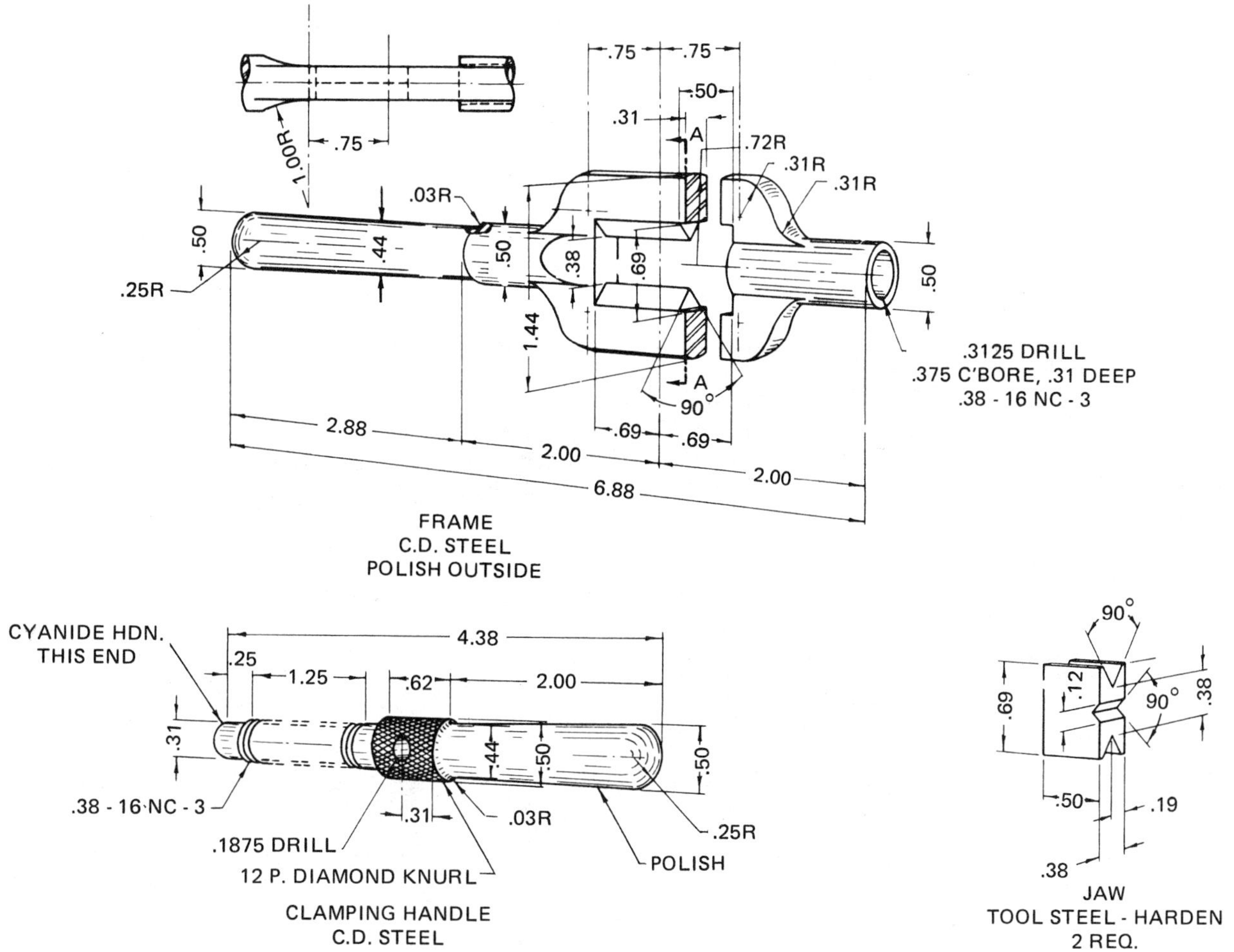

NAME OF SCHOOL CITY OR TOWN	TAP WRENCH DETAILS	SHEET No. 14-C NAME	CLASS DATE

Fig. 14-3

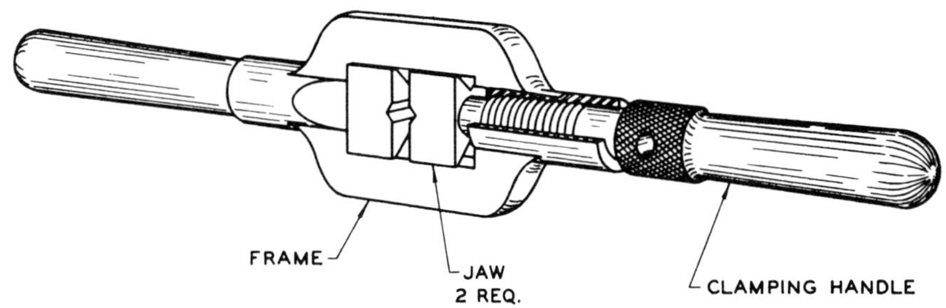

NAME OF SCHOOL CITY OR TOWN	TAP WRENCH ASSEMBLY	SHEET No. 14-D NAME	CLASS DATE

Fig. 14-4

7. Draw the four views of the frame. Dimension the views. Letter the notes, the name of the part, and the material.
8. Draw a horizontal center line 2 1/4 inches above the title strip.
9. Draw a vertical center line 3 inches from the left margin line.
10. Draw the left end view of the clamping handle.
11. Draw the front view 1 1/4 inches from the center of the end view. Dimension the views. Letter the notes, the name of the part, and the material.
12. Draw the front view of the jaw 2 inches above the title strip and 5 inches from the right margin line.
13. Draw the right-side view 1 1/4 inches from the front view. Dimension the views. Letter the name, material, heat treatment, and number required.
14. Check each detail drawing for accuracy of representation and dimensioning.

REVIEW QUESTIONS

1. For what purposes are tap wrenches used?
2. What is meant by cyanide harden?
3. What is tool steel?

THREE-VIEW ASSEMBLY DRAWING

Figure 14-4 shows how the jaws and the clamping handle are assembled in the frame of the tap wrench. The square end of a tap or reamer is secured between the jaws when the clamping handle is tightened.

ASSIGNMENT D

Make a three-view assembly drawing of the tap wrench from the detail drawings.

Procedure

1. Lay out the margin lines on the drawing sheet and letter the title strip.
2. Draw a horizontal center line for the top view 3 1/4 inches below the top margin line.
3. Draw a horizontal center line for the front view and the side view 3 3/4 inches above the title strip.
4. Draw vertical center line for front and top views 7 1/2 inches from left margin line.
5. Draw vertical center line for right-side view 2 7/8 inches from right margin line.
6. Draw front, right side, and top view. Do not draw the front view in section.
7. Letter the name of each part and draw the leaders. Check the accuracy of all lines, dimensions, and notes on the drawing.

REVIEW QUESTIONS

1. When should hidden lines be drawn in assembly drawings?
2. When should the views of assembly drawings be drawn in section?

DETAILING THE C-CLAMP

The C-clamp in figure 14-5 and 14-6 is a heavy duty clamp. The T-shaped cross section is designed to provide a strong rigid body without excessive weight. The body is made of high grade steel which is forged to shape.

A small finish allowance is included on die forgings where surfaces are to be finished. The flat clamping surface on the body is to be finished with a finish mark.

The wing screw is a drop forging. The square threads, the ball on the end, and the neck behind the ball are finished. Several dimensions on the pictorial drawing of the wing screw are fol-

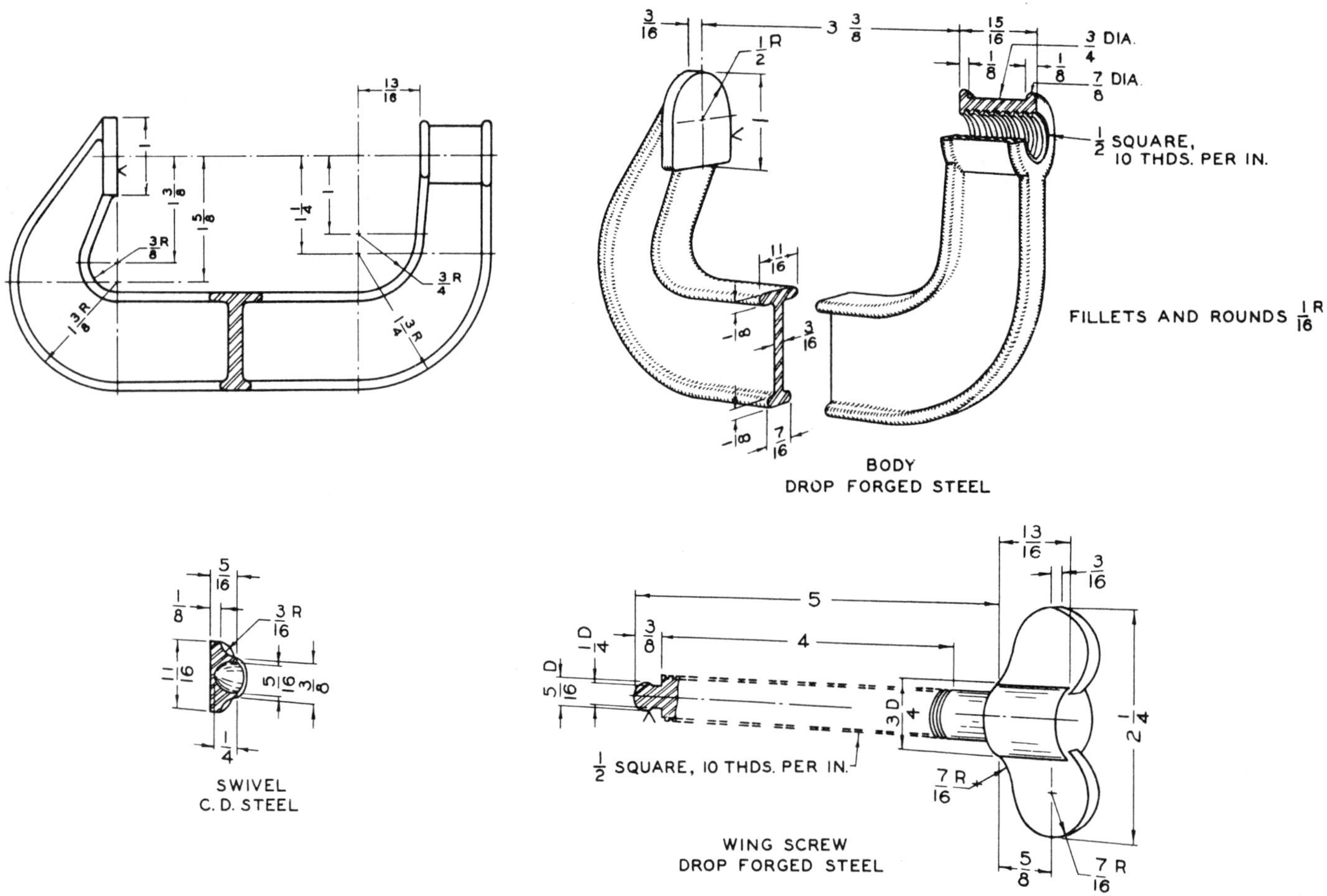

NAME OF SCHOOL CITY OR TOWN	"C" CLAMP DETAILS	SHEET No. 14-E NAME	CLASS DATE

Fig. 14-5

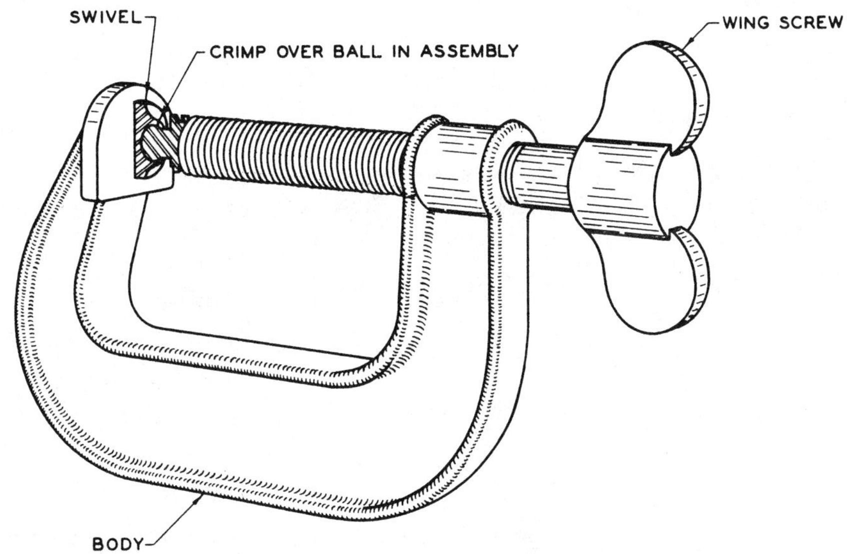

NAME OF SCHOOL CITY OR TOWN	"C" CLAMP ASSEMBLY	SHEET No. 14-F NAME	CLASS DATE

Fig. 14-6

lowed by the letter "D" to indicate clearly that these dimensions are diameters. The letter D is not necessary on these dimensions on the working drawing.

The swivel has an opening or socket to fit the ball on the end of the wing screw. Together they make a ball and socket joint. The purpose of the ball and socket joint is to allow the swivel to adjust itself to the surface of the work which is clamped. The edge of the socket is *crimped* or squeezed in around the ball to hold the swivel on the wing screw.

ASSIGNMENT E

Draw the front view, the left-side view, and the right-side view of the body of the C-clamp. Draw a front view and a right-side view of the wing screw. Draw a right-side view and a front view in full section of the swivel.

Procedure

1. Lay out the margin lines on the drawing sheet and letter the title strip.
2. Draw a horizontal center line 2 inches below the top margin line.
3. Draw vertical center lines 2 1/4 and 6 1/8 inches from the left margin line. Draw another 3 1/4 inches from the right margin line.
4. Lay out the other center lines for the front view as shown in the inserted front view on the detail drawing, figure 14-5.
5. Draw the front, right-side, and left-side views of the body. First draw the circles and arcs in each view. Draw a revolved section in the front view. The details shown clearly in one side view are not repeated in the other side view. Dimension the views and letter the notes.
6. Draw a horizontal center line for the wing screw and the swivel 2 1/8 inches above the title strip.
7. Draw a vertical center line 2 1/2 inches from the right margin line.
8. Draw the right-side view of the wing screw.
9. Draw the front view of the wing screw. The shoulder of the screw should be 4 7/8 inches from the right margin line. To save time, draw several square threads on each end of the screw and connect them with ditto lines.
10. Dimension the views and letter the notes.
11. Draw a vertical center line 4 1/4 inches from the left margin line.
12. Draw the right-side view of the swivel.
13. Draw the front view in full section 2 1/4 inches from the left margin line.
14. Dimension the views and letter the note. Check each view, line, and standard method of representing sections and screw threads on the drawing.

REVIEW QUESTIONS

1. For what purpose are C-clamps used?
2. What is meant by forging? What is a die forging? A drop forging?
3. Why are finish marks required for finished surfaces on drawings of forgings?
4. What is a ball and socket joint?

THREE-VIEW ASSEMBLY DRAWING OF THE C-CLAMP

In the pictorial assembly drawing of the C-clamp, figure 14-6, the end of the screw and the swivel are drawn in section. This is to show clearly the assembly of the ball and socket. Only the swivel need be drawn in section to show the ball and socket assembly.

ASSIGNMENT F

Make a three-view assembly drawing of the C-clamp. Draw the front view, the left-side view, and the right-side view.

Procedure

1. Lay out the margin lines on the drawing sheet and letter the title strip.
2. Draw a horizontal center line for the screw 3 3/4 inches below the top margin line.
3. Draw vertical center lines for the side views 2 1/4 inches from the right and left margin lines.
4. Draw another vertical center line 6 inches from the left margin line.
5. Locate and draw the remaining center lines using the dimensions given on the detail drawings.
6. Draw the front view, the left-side view, and the right-side view. First draw all circles and arcs. Complete the views. The hidden lines are omitted from the views. Details shown clearly in one side view are not repeated in the other side view.
7. Draw several of the square threads at each end of the screw and connect them with ditto lines, or the threads may be drawn their entire length.
8. Letter the names of each part and the note.
9. Check all lines, views, and dimensions for accuracy.

REVIEW QUESTIONS

1. Why is it good practice to make the assembly drawing from the detail drawings?
2. Why should details shown clearly in one side view be omitted from the other side view?
3. Why are hidden lines usually omitted from assembly drawings?
4. What is meant by CRIMP OVER BALL IN ASSEMBLY?

UNIVERSAL V-BLOCK

The universal V-block, figure 14-7, is a precision ground V-block. It is used to hold cylindrical parts for drilling, reaming, milling, and other machining operations. The parts are clamped in the V-block with a clamp and clamping screws. Toolmakers use this V-block for laying out work and for checking and inspecting work.

Aligning Slots

When used for milling operations, the V-block must be accurately aligned on the milling machine table. A long key is used for this purpose. The key fits into one of the aligning slots in the V-block and into one of the T slots in the milling machine table. Because the V-block has three aligning slots, it can be used in three different positions on the milling machine table.

Clamping Recess

The V-block has recessed sides. These recesses provide clamping flanges so that the V-block can be bolted to the milling machine table with strap clamps and T bolts.

The V-block is an alloy steel casting. It is machined and then hardened. After it is hardened, the sides, top, bottom, and V-groove are ground parallel, and the ends are ground square with the other surfaces.

Grinding Recess

The rectangular slot in the bottom of the V-groove is called a grinding recess. It allows the corner of the grinding wheel to run beyond the edges of the sloping surfaces. If the recess is omitted, the grinding wheel leaves a fillet in the bottom of the V-groove.

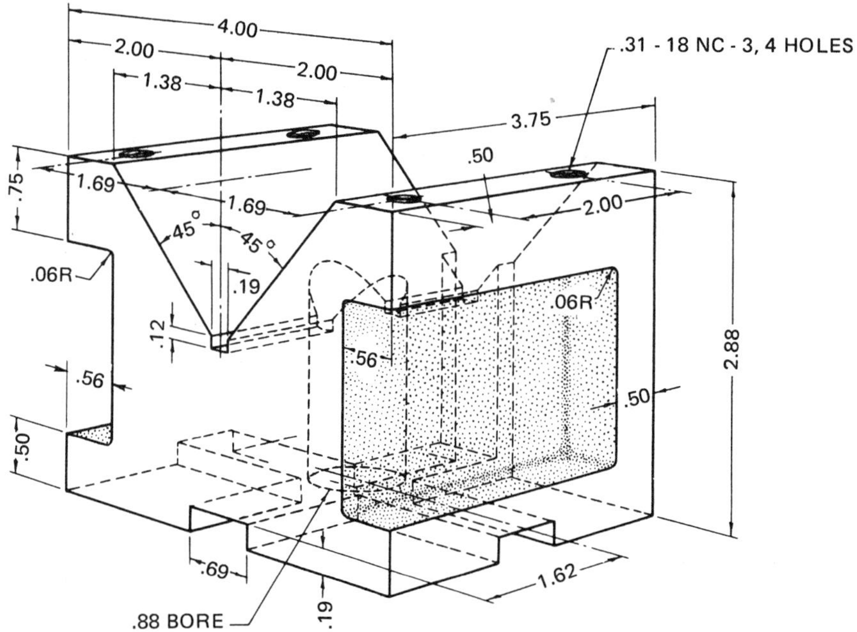

"V" BLOCK
CAST ALLOY STEEL
HARDEN AND GRIND

NAME OF SCHOOL CITY OR TOWN	UNIVERSAL "V" BLOCK DETAILS	SHEET No. 14-G NAME	CLASS DATE

Fig. 14-7

ASSIGNMENT G

Make a three-view working drawing of the universal V-block. Draw the top, front, and right-side views.

Procedure

1. Lay out the margin lines on the drawing sheet and letter the title strip.
2. Draw the top view 3/4 inch below top margin line and 3 1/4 inches from left margin line.
3. Draw the front view 1 inch above the title strip.
4. Draw the side view 3 1/4 inches from the right margin line. The edge at the intersection of the .88 inch bored hole and the V-groove is an elliptical curve in the side view. Figure 14-8 shows how to locate points to develop this elliptical curve.
5. Dual-dimension the views.
6. Place finish marks on all lines representing finished surfaces.
7. Letter the notes. Check the views and dimensions carefully.

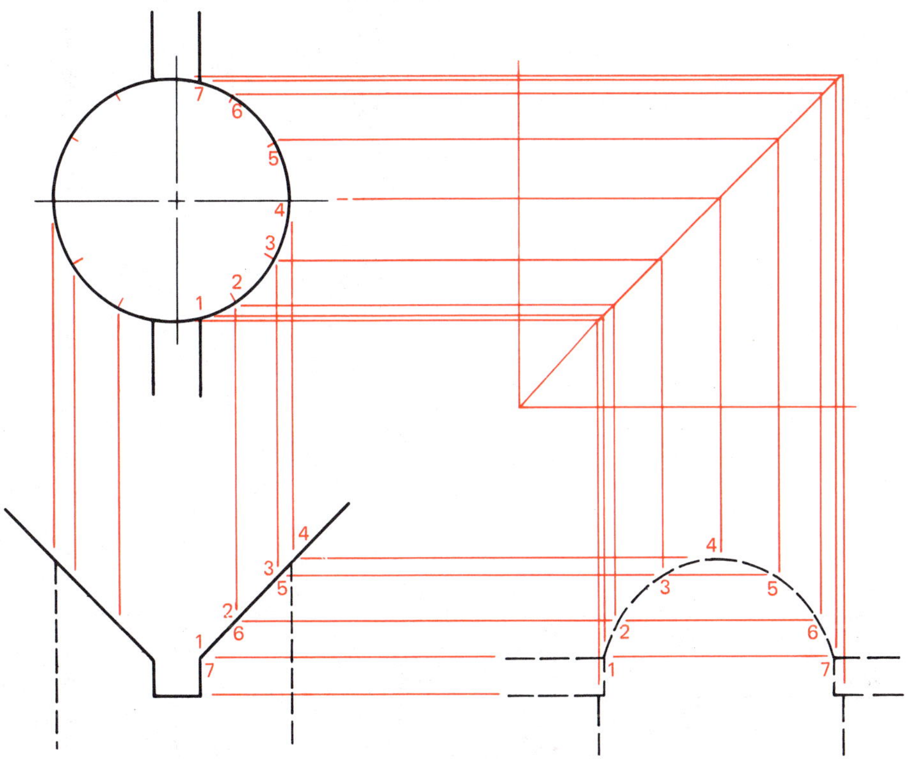

Fig. 14-8 Development of an elliptical curve

REVIEW QUESTIONS

1. For what purposes are V-blocks used?
2. How is the V-block aligned on the milling machine table?
3. How is the V-block clamped to the table of the milling machine?
4. What is the purpose of the grinding recess in the bottom of the V-groove?

V-BLOCK CLAMP

The clamp shown in figure 14-9 uses two knurled clamping screws to hold work in the universal V-block. The clamp is used in the upright position shown on the drawing for clamping work of large diameters. For clamping work of small diameters, the clamp is inverted.

The openings for the clamping screws are designed to permit the clamp to be removed quickly without removing the clamping screws. When the clamping screws are loosened, one end of the clamp is swung out from under the head of one screw. The other end of the clamp can slide out from under the head of the second screw.

To show the shape of the cross section of the clamp, a removed or detail section view is drawn. The removed section is shown in figure 14-9.

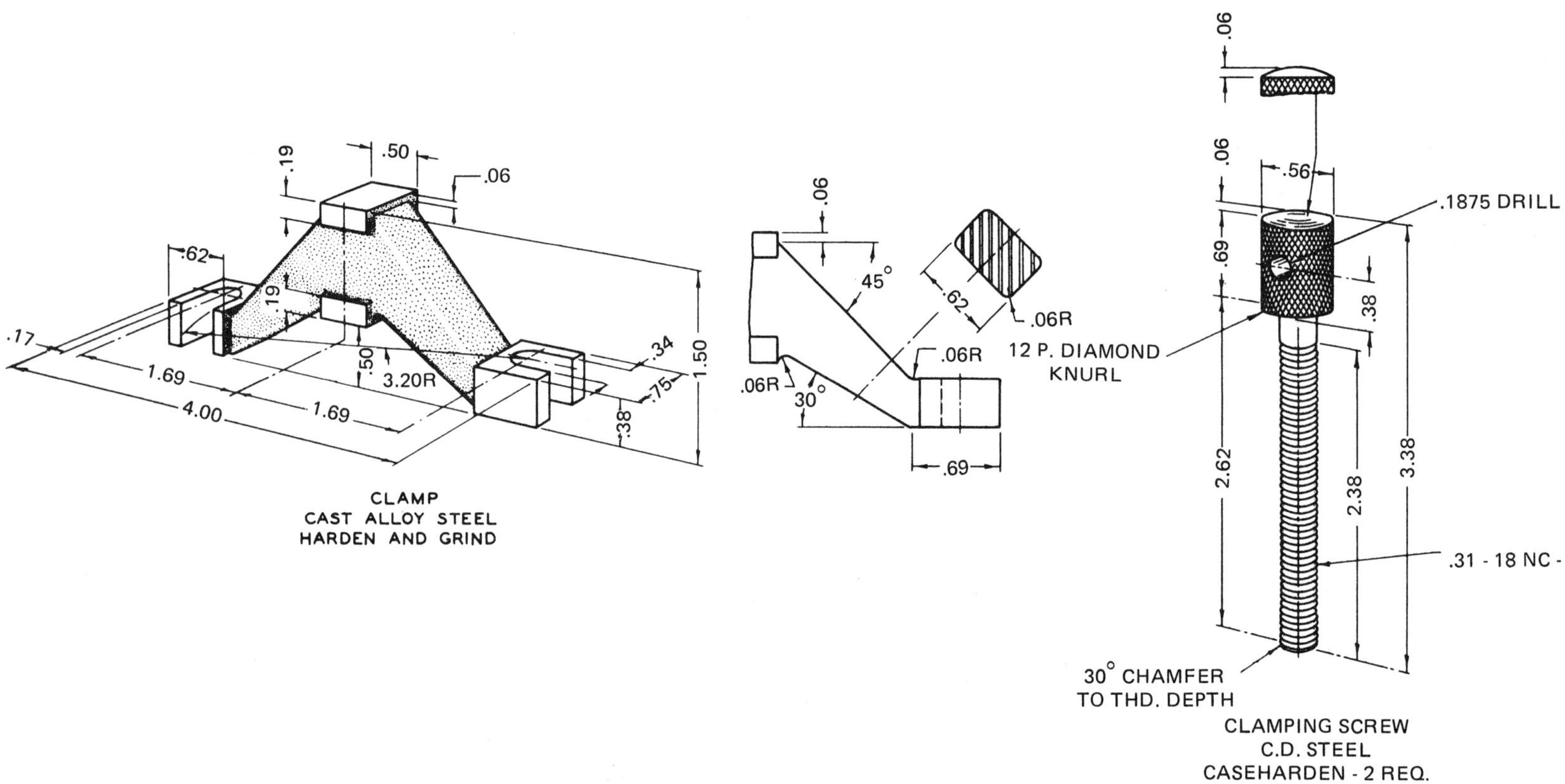

NAME OF SCHOOL CITY OR TOWN	UNIVERSAL "V" BLOCK DETAILS	SHEET No. 14-H NAME	CLASS DATE

Fig. 14-9

The clamp is made of cast alloy steel and is finished by grinding after it is hardened.

ASSIGNMENT H

Make two-view working drawings of the clamp and the clamping screw. Draw the top and front views.

Procedure

1. Lay out the margin lines on the drawing sheet and letter the title strip.
2. Draw the front view of the clamp 3 1/4 inches from the left margin line and 2 3/4 inches above the lower margin line.
3. Draw the top view 2 1/4 inches below the top margin line.
4. Draw the removed section on a center line that is perpendicular to the upper sloping surface of the clamp. Draw this view 2 1/8 inches away from the front view.
5. Dimension the views, add the finish marks, and letter the note.
6. Draw a horizontal center line for the top view of the clamping screw 2 1/2 inches below the top margin line.
7. Draw a vertical center line 4 1/2 inches from the right margin line.
8. Draw the top view.
9. Draw the front view 1 1/8 inches below the top view.
10. Dimension the views and letter the notes.
11. Check all lines, views, dimensions, and notes.

REVIEW QUESTIONS

1. Why is the 3.20-inch radius necessary on one side of the clamp-screw opening?
2. Why are the clamping screw heads knurled?
3. What is the purpose of the .1875-inch drilled hole in the clamping screw?
4. What is a removed or detail section?

V-BLOCK ASSEMBLY DRAWING

The pictorial drawing, figure 14-10, shows the assembly of the universal V-block. The clamp may be drawn with the clamping screws in either the front or second pair of holes, or inverted in either place.

ASSIGNMENT I

Make a three-view working drawing of the universal V-block assembly. Draw the front, top, and right-side views.

Procedure

1. Lay out the margin lines on the drawing sheet and letter the title strip.
2. Place the top view 3/4 inch below the top margin line, and 3 1/4 inches from the left margin line.
3. Place the front and side views 3/4 inch above the title strip.
4. Place the side view 3 1/4 inches from the right margin line. Draw the three views.
5. Letter the names of the parts. Check each part for accuracy.

REVIEW QUESTIONS

1. Give the diameter of the largest cylinder which can be clamped in the V-block.
2. When the clamp is inverted, give the diameter of the smallest part which can be clamped.

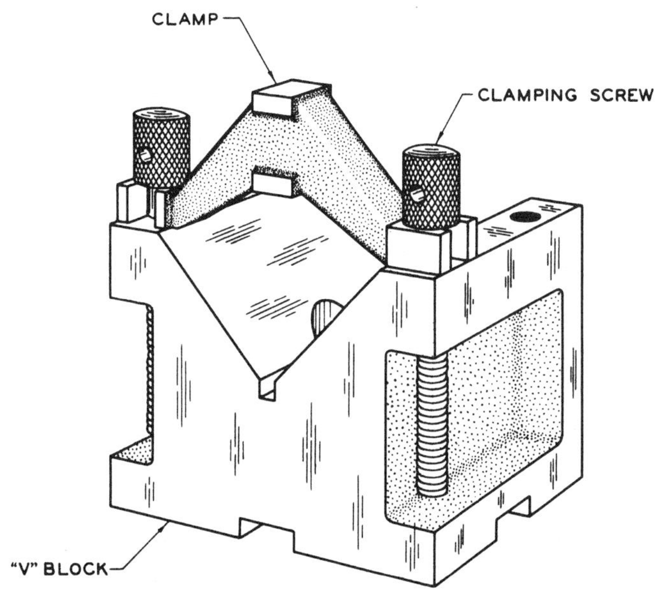

NAME OF SCHOOL CITY OR TOWN	UNIVERSAL "V" BLOCK ASSEMBLY	SHEET No. 14-I NAME	CLASS DATE

Fig. 14-10

MAGIC CHUCK

This is the first of a series of drawings of the magic chuck. The magic chuck holds drills, reamers, taps, boring tools, counterbores, countersinks, and other cutting tools. These various tools can be quickly changed without stopping the machine.

Machine Tapers

The magic chuck body has a tapered shank that fits into a tapered socket in the spindle of the machine. The taper on the shank of the magic chuck body is number 2 in the Morse Taper Series.

The American Standard Machine Tapers is a series that aims to reduce the number of sizes of tapers and to combine the best features of the various taper systems. The first three sizes in this system are from the Brown and Sharpe Taper Series. The next seven sizes are taken from the Morse Taper Series. These are followed by ten larger sizes having a taper of .75 inch per foot. The dimensions of standard tapers are given in tables in the Appendix.

The end of the shank on the magic chuck body is milled on each side to make a flat *tang* or *tongue.* The tang fits into a slot in the bottom of the tapered socket. It prevents the shank from turning in the socket if the friction grip of the taper should not hold.

In the center of the tang a hole is drilled with a combined drill and 60-degree countersink. The edge of the hole in the opposite end of the chuck body is chamfered at 30 degrees to make a 60-degree included angle for this hole. The 60-degree centers of the lathe and grinder fit into these 60-degree holes to align the chuck body when it is machined.

Hardness Specifications

The magic chuck body is made of tool steel and is hardened. To insure proper hardness, the degree of hardness for the shank end is specified on the drawing as Rock. C 44-48, Sclero. 56-63. This means that the hardness, when measured by the Rockwell Hardness Tester, must be between 44 and 48 on the C scale. When the hardness is measured with an instrument called the Shore's Scleroscope, the hardness must register between 56 and 63. The hardness specifications Rock. C 52-58 and Sclero. 67-77 on the chuck end indicate that the degree of hardness on this end must be greater than on the shank end.

Grinding Symbol

After the piece is hardened, it is finished to within very close limits by grinding. The surfaces to be ground are indicated with a 60-degree V, with the letter G inserted in the opening in the V. This symbol for finishing is often used although it is not a recognized standard finish mark.

ASSIGNMENT J

Draw a right end view, a front view partially drawn in full section, and a top view of the magic chuck body, figure 14-11. Draw an enlarged partial sectional view on cutting-plane line A-A.

Procedure

1. Lay out the margin lines on the drawing sheet and letter the title strip.
2. Draw the center lines for the right end view 4 1/4 inches from the right margin line and 3 3/4 inches above the title strip. Draw the right end view.
3. Draw the front view 3 inches from the center of the end view. The right portion of this view is drawn in full section.
4. Draw the center line for the top view 3/4 inch below the top margin line. Draw an exterior top view.
5. Draw the center lines for the partial section view 2 3/4 inches below the top margin line and 2 1/2 inches from the right

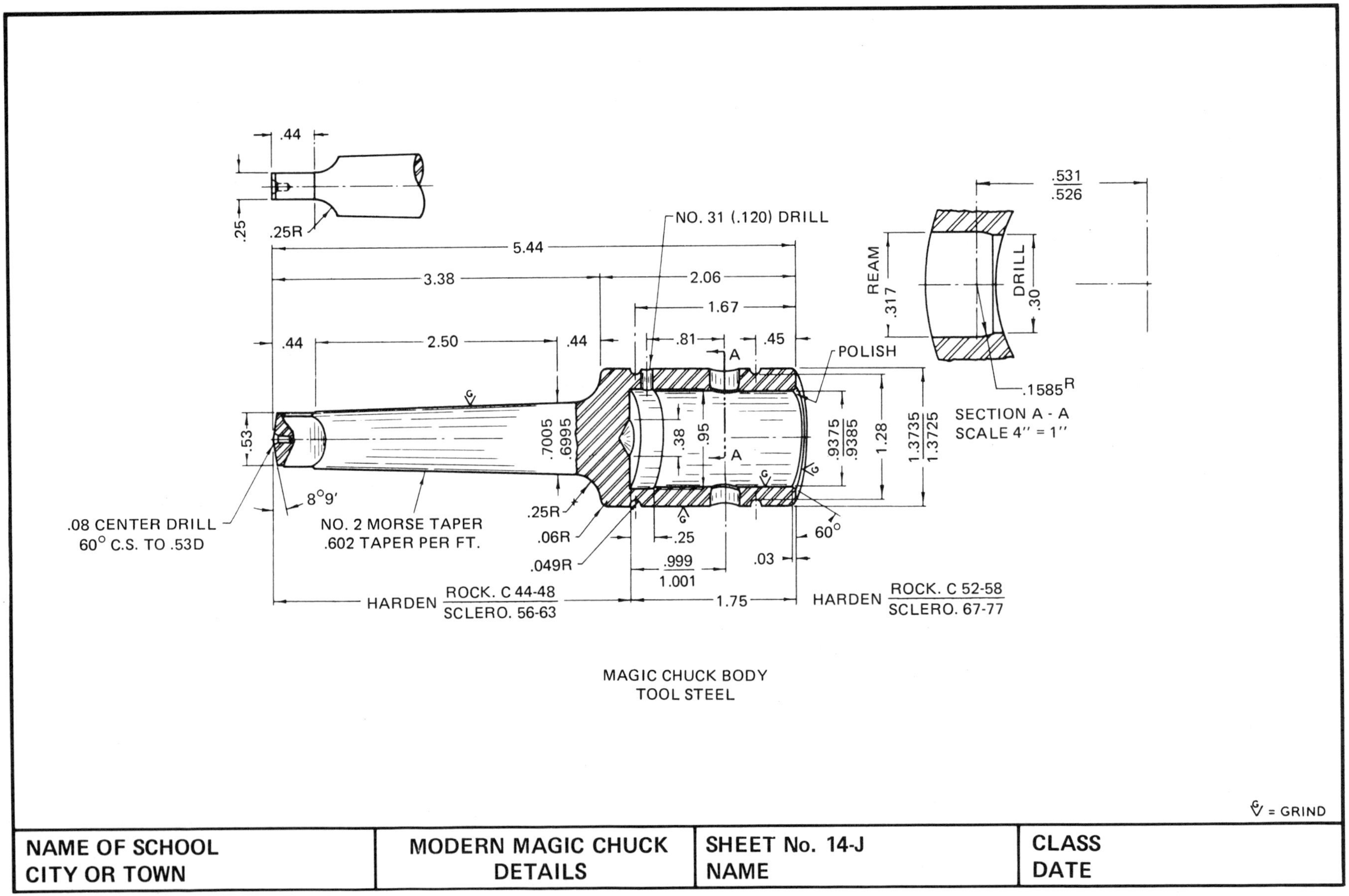
.44
.25
.25R
5.44
3.38
2.06
1.67
.44
2.50
.44
.81
.45
NO. 31 (.120) DRILL
POLISH
A
A
.53
.7005
.6995
.38
.95
.9375
.9385
1.28
1.3735
1.3725
8°9′
.08 CENTER DRILL
60° C.S. TO .53D
NO. 2 MORSE TAPER
.602 TAPER PER FT.
.25R
.06R
.049R
.25
.999
1.001
.03
60°
1.75
HARDEN ROCK. C 44-48 / SCLERO. 56-63
HARDEN ROCK. C 52-58 / SCLERO. 67-77
.531
.526
REAM
.317
DRILL
.30
.1585R
SECTION A - A
SCALE 4″ = 1″
MAGIC CHUCK BODY
TOOL STEEL
G = GRIND
NAME OF SCHOOL
CITY OR TOWN
MODERN MAGIC CHUCK
DETAILS
SHEET No. 14-J
NAME
CLASS
DATE

Fig. 14-11

margin line. Draw a partial section view, enlarged four times, of the section on cutting-plane line A-A.

6. Complete the drawing. Check all lines, view, dimensions, and notes.

REVIEW QUESTIONS

1. For what purposes are machine tapers used?
2. Describe a tapered shank and a tapered socket.
3. What is meant by .602 taper per foot?
4. Name the flat part on the end of the shank. What is its purpose?
5. Why are 60-degree countersunk holes often made in the ends of cylindrical machine parts.
6. In dimensioning, what is meant by the upper limit? The lower limit?
7. What is meant by the tolerance on a dimension?

MAGIC CHUCK COLLET AND LOCKING RING

The collet in figure 14-12 fits into the hole in the magic chuck body. It has a No. 1 Morse taper socket and is used to hold twist drills and reamers having a No. 1 Morse taper shank.

To separate a taper shank from its socket, a long hole, called a drift keyslot is cut through the collet. A wedge-shaped drift, figure 14-13, is driven into this hole to force the taper shank out of the socket.

The drift has a taper of 1.75 inches per foot which makes an angle of approximately 8° 19′. The end of the tang on taper shanks is cut at 8° 19′ to match the angle on the drift. This angle is shown on the tang on the magic chuck body.

The collet, made of cold drawn steel, is carburized to harden.

The locking ring is made from steel tubing. Steel tubing is specified because, if this part is made from solid material, it requires time to remove the excess metal from the hole, material is wasted, and it costs more to manufacture the part. The locking ring fits the outside of the magic chuck body.

The stop rings are made of spring steel and are tempered. They fit into the grooves on the magic chuck body.

ASSIGNMENT K

Make a two-view working drawing of each of the parts shown on figure 14-12.

Procedure

1. Lay out the margin lines on the drawing sheet and letter the title strip.
2. Draw the center lines for the side view of the locking ring 2 inches below the top margin line and 6 inches from the left margin line. Draw the two views allowing 1 3/4 inches between the views.
3. Draw the center lines for the side view of the stop ring 2 inches below the top margin line and 3 inches from the right margin line. Draw the two views allowing 1 1/4 inches between the views.
4. Draw the center lines for the left end view of the collet 3 inches above the title strip and 4 3/4 inches from the left margin line.
5. Draw the two views allowing 2 1/8 inches between the views. Complete the drawing. Check all dimensions, tapers, and notes.

REVIEW QUESTIONS

1. How is a taper socket made? How is it measured for correct size?

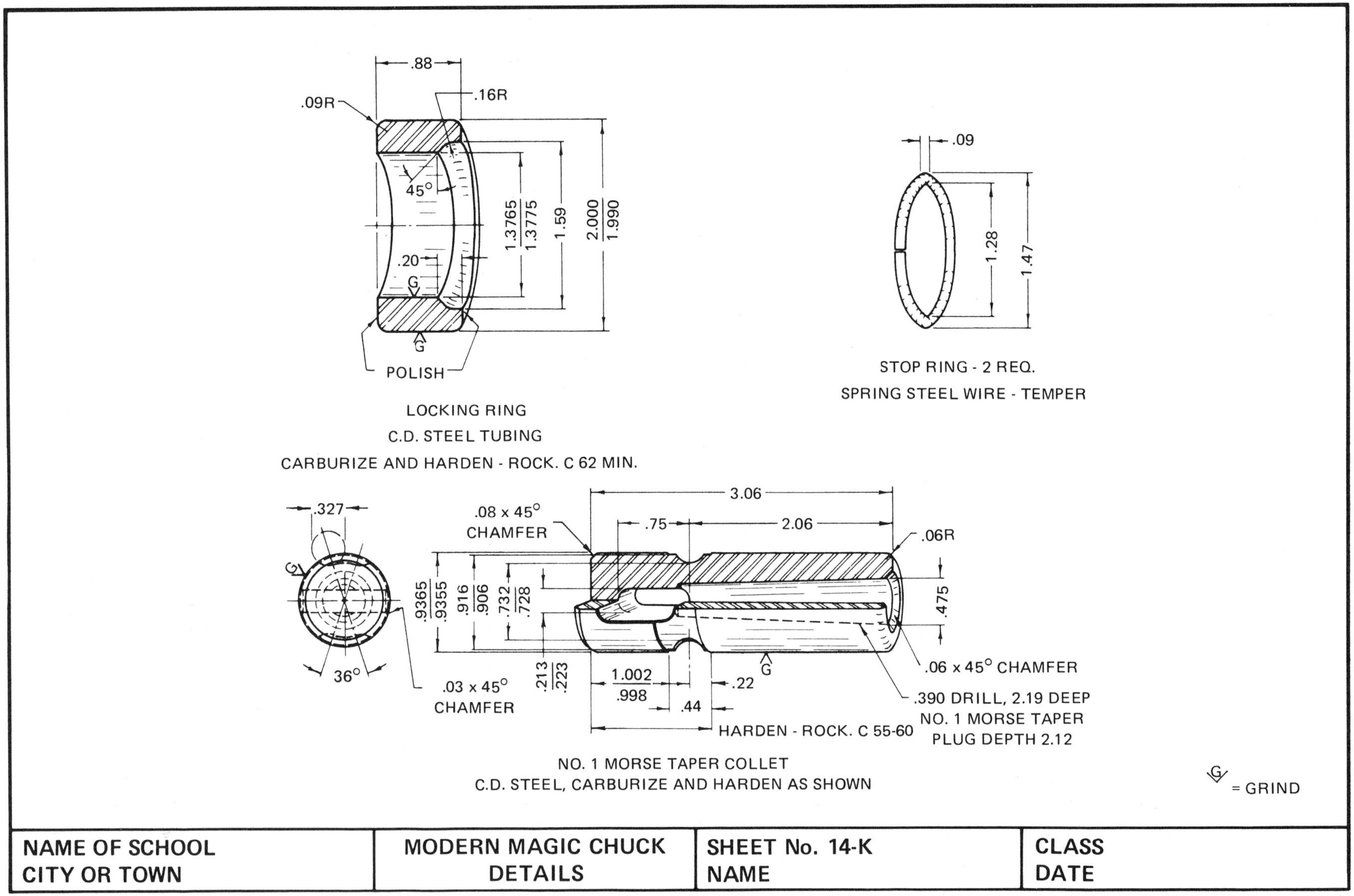
.88
.09R
.16R
45°
.20
G
1.3765
1.3775
1.59
2.000
1.990
POLISH
LOCKING RING
C.D. STEEL TUBING
CARBURIZE AND HARDEN - ROCK. C 62 MIN.
.09
1.28
1.47
STOP RING - 2 REQ.
SPRING STEEL WIRE - TEMPER
.327
36°
.9365
.9355
.916
.906
.732
.728
.08 x 45° CHAMFER
.03 x 45° CHAMFER
3.06
.75
2.06
.06R
.475
.213
.223
1.002
.998
.22
.44
.06 x 45° CHAMFER
.390 DRILL, 2.19 DEEP
NO. 1 MORSE TAPER
PLUG DEPTH 2.12
HARDEN - ROCK. C 55-60
NO. 1 MORSE TAPER COLLET
C.D. STEEL, CARBURIZE AND HARDEN AS SHOWN
G = GRIND
NAME OF SCHOOL
CITY OR TOWN
MODERN MAGIC CHUCK
DETAILS
SHEET No. 14-K
NAME
CLASS
DATE

Fig. 14-12

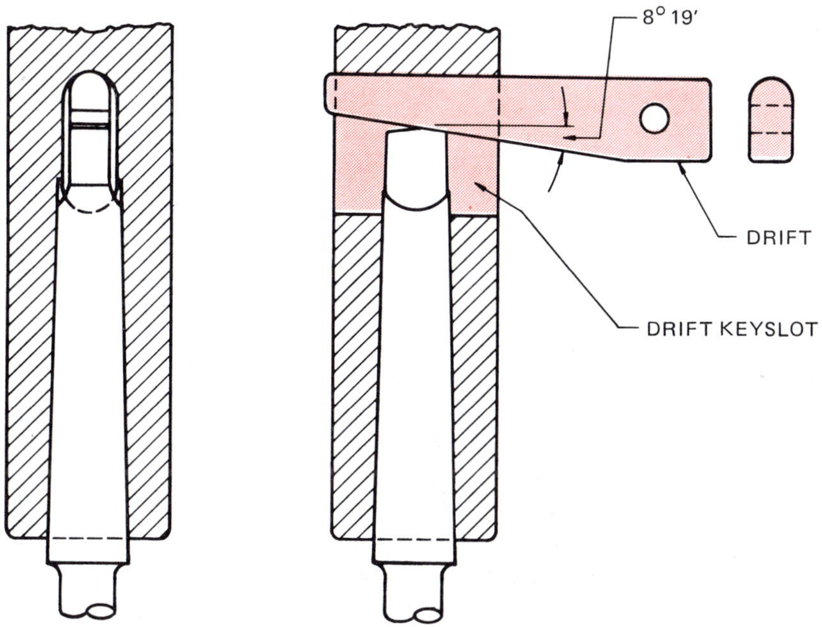

Fig. 14-13 Drift keyslot and drift

2. How should a standard taper be dimensioned? How is a taper shank removed from a taper socket?
3. What is meant by carburizing?

MAGIC CHUCK DETAILS

The collet and its bushing hold straight shank twist drills. A separate bushing must be made for each drill size, and the hole marked with dimension "A" is made to fit the drill.

A .06 inch saw is used to cut the bushing part way through from the top and from the bottom, leaving a section .06 inches thick. When the bushing is wedged into the taper socket, it closes and grips the drill shank firmly. In order to separate the parts, a pin is inserted in the .375 inches diameter hole in the collet and the bushing is driven from the socket. On the bushing and in the socket is a special taper. The collet for the taper bushing fits into the magic chuck body.

ASSIGNMENT L

Make a two-view working drawing of each of the parts shown in figure 14-14.

Procedure

1. Lay out the margin lines on the drawing sheet and letter the title strip.
2. Draw the center lines for the left end view of the bushing collet 4 inches below the top margin line and 2 1/2 inches from the left margin line. Draw the end view.
3. Draw the front view 2 1/2 inches from the center of the end view.
4. Draw the center lines for right end view of bushing 4 1/2 inches below the top margin line and 2 1/2 inches from right margin line. Draw the right end view.
5. Draw the front view 2 1/8 inches from the center of the end view. Complete the drawing.
6. Check all dimensions.

REVIEW QUESTIONS

1. For what purpose is the collet and bushing used? What causes the drills to be gripped tightly in the bushing?
2. How are the parts separated? How are special tapers dimensioned?
3. What is meant by the upper limit on a dimension? The lower limit?
4. What is meant by the tolerance on a dimension?

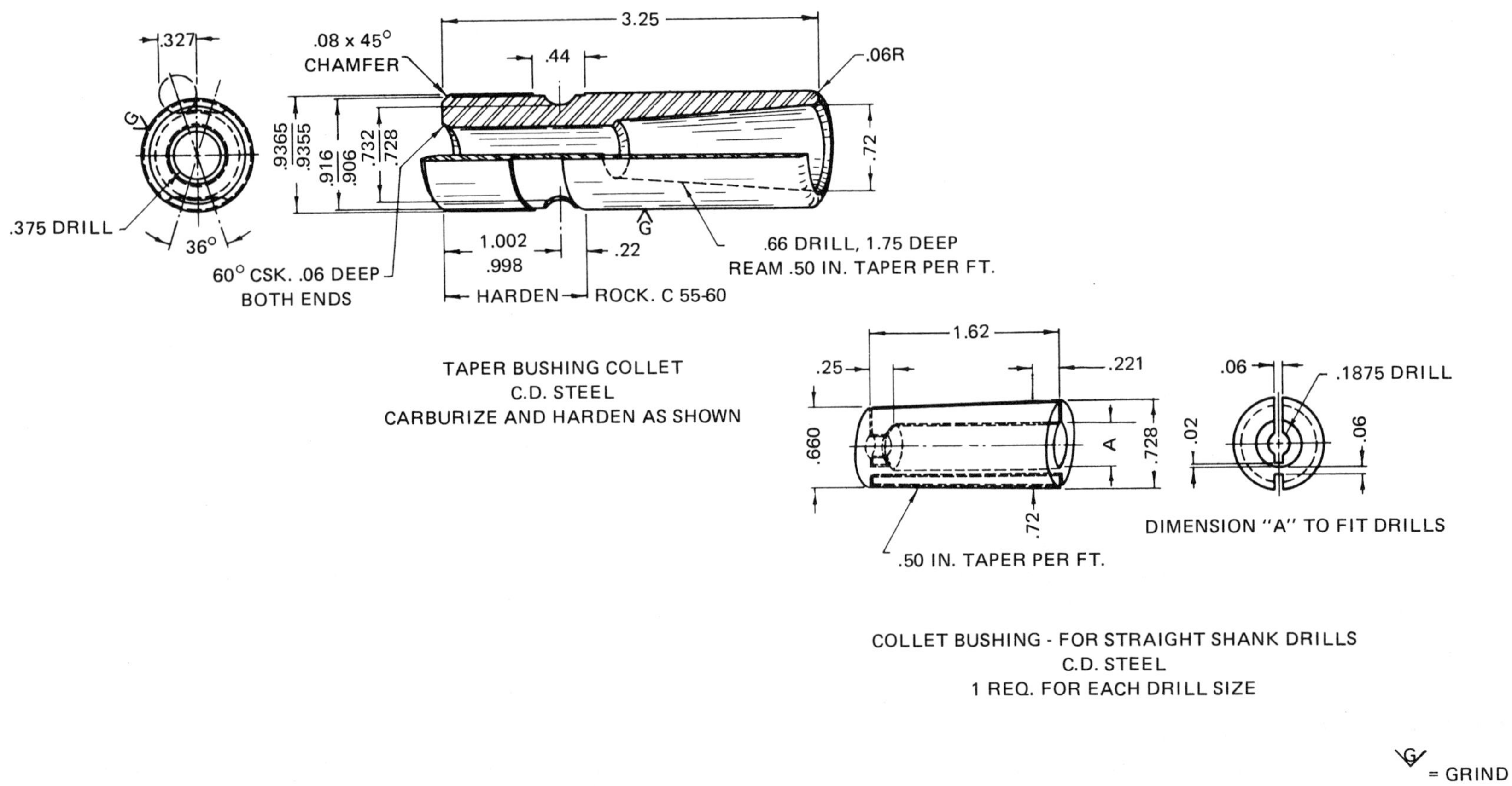

NAME OF SCHOOL CITY OR TOWN	MAGIC CHUCK DETAILS	SHEET No. 14-L NAME	CLASS DATE

Fig. 14-14

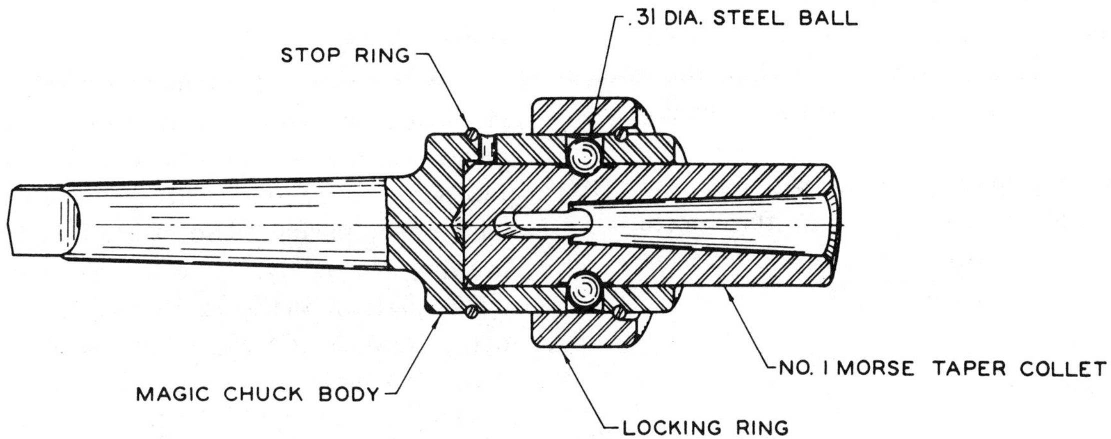

NAME OF SCHOOL CITY OR TOWN	MAGIC CHUCK ASSEMBLY	SHEET No. 14-M NAME	CLASS DATE

Fig. 14-15

MAGIC CHUCK ASSEMBLY DRAWING

Figure 14-15 shows how the magic chuck is assembled. One stop ring is snapped in place in the groove. The two steel balls are inserted in the holes in the body. The locking ring is slipped on the body and keeps the balls in the holes. The second stop ring is then snapped in place. The stop rings limit the travel of the locking ring.

In actual operation, the magic chuck is revolving at high speeds in the machine. As the locking ring is raised, the balls are forced outward by the action of centrifugal force, and the collet and tool can be inserted in the chuck. When the locking ring is lowered, the balls are forced into the collet recesses and lock it in place.

To change the tools, the locking ring is raised with one hand. The balls then disengage from the collet by centrifugal force. These permit the collet and tool to drop out into the other hand. Then another collet and tool can be inserted.

ASSIGNMENT M

Make a two-view drawing of the magic chuck assembly.

Procedure for Making Magic Chuck Drawing

1. Lay out the margin lines on the drawing sheet and letter the title strip.
2. Draw the center lines for the right end view 4 3/4 inches above the title strip and 3 7/8 inches from the right margin line. Draw the right end view.
3. Draw the front view 2 3/4 inches from the center of the end view. This view may be an exterior view with hidden lines to show the interior construction, or it may be partially drawn in full section to show the interior.
4. Complete the drawing. Check all lines, views, and notes.

REVIEW QUESTIONS

1. Calculate the maximum amount of clearance possible between the chuck body and the locking ring. The maximum clearance results if the hole in the locking ring is made the upper limit and the outside diameter of the chuck body is made the lower limit.
2. Calculate the minimum amount of clearance between chuck body and locking ring.

BENCH VISE

A dimensioned assembly drawing of a small bench vise is given in figure 14-16. Ordinarily, assembly drawings are not dimensioned unless the machine or device is small, simple, and has few parts. Dimensions when placed on assembly drawings tend to clutter the drawing. Also, it is difficult to dimension an assembly drawing adequately. It often is impossible to show clearly to which parts of the assembly each dimension applies. Therefore, detail drawings of each part are usually made from undimensioned design assembly drawings. This particular assembly drawing is dimensioned to supply the student with the necessary dimensions so that accurate detail drawings may be made.

Parts List

Assembly drawings, unit assemblies, group assemblies, or subassemblies usually contain a list showing the number and the name of each part included in the assembly. This list is commonly called a *parts list*. The parts are identified in the assembly drawing by a detail number. The detail number is enclosed in a circle which is attached to the part with a leader. A parts list and the identifying numbers are shown on this assembly drawing of the vise.

Material List

A bill of material (or material list) is placed on detail drawings which show more than one part. The material list usually

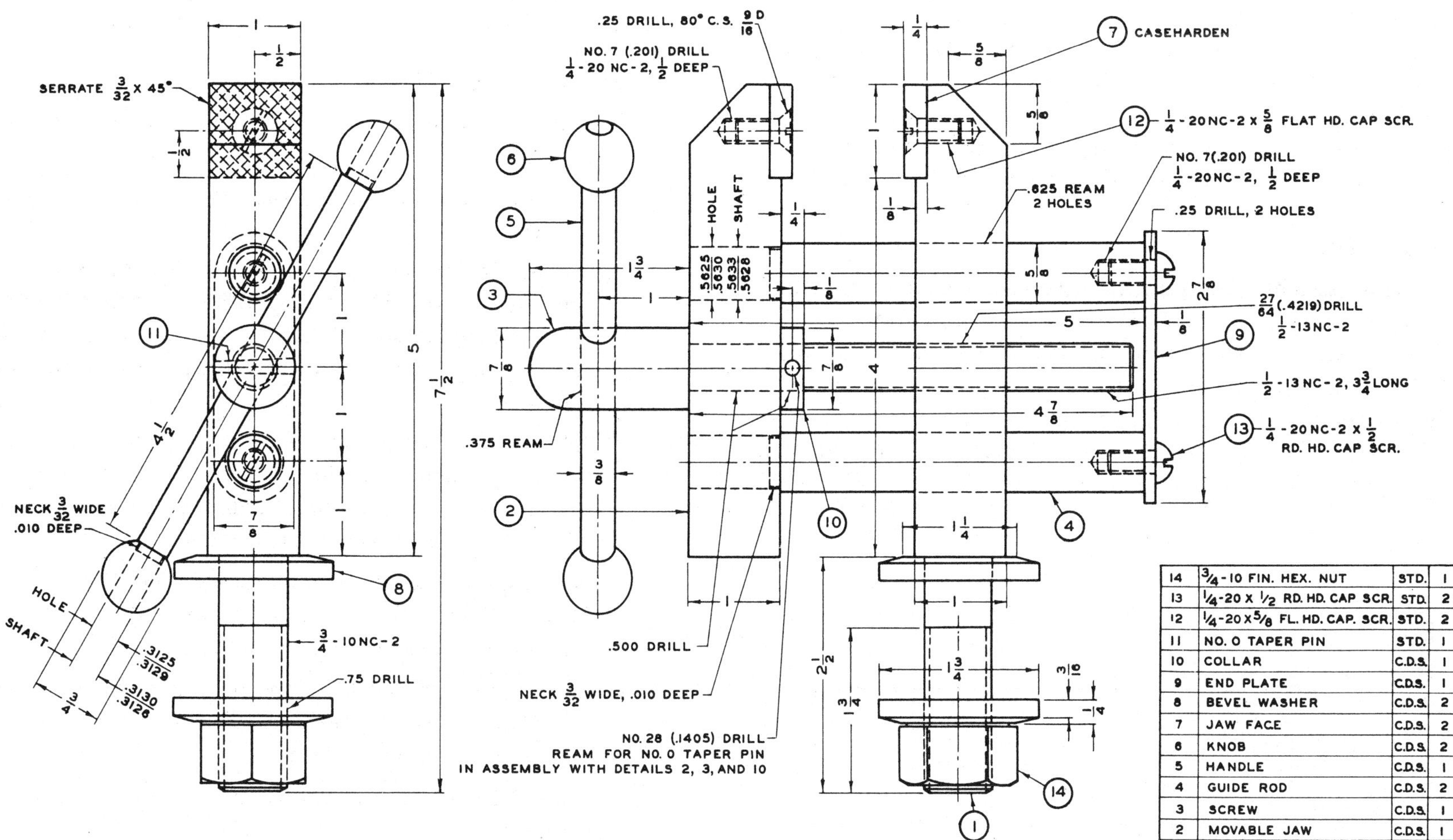

PART NO	NAME OF PART	MAT.	NO. REQ.
14	3/4-10 FIN. HEX. NUT	STD.	1
13	1/4-20 X 1/2 RD. HD. CAP SCR.	STD.	2
12	1/4-20 X 5/8 FL. HD. CAP. SCR.	STD.	2
11	NO. 0 TAPER PIN	STD.	1
10	COLLAR	C.D.S.	1
9	END PLATE	C.D.S.	1
8	BEVEL WASHER	C.D.S.	2
7	JAW FACE	C.D.S.	2
6	KNOB	C.D.S.	2
5	HANDLE	C.D.S.	1
4	GUIDE ROD	C.D.S.	2
3	SCREW	C.D.S.	1
2	MOVABLE JAW	C.D.S.	1
1	STATIONARY JAW	C.D.S.	1

NAME OF SCHOOL CITY OR TOWN	VISE	SHEET No. 14-N/O NAME	CLASS DATE

Fig. 14-16

specifies the part number; the size of the unfinished or raw material with allowance for machining; the pattern number of each casting; a description of standard parts such as bolts, screws, and nuts; an identifying name or number of special parts which may be purchased; along with the number required of each part and the kind of material. The material list is used as a guide when the necessary raw material and standard parts are ordered from the stock department. A typical material list is illustrated in figure 14-17.

Standard Taper Pins

This assembly drawing shows that the collar (detail 10) is fastened to the screw (detail 3) with a No. 0 taper pin (detail 11). After the screw and collar are assembled with the movable jaw (detail 2), the hole for the taper pin is drilled and then reamed with a taper reamer. The taper pin is then driven into the tapered hole to complete the assembly of these parts. The sizes of standard taper pins are given in the Appendix.

PART NO.	SIZE OF MATERIAL	MAT.	NO. REQ.
5	DIE CASTING D-18	ZINC ALLOY	1
4	NO. 2 TAPER PIN	STD.	2
3	1/2-13 HEX. HD. CAP SCR.	STD.	4
2	PATT. NO. 4718-L	C.I.	1
1	5/8 x 6 1/8 LONG	C.D.S.	1

Fig. 14-17 A material list

ASSIGNMENT N

Make a three-view working drawing of the stationary jaw (detail 1). Make a two-view working drawing of the screw (detail 3).

Procedure

1. Lay out the margin lines on the drawing sheet and letter the title strip.
2. Lay out the material list using the dimensions given in figure 14-16.
3. The stationary jaw (detail 1) is drawn lengthwise on the sheet. Find the length and width of a block which is large enough for the three views of detail 1. Allow 1 1/2 inches between the top and front views, and 1 inch between the front and side views.
4. The front view and end view of the screw, detail 3 are drawn. Find the length and width of a block which is large enough for the two views of detail 3, allowing 3/4 inch between the views. Draw light blocks for the views.
5. When the blocks are suitably spaced, draw the required views of each detail. Dimension the views. Add the notes, the detail number, and the name of each part.
6. Fill in the material list. Specify the size of the cold-drawn steel bar stock that is used for each part. Add 1/8 inch to the length for finishing the ends.
7. Complete the drawing. Check all lines, views, and dimensions for accuracy.

REVIEW QUESTIONS

1. What is the purpose of a parts list?
2. What is the purpose of a material list?
3. For what purpose are taper pins used?
4. What is the reason for blocking-in the views on a drawing?

BENCH VISE DETAIL DRAWINGS

Interchangeability of machined parts can only be achieved when the dimensions of mating parts have small tolerances controlled by limit dimensions.

Limits on Cylindrical Fits

Tables have been devised, for the convenience of machine designers, which specify the tolerances on the diameters of holes and mating shafts for eight kinds of fits, ranging from a loose fit (Class 1) to a heavy force fit (Class 8).

The end of the guide rods, detail 4, must be pressed into the holes in the movable jaw, detail 2. The dimensions specified on these parts are for a class 7 fit which is a medium force fit. The lower limit on the hole is .5625 inch and the lower limit on the guide rod is .5628 inch which gives an interference of .0003 inch for a medium force fit. The machinist cannot be expected to make these parts to the exact dimension so he is allowed a .0005 inch variation or tolerance. The upper limits on the hole and guide rod are .5630 inch and .5633 inch which also provides an interference of .0003 inch. The ideal condition is to have each guide rod .0003 inch larger than the hole in which it is assembled for a medium force fit.

To attain this condition in quantity production, the guide rods must be selected to fit the holes as follows:

Diameter of hole	Diameter of Guide Rod
.5625	.5628
.5626	.5629
.5627	.5630
.5628	.5631
.5629	.5632
.5630	.5633

This process of selecting the guide rods to fit the holes so that the desired amount of interference is maintained in each assembly is called *selective assembly.* Selective assembly is essential for all force or tight fits. The interference in force fits is also called *negative allowance.*

The handle (detail 5) must be a tight fit in the knobs, (detail 6). The dimensions specified are for a class 6 fit and are intended to fit the diameter of the handle .0001 inch larger than the diameter of the holes in the knobs. To reach this condition in quantity production, selective assembly is essential.

Neck

The 3/32 inch wide by .010 inch deep necks on the guide rods and the handle are sometimes called a grinding recess. Their purpose is to allow the corner of the grinding wheel to run nearly to the shoulder. This neck makes it possible to have a uniform diameter throughout the length of the part. If these parts were made without a grinding recess, a small fillet would be left in the corner by the edge of the grinding wheel.

ASSIGNMENT O

Draw a front and end view of the guide rod (detail 4). Make a front and end view of the handle (detail 5). Draw a front and side view of the movable jaw (detail 2). Draw the front views in a vertical position.

Procedure

1. Lay out the margin lines on the drawing sheet and letter the title strip.
2. Lay out the material list using the same dimensions given in figure 14-16.
3. Figure the length and the width of the blocks which are large enough for the required views. Allow 3/4 inch between the front and end views of the guide rod and the handle. Allow 1 1/2 inches between the front and side views of the jaw.

4. When the length and width of the blocks have been determined, block-in the views lightly.
5. Shift the blocks if needed until they are suitably spaced. Draw the views.
6. Dimension the views. Add the notes, the detail numbers, and the name of each part.
7. Fill in the material list for each part. Complete the drawing. Check the material list and the individual drawings.

REVIEW QUESTIONS

1. What is meant by the term mass production? Interchangeable parts?
2. What are limit dimensions?
3. What is the upper limit on a dimension? The lower limit?
4. What is the tolerance?
5. What is meant by selective assembly? By blocking-in the views?

BENCH VISE DETAIL DRAWINGS

The remaining parts on the dimensioned assembly drawing, figure 14-16, are drawn on this next detail drawing. Standard parts which are usually purchased are not to be drawn.

Serrate

The jaw faces are serrated by cutting a series of grooves at 45 degrees each way across the face. The serrated surfaces hold the work securely in the vise. The jaw faces are case hardened to resist wear.

ASSIGNMENT P

Draw a front and side view of the end plate (detail 9). Make a front and side view of the knob (detail 6). Draw a top and front view of the bevel washer (detail 8). Make a front and side view of the jaw face (detail 7). Draw a front and side view of the collar (detail 10).

Procedure

1. Lay out the margin lines and the material list on the drawing sheet and letter the title strip.
2. Block-in the views.
3. Draw the views of each part complete with dimensions and notes.
4. Fill in the material list for each part.
5. Complete the drawing. Check the different views, dimensions, and notes.
6. Check the accuracy of all lines and views.

REVIEW QUESTIONS

1. What is meant by serrate?
2. What is its purpose?

UNIT 15
CAREERS IN DRAFTING

This text is designed to provide an understanding of basic drafting concepts and to develop fundamental drawing skills common to many drawing-based career opportunities. The hands-on introduction to the graphic language of industry is valuable to the future machinist, tool-and-die maker, patternmaker, engineer, cartographer, carpenter, architect, interior decorator, and industrial designer. It is also valuable to workers whose job descriptions include interpreting plans, but who are not usually responsible for generating plans.

The following job descriptions identified by the American Institute for Design and Drafting (AIDD) are those that would require only the use of drawing skills.

A *drafter trainee* begins as a tracer who is expected to have an aptitude for drawing and be able to do neat, freehand lettering. The trainee works on revising or redrawing from existing prints or drawings to learn the fundamental drafting requirements of the company.

A *drafter* needs a minimum amount of training on the job, in a certified technical school program, or in both. The drafter is expected to be familiar with drafting standards and procedures. The job includes normal drafting assignments, data gathering, routine calculations, and measurements under regular supervision. Also incorporated under this new title is the experienced drafter known as a *drafting checker.* The checker is thoroughly familiar with manufacturing processes, drafting practices, and company policies, and must approve drawings with a signature after checking them for accuracy, completeness, and manufacturing feasibility.

A *design drafter* is an experienced drafter with an understanding of all drafting standards and practices, as well as mathematics and kinematics which are essential to this position. Often a member of the engineering team, the design drafter may make design changes in consultation with the engineering designer or architect, follow through on project changes, make estimates for material, time and cost, and instruct and supervise other drafters in support activities on design problems and related drawings.

The *design drafting supervisor* is responsible for coordinating all activities within a design drafting section. The position requires establishing work schedules, drafting standards and practices, and work loads. The design drafting supervisor also reviews layouts, designs and calculation, recommends personnel changes, and handles the budgeting and purchasing for the drafting staff.

The *design drafting manager* is an administrator responsible for managing and coordinating the technical, fiscal, and personnel activities of the department. The position entails enforcing compliance with all company rules and procedures, hiring and evaluating personnel, maintaining training programs, making changes to improve the financial or operational conditions, and acting as the liaison between the design drafting department and other departments within the company.

In addition to these job descriptions identified by the AIDD, a *technical illustrator* is responsible for making pictorial representations of objects. Technical illustrations are used in manufacturing and production, marketing and advertising literature, technical manuals for consumers, the aerospace program, and other areas where graphic interpretation of ideas is necessary to help in visualization of the object, its construction, or its use.

Drafting and drafting-related careers such as engineering, architecture, industrial design, and cartography are open to both men and women who have the patience to handle repetitive tasks and the ability to draw, Figure 15-1.

A drafter is expected to be neat, accurate, and precise. The drafter must be able to visualize objects and solutions and draw these visual images on drafting media. It is helpful to have a background in high school mathematics, science, and drafting. Most careers in drafting and drafting-related fields depend on some type of post-high-school education, often in addition to an apprenticeship program in industry.

For more information about careers in drafting, write to the American Institute for Design and Drafting, 349 Price Road, Bartlesville, OK 74003.

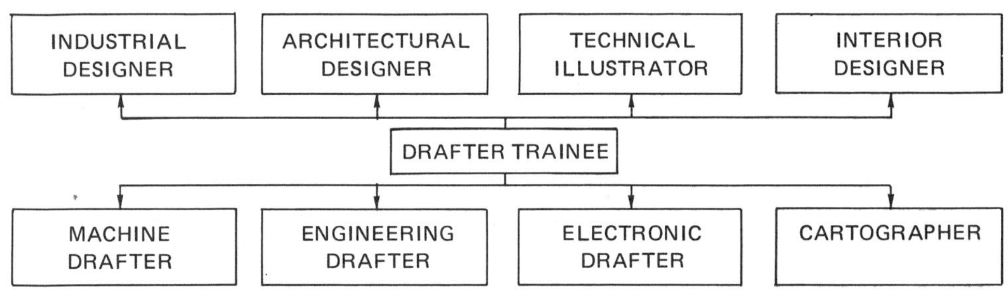

Fig. 15-1 Examples of drafting and drafting-related career opportunities

APPENDIX

TABLES

TABLE I. DECIMAL METRIC EQUIVALENTS

Inches		Milli-	Inches		Milli-	Inches		Milli-
Fractions	Decimals	meters	Fractions	Decimals	meters	Fractions	Decimals	meters
	.00394	.1	9/32	.28125	7.1438	21/32	.65625	16.6688
	.00787	.2	19/64	.296875	7.5406		.66929	17.00
	.01181	.3	5/16	.3125	7.9375	43/64	.671875	17.0657
1/64	.015625	.3969		.31496	8.00	11/16	.6875	17.4625
	.01575	.4	21/64	.328125	8.3344	45/64	.703125	17.8594
	.01969	.5	11/32	.34375	8.7313		.70866	18.00
	.02362	.6		.35433	9.00	23/32	.71875	18.2563
	.02756	.7	23/64	.359375	9.1281	47/64	.734375	18.6532
1/32	.03125	.7938	3/8	.375	9.525		.74803	19.00
	.0315	.8	25/64	.390625	9.9219	3/4	.7500	19.05
	.03543	.9		.3937	10.00	49/64	.765625	19.4469
	.03937	1.00	13/32	.40625	10.3188	25/32	.78125	19.8438
3/64	.046875	1.1906	27/64	.421875	10.7156		.7874	20.00
1/16	.0625	1.5875		.43307	11.00	51/64	.796875	20.2407
5/64	.078125	1.9844	7/16	.4375	11.1125	13/16	.8125	20.6375
	.07874	2.00	29/64	.453125	11.5094		.82677	21.00
3/32	.09375	2.3813	15/32	.46875	11.9063	53/64	.828125	21.0344
7/64	.109375	2.7781		.47244	12.00	27/32	.84375	21.4313
	.11811	3.00	31/64	.484375	12.3031	55/64	.859375	21.8282
1/8	.125	3.175	1/2	.5000	12.70		.86614	22.00
9/64	.140625	3.5719		.51181	13.00	7/8	.875	22.225
5/32	.15625	3.9688	33/64	.515625	13.0969	57/64	.890625	22.6219
	.15748	4.00	17/32	.53125	13.4938		.90551	23.00
11/64	.171875	4.3656	35/64	.546875	13.8907	29/32	.90625	23.0188
3/16	.1875	4.7625		.55118	14.00	59/64	.921875	23.4157
	.19685	5.00	9/16	.5625	14.2875	15/16	.9375	23.8125
13/64	.203125	5.1594	37/64	.578125	14.6844		.94488	24.00
7/32	.21875	5.5563		.59055	15.00	61/64	.953125	24.2094
15/64	.234375	5.9531	19/32	.59375	15.0813	31/32	.96875	24.6063
	.23622	6.00	39/64	.609375	15.4782		.98425	25.00
1/4	.2500	6.35	5/8	.625	15.875	63/64	.984375	25.0032
17/64	.265625	6.7469		.62992	16.00	1	1.0000	25.4001
	.27559	7.00	41/64	.640625	16.2719			

TABLE II. SCREW THREAD DIAMETERS AND NUMBER OF THREADS PER INCH

Diameter	National Coarse N.C.	National Fine N.F.	National Extra-Fine N.E.F.	8-Pitch 8N	12-Pitch 12N	16-Pitch 16N	Acme
No. 0 –.060	—	80	—	—	—	—	—
1 –.073	64	72	—	—	—	—	—
2 –.086	56	64	—	—	—	—	—
3 –.099	48	56	—	—	—	—	—
4 –.112	40	48	—	—	—	—	—
5 –.125	40	44	—	—	—	—	—
6 –.138	32	40	—	—	—	—	—
8 –.164	32	36	—	—	—	—	—
10 –.190	24	32	—	—	—	—	—
12 –.216	24	28	—	—	—	—	—
1/4	20	28	32	—	—	—	16
5/16	18	24	32	—	—	—	14
3/8	16	24	32	—	—	—	12
7/16	14	20	28	—	—	—	12
1/2	13	20	28	—	12	—	10
9/16	12	18	24	—	12	—	—
5/8	11	18	24	—	12	—	8
11/16	—	—	24	—	12	—	—
3/4	10	16	20	—	12	16	6
13/16	—	—	20	—	12	16	—
7/8	9	14	20	—	12	16	6
15/16	—	—	20	—	12	16	—
1	8	14	20	8	12	16	5
1 1/16	—	—	18	—	12	16	—
1 1/8	7	12	18	8	12	16	5
1 3/16	—	—	18	—	12	16	—
1 1/4	7	12	18	8	12	16	5
1 5/16	—	—	18	—	12	16	—
1 3/8	6	12	18	8	12	16	4
1 7/16	—	—	18	—	12	16	—
1 1/2	6	12	18	8	12	16	4
1 9/16	—	—	18	—	—	16	—
1 5/8	—	—	18	8	12	16	—
1 11/16	—	—	18	—	—	16	—
1 3/4	5	—	16	8	12	16	4
1 13/16	—	—	—	—	—	16	—
1 7/8	—	—	—	8	12	16	—
1 15/16	—	—	—	—	—	16	—
2	4 1/2	—	16	8	12	16	4
2 1/16	—	—	—	—	—	16	—
2 1/8	—	—	—	8	12	16	—
2 3/16	—	—	—	—	—	16	—
2 1/4	4 1/2	—	—	8	12	16	3
2 5/16	—	—	—	—	—	16	—
2 3/8	—	—	—	—	12	16	—
2 7/16	—	—	—	—	—	16	—
2 1/2	4	—	—	8	12	16	3
2 5/8	—	—	—	—	12	16	—
2 3/4	4	—	—	8	12	16	3
2 7/8	—	—	—	—	12	16	—
3	4	—	—	8	12	16	2
3 1/8	—	—	—	—	12	16	—
3 1/4	4	—	—	8	12	16	—
3 3/8	—	—	—	—	12	16	—
3 1/2	4	—	—	8	12	16	2
3 5/8	—	—	—	—	12	16	—
3 3/4	4	—	—	8	12	16	—
3 7/8	—	—	—	—	12	16	—
4	4	—	—	8	12	16	2
4 1/4	—	—	—	8	12	—	—
4 1/2	—	—	—	8	12	—	2
4 3/4	—	—	—	8	12	—	—
5	—	—	—	8	12	—	2
5 1/4	—	—	—	8	12	—	—
5 1/2	—	—	—	8	12	—	—
5 3/4	—	—	—	8	12	—	—
6	—	—	—	8	12	—	—

TABLE III. ISOMETRIC STANDARD SIZES

Nominal Size Diam. (mm)			Pitches (mm)														Nom-inal Size Diam. (mm)
Column[a]			Series With Graded Pitches		Series With Constant Pitches												
1	2	3	Coarse	Fine	6	4	3	2	1.5	1.25	1	0.75	0.5	0.35	0.25	0.2	
0.25			0.075	—	—	—	—	—	—	—	—	—	—	—	—	—	0.25
0.3			0.08	—	—	—	—	—	—	—	—	—	—	—	—	—	0.3
	0.35		0.09	—	—	—	—	—	—	—	—	—	—	—	—	—	0.35
0.4			0.1	—	—	—	—	—	—	—	—	—	—	—	—	—	0.4
	0.45		0.1	—	—	—	—	—	—	—	—	—	—	—	—	—	0.45
0.5			0.125	—	—	—	—	—	—	—	—	—	—	—	—	—	0.5
	0.55		0.125	—	—	—	—	—	—	—	—	—	—	—	—	—	0.55
0.6			0.15	—	—	—	—	—	—	—	—	—	—	—	—	—	0.6
	0.7		0.175	—	—	—	—	—	—	—	—	—	—	—	—	—	0.7
0.8			0.2	—	—	—	—	—	—	—	—	—	—	—	—	—	0.8
	0.9		0.225	—	—	—	—	—	—	—	—	—	—	—	—	—	0.9
1			0.25	—	—	—	—	—	—	—	—	—	—	—	—	0.2	1
	1.1		0.25	—	—	—	—	—	—	—	—	—	—	—	—	0.2	1.1
1.2			0.25	—	—	—	—	—	—	—	—	—	—	—	—	0.2	1.2
	1.4		0.3	—	—	—	—	—	—	—	—	—	—	—	—	0.2	1.4
1.6			0.35	—	—	—	—	—	—	—	—	—	—	—	—	0.2	1.6
	1.8		0.35	—	—	—	—	—	—	—	—	—	—	—	—	0.2	1.8
2			0.4	—	—	—	—	—	—	—	—	—	—	—	0.25	—	2
	2.2		0.45	—	—	—	—	—	—	—	—	—	—	—	0.25	—	2.2
2.5			0.45	—	—	—	—	—	—	—	—	—	—	0.35	—	—	2.5
3			0.5	—	—	—	—	—	—	—	—	—	—	0.35	—	—	3
	3.5		0.6	—	—	—	—	—	—	—	—	—	—	0.35	—	—	3.5
4			0.7	—	—	—	—	—	—	—	—	—	0.5	—	—	—	4
	4.5		0.75	—	—	—	—	—	—	—	—	—	0.5	—	—	—	4.5
5			0.8	—	—	—	—	—	—	—	—	—	0.5	—	—	—	5
		5.5	—	—	—	—	—	—	—	—	—	—	0.5	—	—	—	5.5
6			1	—	—	—	—	—	—	—	—	0.75	—	—	—	—	6
		7	1	—	—	—	—	—	—	—	—	0.75	—	—	—	—	7
8			1.25	1	—	—	—	—	—	—	1	0.75	—	—	—	—	8
		9	1.25	—	—	—	—	—	—	—	1	0.75	—	—	—	—	9
10			1.5	1.25	—	—	—	—	—	1.25	1	0.75	—	—	—	—	10
		11	1.5	—	—	—	—	—	—	—	1	0.75	—	—	—	—	11
12			1.75	1.25	—	—	—	—	1.5	1.25	1	—	—	—	—	—	12
	14		2	1.5	—	—	—	—	1.5	1.25[b]	1	—	—	—	—	—	14
		15	—	—	—	—	—	—	1.5	—	1	—	—	—	—	—	15
16			2	1.5	—	—	—	—	1.5	—	1	—	—	—	—	—	16
		17	—	—	—	—	—	—	1.5	—	1	—	—	—	—	—	17
	18		2.5	1.5	—	—	—	2	1.5	—	1	—	—	—	—	—	18
20			2.5	1.5	—	—	—	2	1.5	—	1	—	—	—	—	—	20
	22		2.5	1.5	—	—	—	2	1.5	—	1	—	—	—	—	—	22
24			3	2	—	—	—	2	1.5	—	1	—	—	—	—	—	24
		25	—	—	—	—	—	2	1.5	—	1	—	—	—	—	—	25
		26	—	—	—	—	—	—	1.5	—	1	—	—	—	—	—	26
	27		3	2	—	—	—	2	1.5	—	1	—	—	—	—	—	27
		28	—	—	—	—	—	2	1.5	—	1	—	—	—	—	—	28
30			3.5	2	—	—	(3)	2	1.5	—	1	—	—	—	—	—	30
		32	—	—	—	—	—	2	1.5	—	—	—	—	—	—	—	32
	33		3.5	2	—	—	(3)	2	1.5	—	—	—	—	—	—	—	33
		35[c]	—	—	—	—	—	—	1.5	—	—	—	—	—	—	—	35[c]
36			4	3	—	—	—	2	1.5	—	—	—	—	—	—	—	36
		38	—	—	—	—	—	—	1.5	—	—	—	—	—	—	—	38
	39		4	3	—	—	—	2	1.5	—	—	—	—	—	—	—	39
		40	—	—	—	—	3	2	1.5	—	—	—	—	—	—	—	40
42			4.5	3	—	4	3	2	1.5	—	—	—	—	—	—	—	42
	45		4.5	3	—	4	3	2	1.5	—	—	—	—	—	—	—	45

a Thread diameter should be selected from columns 1, 2 or 3; with preference being given in that order.
b Pitch 1.25 mm in combination with diameter 14 mm has been included for spark plug applications.
c Diameter 35 mm has been included for bearing locknut applications.
The use of pitches shown in parentheses should be avoided wherever possible.
The pitches enclosed in the bold frame, together with the corresponding nominal diameters in Columns 1 and 2, are those combinations which have been established by ISO Recommendations as a selected "coarse" and "fine" series for commercial fasteners. Sizes 0.25 mm through 1.4 mm are covered in ISO Recommendation R 68 and, except for the 0.25 mm size, in AN Standard ANSI B1.10.

(ANSI)

TABLE IV. TAP DRILL SIZES FOR AMERICAN STANDARD THREADS

Diam. of Thread	Threads per Inch	Drill*	Decimal Equiv.
No. 0 – .060	80 NF	3/64	.0469
1 – .073	64 NC	1.5 MM	.0591
	72 NF	53	.0595
2 – .086	56 NC	50	.0700
	64 NF	50	.0700
3 – .099	48 NC	5/64	.0781
	56 NF	45	.0820
4 – .112	40 NC	43	.0890
	48 NF	42	.0935
5 – .125	40 NC	38	.1015
	44 NF	37	.1040
6 – .138	32 NC	36	.1065
	40 NF	33	.1130
8 – .164	32 NC	29	.1360
	36 NF	29	.1360
10 – .190	24 NC	25	.1495
	32 NF	21	.1590
12 – .216	24 NC	16	.1770
	28 NF	14	.1820
1/4	20 NC	7	.2010
	28 NF	3	.2130
	32 NEF	7/32	.2188
5/16	18 NC	F	.2570
	24 NF	I	.2720
	32 NEF	9/32	.2812
3/8	16 NC	5/16	.3125
	24 NF	Q	.3320
	32 NEF	11/32	.3438
7/16	14 NC	U	.3680
	20 NF	25/64	.3906
	28 NEF	Y	.4040
1/2	12 N	27/64	.4219
	13 NC	27/64	.4219
	20 NF	29/64	.4531
	28 NEF	15/32	.4687
9/16	12 NC	31/64	.4844
	18 NF	33/64	.5156
	24 NEF	33/64	.5156
5/8	11 NC	17/32	.5312
	12 N	35/64	.5469
	18 NF	14.5 MM	.5709
	24 NEF	37/64	.5781

Diam. of Thread	Threads per Inch	Drill*	Decimal Equiv.
11/16	12 N	39/64	.6094
	24 NEF	16.5 MM	.6496
3/4	10 NC	16.5 MM	.6496
	12 N	17 MM	.6693
	16 NF	17.5 MM	.6890
	20 NEF	45/64	.7031
13/16	12 N	18.5 MM	.7283
	16 N	3/4	.7500
	20 NEF	49/64	.7656
7/8	9 NC	49/64	.7656
	12 N	20 MM	.7874
	14 NF	20.5 MM	.8071
	16 N	13/16	.8125
	20 NEF	21 MM	.8268
15/16	12 N	55/64	.8594
	16 N	7/8	.8750
	20 NEF	22.5 MM	.8858
1	8 NC	7/8	.8750
	12 N	59/64	.9219
	14 NF	23.5 MM	.9252
	16 N	15/16	.9375
	20 NEF	61/64	.9531
1 1/16	12 N	25 MM	.9843
	16 N	1	1.0000
	18 NEF	25.5 MM	1.0040
1 1/8	7 NC	63/64	.9844
	8 N	25.5 MM	1.0039
	12 NF	26.5 MM	1.0433
	16 N	1 1/16	1.0625
	18 NEF	1 5/64	1.0781
1 3/16	12 N	28 MM	1.1024
	16 N	1 1/8	1.1250
	18 NEF	1 9/64	1.1406
1 1/4	7 NC	1 7/64	1.1094
	8 N	1 1/8	1.1250
	12 NF	29.5 MM	1.1614
	16 N	1 3/16	1.1875
	18 NEF	30.5 MM	1.2008
1 5/16	12 N	1 15/64	1.2344
	16 N	1 1/4	1.2500
	18 NEF	32 MM	1.2598

*To produce approximately 75% full thread

TABLE IV. CONTINUED – TAP DRILL SIZES FOR AMERICAN STANDARD THREADS

Diam. of Thread	Threads per Inch	Drill*	Decimal Equiv.
1 3/8	6 NC	1 13/64	1.2031
	8 N	1 1/4	1.2500
	12 NF	1 19/64	1.2969
	16 N	1 5/16	1.3125
	18 NEF	33.5 MM	1.3189
1 7/16	12 N	34.5 MM	1.3583
	16 N	1 3/8	1.3750
	18 NEF	35 MM	1.3780
1 1/2	6 NC	1 21/64	1.3281
	8 N	1 3/8	1.3750
	12 NF	36 MM	1.4173
	16 N	1 7/16	1.4375
	18 NEF	1 29/64	1.4531
1 9/16	16 N	1 1/2	1.5000
	18 NEF	1 33/64	1.5156
1 5/8	8 N	1 1/2	1.5000
	12 N	39 MM	1.5354
	16 N	1 9/16	1.5625
	18 NEF	40 MM	1.5748
1 11/16	16 N	1 5/8	1.6250
	18 NEF	41.5 MM	1.6339
1 3/4	5 NC	1 35/64	1.5469
	8 N	1 5/8	1.6250
	12 N	1 43/64	1.6719
	16 NEF	1 11/16	1.6875
1 13/16	16 N	1 3/4	1.7500
1 7/8	8 N	1 3/4	1.7500
	12 N	45.5 MM	1.7913
	16 N	1 13/16	1.8125
1 15/16	16 N	1 7/8	1.8750
2	4 1/2 NC	1 25/32	1.7812
	8 N	1 7/8	1.8750
	12 N	1 59/64	1.9219
	16 NEF	1 15/16	1.9375
2 1/16	16 N	2	2.0000
2 1/8	8 N	2	2.0000
	12 N	2 3/64	2.0469
	16 N	2 1/16	2.0625

Diam. of Thread	Threads per Inch	Drill*	Decimal Equiv.
2 3/16	16 N	2 1/8	2.1250
2 1/4	4 1/2 NC	2 1/32	2.0312
	8 N	2 1/8	2.1250
	12 N	55 MM	2.1654
	16 N	2 3/16	2.1875
2 5/16	16 N	2 1/4	2.2500
2 3/8	12 N	2 19/64	2.2969
	16 N	2 5/16	2.3125
2 7/16	16 N	2 3/8	2.3750
2 1/2	4 NC	2 1/4	2.2500
	8 N	2 3/8	2.3750
	12 N	61.5 MM	2.4213
	16 N	2 7/16	2.4375
2 5/8	12 N	64.5 MM	2.5394
	16 N	2 9/16	2.5625
2 3/4	4 NC	2 1/2	2.5000
	8 N	2 5/8	2.6250
	12 N	2 43/64	2.6719
	16 N	2 11/16	2.6875
2 7/8	12 N	71 MM	2.7953
	16 N	2 13/16	2.8125
3	4 NC	2 3/4	2.7500
	8 N	2 7/8	2.8750
	12 N	74 MM	2.9134
	16 N	2 15/16	2.9375
3 1/8	12 N	3 1/16	3.0625
	16 N	3 1/16	3.0625
3 1/4	4 NC	3	3.0000
	8 N	3 1/8	3.1250
	12 N	3 3/16	3.1875
	16 N	3 3/16	3.1875
3 3/8	12 N	3 5/16	3.3125
	16 N	3 5/16	3.3125
3 1/2	4 NC	3 1/4	3.2500
	8 N	3 3/8	3.3750
	12 N	3 7/16	3.4375
	16 N	3 7/16	3.4375
3 3/4	4 NC	3 1/2	3.5000

* To produce approximately 75% full thread

TABLE V. DRILL SIZE DECIMAL EQUIVALENT

Drill No.	Frac.	Deci.
80	—	.0135
79	—	.0145
	1/64	.0156
78	—	.0160
77	—	.0180
76	—	.0200
75	—	.0210
74	—	.0225
73	—	.0240
72	—	.0250
71	—	.0260
70	—	.0280
69	—	.0292
68	—	.0310
	1/32	.0313
67	—	.0320
66	—	.0330
65	—	.0350
64	—	.0360
63	—	.0370
62	—	.0380
61	—	.0390
60	—	.0400
59	—	.0410
58	—	.0420
57	—	.0430
56	—	.0465
	3/64	.0469
55	—	.0520
54	—	.0550
53	—	.0595
	1/16	.0625
52	—	.0635
51	—	.0670
50	—	.0700
49	—	.0730
48	—	.0760
	5/64	.0781
47	—	.0785
46	—	.0810
45	—	.0820
44	—	.0860
43	—	.0890

Drill No.	Frac.	Deci.
42	—	.0935
	3/32	.0938
41	—	.0960
40	—	.0980
39	—	.0995
38	—	.1015
37	—	.1040
36	—	.1065
	7/64	.1094
35	—	.1100
34	—	.1110
33	—	.1130
32	—	.116
31	—	.120
	1/8	.125
30	—	.129
29	—	.136
	9/64	.140
28	—	.141
27	—	.144
26	—	.147
25	—	.150
24	—	.152
23	—	.154
	5/32	.156
22	—	.157
21	—	.159
20	—	.161
19	—	.166
18	—	.170
	11/64	.172
17	—	.173
16	—	.177
15	—	.180
14	—	.182
13	—	.185
	3/16	.188
12	—	.189
11	—	.191
10	—	.194
9	—	.196
8	—	.199
7	—	.201

Drill No.	Frac.	Deci.
	13/64	.203
6	—	.204
5	—	.206
4	—	.209
3	—	.213
	7/32	.219
2	—	.221
1	—	.228
A	—	.234
	15/64	.234
B	—	.238
C	—	.242
D	—	.246
	1/4	.250
E	—	.250
F	—	.257
G	—	.261
	17/64	.266
H	—	.266
I	—	.272
J	—	.277
	9/32	.281
K	—	.281
L	—	.290
M	—	.295
	19/64	.297
N	—	.302
	5/16	.313
O	—	.316
P	—	.323
	21/64	.328
Q	—	.332
R	—	.339
	11/32	.344
S	—	.348
T	—	.358
	23/64	.359
U	—	.368
	3/8	.375
V	—	.377
W	—	.386
	25/64	.391
X	—	.397

Drill No.	Frac.	Deci.
Y	—	.404
	13/32	.406
Z	—	.413
	27/64	.422
	7/16	.438
	29/64	.453
	15/32	.469
	31/64	.484
	1/2	.500
	33/64	.516
	17/32	.531
	35/64	.547
	9/16	.562
	37/64	.578
	19/32	.594
	39/64	.609
	5/8	.625
	41/64	.641
	21/32	.656
	43/64	.672
	11/16	.688
	45/64	.703
	23/32	.719
	47/64	.734
	3/4	.750
	49/64	.766
	25/32	.781
	51/64	.797
	13/16	.813
	53/64	.828
	27/32	.844
	55/64	.859
	7/8	.875
	57/64	.891
	29/32	.906
	59/64	.922
	15/16	.938
	61/64	.953
	31/32	.969
	63/64	.984
	1	1.000

MM	DEC.	MM	DEC.	MM	DEC.	MM	DEC.
1.	.0394	3.2	.1260	6.3	.2480	9.5	.3740
1.05	.0413	3.25	.1280	6.4	.2520	9.6	.3780
1.1	.0433	3.3	.1299	6.5	.2559	9.7	.3819
1.15	.0453	3.4	.1339	6.6	.2598	9.75	.3839
1.2	.0472	3.5	.1378	6.7	.2638	9.8	.3858
1.25	.0492	3.6	.1417	6.75	.2657	9.9	.3898
1.3	.0512	3.7	.1457	6.8	.2677	10.	.3937
1.35	.0531	3.75	.1476	6.9	.2717	10.5	.4134
1.4	.0551	3.8	.1496	7.	.2756	11.	.4331
1.45	.9571	3.9	.1535	7.1	.2795	11.5	.4528
1.5	.0591	4.	.1575	7.2	.2835	12.	.4724
1.55	.0610	4.1	.1614	7.25	.2854	12.5	.4921
1.6	.0630	4.2	.1654	7.3	.2874	13.	.5118
1.65	.0650	4.25	.1673	7.4	.2913	13.5	.5315
1.7	.0669	4.3	.1693	7.5	.2953	14.	.5512
1.75	.0689	4.4	.1732	7.6	.2992	14.5	.5709
1.8	.0709	4.5	.1772	7.7	.3031	15.	.5906
1.85	.0728	4.6	.1811	7.75	.3051	15.5	.6102
1.9	.0748	4.7	.1850	7.8	.3071	16.	.6299
1.95	.0768	4.75	.1870	7.9	.3110	16.5	.6496
2.	.0787	4.8	.1890	8.	.3150	17.	.6693
2.05	.0807	4.9	.1929	8.1	.3189	17.5	.6890
2.1	.0827	5.	.1968	8.2	.3228	18.	.7087
2.15	.9846	5.1	.2008	8.25	.3248	18.5	.7283
2.2	.0866	5.2	.2047	8.3	.3268	19.	.7480
2.25	.0886	5.25	.2067	8.4	.3307	19.5	.7677
2.3	.0906	5.3	.2087	8.5	.3346	20.	.7874
2.35	.0925	5.4	.2126	8.6	.3386	20.5	.8071
2.4	.0945	5.5	.2165	8.7	.3425	21.	.8268
2.45	.0965	5.6	.2205	8.75	.3445	21.5	.8465
2.5	.0984	5.7	.2244	8.8	.3465	22.	.8661
2.6	.1024	5.75	.2264	8.9	.3504	22.5	.8858
2.7	.1063	5.8	.2283	9.	.3543	23.	.9055
2.75	.1083	5.9	.2323	9.1	.3583	23.5	.9252
2.8	.1102	6.	.2362	9.2	.3622	24.	.9449
2.9	.1142	6.1	.2402	9.25	.3642	24.5	.9646
3.	.1181	6.2	.2441	9.3	.3661	25.	.9843
3.1	.1220	6.25	.2461	9.4	.3701		

Drill Size Decimal Equivalents. Left. Number and Letter Drills. Above. Metric Drills.

TABLE VI. AMERICAN STANDARD WRENCH HEAD BOLTS AND NUTS–REGULAR SERIES

	HEADS - Unfinished, Sq. and Hex.; Semi-fin., Hex. Only					NUTS - Unfinished, Sq. and Hex.; Semi-fin., Hex.Only JAM NUTS - Unfinished and Semi-fin., Hex.Only						
Diam. of Bolt	Width			Height		Width			Thickness-Reg.		Thickness - Jam	
	Across Flats	Across Corners Hex.	Across Corners Square	Unfin. Sq. and Hex.	Semifin. Hex.	Across Flats	Across Corners Hex.	Across Corners Square	Unfin. Sq. and Hex.	Semifin. Hex.	Unfin-ished	Semi-finished
1/4	3/8	.433	.530	11/64	5/32	7/16	.505	.619	7/32	13/64	5/32	9/64
5/16	1/2	.577	.707	13/64	3/16	9/16	.649	.796	17/64	1/4	3/16	11/64
3/8	9/16	.649	.796	1/4	15/64	5/8	.722	.884	21/64	5/16	7/32	13/64
7/16	5/8	.722	.884	19/64	9/32	3/4	.866	1.061	3/8	23/64	1/4	15/64
1/2	3/4	.866	1.061	21/64	19/64	13/16	.938	1.149	7/16	27/64	5/16	19/64
9/16	7/8	1.010	1.237	3/8	11/32	7/8	1.010	1.237	1/2	31/64	11/32	21/64
5/8	15/16	1.082	1.326	27/64	25/64	1	1.155	1.414	35/64	17/32	3/8	23/64
3/4	1 1/8	1.299	1.591	1/2	15/32	1 1/8	1.299	1.591	21/32	41/64	7/16	27/64
7/8	1 5/16	1.516	1.856	19/32	9/16	1 5/16	1.516	1.856	49/64	3/4	1/2	31/64
1	1 1/2	1.732	2.121	21/32	19/32	1 1/2	1.732	2.121	7/8	55/64	9/16	35/64
1 1/8	1 11/16	1.949	2.387	3/4	11/16	1 11/16	1.949	2.387	1	31/32	5/8	39/64
1 1/4	1 7/8	2.165	2.652	27/32	25/32	1 7/8	2.165	2.651	1 3/32	1 1/16	3/4	23/32
1 3/8	2 1/16	2.381	2.917	29/32	27/32	2 1/16	2.382	2.917	1 13/64	1 11/64	13/16	25/32
1 1/2	2 1/4	2.598	3.182	1	15/16	2 1/4	2.598	3.182	1 5/16	1 9/32	7/8	27/32
1 5/8	2 7/16	2.815	3.447	1 3/32	1 1/32	2 7/16	2.815	3.447	1 27/64	1 25/64	15/16	29/32
1 3/4	2 5/8	3.031	3.712	1 5/32	1 3/32	2 5/8	3.031	3.712	1 17/32	1 1/2	1	31/32
1 7/8	2 13/16	3.248	3.979	1 1/4	1 3/16	2 13/16	3.248	3.979	1 41/64	1 39/64	1 1/16	1 1/32
2	3	3.464	4.243	1 11/32	1 7/32	3	3.464	4.243	1 3/4	1 23/32	1 1/8	1 3/32
2 1/4	3 3/8	3.897	4.773	1 1/2	1 3/8	3 3/8	3.897	4.773	1 31/32	1 59/64	1 1/4	1 13/64
2 1/2	3 3/4	4.330	5.303	1 21/32	1 17/32	3 3/4	4.330	5.303	2 3/16	2 9/64	1 1/2	1 29/64
2 3/4	4 1/8	4.763	5.834	1 13/16	1 11/16	4 1/8	4.763	5.834	2 13/32	2 23/64	1 5/8	1 37/64
3	4 1/2	5.196	6.364	2	1 7/8	4 1/2	5.196	6.364	2 5/8	2 27/64	1 3/4	1 45/64

Finished Regular Bolt Heads and Nuts, When Specified, Are Made to the Dimensions Given For Semi-finished Bolts and Nuts.

TABLE VII. AMERICAN STANDARD WRENCH HEAD BOLTS AND NUTS—HEAVY SERIES

HEADS - Unfinished, Sq. and Hex. Semi-finished, Hex. Only						NUTS - Unfinished, Sq. and Hex., Semi-fin., Hex.Only JAM NUTS - Unfinished and Semi-fin., Hex. Only						
Diam. of Bolt	Width			Height		Width			Thickness-Heavy		Thickness - Jam	
	Across Flats	Across Corners		Unfin. Sq. and Hex.	Semifin. Hex.	Across Flats	Across Corners		Unfin. Sq. and Hex.	Semifin. Hex.	Unfin-finished	Semi-finished
		Hex.	Square				Hex.	Square				
1/2	7/8	.969	1.167	7/16	13/32	7/8	.969	1.167	1/2	31/64	5/16	19/64
9/16	15/16	1.037	1.249	15/32	7/16	15/16	1.037	1.249	9/16	35/64	11/32	21/64
5/8	1 1/16	1.175	1.416	17/32	1/2	1 1/16	1.175	1.416	5/8	39/64	3/8	23/64
3/4	1 1/4	1.383	1.665	5/8	19/32	1 1/4	1.382	1.665	3/4	47/64	7/16	27/64
7/8	1 7/16	1.589	1.914	23/32	11/16	1 7/16	1.589	1.914	7/8	55/64	1/2	31/64
1	1 5/8	1.796	2.162	13/16	3/4	1 5/8	1.796	2.162	1	63/64	9/16	35/64
1 1/8	1 13/16	2.002	2.411	29/32	27/32	1 13/16	2.002	2.411	1 1/8	1 7/64	5/8	39/64
1 1/4	2	2.209	2.661	1	15/16	2	2.209	2.661	1 1/4	1 7/32	3/4	23/32
1 3/8	2 3/16	2.416	2.909	1 3/32	1 1/32	2 3/16	2.416	2.909	1 3/8	1 11/32	13/16	25/32
1 1/2	2 3/8	2.622	3.158	1 3/16	1 1/8	2 3/8	2.622	3.158	1 1/2	1 15/32	7/8	27/32
1 5/8	2 9/16	2.828	3.406	1 9/32	1 7/32	2 9/16	2.828	3.406	1 5/8	1 19/32	15/16	29/32
1 3/4	2 3/4	3.036	3.655	1 3/8	1 5/16	2 3/4	3.035	3.655	1 3/4	1 23/32	1	31/32
1 7/8	2 15/16	3.242	3.905	1 15/32	1 13/32	2 15/16	3.242	3.905	1 7/8	1 27/32	1 1/16	1 1/32
2	3 1/8	3.449	4.153	1 9/16	1 7/16	3 1/8	3.449	4.153	2	1 31/32	1 1/8	1 3/32
2 1/4	3 1/2	3.862	4.652	1 3/4	1 5/8	3 1/2	3.862	4.652	2 1/4	2 13/64	1 1/4	1 13/64
2 1/2	3 7/8	4.275	5.149	1 15/16	1 13/16	3 7/8	4.275	5.149	2 1/2	2 29/64	1 1/2	1 29/64
2 3/4	4 1/4	4.688	5.646	2 1/8	2	4 1/4	4.688	5.646	2 3/4	2 45/64	1 5/8	1 37/64
3	4 5/8	5.102	6.144	2 5/16	2 3/16	4 5/8	5.102	6.144	3	2 61/64	1 3/4	1 45/64

Finished Heavy Bolt Heads and Nuts When Specified Are Made to the Dimensions Given For Semi-finished Bolt Heads and Nuts.

TABLE VIII. RECOMMENDED MINIMUM THREAD LENGTHS FOR BOLTS

Bolt Length	NOMINAL DIAMETER OF BOLT														
	No. 10 1/4	5/16 3/8	7/16 1/2	9/16 5/8	3/4	7/8	1	1 1/8 1 1/4	1 3/8 1 1/2	1 5/8 1 3/4	1 7/8 2	2 1/4	2 1/2	2 3/4	3
	MINIMUM LENGTH OF THREAD														
3/4	1/2	1/2													
1	3/4	3/4	3/4	3/4											
1 1/4	3/4	3/4	1	1	1										
1 1/2	3/4	7/8	1	1 1/8	1 1/8	1 1/8									
1 3/4	3/4	7/8	1	1 3/16	1 3/8	1 3/8	1 3/8								
2	3/4	1	1 1/4	1 1/4	1 3/8	1 9/16	1 5/8	1 5/8							
2 1/2	3/4	1	1 1/4	1 1/2	1 1/2	1 9/16	1 3/4	2	2						
3	7/8	1	1 1/4	1 1/2	1 3/4	1 3/4	1 3/4	2 1/8	2 1/2	2 1/2					
4	7/8	1	1 1/4	1 1/2	1 3/4	2	2 1/4	2 1/4	2 1/2	2 7/8	3 1/4	3 1/4	3 1/4		
5	7/8	1 3/16	1 1/4	1 1/2	1 3/4	2	2 1/4	2 3/4	2 3/4	2 7/8	3 1/4	3 5/8	4	4 1/8	4 1/4
6	7/8	1 3/16	1 1/2	1 1/2	1 3/4	2	2 1/4	2 3/4	3 1/4	3 1/4	3 1/4	3 5/8	4	4 1/8	4 3/4
8	7/8	1 3/16	1 1/2	1 13/16	2	2	2 1/4	2 3/4	3 1/4	3 3/4	4	4	4	4 1/8	4 3/4
10	7/8	1 3/16	1 1/2	1 13/16	2 1/8	2 7/16	2 1/2	2 3/4	3 1/4	3 3/4	4 1/4	4 3/4	4 3/4	4 3/4	4 3/4
12	7/8	1 3/16	1 1/2	1 13/16	2 1/8	2 7/16	2 3/4	2 3/4	3 1/4	3 3/4	4 1/4	4 3/4	5 1/4	5 3/4	6 1/4
16	1	1 3/16	1 1/2	1 13/16	2 1/8	2 7/16	2 3/4	3 1/4	3 1/4	3 3/4	4 1/4	4 3/4	5 1/4	5 3/4	6 1/4
20	1	1 3/8	1 1/2	1 13/16	2 1/8	2 7/16	2 3/4	3 3/8	4	4 5/8	4 3/4	4 3/4	5 1/4	5 3/4	6 1/4
30			1 3/4	1 13/16	2 1/8	2 7/16	2 3/4	3 3/8	4	4 5/8	5 1/4	5 7/8	6 1/2	6 1/2	6 1/2

TABLE IX. DIMENSIONS OF CAP SCREW HEADS

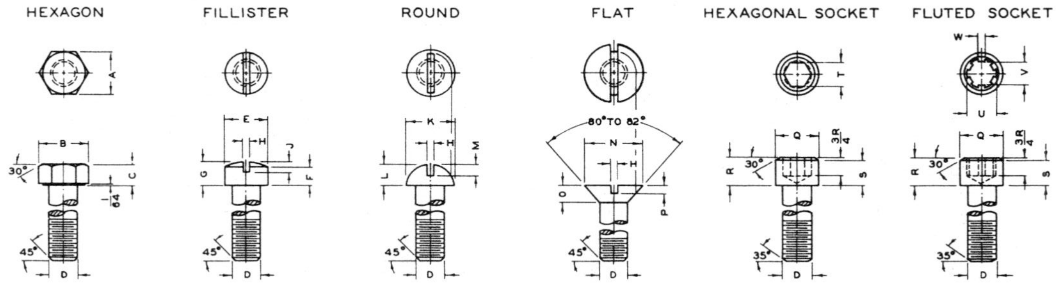

Diam.	A	B	C	E	F	G	H	J	K	L	M	N	O	P	Q	R	S	T	U	V	W	
2-.086															.140 .136	.086 .083	.079	1/16	.074 .073	.064 .063	.016 .015	Max. Min.
3-.099															.161 .157	.099 .096	.091	5/64	.098 .097	.082 .080	.022 .021	Max. Min.
4-.112															.183 .178	.112 .109	.103	5/64	.098 .097	.082 .080	.022 .021	Max. Min.
5-.125															.205 .200	.125 .122	.115	3/32	.115 .113	.098 .096	.025 .023	Max. Min.
6-.138															.226 .221	.138 .134	.127	3/32	.115 .113	.098 .096	.025 .023	Max. Min.
8-.164															.270 .265	.164 .160	.150	1/8	.149 .147	.128 .126	.032 .030	Max. Min.
10-.190															5/16	.190 .185	.174	5/32	.188 .186	.163 .161	.039 .037	Max. Min.
12-.216															11/32	.216 .211	.198	5/32	.188 .186	.163 .161	.039 .037	Max. Min.
1/4	7/16	.505	3/16	3/8	.172 .157	.216 .195	.075 .064	.097 .077	7/16	.191 .175	.117 .097	1/2	.140	.069 .046	3/8	1/4	.229	3/16	.221 .219	.190 .188	.050 .048	Max. Min.
5/16	1/2	.577	15/64	7/16	.203 .186	.253 .230	.084 .072	.115 .090	9/16	.246 .226	.151 .126	5/8	.176	.086 .057	7/16	5/16	.286	7/32	.256 .254	.221 .219	.060 .058	Max. Min.
3/8	9/16	.649	9/32	9/16	.250 .229	.314 .285	.094 .081	.143 .113	5/8	.273 .252	.168 .138	3/4	.210	.103 .069	9/16	3/8	.344	5/16	.380 .377	.319 .316	.092 .089	Max. Min.
7/16	5/8	.722	21/64	5/8	.297 .274	.368 .337	.094 .081	.168 .133	3/4	.328 .302	.202 .167	13/16	.210	.103 .069	5/8	7/16	.401	5/16	.380 .377	.319 .316	.092 .089	Max. Min.
1/2	3/4	.866	3/8	3/4	.328 .301	.412 .376	.106 .091	.188 .148	13/16	.355 .328	.219 .179	7/8	.210	.103 .069	3/4	1/2	.458	3/8	.463 .460	.386 .383	.112 .109	Max. Min.
9/16	13/16	.938	27/64	13/16	.375 .347	.466 .428	.118 .102	.214 .169	15/16	.410 .379	.253 .208	1	.245	.120 .080	13/16	9/16	.516	3/8	.463 .460	.386 .383	.112 .109	Max. Min.
5/8	7/8	1.010	15/32	7/8	.422 .392	.521 .480	.133 .116	.240 .190	1	.438 .405	.270 .220	1 1/8	.281	.137 .092	7/8	5/8	.573	1/2	.604 .601	.509 .506	.138 .134	Max. Min.
3/4	1	1.155	9/16	1	.500 .466	.612 .566	.149 .131	.283 .233	1 1/4	.547 .506	.337 .277	1 3/8	.352	.171 .115	1	3/4	.688	9/16	.631 .627	.535 .531	.149 .145	Max. Min.
7/8	1 1/8	1.299	21/32	1 1/8	.594 .556	.720 .669	.167 .147	.334 .264				1 5/8	.423	.206 .139	1 1/8	7/8	.802	9/16	.709 .705	.604 .600	.168 .164	Max. Min.
1	1 5/16	1.516	3/4	1 5/16	.656 .613	.802 .744	.188 .166	.372 .292				1 7/8	.494	.240 .162	1 5/16	1	.917	5/8	.801 .797	.685 .681	.189 .185	Max. Min.
1 1/8	1 1/2	1.732	27/32												1 1/2	1 1/8	1.031	3/4	.970 .966	.828 .824	.231 .227	Max. Min.
1 1/4	1 11/16	1.949	15/16												1 3/4	1 1/4	1.146	3/4	.970 .966	.828 .824	.231 .227	Max. Min.
1 3/8															1 7/8	1 3/8	1.260	3/4	.970 .966	.828 .824	.231 .227	Max. Min.
1 1/2															2	1 1/2	1.375	1	1.275 1.271	1.007 1.003	.298 .294	Max. Min.

Length of Thread on Hexagon, Fillister, Round and Flat = 2 D + 1/4"

Length of Thread on Hex. and Fluted Socket, For NC Thread = 2D + 1/2", or 1/2 Bolt Length } Whichever is Greater
For NF Thread = 1 1/2D + 1/2, or 3/8 Bolt Length

TABLE X. DIMENSIONS OF MACHINE SCREW HEADS

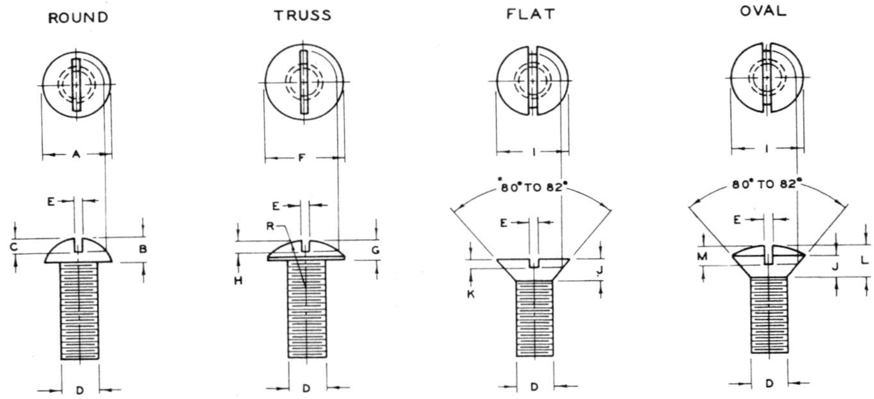

Diam.	A	B	C	E	F	G	H	H[1]	I	J	K	L	M	N	
0-.060	.113	.053	.039	.023					.119	.035	.015	.056	.030	.096	Max.
	.099	.043	.029	.016					.105	.026	.010	.041	.025	.083	Min.
1-.073	.138	.061	.044	.026					.146	.043	.019	.068	.038	.118	Max.
	.122	.051	.033	.019					.130	.033	.012	.052	.031	.104	Min.
2-.086	.162	.069	.048	.031	.194	.053	.031	.050	.172	.051	.023	.080	.045	.140	Max.
	.146	.059	.037	.023	.180	.044	.022	.040	.156	.040	.015	.063	.037	.124	Min.
3-.099	.187	.078	.053	.035	.226	.061	.036	.055	.199	.059	.027	.092	.052	.161	Max.
	.169	.067	.040	.027	.211	.051	.026	.044	.181	.048	.017	.073	.043	.145	Min.
4-.112	.211	.086	.058	.039	.257	.069	.040	.060	.225	.067	.030	.104	.059	.183	Max.
	.193	.075	.044	.031	.241	.059	.030	.049	.207	.055	.020	.084	.049	.166	Min.
5-.125	.236	.095	.063	.043	.289	.078	.045	.070	.252	.075	.034	.116	.067	.205	Max.
	.217	.083	.047	.035	.272	.066	.034	.058	.232	.062	.022	.095	.055	.187	Min.
6-.138	.260	.103	.068	.048	.321	.086	.050	.080	.279	.083	.038	.128	.074	.226	Max.
	.240	.091	.051	.039	.303	.074	.037	.067	.257	.069	.024	.105	.060	.208	Min.
8-.164	.309	.120	.077	.054	.384	.102	.058	.110	.332	.100	.045	.152	.088	.270	Max.
	.287	.107	.058	.045	.364	.088	.045	.096	.308	.084	.029	.126	.072	.250	Min.
10-.190	.359	.137	.087	.060	.448	.118	.068	.120	.385	.116	.053	.176	.103	.313	Max.
	.334	.123	.065	.050	.425	.103	.053	.105	.359	.098	.034	.148	.084	.292	Min.
12-.216	.408	.153	.096	.067	.511	.134	.077	.155	.438	.132	.060	.200	.117	.357	Max.
	.382	.139	.072	.056	.487	.118	.061	.139	.410	.112	.039	.169	.096	.334	Min.
1/4	.472	.175	.109	.075	.573	.150	.087	.190	.507	.153	.070	.232	.136	.414	Max.
	.443	.160	.082	.064	.546	.133	.070	.172	.477	.131	.046	.197	.112	.389	Min.
5/16	.590	.216	.132	.084	.698	.183	.106	.230	.635	.191	.088	.290	.171	.518	Max.
	.557	.198	.099	.072	.666	.162	.085	.208	.600	.165	.058	.249	.141	.490	Min.
3/8	.708	.256	.155	.094	.823	.215	.124	.295	.762	.230	.106	.347	.206	.622	Max.
	.670	.237	.117	.081	.787	.191	.100	.270	.722	.200	.070	.300	.170	.590	Min.
7/16	.750	.328	.196	.094	.948	.248	.142		.812	.223	.103	.345	.210	.625	Max.
	.707	.307	.148	.081	.907	.221	.116		.771	.190	.066	.295	.174	.589	Min.
1/2	.813	.355	.211	.106	1.073	.280	.161		.875	.223	.103	.354	.216	.750	Max.
	.766	.332	.159	.091	1.028	.250	.131		.831	.186	.065	.299	.176	.710	Min.
9/16	.938	.410	.242	.118	1.198	.312	.179		1.000	.260	.120	.410	.250	.812	Max.
	.887	.385	.183	.102	1.149	.279	.146		.950	.220	.077	.350	.207	.768	Min.
5/8	1.000	.438	.258	.133	1.323	.345	.196		1.125	.298	.137	.467	.285	.875	Max.
	.944	.411	.195	.116	1.269	.309	.162		1.069	.253	.088	.399	.235	.827	Min.
3/4	1.250	.547	.320	.149	1.573	.410	.234		1.375	.372	.171	.578	.353	1.000	Max.
	1.185	.516	.242	.131	1.511	.368	.182		1.306	.319	.111	.497	.293	.945	Min.

* Flat Head Machine Screws Also Are Made With 100° Angle

TABLE X. CONTINUED – DIMENSIONS OF MACHINE SCREW HEADS

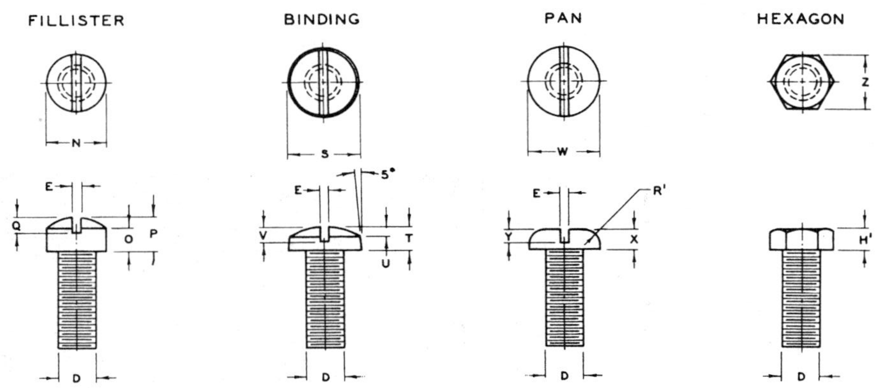

Diam.	O	P	Q	R	R[1]	S	T	U	V	W	X	Y	Z	
0-.060	.045	.059	.025											Max.
	.037	.043	.015											Min.
1-.073	.053	.071	.031											Max.
	.045	.055	.020											Min.
2-.086	.062	.083	.037	.129	.035	.181	.046	.018	.030	.167	.053	.033	.125	Max.
	.053	.066	.025			.171	.041	.013	.024	.155	.045	.023	.120	Min.
3-.099	.070	.095	.043	.151	.037	.208	.054	.022	.036	.193	.060	.037	.187	Max.
	.061	.077	.030			.197	.048	.016	.029	.180	.051	.027	.181	Min.
4-.112	.079	.107	.048	.169	.042	.235	.063	.025	.042	.219	.068	.041	.187	Max.
	.069	.088	.035			.223	.056	.018	.034	.205	.058	.030	.181	Min.
5-.125	.088	.120	.054	.191	.044	.263	.071	.029	.048	.245	.075	.045	.187	Max.
	.078	.100	.040			.249	.064	.021	.039	.231	.065	.032	.181	Min.
6-.138	.096	.132	.060	.211	.046	.290	.080	.032	.053	.270	.082	.050	.250	Max.
	.086	.111	.045			.275	.071	.024	.044	.256	.072	.038	.244	Min.
8-.164	.113	.156	.071	.254	.052	.344	.097	.039	.065	.322	.096	.058	.250	Max.
	.102	.133	.054			.326	.087	.029	.054	.306	.085	.043	.244	Min.
10-.190	.130	.180	.083	.283	.061	.399	.114	.045	.077	.373	.110	.067	.312	Max.
	.118	.156	.064			.378	.102	.034	.064	.357	.099	.050	.305	Min.
12-.216	.148	.205	.094	.336	.078	.454	.130	.052	.089	.425	.125	.077	.312	Max.
	.134	.178	.074			.430	.117	.039	.074	.407	.112	.060	.305	Min.
1/4	.170	.237	.109	.375	.087	.513	.153	.061	.105	.492	.144	.087	.375	Max.
	.155	.207	.087			.488	.138	.046	.088	.473	.130	.070	.367	Min.
5/16	.211	.295	.137	.457	.099	.641	.193	.077	.134	.615	.178	.109	.500	Max.
	.194	.262	.110			.609	.174	.059	.112	.594	.162	.092	.491	Min.
3/8	.253	.355	.164	.538	.143	.769	.234	.094	.163	.740	.212	.130	.562	Max.
	.233	.315	.133			.731	.211	.071	.136	.716	.195	.113	.552	Min.
7/16	.265	.368	.170	.619										Max.
	.242	.321	.135											Min.
1/2	.297	.412	.190	.701										Max.
	.273	.362	.151											Min.
9/16	.336	.466	.214	.783										Max.
	.308	.410	.172											Min.
5/8	.375	.521	.240	.863										Max.
	.345	.461	.193											Min.
3/4	.441	.612	.281	1.024										Max.
	.406	.542	.226											Min.

Length of Thread: Screws 2″ or Shorter are Threaded to Head. Screws over 2″ are Threaded 1 3/4″ Minimum. Screw Heads Also are Made With Cross Recess Slots.

TABLE XI. DIMENSIONS OF MACHINE SCREW AND STOVE BOLT NUTS

SQUARE

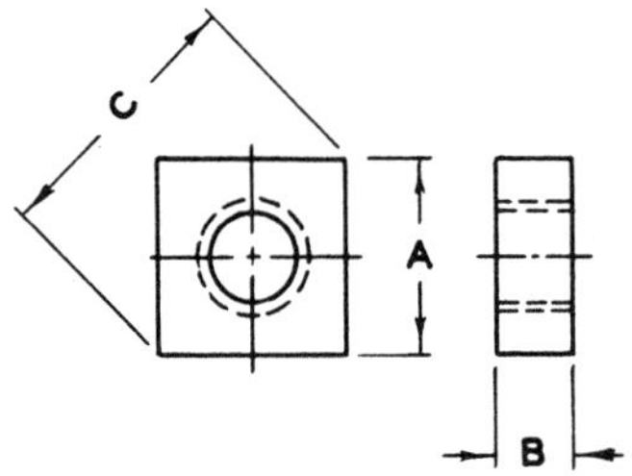

HEXAGON

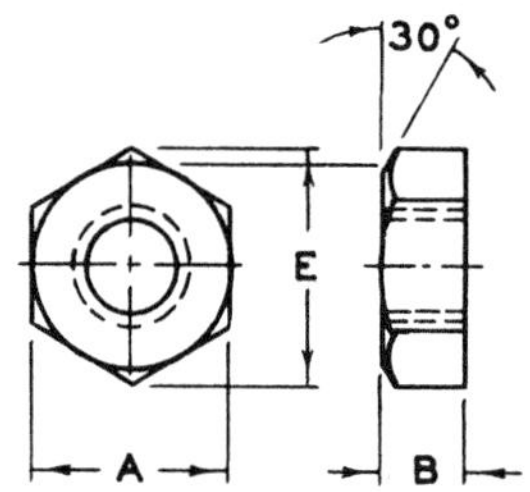

Diameter of Thread	A			B			C	E
		Max.	Min.		Max.	Min.		
0-.060	5/32	.1562	.150	3/64	.050	.043	.2209	.1804
1-.073	5/32	.1562	.150	3/64	.050	.043	.2209	.1804
2-.086	3/16	.1875	.180	1/16	.066	.057	.2651	.2165
3-.099	3/16	.1875	.180	1/16	.066	.057	.2651	.2165
4-.112	1/4	.250	.241	3/32	.098	.087	.3535	.2886
5-.125	5/16	.3125	.302	7/64	.114	.102	.4419	.3608
6-.138	5/16	.3125	.302	7/64	.114	.102	.4419	.3608
8-.164	11/32	.3438	.332	1/8	.130	.117	.4861	.3968
10-.190	3/8	.375	.362	1/8	.130	.117	.5303	.4329
12-.216	7/16	.4375	.423	5/32	.161	.148	.6187	.5051
1/4	7/16	.4375	.423	3/16	.193	.178	.6187	.5051
5/16	9/16	.5625	.545	7/32	.225	.208	.7955	.6494
3/8	5/8	.625	.607	1/4	.257	.239	.8839	.7216

TABLE XII. DIMENSIONS OF SQUARE HEAD AND HEADLESS SET SCREWS

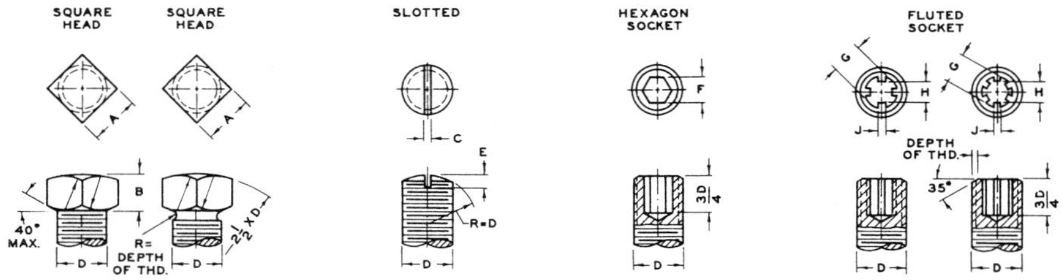

Diam.	A	B	C	E	F	G	H	J	K	L	M	N	
5-.125			.023	.031	1/16	.071 .070	.053 .052	.022 .021	1/16	.083 .078	.06	.03	Max. Min.
6-.138			.025	.035	1/16	.079 .078	.056 .055	.022 .021	.069	.092 .087	.07	.03	Max. Min.
8-.164			.029	.041	5/64	.098 .097	.082 .080	.022 .021	5/64	.109 .103	.08	.04	Max. Min.
10-.190			.032	.048	3/32	.115 .113	.098 .096	.025 .023	3/32	.127 .120	.09	.04	Max. Min.
12-.216			.036	.054	3/32	.115 .113	.098 .096	.025 .023	7/64	.144 .137	.11	.06	Max. Min.
1/4	1/4	3/16	.045	.063	1/8	.149 .147	.128 .126	.032 .030	1/8	5/32	1/8	1/16	Max. Min.
5/16	5/16	15/64	.051	.078	5/32	.188 .186	.163 .161	.039 .037	11/64	13/64	5/32	5/64	Max. Min.
3/8	3/8	9/32	.064	.094	3/16	.221 .219	.190 .188	.050 .048	13/64	1/4	3/16	3/32	Max. Min.
7/16	7/16	21/64	.072	.109	7/32	.256 .254	.221 .219	.060 .058	15/64	19/64	7/32	7/64	Max. Min.
1/2	1/2	3/8	.081	.125	1/4	.298 .296	.254 .252	.068 .066	9/32	11/32	1/4	1/8	Max. Min.
9/16	9/16	27/64	.091	.141	1/4	.298 .296	.254 .252	.068 .066	5/16	25/64	9/32	9/64	Max. Min.
5/8	5/8	15/32	.102	.156	5/16	.380 .377	.319 .316	.092 .089	23/64	15/32	5/16	5/32	Max. Min.
3/4	3/4	9/16	.129	.188	3/8	.463 .460	.386 .383	.112 .109	7/16	9/16	3/8	3/16	Max. Min.
7/8	7/8	21/32			1/2	.604 .601	.509 .506	.138 .134	33/64	21/32	7/16	7/32	Max. Min.
1	1	3/4			9/16	.631 .627	.535 .531	.149 .145	19/32	3/4	1/2	1/4	Max. Min.
1 1/8	1 1/8	27/32			9/16	.709 .705	.604 .600	.168 .164	43/64	27/32	9/16	9/32	Max. Min.
1 1/4	1 1/4	15/16			5/8	.801 .797	.685 .681	.189 .185	3/4	15/16	5/8	5/16	Max. Min.
1 3/8	1 3/8	1 1/32			5/8	.869 .865	.744 .740	.207 .203	53/64	1 1/32	11/16	11/32	Max. Min.
1 1/2	1 1/2	1 1/8			3/4	.970 .966	.828 .824	.231 .227	29/32	1 1/8	3/4	3/8	Max. Min.
1 3/4					1	1.275 1.271	1.007 1.003	.298 .294	1 1/16	1 5/16	7/8	7/16	Max. Min.
2					1	1.275 1.271	1.007 1.003	.298 .294	1 7/32	1 1/2	1	1/2	Max. Min.

SET SCREW POINTS

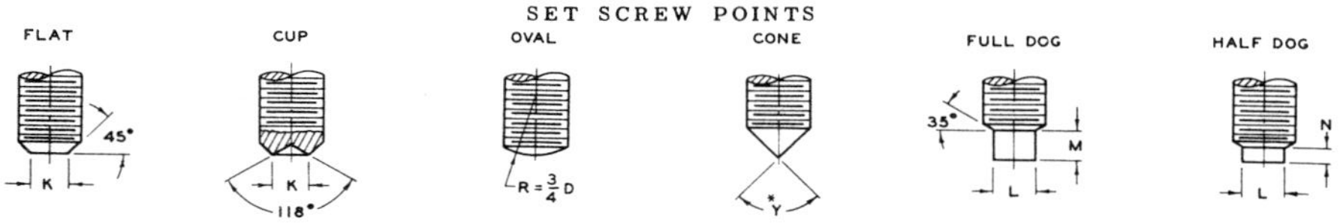

*Angle Y = 118° When Length Equals Diam. or Less, Y = 90° When Length Exceeds Diam.

TABLE XIII. DIMENSIONS OF AMERICAN NATIONAL TAPER PIPE THREADS

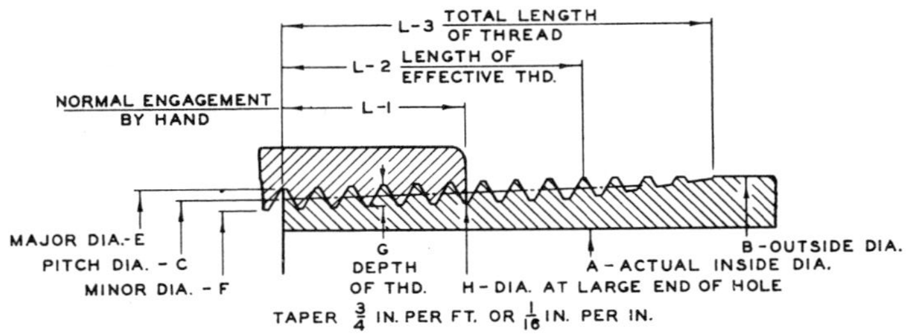

Nominal Pipe Size	Thds per In	A	B	C	E	F	G	L-1	L-2	L-3	H Reamer Dia.	Tap Drill Size
1/16	27	.181	.3125	.271	.301	.242	.0296	.160	.261	.390	.252	1/4
1/8	27	.269	.405	.364	.393	.334	.0296	.180	.264	.392	.345	11/32
1/4	18	.364	.540	.477	.522	.433	.0444	.200	.402	.595	.445	7/16
3/8	18	.493	.675	.612	.656	.568	.0444	.240	.408	.601	.583	19/32
1/2	14	.622	.840	.758	.816	.701	.0571	.320	.534	.782	.721	23/32
3/4	14	.824	1.050	.968	1.025	.911	.0571	.339	.546	.794	.932	15/16
1	11 1/2	1.049	1.315	1.214	1.283	1.144	.0696	.400	.683	.985	1.169	1 5/32
1 1/4	11 1/2	1.380	1.660	1.557	1.627	1.488	.0696	.420	.707	1.009	1.514	1 1/2
1 1/2	11 1/2	1.610	1.900	1.796	1.866	1.727	.0696	.420	.724	1.025	1.753	1 23/32
2	11 1/2	2.067	2.375	2.269	2.339	2.200	.0696	.436	.757	1.058	2.227	2 3/16
2 1/2	8	2.469	2.875	2.720	2.820	2.620	.100	.682	1.138	1.571	2.662	2 5/8
3	8	3.068	3.500	3.341	3.441	3.241	.100	.766	1.200	1.634	3.289	3 1/4
3 1/2	8	3.548	4.000	3.838	3.938	3.738	.100	.821	1.250	1.684	3.789	3 3/4
4	8	4.026	4.500	4.334	4.434	4.234	.100	.844	1.300	1.734	4.287	4 1/4
5	8	5.047	5.563	5.391	5.491	5.291	.100	.937	1.406	1.841	5.349	5 5/16
6	8	6.065	6.625	6.446	6.546	6.346	.100	.958	1.513	1.947	6.406	6 3/8
8	8	7.981	8.625	8.434	8.534	8.334	.100	1.063	1.713	2.147	8.400	
10	8	10.020	10.750	10.545	10.645	10.445	.100	1.210	1.925	2.359	10.521	
12	8	12.000	12.750	12.533	12.633	12.433	.100	1.360	2.125	2.559	12.518	

TABLE XIV. AMERICAN STANDARD HOLDING TAPERS

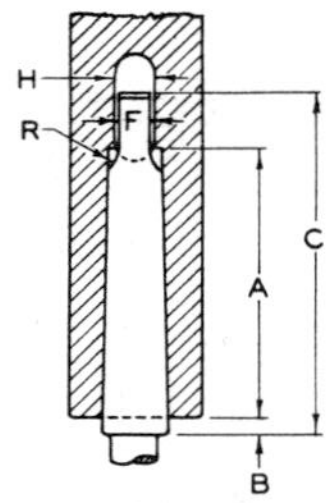

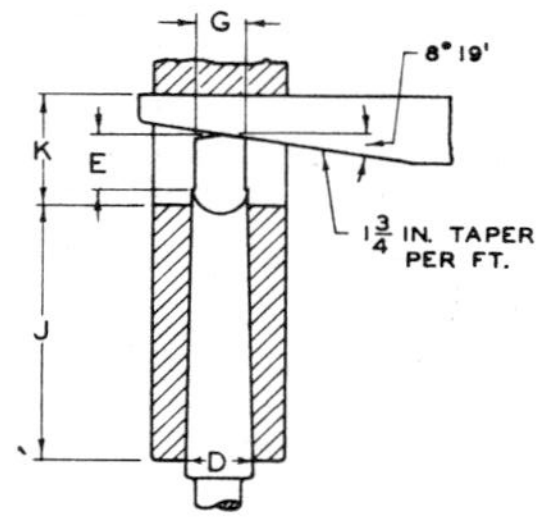

No. of Taper	Taper per Foot	D	A	B	C	E	F	G	R	H	J	K
.239	.502	.2392	1 1/16	3/32	1 9/32	3/16	.125	11/64	3/16	.141	15/16	3/8
.299	.502	.2996	1 5/16	3/32	1 19/32	1/4	.1562	7/32	3/16	.172	1 11/64	1/2
.375	.502	.3752	1 5/8	3/32	1 31/32	5/16	.1875	9/32	3/16	.203	1 15/32	5/8
1	.5985	.4750	2 3/16	1/8	2 9/16	3/8	.2031	11/32	3/16	.213	2 1/16	3/4
2	.5994	.700	2 21/32	3/16	3 1/8	7/16	.250	17/32	1/4	.260	2 1/2	7/8
3	.6023	.938	3 5/16	3/16	3 7/8	9/16	.3125	23/32	9/32	.322	3 1/16	1 3/16
4	.6232	1.231	4 3/16	1/4	4 7/8	5/8	.4687	31/32	5/16	.478	3 7/8	1 1/4
4 1/2	.6240	1.500	4 5/8	1/4	5 3/8	11/16	.5625	1 13/64	3/8	.573	4 5/16	1 3/8
5	.6315	1.748	5 5/16	1/4	6 1/8	3/4	.625	1 13/32	3/8	.635	4 15/16	1 1/2
6	.6256	2.494	7 13/32	5/16	8 9/16	1 1/8	.750	2	1/2	.760	7	1 3/4
7	.6240	3.270	10 5/32	3/8	11 5/8	1 3/8	1.125	2 5/8	3/4	1.135	9 1/2	2 5/8
200	.750	2.000	4 3/4							.656	4 1/2	1 9/16
250	.750	2.500	5 1/2			KEY DRIVE				.781	5 3/16	1 15/16
300	.750	3.000	6 1/4							1.031	5 15/16	2 3/16
350	.750	3.500	6 15/16			WITH				1.031	6 3/4	2 3/16
400	.750	4.000	7 11/16							1.031	7 1/2	2 3/16
450	.750	4.500	8 3/8			SHANK				1.031	8	2 3/4
500	.750.	5.000	9 1/8							1.031	8 3/4	2 3/4
600	.750	6.000	10 9/16			RETAINED				1.281	10 1/8	3 1/4
800	.750	8.000	13 1/2							1.781	12 7/8	4 1/4
1000	.750	10.000	16 5/16			BY KEY				2.031	15 3/4	5
1200	.750	12.000	19							2.531	18 1/2	6

TABLE XV. MORSE TAPERS

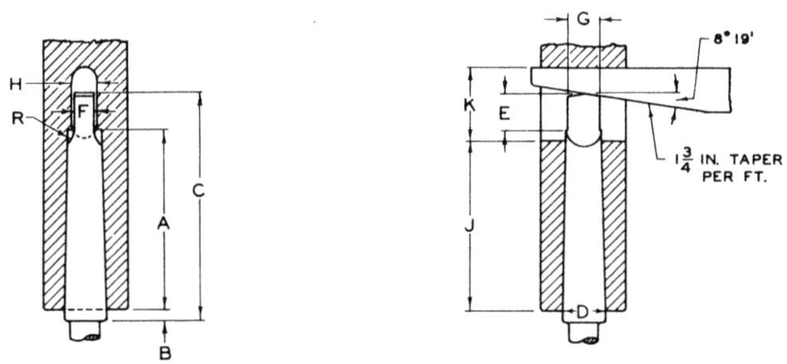

No. of Taper	Taper per Foot	D	A	B	C	E	F	G	R	H	J	K
0	.6246	.356	2 1/32	1/8	2 11/32	1/4	5/32	.235	5/32	.160	1 15/16	9/16
1	.5986	.475	2 3/16	1/8	2 9/16	3/8	13/64	.343	3/16	.213	2 1/16	3/4
2	.5994	.700	2 5/8	3/16	3 1/8	7/16	1/4	17/32	1/4	.260	2 1/2	7/8
3	.6023	.938	3 1/4	3/16	3 7/8	9/16	5/16	23/32	9/32	.322	3 1/16	1 3/16
4	.6233	1.231	4 1/8	1/4	4 7/8	5/8	15/32	31/32	5/16	.478	3 7/8	1 1/4
5	.6315	1.748	5 1/4	1/4	6 1/8	3/4	5/8	1 13/32	3/8	.635	4 15/16	1 1/2
6	.6256	2.494	7 3/8	5/16	8 9/16	1 1/8	3/4	2	1/2	.760	7	1 3/4
7	.6240	3.270	10 1/8	3/8	11 5/8	1 3/8	1 1/8	2 5/8	3/4	1.135	9 1/2	2 5/8

TABLE XVI. BROWN AND SHARPE TAPERS

No. of Taper	D	A	B	C	E	F	G	H	J	K
1	.239	1 1/16	3/32	1 9/32	3/16	1/8	.170	.135	15/16	3/8
2	.299	1 5/16	3/32	1 19/32	1/4	5/32	.220	.166	1 11/64	1/2
3	.375	1 5/8	3/32	1 31/32	5/16	3/16	.282	.197	1 15/32	5/8
3	.385	1 7/8	3/32	2 7/32	5/16	3/16	.282	.197	1 23/32	5/8
3	.395	2 1/8	3/32	2 15/32	5/16	3/16	.282	.197	1 31/32	5/8
4	.402	1 3/8	3/32	1 3/4	11/32	7/32	.320	.228	1 13/64	11/16
4	.420	1 13/16	3/32	2 3/16	11/32	7/32	.320	.228	1 41/64	11/16
5	.523	1 7/8	3/32	2 9/32	3/8	1/4	.420	.260	1 11/16	3/4
5	.533	2 1/8	3/32	2 17/32	3/8	1/4	.420	.260	1 15/16	3/4
5	.539	2 1/4	3/32	2 21/32	3/8	1/4	.420	.260	2 1/16	3/4
6	.599	2 1/2	3/32	2 31/32	7/16	9/32	.460	.291	2 19/64	7/8
6	.635	3 3/8	3/32	3 27/32	7/16	9/32	.460	.291	3 11/64	7/8
7	.725	3 1/8	3/32	3 5/8	15/32	5/16	.560	.322	2 29/32	15/16
8	.898	3 11/16	1/8	4 1/4	1/2	11/32	.710	.353	3 29/64	1
8	.917	4 1/8	1/8	4 11/16	1/2	11/32	.710	.353	3 57/64	1
9	1.067	4 1/8	1/8	4 3/4	9/16	3/8	.860	.385	3 7/8	1 1/8
9	1.077	4 3/8	1/8	5	9/16	3/8	.860	.385	4 1/8	1 1/8
10	1.260	5 1/8	1/8	5 27/32	21/32	7/16	1.010	.447	4 27/32	1 5/16
10	1.289	5 13/16	1/8	6 17/32	21/32	7/16	1.010	.447	5 17/32	1 5/16
10	1.312	6 11/32	1/8	7 1/16	21/32	7/16	1.010	.447	6 1/16	1 5/16
11	1.498	6 1/16	1/8	6 25/32	21/32	7/16	1.210	.447	5 25/32	1 5/16
11	1.531	6 7/8	1/8	7 19/32	21/32	7/16	1.210	.447	6 19/32	1 5/16
12	1.797	7 1/4	1/8	8 1/16	3/4	1/2	1.460	.510	6 15/16	1 1/2
13	2.073	7 7/8	1/8	8 11/16	3/4	1/2	1.710	.510	7 9/16	1 1/2
14	2.344	8 3/8	1/8	9 9/32	27/32	9/16	1.960	.572	8 1/32	1 11/16
15	2.615	8 7/8	1/8	9 25/32	27/32	9/16	2.210	.572	8 17/32	1 11/16
16	2.885	9 3/8	1/8	10 3/8	15/16	5/8	2.450	.635	9	1 7/8
17	3.156	9 7/8								
18	3.427	10 3/8								

TABLE XVII. STANDARD TAPER PINS

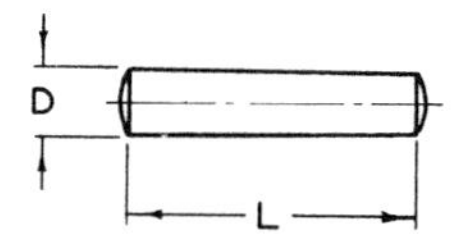

DIAMETER AT SMALL END = D - (L x 0.0208)

No. of Taper Pin	D	L Range of Lengths	No. of Taper Pin	D	L Range of Lengths
6/0	.078	3/8 to 3/4	3	.219	3/4 to 1 3/4
5/0	.094	1/2 to 1	4	.250	3/4 to 2
4/0	.109	1/2 to 1	5	.289	1 to 2 1/4
3/0	.125	1/2 to 1	6	.341	1 1/4 to 3
2/0	.141	1/2 to 1 1/4	7	.409	2 to 3 3/4
0	.156	1/2 to 1 1/4	8	.492	2 to 4 1/2
1	.172	5/8 to 1 1/4	9	.591	2 3/4 to 5 1/4
2	.193	3/4 to 1 1/2	10	.706	3 1/2 to 6

ACKNOWLEDGMENTS

The author wishes to acknowledge the following companies which have supplied material and information:

Bruning, Clevelend, Ohio – figures 1-10, 1-18, 1-19, 1-20, 1-23, 1-24, 1-25, 1-26, 1-33, 1-34
Teledyne Post, Chicago, Illinois – figures 1-1, 1-2, 1-3, 1-9, 1-11, 1-15, 1-16, 1-17, 1-21, 1-27, 1-28, 1-30, 1-31, 1-32

The Delmar Staff

Source Editor: Mark W. Huth
Associate Editor: Vincent A. DeSantis
Photo Editor: Sherry Patnode
Copy Editor: Kathleen Beiswenger

Classroom Testing

The material in this textbook was classroom tested by the students in Hilton Central High School, Hilton, New York.

INDEX